Performance of Materials in Buildings

Performance of Materials in Buildings

A study of the principles and agencies of change

Lyall Addleson
and
Colin Rice

Butterworth-Heinemann Ltd
Linacre House, Jordan Hill, Oxford OX2 8DP

 A member of the Reed Elsevier plc group

OXFORD LONDON BOSTON
MUNICH NEW DELHI SINGAPORE SYDNEY
TOKYO TORONTO WELLINGTON

First published 1991
Paperback edition 1994
Reprinted 1995.

British Library Cataloguing in Publication Data
Addleson, L.
  Performance of materials in buildings:
  A study of the principles and agencies of change.
  I. Title II. Rice, Colin
  691

ISBN 0 7506 1961 9

Library of Congress Cataloguing in Publication Data
Addleson, Lyall.
  Performance of materials in buildings: a study of the principles
  and agencies of change/Lyall Addleson and Colin Rice.
  p. cm.
  Includes bibliographical references and index.
  ISBN 0 7506 1961 9
  1. Building materials. 2. Corrosion and anticorrosives.
  3. Weathering of buildings. I. Rice, Colin. II. Title.
  TA407.A443 1991   91–704
  691—dc20   CIP

Composition by Genesis Typesetting, Laser Quay, Rochester, Kent

Printed in Great Britain by Ipswich Book Co. Ltd., Ipswich, Suffolk.

Contents

Preface   xiv
Glossary   xvi

Introduction   1
General considerations   1
1 Scope   1
2 Objectives   1
3 Approach   2
4 Illustrations   3

Changes and their effects   4
1 Change and complexity   4
2 Innovation and the risk of failure   5
3 Multi-layer construction   7
4 Durability   9

Part 1 Principles for building   15

Need   15
1 Background   15
2 Applying the lessons of failures   15
3 Simple *aide-mémoire*   15

Development   15
1 Early beginnings   15
2 Trials on general practitioners   16
3 Applying everyday experience   16
4 Basis   17
5 Scope   17

The principles   17
1 High>low   17
2 Separate lives   18
3 Creative pessimism   19
4 Continuity   19
5 Balance   19

Rules and precautions   19
1 General considerations   19
2 Rules   20
3 Precautions   20

Application   20
1 Scope   20
2 High>low   20
3 Separate lives   24
4 Creative pessimism   25
5 Continuity   27
6 Balance   28

Compensation   30
1 Scope   30
2 Background   30
3 Application in theory   31

Part 2 Strength of materials   35

2.1 Cracking   37
General considerations   37
1 Broad-brush approach   37
2 Significance   40
3 Structural weakening   40
4 Occurrence   41

Classification and recognition   44
1 Movement/restraint and force relationship   44
2 Crack initiation   46

Effects of variations   50
1 Overview   50
2 Influencing factors   50
3 Stress concentrations – 1: general considerations   57
4 Stress concentrations – 2: typical locations   59

Precautions   66
1 Generally   66
2 Movement joints   67
3 Seeing the problem as a whole   69
   Sources   69

Soils   70
1 Scope   70
2 General considerations   70
3 Settlement   72
4 Clay soils   74
5 Special conditions   80
   Sources   81

Fill and hardcore   82
1 General considerations   82
2 Fill   82
3 Hardcore   86
   Sources   87

2.2 Strength and the use of materials   88
Introduction   88
Strength and the use of materials in construction   88
Shaping and forming   88
1 General principles and objectives   88
2 Shaping by cutting   89
3 Shaping by bending   91
4 Shaping by moulding   94
5 Finishing   95

Jointing, fixing and sealing   95
1 Jointing theory   95
2 Mechanical fixings and fasteners   102
3 Adhesion   108
4 Sealing   117

Strength and materials in use   120
1 Abrasion   120
2 Impact   122
   Sources   126

Part 3 Water and its effects   129

3.1 General considerations   131
Scope   131
Basic properties of water   131
 1 Chemical formula   131
 2 Instability of water   132
 3 Transportation by water   134
 4 Derivation of energy   134
 5 Surface tension and capillarity   134
 6 Thermal properties   135
 7 Density and freezing   135
 8 Electrolytic dissociation and dielectric constant   135
 9 Compressibility   135
 10 Chemical composition of fresh water and sea water   135
    Sources   137

The water cycle   137
1 Background   137
2 Types of cycle   138
3 Equilibrium and distribution   139
4 Microcosm – land areas   139
5 'Local cycle' on land   139
6 Prevailing wet and dry conditions   140
7 Significance of the cycle   143
  Sources   144

Weathering   144
1 Background   144
2 Chemical weathering   145
3 Physical weathering   148
4 Effects of biological agencies   148
5 Significance   148
  Sources   150

Dampness   150
1 Background   150
2 Meaning and use   150
3 Causes of dampness   151
4 Measurement   152
  Sources   154

Flow   154
1 Definition and scope   154
2 Basic exposure   155
3 Soffits   155
4 Profile of front edges   156
5 Eddies   157
6 Returns of projections   157

Assessment of severity of conditions   157
1 Scope   157
2 Precedent   158
3 Difficulties   159
4 Examples   159

3.2 Exposure   164
Introduction   164
Atmosphere   165
1 Generally   165
2 Expression of humidity   165

3 Variations   166
4 Condensation   167
5 Pollution   168
  Sources   180

Rain   180
1 The meaning of 'rain penetration'   181
2 The significance of the wind   181
3 The basic exposure of a site   186
  Sources   195

Ground   196
1 Significance   196
2 Variable factors   197
3 Pressure   197
4 Capillarity and vapour   197
5 Pollutants   198
  Sources   200

During construction   200
1 Background   200
2 Drying and wetting   200
3 Quantities used   201
4 Drying-out time   202
5 Value of drying-out guidance   202
6 Pollutants   202
  Sources   202

Water supply   203
1 Relevance and scope   203
2 Substances in the water   203
3 Hardness   207
4 pH value (acidity and alkalinity)   209

Solution of building materials   211
1 Context   211
2 Sources and effects   211

Faulty services   212
1 Generally   212
2 Design faults   212

Maintenance and user requirements   212
1 Generally   212
2 Pollution   213

Entrapped moisture   214
1 Scope   214
2 Sources of the moisture   214
3 Reasons for the entrapment   215
4 Flat roof precedent   217
  Sources   222

3.3 Moisture content   223
Scope   223
  Sources   223

Definitions and their application   224
1 Moisture content   224
2 Equilibrium moisture content (emc)   225
3 Moisture movement   226
4 Moisture gradient   229
5 Moisture transmission   230

General influences   230
1 Capillarity   230
2 Absorption   231
3 Saturation coefficient   233
4 Permeability   233

The presence of water in materials   234
1 Water in chemical combination   234
2 Free water   234
3 Sorbed water   234

Exposure   236
1 Generally   236
2 Liquid water   237
3 Water vapour   238
4 Interaction between thermal and moisture movements   239

Timber   239
1 Generally   239
2 Effects of conversion   243
3 Equilibrium moisture content (emc)   246
4 Control of hygroscopicity   251
5 Other distortions   252
6 Precautions   253
  Sources   253

Timber-based products   254
1 Generally   254
2 Plywood   256
3 Fibre building board   257
4 Chipboard   259
  Sources   261

Cement-based products   262
1 Generally   262
2 Influencing factors   263
3 Notes, rules and precautions   268
  Sources   276

3.4 Exclusion   278
Principles of exclusion   278
1 Introduction   278
2 Background   281
3 The three methods of exclusion   282
4 Joints and junctions   292

Methods of exclusion   298
1 Introduction   298
2 Roofs   298
3 Walls   311
4 Structures at or below ground level   325
5 Junctions   335
  Sources   335

3.5 Efflorescence   348
Background   348
1 The phenomenon   348
2 Scope   348

General considerations   349
1 Effects of soluble salts   349
2 Variable factors   351

3 Tests   351
4 Apparent transient nature of efflorescence   353
5 Materials commonly affected   353
6 Importance of sources of salts   354

Nature of soluble salts   354
1 Composition   354
2 Solubility   354
3 Significance and effects   354

Sources of soluble salts   356
1 Salts orginally present in a material   356
2 Salts derived from decomposition of a material   361
3 Salts derived from external sources   362

Physical processes   365
1 Background   365
2 The effects of solubility   366
3 The effects of pore structure   366
4 The effects of crystal form of the salts   368
5 The effects of hygroscopicity of salts   368
6 The effects of precipitation   368
7 The effects of the distribution of salts in a material   368
8 The effects of the transfer of water in a material   369

Lime staining   371
1 Background   371
2 Mechanism   371
3 Precautions   372

Precautions   373
1 Generally   373
2 Selection of materials   373
3 Details of design   374
4 Handling, storage and protection of materials   374
5 Drying out   375
6 Removal of efflorescence   375
  Sources   376

3.6 Chemical attack   378
Background   378
1 Scope   378
2 Effects of changes in materials and constructions   379
3 Variations and their effects   380

Unsound materials   382
1 Meaning and scope   382
2 Basic mechanisms   382
3 Plasters and mortars   383
4 Bricks   384
  Sources   385

Acids   385
1 Scope   385
2 Acid gases and stonework   385
  Sources   389
3 Acid gases and brickwork   389
4 Acids and cement-based products   390
  Sources   391

Alkalis   391
1 Scope   391

2 Paint films 391
Sources 396
3 Reactive aggregates 396
Sources 402
4 Staining of limestone 402
Sources 403
5 Staining of brickwork 403
6 Glass 404

Sulphates 405
1 Background 405
2 General considerations 405
3 Basic mechanisms 407
4 Concrete 410
Sources 416
5 Brickwork 417
Sources 419
6 Rendering 420
Sources 421
7 Domestic chimneys 421
Sources 422

Fungi and insects 422
1 Background 422
2 Fungi 423
3 Insects 429
Sources 430

Wood-rots 431
1 Scope 431
2 Types of rot 431
3 Resistance of timbers 432
4 Precautions 438
Sources 450

Ultraviolet radiation 451
1 Background 451
2 General considerations 452
3 Exposure 452
4 Actions and effects 453
5 Increasing durability 456

3.7 Corrosion 457
Introduction 457
1 The scale of the problem 457
2 Inevitability and complexity 457
3 Minimizing the effects of corrosion 458
4 Importance of durability and maintenance 458
5 Initiation and rate of attack 459
6 Scope 459

General considerations 459
1 Scope 459
2 Definition 460
3 Surface films 461
4 Basic mechanisms 461
5 Corrosion classification 467
6 Initiation of attack 470
7 Rate of attack 473
8 Forms of corrosion 476

Effects of corrosion   479
1 Structural soundness   479
2 Distortion or cracking of other building materials   479
3 Entry of water into the building   479
4 Changes in appearance   480

Exposure   480
1 Generally   480
2 External atmospheres   480
3 Internal atmospheres   481
4 Other building materials   482
5 Contact with conductive water   483
6 Contact between dissimilar metals   483

Resistance to corrosion   489
1 Scope   489
2 Ferrous metals   490
3 Non-ferrous metals   498

Protection   506
1 Treatment of the environment   506
2 Protective coatings – 1: generally   508
3 Protective coatings – 2: metals   509
4 Protective coatings – 3: organic   514
5 Vitreous enamelling   524
6 Concrete   524
7 Cathodic protection   529

Precautions   531
1 Generally   531
2 Selection   532
3 Design   532
4 Contact   536
5 Protection   537
6 Maintenance   540
  Sources   540

3.8 Frost action   542
Introduction   542
1 Definition and influencing factors   542
2 Frost resistance   542
3 Deleterious effects   542
4 Guidance and aids available   543

General considerations   543
1 Basic mechanisms   543
2 Freezing in materials   544

Practical considerations   549
1 Effects on materials   549
2 Likely positions   550
3 Assessment of resistance   551
4 Precautions   553

Winter building precautions   556
1 Background   556
2 *In-situ* cement-based products   558
3 Finishes   559
4 Flat roof work   560
  Sources   560

3.9 Flow and changes in appearance   562
Introduction   562
1 Definitions   562
2 Principles relevant   562

Basic mechanisms of flow and staining   562
1 Flow of water on surfaces   562
2 Dirt and washing   564
3 Drying out   565
4 Differential flow   566

Factors affecting flow and staining   566
1 The surface   566
2 The environment   575

Practical precautions   576
1 Objectives   576
2 Saturation staining   576
3 General staining   577
4 Summary   580
Sources   580

Index   581

Preface

Many things have changed in buildings since *Materials for Building* was published in book form in 1972. The principles and factors associated with the performance of materials in constructions have not. In the intervening period there have, of course, been changes of emphasis or detail; new problems have required a reassessment of existing knowledge and, where necessary, the acquisition of new knowledge. In revising the first three volumes of *Materials for Building* account has been taken of the latter but, in addition, the concept of principles *for* rather than of building has been introduced. To better reflect the underlying theme of performance of materials in building constructions, the title has been changed accordingly. The structure of *Materials for Building* has been retained and presented as one volume in a new format but excluding the chapters on basic science and structural mechanics.

A study of this kind, drawing on the work of researchers (notably at the Building Research Establishment – BRE) and others, is an interpretation of that work for the non-specialist in building, architects in particular. It is certainly not intended for bedtime reading but it can be used at different levels and at different times by practitioners and students alike. To this end, the *principles for building* have been developed. They are intended to provide a methodology for fighting through the maze called complexity and as an *aide-mémoire*. Their successful application demands an understanding of fundamentals.

Those chapters that deal essentially with the practical application of aspects such as cracking, strength and use, moisture content, exclusion, efflorescence, chemical attack, corrosion and frost action are so arranged that they proceed from the general to the particular and, where appropriate, conclude with the relevant precautions that should be taken. Taken together with the comprehensively itemized contents pages and the index, it should be possible for readers to identify particular aspects or problems in either general or detailed terms. As a further aid, there is much cross-referencing to related or closely associated aspects. Those wanting even more detail and/or clarification can turn to the comprehensive lists of the works on which the individual chapters or parts of chapters have been based.

The principles for building apart, there has been a major overhaul of two particular chapters: *2.1 Cracking* and *3.4 Exclusion*. Both have problems that appear to be at the heart of much agony in the building industry; both are eminently suitable for applying the principles for building. In each case the underlying principles and fundamentals have been extracted and classified. In common with other aspects, neither contains copybook solutions. Instead, the logical structure used should help to provide the guidance needed to resolve particular problems.

The problems associated with the changing appearance of external building surfaces have also been taken back to fundamentals. The discussion of these concludes the part on water and its effects. The Glosssary has been retained and extended. The many photographs in *Materials for Building* have been replaced by more meaningful diagrams.

Photographs have been reserved for illustrating specific points and those of defects confined to chapters where they are most relevant.

We are most grateful to all those authors, publishers and organizations who have given permission to reproduce or use their copyright material. This work would certainly not have been possible without this material. Rather than list here all those who have contributed in this way, the relevant sources are given in abbreviated form where the material has been used. It should therefore be taken that in *every case* 'By courtesy of' is implied.

Importantly, all material from the Building Research Establishment, the Department of the Environment and other Government departments is Crown Copyright and is reproduced by permission of the Controller of HMSO. All material from BSI (Standards or Codes of Practice) is reproduced by permission of the Director.

We are also most appreciative of the help we have obtained from individuals at the BRE and other research and development organizations such as PSA, TRADA, BDA and C&CA (now BCA). Finally our thanks to our publishers for their assistance and guidance.

Lyall Addleson and Colin Rice
London, August 1991

Glossary

Abrasion   The wear or removal of the surface of a solid material as a result of relative movement of other solid bodies in contact with it

Absorption   The more or less uniform penetration of one substance into the body of another. Cf. *Adsorption*

Acids   Compounds of acidic radical with hydrogen which can be replaced by a metal (usually sodium) either wholly or in part. Acids turn blue litmus red. Cf. *Alkalis*

Adhesion   The property of matter by which close contact is established between two or more surfaces when they are brought intimately together. Force is required to separate the surfaces. Cf. *Cohesion*

Adsorption   A special type of absorption in which only the surface functions as the absorbing medium. Cf. *Absorption*

Alkalis   Soluble bases or hydroxides. A base is a compound which reacts with an acid to yield a salt and water only. Alkalis turn red litmus blue. Cf. *Acids*

Anisotropic   Different physical properties in different directions. Cf. *Isotropic*

Atom   The smallest uncharged particle of an element which can enter into, or be expelled from, chemical combination. The particle is said to be uncharged in order to distinguish it from an ion. An atom consists of one or several of each of three sub-atomic units, the electron, the proton and the neutron. Cf. *Ion* and *Molecule*

Atomic weight   The atomic weight of an element is the weight of the atom of the element as a multiple of the weight of an atom of hydrogen. Cf. *Molecular weight* and *Equivalents*

Capillarity   The capacity of a liquid to move upwards (or downwards) in fine-bore tubes (or narrow spaces) due to surface tension effects

Capillary   Circular tube of narrow bore. Term is also applicable to pores in material or narrow spaces between materials (e.g. in joints)

Capillary pressure   The pressure developed by capillarity

Carbonation   The process of saturation of a liquid with carbon dioxide or converting a compound to carbonate by means of carbon dioxide

Chemical change   A change that is (in building terms particularly) permanent and that results in completely new substances being formed that are completely different from the original substances and with entirely new chemical properties. Cf. *Physical change*

Climate   Basically, the sum total of all the weather processes at a given place, and thus comprises average conditions and the regular sequence of the weather

Cohesion   Forces between the particles of any given mass by virtue of which it resists physical disintegration. Cf. *Adhesion*, which is associated with forces at interfacial surfaces

Cold working   The operation of shaping metals at or near atmospheric temperature by rolling, pressing, drawing, stamping or spinning. Cf. *Work hardening*

Colloid   A substance consisting of a continuous medium and particles dispersed therein. More generally known as a *disperse system*

Compound Has a fixed composition and formed as a result of chemical forces uniting two or more elements. The properties of a compound are different from those of the combining elements. Cf. *Mixture*

Condensation A process of forming a liquid from its vapour. When moist air is cooled below its *dewpoint* it condenses if there are extended surfaces (e.g. of building materials for surface or interstitial condensation) or nuclei (e.g. dust particles or ions for mist, fog and cloud) present

Covalent bond The sharing of two electrons by two atoms

Creep The time-dependent part of strain resulting from stress – the additional deformation that takes place in materials subjected to a constant load over a period of time

Crystal A body, generally solid, where atoms are arranged in a definite pattern, the crystal faces being an outward expression of the regular arrangement of the atoms. Cf. *Space lattice*

Deformation The changed shape of a material due to the application on or inducement of a force in a material

Dewpoint The temperature at which air would become saturated if cooled at constant pressure

Ductility The capacity of a material, usually a metal, to be drawn out plastically before breaking. Cf. *Malleability*

Durability The quality of maintaining a satisfactory appearance and satisfactory performance of required functions. Cf. *Maintenance*

Efflorescence Deposits of soluble salts at or near the surface of a porous material as the result of evaporation of the water in which the salts are dissolved

Elasticity That property of a material which enables it to return to its original shape and form once the stress causing the deformation has been removed

Electrolyte A substance (other than a metal) which, when fused or dissolved in water, conducts an electric current

Electron One of the three sub-atomic units. It is the constituent of negative electricity. Cf. *Proton* and *Neutron*

Elements The basic 'building units' from which all matter is made. They are substances which cannot be broken down or decomposed by chemical means

Equilibrium moisture content The moisture content of a material that it will achieve when it is in equilibrium with the moisture content of the surrounding air. (Principle of balance applies)

Equivalents *Element*: The equivalent of an element is the number of parts by weight of the element which combine with or displace, directly or indirectly, one part by weight of hydrogen. Cf. *Atomic weight* and *Molecular weight*

Compound: The equivalent of a compound is the number of parts by weight of the compound which react or yield, directly or indirectly, one part by weight of hydrogen. Cf. *Atomic weight* and *Molecular weight*

Evaporation A process whereby the quantity of a liquid exposed to air is progressively reduced until it eventually disappears

Exfoliation The splitting off of sheets of rock from stonework. Also loosely applied to splitting off of sheets from brickwork

Fatigue A phenomenon whereby materials fracture when subjected to a fluctuating or repeated load which is within the stress limit for static loading. Cf. *Impact*

Force (1) The physical agent which causes a change in momentum. (2) The physical agent which produces an elastic strain in a body. Cf. *Weight* and *Mass*

Frost splitting Splitting off of thin layers of the surfaces of materials (such as brick or stone) as the result of frost action beneath the surface

Fungi   Large group of very simple plants that take the form of a fine web of microscopic root-like threads which grow over and into the materials from which they extract the food they need. The reproductive parts are contained in fruiting bodies that release spores (seeds in higher plants) into the air in very large numbers

Gas   One of the states of matter. A gas has no definite volume or shape, but fills any vessel into which it is put, irrespective of shape or size. Cf. *Liquid* and *Solid*

Gel   The apparently solid, often jelly-like, material formed from a colloidal solution on standing. Properties, even when containing appreciable quantities of water, are more like those of solids than liquids

Hardness   The resistance of a material to permanent deformation of its surface

Impact   The sudden application of a load on a material that results in stresses that are momentarily higher than those due to the same static load. Cf. *Fatigue*

Impervious   Said of materials that have the property of satisfactorily resisting the passage of water

Ion   A charged atom, molecule or radical whose migration effects the transport of electricity through an electrolyte, or, to a certain extent, through a gas

Ionic bond   Formed by the complete transference of an electron from one atom to another

Isotropic   The same physical properties in different directions. Cf. *Anisotropic*

Liquid   One of the states of matter. A liquid has a definite volume, but no definite shape, taking on the shape of its containing vessel. Cf. *Solid* and *Gas*

Load   Imposition of some weight or force on a structural member or element

Maintenance   Work undertaken in order to keep or restore every facility, that is, every part of a site, building and contents, to an acceptable standard. Cf. *Durability*

Malleability   The ability of a material, usually a metal, to be beaten into sheets without rupturing. Cf. *Ductility*

Mass   The quantity of matter in a body. Cf. *Weight* and *Force*

Meniscus   The curved, upper surface of a liquid in a tube or other container due to surface tension effects

Mixture   Consists of two or more substances whose individual properties remain unaltered. A mixture has properties which vary from point to point and is said to be heterogeneous. Cf. *Compound*

Modulus of elasticity (Young's modulus)   The ratio of the direct stress to the strain produced by that stress

Moisture content   The amount of moisture that a material contains at a given time and is expressed as a percentage of its (oven-dry) dry mass

Moisture gradient   The variation of moisture content between the outer and inner parts of a material

Molecular weight   The molecular weight of an element or compound is the weight of a molecule of the substance as a multiple of the weight of an atom of hydrogen. Cf. *Atomic weight* and *Equivalents*

Molecule   The smallest particle of a substance capable of independent physical existence and built up of groups of atoms of the elements

Mould growth   The development of exceedingly fine interwoven hollow threads (hyphae) arranged to form mycelium and found on most building surfaces. The growth is initiated and sustained by consistently high moisture contents

Neutron   One of the three sub-atomic units, a component of the nucleus and with no electrical charge. Cf. *Electron* and *Proton*

Oxidation   The reactions in which oxygen (or an element chemically similar to oxygen, such as sulphur or chlorine) is added to an element or compound, or in which hydrogen is removed from a compound. Cf. *Reduction*

Physical change   A change that lasts as long as the cause of the change persists. In general, a physical change is limited to a change in shape or appearance of the material concerned. Cf. *Chemical change*

pH value   The logarithm to base 10 of the hydrogen ion concentration with the negative sign omitted. Denotes the degree of acidity of a solution. Pure water has a pH value of 7.0, acids have a value *below* 7.0 and alkalis a value *above* 7.0

Plastic   Adjective used in connection with wet mixes of mortar, plaster, renders and concrete and implies easy to trowel or spread

Plasticity   The property in a mortar, plaster, render or concrete implying ease of trowelling or spreading

Plasticizer   (1) An admixture in mortar or concrete which can increase the workability or plasticity of a mix with a low water content in the mix. (2) A non-volatile substance mixed with the medium of paint, varnish or lacquer to improve flexibility of the hardened film

Plastics   Artificial materials generally of synthetic organic origin and which are plastic at some stage of their manufacture, during which heat and/or pressure are used

Pores   The spaces between the particles of which a material is composed

Porosity   The ratio of the volume of voids in a material to that of the overall volume of the material and expressed as a percentage

Proton   One of the three sub-atomic units, a component of the nucleus and carries a positive electric charge. Cf. *Electron* and *Neutron*

Reduction   The reactions in which oxygen (or an element similar to oxygen, such as sulphur or chlorine) is removed from an element or compound, or in which hydrogen is added to a compound. Cf. *Oxidation*

Resilience   (1) Technically, the amount of energy stored in a material. (2) Non-technically, the power of a strained body to 'spring back' on removal of the acting force causing the straining

Salts   Formed as a result of a chemical reaction either between an acid and an alkali or an acid and a metal

Solid   One of the states of matter. A solid has a definite volume and shape. Solids may be either crystalline or amorphous (non-crystalline). Cf. *Gas* and *Liquid*

Solubility   (of a solid in water): The number of grams of a solid which dissolves, at a given temperature, in 100 grams of water to give a saturated solution at that temperature

Solute Solution   The dissolved solid in a solution. Cf. *Solvent* (1) A *saturated* solution is a solution which, at a given temperature, is in equilibrium with undissolved solid. (2) A *colloidal* solution, or sol, is heterogeneous and is a system in which one of the components is dispersed throughout the other as small particles or droplets. (3) A *supersaturated* solution is a solution which contains more solute in a given weight of solvent than is required to form a saturated solution at the same temperature. Usually an unstable solution. (4) A *true* solution is a molecularly homogeneous mixture of two or more substances

Solvent   The liquid which holds the solid in solution

Space lattice   The regular geometrical pattern in which the structural units of a *crystal* are arranged

Strain   A measure of the deformation produced in a member by an acting force and relates change in form (length, width, depth or volume) with the original form of the member, that is, prior to loading. Defined as a *ratio*

Strain hardening   Increase in the resistance to deformation (that is, hardness) produced by earlier deformation. Cf. *Work hardening*

Strength   The ability of a material to sustain loads without undue distortion or failure

Stress (intensity of stress)   The intensity of internal forces called into play by the external forces. The intensity is expressed as units of force per unit area

Surface tension   Property of liquid surfaces to assume minimum area and, in so doing, liquid surfaces exhibit certain features resembling the properties of a stretched elastic membrane

Temperature gradient   The variation of temperature between the outer and inner parts of a material or construction

Thermal pumping   A physical process in which air or water is drawn through a material bounding an air-filled space by the vacuum caused as a result of pressure changes due to changes in temperature

Thermoplastic plastics   Class of plastics which can be softened and re-softened indefinitely by the application of controlled heat and pressure

Thermosetting plastics   Class of plastics which undergoes a chemical reaction during the hardening process and cannot subsequently be reshaped by the application of heat and pressure

Thixotropy   The property shown by certain gels of liquefying on being shaken and of re-forming on standing

Valency   The property of an atom which enables it to enter into chemical combination with other atoms

Vapour pressure   Represents the partial pressure exerted by the molecules of water vapour alone (that is, in the absence of other gases normally present in the atmosphere) and expressed as N/m^2 or millibars

Viscosity   A property of liquids which appears as a dissipative resistance to flow

Weight   The effect of gravitational force acting on a body. Cf. *Mass* and *Force*

Work hardening   The increase in strength and hardness (that is, the resistance to deformation) produced by working metals. It is most pronounced in cold working. Cf. *Strain hardening*

Introduction

General considerations

1 Scope

(a) The study

This is a study of the way materials perform or are expected to perform in constructions that make up the fabric of buildings. The concept of principles *for*, rather than *of*, building is introduced. There are five principles, and these are explained in detail, as are the concepts of rules and precautions which arise from the principles. Guidance is given on the way conflicts between the principles, rules, precautions and practical necessity should be resolved by compensation. Taken together, these are intended to provide a methodology for the resolution of the technical problems that may be encounted in practice.

Unlike other studies of building materials, the factors against which materials have to perform in practice are considered first. The factors considered are the *strength of materials* and *water and its effects*.* In each chapter the relevant properties and performance of materials and constructions are described on a comparative basis. Throughout, the relevant fundamentals of building are explained and applied.†

(b) The introduction

This introduction sets the scene for the study. In it the relevant background is discussed with particular attention given to the effects of change in forms of construction and the durability of materials.

2 Objectives

The objectives of this study are:

1. To provide general guidance for those wishing to have an understanding of the problems associated with the performance of materials in modern building construction and the way these problems may be solved in principle;
2. To analyse systematically the factors that influence both the properties and performance of materials in respect of strength and the effects of water;
3. To gather in and then interpret and relate systematically, for use by general practitioners and students rather than specialists, a considerable amount of diversified information (on research, development and experience in the field) from authoritative sources‡ on the subject of the performance of construction as a whole;
4. To introduce and explain the concept of the *principles for building*, the need to differentiate between principles, rules and precautions and how conflicts between these and practical necessity should be resolved by compensation.

*Heat and fire and their effects are covered in *Materials for Building*, Vol. 4.

†Reference should be made as relevant to Part 1 of *Materials for Building* (Vol. 1) that includes a description of those physical and chemical aspects related specifically to the properties and performance of building materials.

‡Such as the Building Research Establishment, the Construction Industry Research and Development Association, the Property Services Agency, the British Standards Institution, Universities, Polytechnics and trade research and development associations.

3 Approach

(a) Use and interpretation of science

In the treatment of the subject matter of this study what is commonly known as the scientific method has been adopted. That is, an attempt has been made to follow through all aspects of each subject/factor in a systematic way. In another respect, a grasp of scientific knowledge or principles can be useful in gaining an understanding of particular phenomena associated with building constructions and their perform-ance. Such knowledge can also be helpful in predicting the likely performance or the risk of failure of novel forms of construction aimed to meet new requirements.

The scientific aspects and principles relevant to the use and performance of materials are explained as interpreted and simplified by two architects, both with practical experience of building generally and the investigation of building failures in particular and one with much teaching experience. It should go without saying that scientists and/or specialists should always be consulted on those problems that are outside the scope of the general practitioner. This study should help the latter to recognize when specialist help is required and to understand better the advice that particular specialists might give.

(b) Division and coverage

The study starts in Part 1 with the background to and an explanation of *principles for building* and the related *rules* and *precautions*. The application of these in typical cases is described. The concept of *compensation is* explained and applied.

In discussing the strength of materials (Part 2) emphasis is given to cracking and the strength of materials in use. It has been assumed that the relevant elementary structural behaviour and strength properties of materials are understood.*

Water and its effects takes up by far the greater part of this study. This reflects, in part at any rate, the fact that 50% of problems and/or failures in modern building have been attributed by BRE to the effects of water in its gas, liquid or solid state.†

In view of the novelty of the concept of the *principles for building*, the related *rules* and *precautions* and the need for *compensation* to resolve the technical and other conflicts in particular cases, it has been neither possible nor desirable to restrict the use of the term 'principle' to one of the five principles for building. Where relevant, a particular principle is printed in bold type in the text. However, the context should make it clear whether or not it is these principles to which reference is being made. The same applies to rules and precautions.

(c) Variations and their effects

It is inevitable that different 'samples' (to use an appropriate term from statistics) of a building material, like samples of any other natural or manufactured material, will vary to some extent in their properties. In building, the way materials in particular cases are designed, assembled and exposed (to the weather, use and care) will be another important determinant of variation. The need to recognize the inevitability of variations of all kinds is explained under 'creative pessimism' in the *principles for building* later in this introduction. For the moment, suffice it to say that, in a study of this kind, it is impossible to take into account

*Chapter 2.1 and 2.2 of *Materials for Building*, Vol 1 contain the fundamentals of structural mechanics and the related properties of materials.

†BRE Digest 176, April 1975.

specific variations or their effects that are likely to occur in particular cases in practice.

This difficulty should help to underline the need for a proper analysis of performance and other requirements in relation to the site followed by a well-reasoned synthesis that takes all factors into account. In this, it should also be recognized that nowadays there are seldom universal or copybook answers to problems that are encountered in buildings, more so when requirements or expectations change rapidly.

Accordingly, in this study, the effects of variations impose certain limitations. Summarized, these are:

1. Explanations of the structure and properties of matter and materials assume ideal conditions. In other words, the best possible case is being explained. Any changes or impurities which may arise in practice should be regarded as a variation and, unless there are special circumstances, should, in general, be taken to be worse than the ideal.
2. The physical and chemical properties quoted are average properties for the particular materials. The properties assume the use of the materials under what are usually described as 'normal working conditions' unless, of course, otherwise stated.*
3. Unless otherwise stated, descriptions of the use of materials also assume ideal conditions. Such conditions include the application of good-quality workmanship and control and the use of materials of good average quality.

In all cases, where applicable, a British Standard is used as the criterion for either the properties of materials or the quality of workmanship and the materials involved.

4 Illustrations

Illustrations by way of diagrams, charts, tables and photographs are numbered separately in each chapter. In addition, each of the four forms of illustration, as described below, is prefixed by its first letter. For example, Diagram D2.3/2 means Diagram 2 in Chapter 2.3.

(a) Diagrams, sketches, graphs and maps

Wherever possible, diagrammatic drawings are used to support the text. For clarity, the scale used has sometimes been exaggerated. Generally, details of methods of construction are not given although, where appropriate, practical examples are included. Sketches and drawings, including graphs, maps, etc., are referred to as diagrams and prefixed 'D'.

(b) Properties and other numerical data

Bar charts have generally been used in preference to tables to set out the properties of materials. Most charts include the numerical value of the properties in case these are needed in predictions. Sometimes graphs are presented as comparative graphs. Another form of chart, the conversion chart, based on the principle used in nomograms (e.g. thermometers with both Fahrenheit and Celsius scales), are used where appropriate. All these are referred to in the text as charts and are prefixed 'C'.

(c) Tables

Tables are used where it is more convenient to summarize certain information and some tables include drawings. Tables are prefixed 'T'.

*In this, it is important to note that many tests on materials have been (and continue to be) made using 'traditional' constructions and environmental conditions. For example, a test may ignore the effects of or exclude thermal insulation.

(d) Photographs

Generally, photographs illustrating failures or what should not be done have been restricted to those parts where such illustrations would be most informative (e.g. in cracking, efflorescence, chemical attack, corrosion, frost action and changes in appearance). Otherwise, photographs showing 'the correct approach' have been used, but selectively. Photographs are prefixed 'P'.

Changes and their effects

1 Change and complexity

(a) Climatic modification

It is a truism that change is an essential characteristic of life and, naturally enough, also of the buildings designed and constructed to meet particular requirements of life at a particular time. From the earliest times, buildings have been designed and constructed with the express aim of providing shelter from the external climate (or environment, as more commonly termed nowadays). The modification of the external environment was achieved by using certain types and arrangements of materials essentially in the external fabric of buildings. In time, the nature and characteristics of the external fabric were changed to provide greater control of the external environment.

Mechanical aids have, to some extent, always been required to supplement any shortcomings in materials and forms of construction used to provide protection and comfort inside buildings and overall pleasure. Initially, there was the simple fire without a proper flue, but in time that changed to centrally controlled heating systems now common, in the Western world at least. In some types of building the need arose for mechanical ventilation that culminated in air-conditioning systems. With the arrival of the electronic age, the comfort-providing systems have become more sophisticated but, in addition, many buildings, even some domestic ones, have to house a wide variety of electronic gadgetry, from telephones and fax machines to computers in addition to lighting, kitchen and other equipment.

The history of building has one simple message. Materials provide the basis for changing the effects of the external environment so that the resulting internal environments can meet needs and aspirations at a particular time within the relevant economic, social and political restraints. It is also clear that, generally, forms of construction, the materials in them and the shapes of the buildings have changed, sometimes markedly, in response to changes in functional and other requirements. Consequently, the success or otherwise with which new or changing needs and aspirations have been met has depended (and continues to depend) in no small measure upon the imagination, skill and care of all involved (i.e. designers, builders and users alike) in the selection and use of materials in the constructions that make up the building fabric.

(b) Pressures and their effects

At various times in recent decades social and political pressures have required designers and builders to build more buildings faster. This has resulted, in part at any rate, on makers and suppliers of materials being encouraged to develop new materials. Prefabrication of component parts, if not most, of a building always seemed to be the obvious answer. Speed of building had, in many cases, to take into account more sophisticated control of the internal environment. The scale of building changed in form, size and number. Tall buildings became commonplace. The way in which buildings are procured, designed and built has undergone much change. These were some of the answers for densely packed developments of many different types of building that also had to contain an ever-increasing number of different types of equipment.

Often, mainly for economic reasons, the whole process went into reverse. The intense building activity changed to one of virtual inactivity. As a result of this, the building industry lost much of its impetus and skills. When the activity recovered, there were new requirements to meet or new problems to resolve. The stop–go did neither the building industry nor buildings much good – indeed the opposite. When buildings designed and constructed under these conditions, particularly during the 1950s, 1960s and 1970s, were seen by society in general not to be performing as anticipated, the external facades of a large number of new buildings began to use buildings of a former age as models for their appearance. This has often taken place without regard to a fundamental change in the arrangements of materials in constructions.

(c) The growth of complexity

Inevitably, changes in the performance requirements of building to meet the ever-changing and more sophisticated user requirements have led to buildings not simply looking different but becoming complex assemblies of materials and equipment. Complexity abounds, or appears to abound, in other ways. For example:

1. The range of materials available has increased;
2. The number of specialists needed to assemble many materials has grown;
3. There is a greater need for the application of science and technology in building;
4. Design and constructional management techniques have had to be reassessed; and
5. Starting with asbestos, there have been worries recently about the health hazards or potential health hazards from the use of certain materials and 'sick' buildings in general.

As valuable as computers may be, their use has somehow added to the complexity.

2 Innovation and the risk of failure

For reasons given above, innovation is a natural and therefore inescapable part of building generally and in the selection and use of materials in particular. The problems inherent in any innovation might not be as difficult as they appear to have been in building in recent decades if the innovation could have taken place leisurely, without the restraints of cost or without regard to the appearance of the final result. The reality in the building industry since the end of World War II is rather different, as was the traditional approach to innovation.

(a) The traditional approach

What is sometimes loosely, if not fondly, referred to as 'traditional' construction was developed comparatively slowly over many years, sometimes centuries. There was, or there seemed to be, time to advance or develop step by step, correcting such failures as were found on the way, as part of an evolutionary process. Importantly, at a given time, all involved understood what a particular form of construction would achieve in practice, how it could be included in a design, how it was to be built and what was required in its care.

In short, any inadequacies that particular materials or parts of a construction might have were recognized and understood by all concerned. Slight changes that were made were part of an evolutionary process. As importantly, choice of suitable materials and constructional details was, by today's standards, limited. There were few, if any, dangers attached to the copybook approach or the use of standard details.

(b) *The scientific approach*

In contrast, innovation since World War II, whether in materials or forms of construction, has adopted a somewhat different approach, almost perforce. For reasons explained earlier, there has not been the time to indulge in lengthy trials, correcting errors at each stage, as would have been done in the traditional approach. However, trial and error of some kind is still necessary.

Accordingly, the application of science was seen by many as a means of predicting the likely performance of certain combinations of materials, including new or innovative ones, under given conditions, thereby reducing the time necessary for trials. However, in many cases the fundamental data in respect of materials, particularly the variations that could be expected in the environment in and around buildings, required for the predictions were not available. Often existing information was interpreted, if there was insufficient time for research. The new approach became known as 'research and development'. In some cases, existing details were modified, as opposed to being fundamentally rethought, to include extra requirements such as thermal insulation. By so doing, the performance of many, if not most, traditional constructions was changed.

(c) *Risk of failure*

For the present purposes, failure is defined as a shortcoming in expected or required performance. In the application of the scientific method to building there is, or appears to be, an inherent risk of failure in the outcome. The accuracy with which performance in use, and hence the risks, may be predicted is largely dependent on the accuracy and reliability of the data and the prediction method used. Generally, the less accurate or reliable either is, the greater the risk of failure. Structural engineering has long recognized this. Accordingly, factors of safety were introduced to take care, so to speak, of inaccuracies, unreliabilities and unknowns. The same approach has not been applied systematically to other engineering or science-orientated aspects of building. Consequently, risks of failure abound, but such a concept is not readily recognized or accepted in the building industry generally. As already noted, the concept of risk of failure needs to be applied to any construction or constructional details that differs from its traditional model.

(d) *Assessment and quality control*

Attempts have been made over the past 20 years or so to reduce some of the agonies or worries associated with the selection, use and likely performance of new materials and constructional techniques, first, with the establishment of what is now the British Board of Agrément. The BBA certifies the suitability of materials for use in specified circumstances.

More recently, testing laboratories have begun to run certification schemes. At another level, there has been great activity in quality control.* Few, if any, of the certificates issued by any testing laboratories or quality control agencies are without qualification. What they do is to gradually fill some of the gaps in knowledge, thereby helping to assist in the prediction of the risk of failure in particular cases.

3 Multi-layer construction

(a) Background

Almost all modern constructions, including most that have a traditional appearance, are multi-layer in character. Each separate layer usually performs a specific function. Multi-layer constructions are not entirely new. What is new about modern versions are their characteristics. A brief look at the development of the use of layers of construction is therefore appropriate.

In traditional constructions the functional requirements of a wall or roof were achieved by essentially homogeneous construction. All masonry constructions in earth, mud, brick and stone are obvious examples of traditional homogeneous constructions. However, even in timber-framed buildings the infill was essentially a homogeneous mass. Additional layers, such as plaster on the inside of walls, performed a secondary role.

Exceptionally, of course, rendering used on the external surfaces of walls, particularly in parts of the country severely exposed to driving rain, provided the primary layer for the exclusion of rainwater. Other examples of additional layers in walling include tile hanging and timber (shiplap) boarding.

The position is somewhat different with roofs. Traditionally, these, whether pitched or flat, had a timber frame of some kind. The frame could only provide the structural function. Another material was needed to exclude the weather. That material, whether tiles, slates, thatch, metal or a bituminous-based weatherproofing layer, needed to have a secondary support by way of timber battens or boards. Nevertheless, the number of layers involved were few. This reflected the limited extent to which other functions, such as thermal insulation, were required.

(b) Framed and clad buildings

The principles underlying a roof where structural and weatherproofing functions were performed by two different materials were adopted, whether consciously or otherwise, in the development of the modern glass/metal clad framed building. In these, the frame provides the structural support and relatively thin materials – the glass and/or the metal – the weatherproofing functions. Fire resistance, on the other hand, was initially provided, in the UK at any rate, by a masonry or concrete upstand between the head and sill of the windows. Later it was possible in certain circumstances for the requisite fire resistance to be provided by thinner board-type materials. (The quest for thinner materials has characterized the development not only of this type of

*BS 5750, *Quality systems* is the source and guide for a number of schemes in the field of quality control.

framed building but also of other framed and even load-bearing constructions in masonry. The aim has been to try to reduce the heaviness and thereby the cost of the basic structure.) The need for thermal insulation, such as it was in the early days, was achieved by the addition of another suitable layer of material.

(c) The cavity wall

The cavity wall is a key example of the change from homogeneous to multi-layer construction in modern building. In its development, air, as a functional layer, was introduced principally to increase the resistance of a solid brick wall to rainwater without the need for a rendered outer layer. This adopts the principle of building known as 'two lines of defence' for the exclusion of rainwater.

Of necessity, the thickness of a solid wall had to be increased by 50 mm, a width of cavity that, by trial and error, was found to be practical for a bricklayer to achieve. The functioning of the cavity wall was briefly as follows:

1. The air in the cavity provided the separating layer between the two leaves of brickwork.
2. The outer leaf was intended to provide the first line of defence for the exclusion of rainwater and the air the second in that any water penetrating the outer leaf, as it was expected it might sometimes, could run down the cavity face of the outer leaf to drain out at weepholes.
3. The inner leaf provided the permanently dry zone. It would normally be covered with another layer such as plaster for appearance.
4. The leaves had to be tied together in places so that they performed the required structural function.*
5. That the air also provided additional thermal insulation was a hidden bonus, not a primary requirement.

The need for better thermal insulation and, to some extent, the desire for lighter structures led to the replacement of the inner leaf of brickwork with lightweight concrete blockwork. As requirements for thermal insulation increased, the relatively poor thermal insulation provided by the cavity was, initially at any rate, improved by filling it completely with a thermal insulant. In existing buildings, notably houses, this was convenient and in a new building there was no need to increase the thickness of the wall to accommodate an extra layer of thermal insulation. The justification for this approach was the apparent ability of solid fairfaced brickwalls 230 mm (9 in) thick to exclude rainwater in the less severely exposed areas of the country or for similar walls but with rendering outside to do so in the more severely exposed areas. In addition, the insulants were claimed to have certain water-excluding properties.

Cavity filling was reasonably successful in excluding rainwater in older properties. In new buildings the position was different, mostly because of changes in the composition of mortar (i.e. having less lime and considerably more cement), bricks (e.g. perforations instead of frogs and/or greater use of bricks with low water absorption), pointing (e.g. recessed joints) and the quality of workmanship (i.e. poorer quality). There has been a move therefore to restore the unobstructed cavity for its water-exclusion role and to increase the width of the wall to accommodate the thickness of the thermal insulant or, alternatively, to use better-insulated blockwork without increasing the overall thickness of the wall significantly.

*The special needs of the ties and the precautions that need to be taken to ensure that water does not bridge the cavity are covered more fully in 3.4 *Exclusion*.

(d) Difficulties and problems

Stated simply, the difficulties and problems associated with modern multi-layer constructions, such as rainwater penetration, condensation and cracking, can be attributed to a single predominant cause – the introduction of a separate layer of thermal insulation. Other important contributory causes include the use of adapted traditional details, the greater utilization of relatively thin sections (some with integral finishes), the speed of construction and lower standards of workman-ship and site control.

Looked at in another way, the problems reflect the difficulties that designers and builders alike have had in recognizing the fundamental differences between traditional and modern multi-layer constructions. The latter are complex. Briefly:

1. To function as intended, the component parts must be continuous. Discontinuities are often less apparent to the untrained eye, or, for many constructions, continuity is difficult to achieve. Nevertheless, discontinuities are more critical in terms of performance than they were in traditional constructions.
2. There is greater scope, largely but not entirely because of misunderstanding, for component parts to be omitted or put in the wrong place.
3. As the component parts tend to be thinner, they require more care in handling and application. Integral finishes also require special care to prevent damage. That care is especially important as it is usually difficult, in some cases virtually impossible, to repair the damage *in situ* as it is with site-applied finishes.

Clearly, all these factors place greater demands on the techniques of modern constructions. In addition, they are, or should be, part of its excitement, imaginative impetus and aesthetic.

4 Durability

(a) Scope

The factors associated with durability and maintenance of materials are dealt with at length and historically in the Introduction to *Materials for Building*, Vol. 1 (pp. 10–16). In this study, that detail is not repeated. Instead, emphasis is given to the element of change. For convenience, Table T1/1 includes those aspects that primarily influence durability.

(b) Inherent nature of materials to change

All materials change in appearance or composition to some extent when exposed to the physical and chemical conditions that arise or prevail around and within constructions. The desire, so to speak, of all materials to want to change is inherent in them. The principle of change can be termed *reversion*, and is a general observation by the authors of this study based on the second law of thermodynamics and is explained later in this chapter. (It is also included later in the *principles for building*.) By way of summary, materials can be said, therefore, to have an inherent tendency to change, the rate of change being dependent on the characteristics of a material and the severity of its exposure to deteriorating factors.

Table T1/1

Deterioration		Substances involved		Materials commonly affected
Cause	Effects	Type	Source	
Atmospheric gases (naturally occurring and 'man-made')	Corrosion Erosion Disintegration	WATER + Acids Alkalis Salts	1. Natural constituents of air (e.g. oxygen, carbon dioxide and chlorides, the latter at or near sea) 2. Fuel-burning and industrial processes for sulphur dioxide, carbon dioxide, chlorides, other gases such as the chlorofluorocarbons (CFCs). *Note*: The effects on materials of climatic change caused by the build-up of CO_2 and CFCs in the atmosphere has yet to be assessed by BRE and others	Building stones; metals; some types of paint; some bricks; cement- and lime-based products; plastics (subject to type)
Cleaning and maintenance	Physical and chemical changes	WATER + Abrasives Acids Alkalis	Cleansing agents	All building materials, their surfaces in particular. Careful selection of cleansing agents in relation to materials essential
Efflorescence (evaporation/ crystallization of salts)	1. Appearance when white crystalline growth occurs 2. Crumbling when crystallization occurs below surface 3. Loss of adhesion (paint films notably)	Soluble salts	1. Inherent in many materials (bricks and building stone notably) 2. From other materials in close contact resulting from chemical action (e.g. lime staining from limestones, mortars and concrete) 3. Atmospheric reactions on certain materials involving a chemical change	Porous materials, mainly bricks, sedimentary stones, concrete and some paints
Electrolytic action	Corrosion of metals	Electrolytic cell (not a substance) in the presence of moisture essential. Extent of action dependent on amount of water-soluble matter present	1. Local difference in composition of the same metal 2. Embedment of a metal in the soil 3. Dissimilar metals in contact 4. Stray or induced electrical currents	All metals commonly used in buildings

Table T1/1   *(Continued)*

Deterioration		Substances involved		Materials commonly affected
Cause	*Effects*	*Type*	*Source*	
Frost action	Physical change occurring mainly in exposed and wet conditions and normally resulting in surface disintegration and cracking. May cause heave of concrete ground floors on chalk	WATER	Rain mainly, sometimes groundwater (e.g. walls below dpc)	Porous materials (bricks and sedimentary stones notably), especially when used externally and below ground
Fungal attack	1. Disfigurement/ deterioration of surfaces – mould growth 2. Decay – wood-rots	Various fungus spores leading to growth of mycelium, the growth initiated and sustained in the presence of moisture. (Dry rot can grow and/or be sustained by water transported from wetter areas)	1. Spores normally in the atmosphere or transported via the material 2. Food for growth from the material affected 3. Moisture in the atmosphere (implies air high humidities); condensation; rain penetration; rising damp; faulty services (e.g. leaking drains or water pipes)	1. For mould growth – all organic materials 2. For wood-rots – timber (depending on species; heartwood usually more resistant than sapwood)
Incompatible materials (in contact with or in association with one another)	Corrosion Decay	Various but all produced under *damp* conditions	Dissimilar metals Copper and brass Oak and other timbers Cement and lime Bricks containing much soluble sulphate Building stones (some types) Magnesium oxychloride	Copper/zinc Aluminium/copper or brass Vulcanized rubber Lead, iron, steel and aluminium Lead, zinc, some aluminium alloys and glass Cement renderings With other types of stone and some bricks Most metals except lead
Induced stresses	Cracking Distortion Loss of adhesion	Physical changes	1. Loading 2. Moisture and/or thermal movements (including drying-out shrinkage) 3. Unbalanced composites	Most building materials, particularly timber and timber-based products, cement-based products, composites (e.g. with bonded finishes or bonded thermal insulation)
Industrial processes	Mainly decomposition; usually of a specific type depending on chemicals/materials involved	WATER + various substances	1. During processes 2. Storage 3. Spillage 4. Fumes	Most building materials – careful selection essential in particular cases

Table T1/1   *(Continued)*

Deterioration		Substances involved		Materials commonly affected
Cause	Effects	Type	Source	
Moisture content changes	Physical effects only: 1. Contraction during drying out (irreversible) resulting in cracking, surface crazing, warping 2. Expansion on exposure to moisture (irreversible – clay bricks) resulting in cracking (of brickwork) 3. Expansion or contraction (reversible)	WATER (liquid or vapour)	1. Inherent in the material or building process (e.g. wet trades) 2. Atmosphere, condensation, rainwater, ground	Porous materials but excluding expanded or foamed plastics
Soil and groundwater action (*excluding* sulphate attack – see below)	Corrosion Erosion Expansion Disintegration	WATER + Soluble sulphates Acids	1. Clays soils, acid soils, groundwater 2. Ashes and industrial wastes used as fill for 'made ground' or hardcore	1. Cement- and lime-based products used in foundations, retaining walls, floor slabs in contact with the ground but also other elements connected by capillary paths 2. Metals used in service pipes, sewers, etc. in contact with the ground
Sulphate attack (*excluding* soil and groundwater action – see above)	Expansion usually resulting in cracking Disintegration	WATER + Soluble sulphates	Bricks Stones Atmospheric action	Cement- and lime-based products (mortars, renders, concrete)
Temperature changes	Distortion Cracking	Heat energy related to surface characteristics and thermal capacity of materials with location and amount of thermal insulation layer an important determinant of effects	1. Externally – air temperature and solar radiation 2. Internally and externally – air temperature and local heat sources	All building materials
Ultraviolet radiation	1. Surface degradation (e.g. fading, yellowing, erosion and accumulation of dirt) 2. Loss of light transmission (in transparent or translucent plastics)	Solar radiation	Mostly plastics but also some paint films and some timbers	

Table T1/1 *(Continued)*

Deterioration		Substances involved		Materials commonly affected
Cause	*Effects*	*Type*	*Source*	
	3. Loss of plasticizers or degradation of the basic material resulting in brittleness			
Unsound materials	Expansion Disintegration	WATER + Unburnt or partially burnt particles	1. Lime-burning processes 2. Industrial waste (e.g. coal, blastfurnace slag)	Mortars, plasters, lightweight concrete blocks and clay bricks
Wastes – domestic and industrial	Chemical changes	WATER + Acids Alkalis	1. Sanitary installations 2. Sewage-disposal systems 3. Industrial processes	Materials used in all waste-disposal systems. Concrete, mortars and metals in particular may be affected
Water supply	Corrosion Choking	WATER + Mainly acids but also alkalis and carbonates in hard waters	Nature of water supply (pH values; hardness category important determinants)	Metal pipes, appliances, boilers, etc. in plumbing and heating systems

(c) Delaying change and deterioration

In practice, the aim is, or should be, to adopt appropriate measures that will delay the rate of change and deterioration of materials for as long as is required practically and economically. Such measures are basically of two kinds: to use materials that (1) are inherently durable (i.e. those that need little or no maintenance); and (2) are inherently prone to early deterioration and then to provide either for their low durability to be increased by protective coatings (some of which may require regular renewal) or for the periodic replacement of the material itself.

As to regular renewal of protective coatings, the on-going repainting of the Forth Bridge is the outstanding example. Regular renewal of protective coatings is accomplished with relative ease compared with the periodic replacement of layers of materials. It is important that provision is made in design so that replacement can be carried out easily and without disturbing other component parts of the structure.

(d) The life of materials

An assessment of the durability of materials and any maintenance attached to them could be made relatively easily with truly traditional constructions. As explained earlier, the likely performance of a given construction was generally known and any weaknesses it may have were recognized. In addition, the durability of the relatively few materials used could be assessed succinctly, as was done in BS Code of Practice, CP3 – Chapter IX: 1950, *Durability*. Some buildings were expected to last 'forever'; most were considered to be capable of having an economic life of about 60 years. That period appears to have been based on the economics of loan periods to finance new buildings, in the public sector especially, rather than technical issues.

Nowadays the position is somewhat different. There appears to be a greater need economically for buildings to be capable of being designed with lives classified broadly either as long-life or short-life. Obviously, in practice they need to be specified in particular cases. However, and additionally, other classifications or concepts of life have emerged. These include: design life, service life, economic life and life to first maintenance.

Whatever life may be required, once it has been defined there still remains the problem for designers and others to have available the relevant data on durability of the materials involved or likely to be involved and, importantly, the related life-cycle costs.* In all of this there is a large gap that needs extensive and time-consuming research to fill. In the meantime, the best that designers and others can do, as with other problems, is to interpret or extrapolate from such data as are currently available. Special care is needed in any assessment for account to be taken of the degree to which the materials involved may be exposed. Each case is, therefore, likely to be different.

(e) The concept of reversion

As stated earlier, the concept of reversion is a general observation by the authors of this study. It is based on the second law of thermodynamics, namely, that the energy put into a part of a system in one form to do work constantly loses its ability to do that work. The unavailability of energy in a form that can do work is known as *entropy*. The amount of entropy – or disorder – within a system is always increasing. It can only be reversed by adding more useful energy into the system. In terms of materials this is the basic physical law behind decay and deterioration, particularly of refined materials whether natural such as timber, whose creative energy came from the sun, or man-refined such as steel, which has to be refined from its oxidized (i.e. 'natural') state as ore by the application of large amounts of heat energy.

This law also applies to natural unrefined materials such as stone. Geologically, all rocks find their source in the energy of the sun. Igneous rock was formed by the crystallization of the molten magma as it cooled; metamorphoric rock by the application of heat and pressure derived from the movements of the earth's crust; and sedimentary rocks by the laying down of material eroded from higher parts of the crust, themselves only existing because of movements of the earth's crust.

The weathering processes (wind, rain and ultraviolet radiation) cause materials to revert again to some more primitive order. The effect of this can be seen in the natural environment by comparing the landscape of an area that is ancient in geological terms, such as the red heart of Australia which has weathered to a dusty flatness, with a young mountain range such as the Swiss Alps.

The rate of change, of course, varies considerably between materials, and has to be known for different materials, as explained earlier. Thus it is not true to say that metals are more durable than timber if the performance of mild steel is compared with a dense hardwood such as Greenheart.

*Inevitably, it has been necessary to develop the processes involved in life-cycle costing. See, for example, Flanagan, Roger and Furbur, David J., *Life cycle costing for construction*, Report sponsored by the Quantity Surveyors Division of the Royal Institute of Chartered Surveyors, RICS, July 1983.

Part 1

Principles for building

Need

1 Background

There can be little doubt that the track record of the performance of materials during the past 30 years or so has been poor. The same can be said of other aspects of building. Failures of materials, constructions and other parts of buildings, dissatisfaction with the environment and other criticisms about modern building have been frequently reported in the press, discussed at conferences or debated in Parliament. Litigation of building cases, as in other spheres, abounds.

2 Applying the lessons of failures

This is not the place to rehearse the reasons for the failures. That has been done by others, including Lyall Addleson.* Instead, this study draws on and applies the lessons of the failures so that the performance of materials in building constructions may be better achieved in the future. However, it is unlikely that everything included in this study on technical factors that influence the selection, use and thereby the performance of materials can be retained in the memory. Among other things, the complexity of all aspects of modern building will almost certainly cause some mental blockage.

3 Simple aide-mémoire

Accordingly there is, or appears to be, a pressing need for some simple *aide-mémoire*. The five principles, and there are only five to remember, are aimed to provide the basis for just that. Apart from their use as an analytical tool in design they should also serve as a useful checklist at meetings or on-site. It should be noted that an understanding of the fundamentals of building is required if the full potential of the application of the principles is to be achieved. The fundamentals of building as they relate to this study are given where appropriate.

It is recognized that the principles or their application cannot be regarded as a panacea for the resolution of all building problems. They are meant simply to supplement other aids to an improvement of the performance of materials in constructions. At the very least, the application of the principles should highlight areas of possible weakness in a design.

Development

1 Early beginnings

The seeds of the five principles were sown in the authors' desktop analytic work in connection with defects. That work was done initially

*Addleson, Lyall, *Building Failures, a guide to diagnosis, remedy and prevention*, 2nd edn, Butterworths, London, 1989. This includes references to other studies of the reasons for failures.

on the design drawings with the aid of tracing paper overlays and coloured fibre pens, the latter to track the likely path(s) of water (in liquid and vapour form), heat, the wind and so on, all before embarking on predictions. The results of the overlaying provided what engineers sometimes refer to as 'a first approximation'. With more experience it became apparent that there were principles, other than those of building, that could be applied to assist the analysis.

2 Trials on general practitioners

All-day workshops on the physics of building conducted by Lyall Addleson for certain public sector architectural departments and private firms in connection with the continuing education programme of the Institute of Advanced Architectural Studies in York provided an opportunity to test the approach. First, Lyall Addleson introduced the principles and the overlay technique. For the remainder of the day the participants, mostly architects but also including some surveyors and clerks of works, working in small groups, first analysed details that had been produced in their offices. Each group subsequently presented the results of their findings in an open forum. In successive workshops it was possible to improve on the technique and make the principles a little firmer.

At the end of each session one thing was clear: the participants recognized that in their detailed design of construction they had often overlooked some elementary aspects of building construction. Some, if not most, were surprised to find how a few simple principles could assist in exposing weaknesses or potential weaknesses in the detailed design of a construction and without the aid of calculators or computers.

3 Applying everyday experience

At another level, the workshops, and later lectures to architects and others, suggested to the authors that complexity of choice was a root cause of the difficulties that architects and others were experiencing when it came to dealing with materials. The choice related to aspects such as requirements, properties and methods, in particular. How could this problem be resolved simply within the context of the principles?

The answer seemed to lie in applying or adapting experience in everyday life, such as when shopping for something. Among other things, this involves making a choice. This analogy with the choice of materials is not as remote as it might first appear. With materials, the object is also to find the 'best buy' within certain limits. In everyday life the problem is resolved by reducing choice. This is normally done, mostly subconsciously it seems, by a process of elimination, and is achieved by the application and eventually the synthesizing of *objectives, constraints* and *control*.

Consider, for example, buying an item of clothing. One of the objectives is clear: wanting to buy the item. Others might be style, pattern or colour. One of the constraints is also clear: it is size. Another would almost certainly include cost. Control is slightly more difficult to define. Put simply, it is associated with the need to be on one's guard to avoid things going wrong, as they almost certainly will do if certain precautions or checks are not carried out. In the present example and at its simplest level, control would include making certain that the item fits. Equally important, but probably more important in other cases, is whether the quality is adequate for its intended purpose. Are the salesmen or the technical literature to be trusted, or should further checks be made?

4 Basis

Objectives, control and constraints form the basis of the five principles applicable to building, that is, *the principles for building.* The principles should not change with time.* They are, or should be, relatively easy to remember. However, their introduction into the vocabulary of building means that a distinction needs to be made between them and the *rules* and *precautions* that stem, whether consciously or not, from the application of the principles. The concept of *compensation* has to be recognized and applied.

5 Scope

The principles and their relationship with rules and precautions are discussed. The detailed application of all the principles, rules and precautions are explained. The concept of compensation is explained and applied to simple cases.

The principles

The principles are illustrated in Chart C1/1. It will be seen that they have not been grouped in the order of objectives, constraints and control. As an analytical tool or even as a checklist, the order is not important. In any event, there are only five principles. However, as control should be applied to both objectives and constraints it has been placed between the two. Objectives and constraints each have two principles; control just the one.
  From the explanation of each of the principles given below it should be recognized that there is close connection between **high>low** (a constraint) and **balance** (an objective) and between **separate lives** (a constraint) and **continuity** (an objective) with **creative pessimism** (the control) related to both. The connection between **high>low** and **continuity** and between **separate lives** and **balance** is not as close. The connections are illustrated in Chart C1/2.

1 High>low

The principle of **high>low** is derived from thermodynamics, and refers to the natural order of things in two respects: first, gravity and second, the phenomenon whereby a large quantity, whether of water, heat or pressure, in one place (or on one side of a construction) tends to move or want to move towards places with a lesser quantity. Although the latter is a process that occurs during balancing (Fifth Principle), the principle of **high>low** can stand on its own. (Given the pervasiveness of the constraint of **high>low**, the objective of **balance** is to hold, where necessary, different conditions of equilibrium.) The principle includes reversion, explained in Chapter 1 under '4 Durability' earlier (p. 9).

*In this chapter the term 'principle' or 'principles' relates to the principles for building, unless otherwise stated.

1 HIGH > LOW	<ul><li>Gravity (water flow/structure)</li><li>Temperature gradient</li><li>Vapour pressure gradient</li><li>Air pressure gradient</li><li>Reversion (steel > iron)</li></ul>	[DAMP > DRY] [HOT > COLD] [MOIST > DRY] [HIGH > LOW] [PROCESSED > ORIGINAL]	CONSTRAINTS
2 SEPARATE LIVES	<ul><li>Differential movements</li><li>Differential durability</li><li>Incompatible materials</li><li>The process of assembly</li></ul>		
3 CREATIVE PESSIMISM	Allowances for UNCERTAINTY because the properties of all materials, their assembly and performance and the way a building is used are neither totally ideal nor totally predictable		CONTROL
4 CONTINUITY	<ul><li>Structure</li><li>Thermal insulation</li><li>Fire protection</li><li>Cavity separator</li><li>Damp-/waterproofing</li></ul>		OBJECTIVES
5 BALANCE	<ul><li>With surroundings</li><li>Laminates</li><li>Between the parts and the whole</li></ul>	[temperature/moisture] [plywood/facings]	

Chart C1/1 *Principles for building*

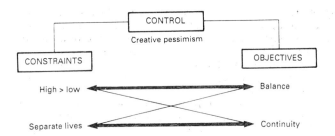

Chart C1/2 *Connections and relationship between the principles in Chart C1/1*

2 Separate lives

As a result of their response to changing conditions, materials tend to be in a state of constant movement. In practice, the amount of movement to be expected in particular cases will, of course, be the determining factor. Apart from, or in addition to, the physical processes, some materials will react with others if the two different materials are in contact and there is another agency present, such as moisture. Therefore there may be either physical or chemical reasons for separating some materials from others (in other words, allowing the respective materials to have **separate lives**). The principle includes the process of assembling buildings.

3 Creative pessimism

The principle of **creative pessimism** follows naturally from the inevitability of variations (see '3 Approach (c) Variations and their effects', p. 2) and uncertainty generally. Importantly, caution is signalled. It is in this sense that it is used in the principles. Furthermore, the principle underlines the need for all concerned with building to recognize and allow for the inherent and almost inevitable shortcomings at a given time in knowledge of materials, of the performance of elements and how buildings are designed, built and used. **Creative pessimism**, as here interpreted, helps to focus on reality. That is why it sits between objectives and constraints.

4 Continuity

The term and the principle it describes should be self-explanatory. The thinking behind it is that no material can be expected to perform its intended function fully if it is not continuous. Structural continuity is probably the most obvious example of the absolute necessity for **continuity**. The same applies to other functional requirements such as damp- and waterproofing, thermal insulation, sound insulation, fire protection and cavity separation. However, in some cases, continuity of performance is achieved by a discontinuity of materials.

5 Balance

A characteristic of constructions is that they create a different environment internally to that outside. A characteristic of environments is that they are changing virtually all the time. The objective of **balance** is to hold or maintain the natural equilibrium of a construction relative to the environment within and outside a building. When either or both environments change, the related construction becomes or tends to become unbalanced to some extent. In trying to achieve what is in effect a new state of equilibrium with its environment, movements and/or adjustments of materials within a construction take place. These may be facilitated or their worst effects reduced in a number of different ways, such as provision for movements, accommodating movements or changes and providing additional heating or cooling.

The objective of **balance** also needs to be considered in another way. Certain composite materials or elements require arrangements to ensure that their inherent balance is maintained so as to avoid or reduce their distortion or deformation.

Rules and precautions

1 General considerations

Rules and precautions result from the practical experience gained from empirical or science-based approaches to the resolution of particular problems in design and/or construction. The resolution may or may not have involved the application of the principles directly or indirectly. Rules and precautions, unlike the principles, are constantly changing, sometimes rapidly. In some cases, further research into a particular subject or other aspects of it has led, understandably, to a change in advice that might have been given earlier. In view of the rapid way in which requirements in buildings have changed, it is not surprising that advice and recommendations, whether from research organizations or British Standards, are being reviewed continually.

2 Rules

Rules are essentially (but not exclusively) design orientated. Examples, with the relevant principle noted, include:

- A cavity in a cavity wall should be at least 50 mm wide (**continuity**);
- A cavity for ventilation in a roof of cold deck design should be at least 50 mm high* (**continuity**);
- Mortar should not be stronger than the units it bonds together (**balance**);
- Damp-proof courses should be at least 150 mm above the adjacent external ground level or roof level †(**high>low/continuity**);
- Damp-proof courses need to be bedded on mortar (**separate lives**);
- Asphalt needs to be laid on an underlay (**separate lives**); and
- A fall of 1 in 80 is needed for adequate drainage of a flat roof (**high>low**).

3 Precautions

Precautions are essentially, but not exclusively, site orientated. They may also be maintenance orientated. Examples include:

- Using appropriate measuring devices, together with appropriate site supervision and control of the relevant procedures to achieve the specified accuracy of line and level;
- Using cavity battens to avoid mortar droppings;
- Protecting materials;
- Ensuring the continuity of damp-proofing, thermal insulation, etc.
- Cleaning surfaces before applying adhesives or bond-dependent materials (such as sealants);
- Using appropriate containers and/or weighing machines together with adequate site supervision and control in the measuring mixes; and
- Maintaining solar protection of waterproof membranes on roofs, protective coatings to ferrous metals, etc. in maintenance programmes.

Application

1 Scope

Having described the principles generally above, attention is now given to explaining how they may be applied to practical examples. The format differs, depending on the categoties in each principle. Rules and precautions related to particular categories are included as appropriate.

2 High>low

(a) Gradients and their control

By its name, this principle implies that there is a gradient between the relative high and low values. In the simplest case this is represented by

*The rules for the area of ventilation slots at each end of the cavity have changed in time – see '5(d) Cavity rule changes' later, p. 28.

†In those parts of the UK more severely exposed to driving rain, experience is pointing to a need to raise the minimum to at least 200 mm.

a straight line; in a complex case such as a multi-layer construction with thermal insulation the gradient would be shallow in places and steep in others. For the present purpose, it is sufficient to refer simply to gradient(s).

By way of an example, the raising of performance standards may be described as increasing the steepness of the gradient between inside and external environments. Thus increasing thermal performance is reflected in better heated (and cooled) buildings and causes the temperature gradient between the inside and outside to be greater. This, in turn, leads to the need for greater control of the flow of heat across that gradient. Steeper gradients always require better methods of control. This concept applies to other aspects such as structural performance and vapour resistance.

In almost all cases, the controlling function is performed by a layer of material. In thermal insulation the layer may be the mass of the construction or, as is more common in modern multi-layer construc-tions, a thermal insulating board or quilt. This layer controls the flow of heat across the construction. Thus the temperatures on each side of the layer will tend to find their own equilibrium, with the bulk of the temperature gradient taking place across the thickness of the insulant.

The rate of flow of heat depends on the size of the gradient. As the gradient becomes steeper, the more critical the continuity of the controlling layer becomes for it to be completely effective. Consider, for example, different conditions of a dam. An empty dam leaks no water. A hole at the bottom of a full dam is much more serious as far as leakage is concerned than one of the same size at the top. The same applies to heat loss and thermal insulation. The insulant can be thought of as the dam and the heat energy as the water. Therefore it becomes more critical to achieve full continuity of the insulant as the temperature gradient across it increases.

(b) Structure

Unlike thermal insulation, the effects of discontinuities in structure are visible and sometimes catastrophic. In the development of structure there has been a working away from the adherence to the principle of **high>low** and finding ways of compensating successfully for the risks involved. This is reflected in the fact that factors of safety form part of the vocabulary of structural engineers in a way that is not generally true for architects and other designers.

The rule for structure that flows from the principle of **high>low** is that forces should be taken to ground as directly as possible. The buildings which express this most purely are the Pyramids in Egypt. However, the primary task of structure is to enclose space. As spans increase, there is inherently a growing conflict with the rule of taking forces to the ground as directly as possible. Working with the principle means that each element should rest on top of another: the result is a building which relies only on the ability of materials to carry compressive rather than tensile forces. The objective of continuity of structure is achieved by continuity of material, but the material can be of discrete units without the need for mechanical fixings or adhesives. Working against the principle requires compensation, and in structural terms this means exploiting the capacity of the materials to carry tensile forces. Where structural members are in tension, whether as beams, ties, rods or catenaries, there must be complete continuity of material to achieve continuity of function.

Of the structural materials available to early builders (stone, brick, rammed earth and timber), only timber is capable of carrying significant tensile forces. However, for building important structures, such as temples and tombs, only stone had sufficient durability. Since stone has relatively high compressive strength but very little strength in tension, it therefore had to be used following the principle. This meant that buildings were usually massive structures until ways were found of spanning over spaces with arches and vaults.*

An arch works purely by compression. Any significant tension forces in it causes hingeing between the individual units and thereby collapse. The arch was used widely as a structural form by the Romans and was then developed further during the Byzantine, Early Christian and Gothic periods. It typifies the important point that continuity of structural function can be achieved with small discontinuous units. In that mortar is used, it is required only to achieve a better contact between the units rather than as a bonding agent.

Greek architecture did not use the arch. The stonework is said to follow the model of timber construction. Although the entablatures laid across the tops of the columns do, in theory, have tensile forces in part of the members, these are sufficiently large for this to be insignificant.

Structures based on materials which can carry tensile forces have a history that goes back as far as the earliest stone buildings. Timber was the only material capable of carrying tensile forces but, due to its limited durability, no record survives. In buildings with masonry walls, timber was often used for the construction of floors and roofs where the problem of spanning is most easily solved by using a material which can carry tensile forces. The alternative was to create vaults of stone or brick.

Larger spans became possible with the advent of the new materials, iron and steel. Being very strong in tension (as well as in compression), they could be designed to span much more economically. A steel beam works by carrying forces of both compression and tension. The I-section presents the most economical use of material to carry these forces where they are at their greatest.

Shell structures in which there are only forces of compression represent the peak of achievement in the economical design of masonry. Steel can be used most economically when it is used in tension only. Similarly, fabrics of adequate durability can be employed in tension and have negligible thickness. If the history of structure is seen as the development towards spanning further and more economically, then buildings using these techniques represent the leading edge of structural achievement. The important point is that members in tension rely on total physical continuity of the members, whereas in structures where one unit rests on another this continuity is not necessary. For structures which rely on tensile forces, particular attention has to be given to the design of the joints to achieve this continuity. This reflects the need for compensation.

(c) Water flow – 1: Shedding

A property of liquids is that their flow under the influence of gravity takes the shape of their container. Applied to the way that water runs over buildings, this means that rainfall will always try to find the shortest route to the ground.

The rule which follows the principle here is that the flow of water over the face of a building should be controlled. A simple example of construction that follows this rule is overlapping elements such as tiling, slating or weatherboarding, in which the laps shed the water away from the joints.

*See Mainstone, Rowland J., *Developments in Structural Form*, Penguin Books, Harmondsworth, 1975 for an excellent history of structure. Mainstone cites an early example of an arch made of brick with a span of 25 m in a building in Baghdad which may date from the third century AD.

(d) Water flow – 2: Drainage

Roof drainage is an instructive example. If water is to drain from the surface there must be a high point and a low point:

- *The rule* for adequate drainage is that the fall must be at least 1 in 80 and that must be achieved. **Creative pessimism** dictates that the falls intended – i.e. shown on drawings or specified – will not necessarily be obtained. There is a level of accuracy to which any part of a building can be built using normal methods.
- So a *further rule* has emerged.* Double the fall to be obtained – i.e. an achieved fall of 1 in 80 is allowed for by specifying a fall of 1 in 40. It is then for the contractor to take such *precautions* that will ensure that normal building accuracy (or special building accuracy, if that has been specified) is reached.

(e) Temperature gradients

The background to gradients and their control has been explained earlier, using thermal insulation as an example ('Gradients and their control', p. 20). The reason the gradients matter is partly to do with heat loss and partly with condensation. For the control of interstitital condensation the *rule* is that the temperature gradient should increase towards the cold side – the opposite is true for the vapour pressure gradient.

(f) Air pressure gradients – 1: moisture vapour

As moisture vapour causes a partial pressure (known as vapour pressure) there will be a gradient across a construction whenever air on one side contains more moisture than the other. The overall *steepness* of that gradient will depend on the difference in the vapour pressure on each side of the construction; the *shape* on the vapour-resisting properties of the materials that make up the construction.

High>low also applies to vapour resistance in that there should be a successive reduction of the vapour resistance of the construction (i.e. from high to low) from the zone of high pressure to that of low pressure. This translates to the general *rule* that the vapour control layer should be on the warm side of the thermal insulant.

(g) Air pressure gradients – 2: wind

The action of the wind on the external surface of a construction results in an air pressure. As such pressure inside is usually different, an air pressure gradient arises. A positive air pressure externally assists in driving rain through gaps in the construction. In a construction with a cavity ventilated to the outside air the air pressure on the outside and within the cavity tends to equalize. Provision for pressure equalization to take place is a form of balancing (Fifth Principle). Its achievement can help greatly to reduce the risk of rainwater penetration through constructions – see *3.4 Exclusion*, p. 291.

(h) Reversion

The concept of reversion is described in Chapter 1, under '4 Durability, (e) The concept of reversion' (p. 14). The *rule* that flows from this concept is that reversion has to be acknowledged and considered in the light of knowledge of different materials. Techniques that attempt to increase the durability of materials such as the preservative treatment of timber or galvanizing of steel are actions of compensation that buy time rather than confound the basic principle.

*PSA Technical Guide to Flat Roofing, Property Services Agency, March 1987, p. 27, '2.9 Falls and drainage'.

3 Separate lives

(a) Systems approach

One of the effects of multi-layer construction is that constructions should be thought of a system. There is an interaction between the parts that make up the element and between elements that make up the whole. The result of the failure to do this was seen most vividly in the case of flat roofs soon after the introduction of additional thermal insulation. The effect of this was to impose stresses on the waterproofing membranes then available that had not been anticipated. Foamed plastic insulants in particular, because of their relatively large thermal movements, resulted in excessive stresses in the membranes and then their tearing.

(b) Differential movements

Differential movements may arise because of structural movements or changes in temperature and moisture (thermal and moisture movements, respectively). The *rule* that flows from the principle is that the construction should make allowance for the movements to take place. That allowance may be provided either by expecting the material to be able to accommodate stresses induced when restrained or by allowing the movement to take place with minimum stress inducement (both are discussed in detail in *2.1 Cracking*, p. 44 ff).

Alternatively, the problem may be designed out. Consider the perimeter of a flat roof. Basically, there are two approaches (Diagram D1/1):

- One is to let the supporting wall and the roof have their own lives. This means that the roof sits on the wall, preferably oversailing it.
- The other is to allow the wall to project above the roof that now abuts the wall. In the second example, linking the roofing and the wall should be avoided. To do this means flying in the face of **continuity**. But we have developed techniques – a *rule* – on how to achieve continuity of performance, in this case of the waterproofing, by using flashings. The *precaution* required is to ensure that there is nothing that will cause or create continuity of the membrane over the discontinuity in the structure. **Creative pessimism** says make sure

(a) No conflict with membrane continuity

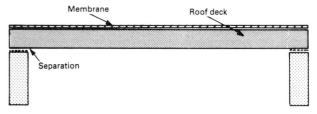

(b) Conflict: membrane **continuity** and **separate lives**

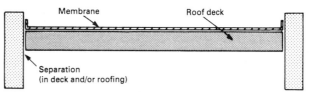

Diagram D1/1 *Flat/wall junction and the application of the principle of* **separate lives**.

that the junction between wall and roof is properly supervised. In the first alternative there is a need to ensure that there is a 'sliding layer' between the top of the wall and the underside of the roof. **Creative pessimism** offers the same advice as before.

(c) Incompatible materials

Where incompatible materials are needed next to one another they need to be separated. Examples include:

- Metals between which there is a risk of bimetallic contact, as in copper/aluminium – see *3.7 Corrosion*, p. 483.
- Certain materials such as bitumen when brought into contact with some plastics roof membranes. Plasticizer migration from plastics such as polystyrene and PVC occurs when in contact with other materials.
- Dust, dirt and debris interfere with or inhibit proper bonding of laps in single membranes. Special *precautions* are needed to ensure that the near-laboratory conditions implied in recommendations are actually achieved on-site.

(d) Differential durability

Differential durability is part of reversion and relates closely to durability. Given that buildings are made up of a number of different materials and components, the *rule* that flows from this principle is to design:

- So as to allow materials of shorter life to be replaced with reasonable ease (e.g. double-glazed units in curtain walling in particular); or
- By choosing materials with similar durability in the same areas of the building (e.g. frame, glazing and cladding in curtain walling – but see note above); or
- By protecting shorter-lived materials so that their effective life is equivalent to the longer-lived ones (e.g. the Forth Bridge approach – see 'Changes and their effects, 4 Durability', p. 9. An example other than timber preservation and galvanizing of steel is the protection of extruded polystyrene in the inverted roof from solar radiation).

4 Creative pessimism

(a) The assembly of buildings

Buildings can be said to be assembled in two main stages: first, on the drawing and, second, on-site. What can be achieved on-site obviously depends on the skills available to assemble the materials to be used but, as importantly, it is, in the first instance, dependent on the nature of the drawing board assembly. The latter should take into account the availability of materials and the skills required for their site assembly.

As constructions become more multi-layered and with many more prefabricated components, so the need increases for designers to be satisfied in advance that all the layers and components will fit together properly on-site and not prejudice one or other functional requirements. The latter will usually mean that one or other of the four principles has been breached. Designers need to recognize that they cannot, as was the case in years gone by, rely on details being 'sorted out on-site' by craftsmen.* Accordingly, there is a greater need for the use of three-dimensional techniques such as drawings, models or, in some cases, full-size mock-ups.

*It is, of course, still possible for a craftsman on-site to be able to assist in sorting out details. However, and importantly, access to craftsmen in general has become more difficult, if not impossible, due to significant changes in contractual procedures. Building contractors, the larger firms in particular, have become no more than managers of the building work, the bulk of which is subcontracted.

However important it is for the design to be capable of being built, there is still a need for the site assemblers to be aware of the relevance or importance of the arrangements of materials within and the functions of a given construction. Site managers, on the other hand, need to exercise greater control if not to adopt different skills for the purpose.

(b) Factors of safety

In the days when choice of materials or the way they should be used was limited the likely performance of a design was known in advance with almost absolute certainty by most, if not all, involved. Performance of materials in use was common knowledge. Designers could proceed with what may be described as 'uncreative optimism'.

Engineers have long recognized that certainty should be constrained by the degree of knowledge available at any given time on properties of materials and related factors such as conditions of use or of the environment. Accordingly, in their predictions they incorporated factors of safety, the factors being governed by the reliability of the relevant data. (The greater the reliability of the data, the smaller the factor of safety.) By so doing, the creativity of engineers is laced with a degree of pessimism, hence **creative pessimism**. The control that pessimism exercises does not nor need not destroy creativity.

The traditional engineering approach should be adopted, therefore, in any aspect of design that involves the use of predictions with numerical data. Care should be taken for designers not be illuded by the apparent accuracy of values to umpteen decimal places that calculators and computers normally produce. In particular, the results of predictions of thermal insulation (but perhaps more importantly the prediction of the risk of condensation) should be regarded as giving basic guidance only. As yet none of the methods of prediction include factors of safety.

Importantly, the same concept is not restricted to numerical aspects. It applies equally to all instances where assumed or specified behaviour, whether of materials, people or the environment, cannot be guaranteed.

(c) Comparing like with like

Traditionally, it was not uncommon for details or forms of construction to be copied from existing examples. This could be done with confidence in the success of the outcome, for the reasons explained in (*b*) above. Looked at in another way, the best traditional constructions and details could be said to have included in them, albeit fortuitously, a large factor of safety or, more simply, that they contained much 'fat' to cope with unknowns, including the quality of workmanship achieved in particular cases.

Designers need, therefore, to take care before they copy existing examples of either constructions as a whole or details in particular. In the final analysis they should assure themselves that they are comparing like with like. In other words, the functional requirements and the conditions of exposure should be similar in both cases (i.e. the existing to be copied and the proposed). In this they should remember that the less 'fat' left in any particular arrangement of materials, the less likely are they to cope with small changes in one or other factor. It is for this reason, among others, that few forms of present-day constructions or their details can be described as traditional. If nothing else, they contain a layer of insulation.*

*It is recognised that the term 'traditional construction' is often used to make a distinction between what is usually referred to as 'high tech'. It might be less confusing in the absence of suitable qualifications whenever either are used if both were given other descriptions.

5 Continuity

(a) Scope

As an objective, **continuity** is applicable to all the categories of performance. It might be better to think of it in terms of performance first when considering, for example, structural stability, water exclusion, thermal insulation, sound insulation, fire protection, damp-proofing and waterproofing. Obviously, when these are translated into materials, then the materials have usually to be continuous, although there are exceptions to this where continuous discontinuity applies, as noted below.

(b) Importance in new fields

The lack of conclusive evidence or knowledge or the fortuitous results of poor workmanship should not be used as an excuse to compromise the continuity of newer layers such as thermal insulation. The importance normally attached to achieving continuity of structure should apply just the same. It should be generally understood, from common knowledge, that breaks in the thermal insulant are likely to lead to cold bridging and possible condensation (surface and/or interstitial). As explained under **high>low** earlier ('2(e) Temperature gradients', p. 23), it should be recognized that the greater the degree of insulation, the greater the risk of the cold bridging and condensation occurring.

The fact that the risk of either happening cannot yet be predicted with confidence would have **creative pessimism** say that such guidance as is available should be laced with a factor of safety. If it is not possible to adapt existing constructions to ensure **continuity**, then a completely new solution should be devised. That is what a creative pessimist would do. The same considerations apply to the closely related vapour control layer.

(c) Continuous discontinuity

Often, a *continuity of function*, such as water exclusion, can and sometimes must be achieved by suitably devised discontinuity. There are two cases of this, both of which are discussed fully in *3.4 Exclusion* (p. 290 and p. 295). *First*, there is the use of small units such as tiles or slates in roofs or wall cladding where the continuity is achieved by overlapping the units. Rules and precautions will guide as to the size of overlap and other features. In this respect, there have been changes in codes of practice that reflect current thinking.* *Second*, there is the use of flashings to waterproof a discontinuity imposed to satisfy the needs of **separate lives**.

In another respect, discontinuity in the form of barriers in cavities required by Building Regulations in respect of fire may be in conflict with the continuity of air paths required for ventilation in the same cavities in a roof of cold deck design. There are obviously ways around this by installing ventilators that would have to penetrate the (continuity of the) waterproofing. **Creative pessimism** cautions that the waterproofing of each ventilator presents a risk of failure that is better avoided by an alternative roof design (i.e. a roof of warm deck design).

Apparent continuity/discontinuity conflicts can sometimes work together. For example, in rainscreen cladding the need to break down the size of cavities behind the screen (to limit the size of the pressure gradients between different parts of the cavity) coincides with similar rules for discontinuity to satisfy fire regulations on limiting the spread of flame and smoke.

*BS 5534: 1978, *Slating and tiling*: Part 1: *Design*.

(d) Cavities: Rule changes

Cavities in cold deck roof designs must be continuous and have inlets/outlets at opposite sides of the roof if there is to be sufficient ventilation of the void. **Creative pessimism** says that there are bound to be obstructions in the void. This means ensuring that there are no obstructions. *Two rules* have emerged:

- One is that proper ventilation of the void requires a cavity height of at least 50 mm;
- The other is that the area of ventilation slots on opposite sides of the roof should equal a certain area or, alternatively, a certain percentage of the roof area.

As to the first, the rules have changed over time. Early on, an area of 0.3% was recommended; this was later increased to 0.4%; now, as a result of further work by BRE, an area of 0.6% is advised in certain cases.* This is perhaps the best example of a change in the rules while the principles remain unchanged.

(e) Performance

On a more philosophical level, continuity could be applied to the performance of the building's functions during its life.

6 Balance

(a) Composite materials

A risk of an imbalance that would result in distortion occurs in almost all composites made up of equal bonded layers of material, whether the materials are the same or not. Plywood is probably the oldest example of a composite made up of layers of the same or similar material bonded to each other. Importantly, practical experience has provided the rule that **balance** is achieved by having an equal number of layers on either side of a core (i.e. together making up unequal layers). Other more recent examples where a balancing layer is required include:

- Laminate-faced plywood, blockboard or chipboard composites. In these, the laminate finish is balanced by another laminate at the back of the board. Both are bonded with an adhesive to the core material.
- Finishes such as tiles and mosaic bonded to plywood, blockboard or chipboard are used as cladding, often on the external walls and exposed to the weather. Although it might be possible to devise a fixing system that can restrain the boards, it is usually more prudent to rely on a balancing layer.
- Coated sheet, usually metal, materials to which a thermal insulant has been bonded either by spraying a foamed plastic or adhering a board. Imbalance is caused by thermal movements. A balancing layer and/or suitable restraint are required.

(b) Composite constructions

The two thermal design concepts of a flat roof (more commonly known as the cold and warm deck) provide good examples of balance/imbalance thermally. The cold deck has only one variant; the warm deck has two, one of which is known as the inverted, upside-down or protected membrane roof. Diagram D1/2 compares the layering of the three. The differences in balance relate to thermal movement, the

*BRE IP13/87, Beech, J. C. and Uberoi, *Ventilating cold deck flat roofs*, October 1987.

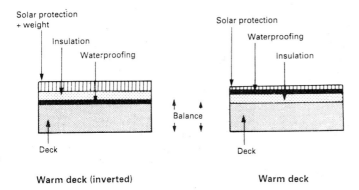

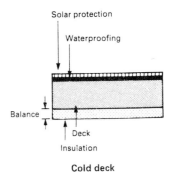

Diagram D1/2 *Flat roof deck types and the application of the principle of* **balance**

location of the thermal insulation layer being the main determining factor. By its nature and function it reduces the passage of heat through. Changes in the thermal environment externally are usually the most influential as to the effects that heat has on movement of the constituent layers. Accordingly, the differences between the three variants of design may be summarized as follows:

- *Cold deck*: Only the insulation, the lowest layer, is in balance. All layers above it are subject, to a lesser or greater degree, to 'extreme' thermal movements. The degree of movement in the waterproof membrane will largely depend on the thermal capacity of the layer(s) below it. Slow-response materials such as concrete will help to reduce the movement; quick-response ones such as timber and air, the opposite. The rule that the waterproofing membrane of all insulated roofs should have solar protection applies.
- *Warm deck – conventional*: In contrast to the cold deck, the layer(s) below the insulation are balanced thermally. Above the layer(s) the effect of the imbalance is to cause comparatively large movements in the thermal insulant, the actual amount depending on the movement characteristics of the material. The stresses created in the layer of thermal insulation because it is bonded to its substrate are transferred to the waterproofing membrane. This means that the latter has to be able to resist the stresses induced, a characteristic not possessed by most membranes used when additional thermal insulation was first introduced into flat roofs. The rule for the solar protection of the membrane is absolute: there must be such protection. The precaution that follows is that such protection must be maintained throughout the service life of the membrane.
- *Warm deck – inverted*: The membrane is thermally the most sensitive of the layers. By transferring the thermal insulant so that it is above rather than below the membrane, the latter is in balance or protected (hence the description 'protected membrane roof' sometimes used). Importantly, the thermal insulant has to be separate from the membrane (**separate lives**) because it is unbalanced and therefore must be free to move to maintain its thermal balance. In addition, it needs to be protected from the effects of sunlight because of the nature of the materials used (i.e. specially formulated foamed plastics or other materials that absorb little water). These are also

light in weight and so have to be weighed down to prevent them being blown away by the wind. The solar protection has therefore to perform this function as well. The choice of material (whether pebbles or paving) will be dependent on whether protection from impact damage by foot and other traffic is required.

(c) Parts and the whole

Balance may also be applied to non-technical issues such as the distribution of resources within a project and the balance of functional objectives between various elements of a building. In deciding on the balance to be achieved in particular cases, consideration should be given to the question of compensation (see below).

Compensation

1 Scope

Almost inevitably, principles cannot always be applied absolutely or the technical truths contained in them cannot always be met fully in practice. It is normal for objectives to be in conflict with constraints (e.g. **continuity** and **separate lives** or **high>low** and **balance**). There may be other pressing demands, such as cost. The probability of a failure of some kind occurring is signalled whenever one or more of the principles is breached in design or is likely to be breached during construction or during the life of the building.

Conflicts that arise should be resolved by *compensation* – not by compromise. The latter implies settlement of differences by partial waiving of theories or principles. So far as the performance of materials is concerned, there is no place for the waiving of theories or principles, the principles for building in particular. Making up for differences or conflicts that may arise is essential – hence compensation. The main purpose of compensation is to reduce the risks of failure that might otherwise occur.

2 Background

The need for compensation as applied in this study arises for two main reasons, both of which have been explained earlier. Briefly, the two reasons result from the following.

(a) Multi-layer construction

Almost all modern constructions are characterized by the inclusion of many layers ('Changes and their effects, 3 Multi-layer construction', p. 7). The nature of traditional construction needs definition. External appearances can be deceptive, as Diagram D1/3 illustrates in the example of the development of the cavity wall, previously described in detail ('Changes and their effects, 3 Multi-layer construction, (c) The cavity wall', p. 8).

Each layer in a multi-layer construction is normally intended to perform a specific function (e.g. exclusion of water, thermal insulation,

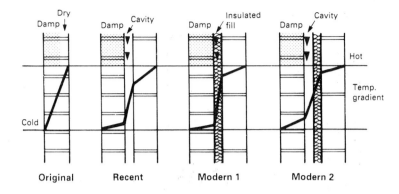

Diagram D1/3 *Stages in the development of traditional cavity wall – an example of multi-layer construction. Note the changes in the damp zones and temperature gradients*

fire resistance or appearance). In almost all cases, the need for compensation arises because of the inclusion of additional thermal insulation – traditional constructions had very little of that. **Continuity** (of damp-proofing, insulation and the vapour control layer in particular) needs to be attained for successful performance.

(b) The application of science

The application of science in order to resolve new and complex problems in buildings has given rise to the need for the consideration and application of the concept of the *risk of failure* ('Changes and their effects, 2 Innovation and the risk of failure', p. 5). This risk increases if tried and tested traditional rules and precautions are ignored. Changes in materials and methods of construction can also be important – examples include firing and design of bricks and mortars. The traditional absorbent 'overcoat' wall may become an impervious 'raincoat' even when apparently traditional materials such as bricks are used (Diagram D1/4). The copying of traditional models without compensating for fundamental differences (e.g. the inclusion of thermal insulation in the 'copy') introduces a high risk of failure. The rule is: only copy like for like; otherwise compensate.

3 Application in theory

Consider the flow diagram D1/5. It starts at the top with the principles and finishes at the bottom with two categories of risk of failure, lowest and highest. An imaginary dividing line runs vertically between the two. On one side of this (i.e. the lowest risk of failure) is the path from the principles to the rules and precautions; on the other (i.e. the highest risk of failure), the need to compensate.

In the simplest and direct case, if all the principles for building can be adopted, the next step is to apply the relevant rules and precautions. If these can be adopted in their entirety then the end result will have the lowest risk of failure for that case. If not, then the end result will carry the highest risk of failure.

On the other hand, if one or more of the principles for building cannot be applied then there is a need to consider compensation. The

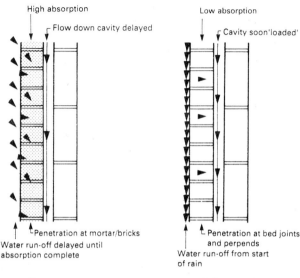

Diagram D1/4 *The effects of changes in the external leaf of a cavity wall (from 'overcoat' to 'raincoat') and the performance of that leaf in relation to the exclusion of rainwater*

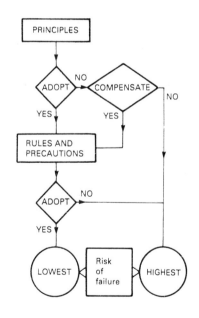

Diagram D1/5 *Flow diagram showing the relationship between principles, rules, precautions and compensation and the risks of failure resulting from taking different routes*

choice is whether to compensate or not. If the decision is not to compensate then the end result will have the highest risk of failure. If compensation is chosen, the relevant rules and precautions need to be considered and, if adopted, the end result will have the lowest risk of failure.

To summarize: in general, the lowest risk of failure (i.e. without quantifying its degree) results from one of two alternative actions: either adopting the principles and rules and precautions in their entirety or compensating for those principles that cannot be applied and adopting the relevant rules and precautions. The opposite occurs again from one or two alternative actions: either adopting the principles but not some or all of the rules and precautions or not compensating for those principles that are not adopted.

(a) Compensating at the jamb of a cavity wall

Adding thermal insulation in the cavity created a new problem. Resolving the conflict between water penetration and heat loss is illustrated is Diagram D1/6 and described briefly below:

- In the traditional example the window frame would be built in as the work proceeded. Providing continuity of the damp-proofing at the joint between the frame and the masonry is relatively simple as the dpc can be tucked into a groove in the frame.
- The problem of water penetration is resolved by compensating the continuity of the outer leaf with a dpc much as in a traditional detail. Because the frame is not built in as the work proceeds, the joint between the frame and the masonry has to be sealed (Compensation 1). The sealant, compared with the dpc, has a relatively short life. Replacement of the sealant during the life of the building is required to compensate for this.

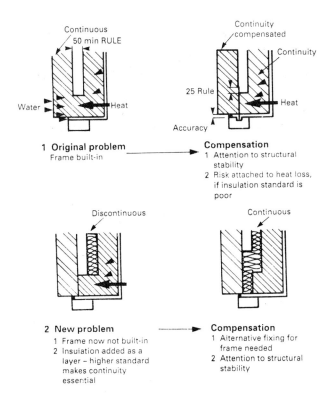

Diagram D1/6 *Application of compensation to the window jamb of a cavity*

• The problem of cold bridging, the risk of it occurring having increased because of the additional thermal insulation, needs to be resolved because the thermal insulation is not continuous. The thermal insulation is made continuous by offsetting it into a cavity at the jamb. The dpc has to be extended in width (Compensation 2). The fixing for the window frame has to be modified.*

(b) Examples of compensating in flat roofs

1. If it is not possible to provide adequate drainage (**high>low**) then the risk of water penetrating at laps – built-up felt or single membranes – is greatly increased. Options to reduce the risk include:
   • Adopting a membrane that is jointless such as asphalt or a sprayed-on polymeric membrane;
   • Ensuring that workmanship and its supervision is above the norm; or
   • Ensuring that maintenance checks are made more frequently than might normally be the case and that such repairs as are found to be necessary are actually carried out.
2. If adequate ventilation of the roof void in a cold deck roof design cannot be assured – because of problems of **continuity**, for example – the risk of failure by excessive condensation is so high that a warm deck roof design is essential. No amount of technological gymnastics will show otherwise.

*Special cavity closers are available to facilitate the practical problems of achieving continuity of the damp-proofing and insulation and the integrity of the fixing.

3. If the conflict between **separate lives** and **continuity** cannot be resolved properly at the perimeter parapets through continuity of a function by discontinuity there is a high risk of cracking of the membrane. Alternatives include:
   - Choosing a membrane that is more likely to withstand the stresses induced;
   - Reducing the area of the roof by incorporating movement joints at smaller intervals than would be the norm; or
   - Having the roof inspected and maintained more frequently than would be normal.

Part 2

Strength of materials

Strength is both a basic property of all solid materials and one of the most important properties of materials used in buildings. A material must, above all else, be able to withstand the anticipated loads imposed on it during its working life. In short, it can only perform its functions properly if, during use, it is able to carry loads without undue distortion or rupture. Such loads include the material's self-weight.*

This ability to carry loads is dependent on two primary, interrelated factors:

- The inherent strength of the material itself, measured by its mechanical properties; and
- The way in which this inherent strength is employed through the size and shape of the material for given methods of loading. The employment of the mechanical properties of a material forms the subject of *structural mechanics*. Hence mechanical properties of materials and structural mechanics are intimately linked.

Loading and its effects on the mechanical behaviour of materials is obviously of first importance in any consideration of the structure of buildings. However, it also plays an important part in understanding the causes of cracking of materials in use, whether forming part of the basic structure of the building, or non-structural elements, such as non-load-bearing partitions, wall and floor finishes.

The inherent strength of materials also influences the way in which materials may be used in forming and fixing operations. Finally, strength plays an important part in the resistance of materials to abrasion and impact as encountered in buildings.

In this study it is the latter applied use of structural principles which is discussed. It is important to emphasize that structural design is outside the scope of this book: for this, standard works must be consulted. An innovatory feature in the chapter on cracking is the classification and guide to the recognition of cracks. Emphasis is given to the effects of variations. Problems associated with soils, fills and hardcore are dealt with separately. The chapter on the strength and the use of materials in constructions relates the assembly of materials to strength. Shaping and forming and strength in use generally are also considered.

*This aspect is sometimes overlooked with what are described as 'non-load-bearing' elements. In the strict structural sense, no material or construction can be said to be non-load-bearing. This is of particular significance in modern multi-layer constructions.

Cracking

General considerations

1 Broad-brush approach

Put simply, cracking is the result of the overstressing of materials due to one or more of the movements given in Table T2.1/1. The way any of these movements may induce the stresses is complex. Materials vary in their properties as regards strength and in other ways. Individual materials are seldom completely homogeneous. These aspects influence the resistance to cracking that a material will provide. In addition,

Table T2.1/1 *Principal causes and effects of movements responsible for cracking*

Cause	Effect	Duration, frequency	Examples of materials or components affected	Notes
1. *Externally applied loads*				
(a) Dead and live loading: Elastic deformation under service	Normally insignificant in vertical members but horizontal members may deflect	Continuous or intermittent under live loads; long term under dead loads	Suspended floor and roof slabs, beams, edge beams or spandrels and claddings; all other elements that support or 'contain' the cladding	For claddings: needs consideration relative to their fixings and bearings and to possible compression of 'contained' claddings: deflections in pre-stressed concrete may be large.
				Deflection of beams and compression of elements 'contained' below them needs consideration
(b) Creep	Contraction of vertical and deflection of horizontal members	Long term	Reinforced and pre-stressed concrete components as above	For claddings: needs consideration as above. May also be significant where load bearing concrete walls or columns have cladding such as mosaic or tiling directly bonded
(c) Wind loading on cladding	Deflection (bowing or hogging)	Intermittent – frequency important	Lightweight cladding, including fixed and opening glazing; sheet siding	Extent of deflection depends on exposure for a given stiffness. Deflection is commonly designed not to exceed 1/240 of the span in order to avoid damage to sealants and glazing

Table T2.1/1   *(Continued)*

Cause	Effect	Duration, frequency	Examples of materials or components affected	Notes
2. *Restraint of internal movements*				
(a) Temperature changes	Expansion and contraction	Intermittent, diurnal, seasonal	All	Temperature gradients and the thermal behaviour of non-homogeneous materials or layers of materials may be significant.
				Extent of movement is influenced by thermal coefficient, exposure, colour, thermal capacity, insulation provided behind membranes or claddings
(b) Moisture content changes:				
1. Initial moisture absorption	Irreversible expansion	Relatively short term, due to absorption of moisture after manufacture	Brick and other ceramic products	Depends on age of product: most movement occurs within first 3 months of product's life
2. Initial moisture release	Irreversible contraction	Relatively short term	Mortar, concrete, sand-lime bricks; unseasoned timber. Ceramic products	May require measures to control or distribute cracking in masonry
3. Alternate absorption release or moisture in service	Expansion and contraction	Periodic – e.g. seasonal	Most porous building materials, including cement-based and wood or wood-based products	In claddings, humidity gradient or non-homogeneity may be significant; laminates of dissimilar materials, particularly if their construction is asymmetrical, need consideration. Large movements in timber across the grain
			Shrinkable clay soils	Proximity of trees; effects of felled trees
3. *Vibrations* (from traffic, machinery, wind forces)				
	Possible loosening of fixings, disturbance of glazing areas	Intermittent and at high frequency	Lightweight cladding, glazing, sheet siding	Authenticated cases of cracking directly attributable to vibrations are rare. Situation may change
4. *Chemical changes*				
(a) Corrosion	Expansion – permanent	Continuous – over months to years	Iron and other ferrous metals in masonry (ties, lintels) or externally exposed reinforced concrete	Depends on protection (adequate cover in reinforced concrete) or resistance of metal

Table T2.1/1   *(Continued)*

Cause	Effect	Duration, frequency	Examples of materials or components affected	Notes
(b) Sulphate attack	Expansion – permanent	Continuous – over months to years	Portland cement products in constructions where soluble sulphates (e.g. high-sulphate bricks, fill or hardcore) and persistent dampness present	Signifiicant for: 1. Cladding where the construction affected has cladding such as mosaic or other tiling or rendering bonded directly to it 2. Ground-floor slabs on colliery shale or other materials containing sulphates
(c) Carbonation	Contraction – permanent	Continuous – over months to years	Porous Portland cement products such as dense and lightweight concrete, asbestos-cement, reinforced fibre products (i.e. asbestos substitutes)	Generally, not significant except asbestos (now fibre-reinforced) sheets painted one side only
(d) Akali silica reaction	Expansion – irreversible	Over many years	Concretes containing reactive aggregates and sufficient alkali	Concrete that remains wet for long periods (e.g. slabs, earth retaining structures)
(e) Moisture expansion of ceramics	Expansion – permanent	Over many years	Fired clay bricks and tiles	1. Chemical reaction with atmospheric moisture 2. Slender brick walls (especially cladding and parapets) 3. Tiled floors and walls
5. *Physical changes*				
(a) Ice or crystalline salt formation	Expansion in building materials Frost heave in soils	Intermittent, dependent on weather conditions and moisture content of materials/soils	Porous natural stones, very exposed brickwork	In walling, damage usually confined to surface spalling or erosion Cracking in exposed ground-floor slabs
(b) Loss of volatiles	Contraction, loss of plasticity	Short or long term, depending on materials, exposure	Some sealants, some plastics	Contributes to age-hardening of some sealants. May lead to embrittlement and distortion of some plastics
(c) Crypto-efflorescence	Expansion; internal damage; spalling. Salt staining	Intermittent – related to the weather	Porous materials – notably fired clay bricks, tiles and natural stones which contain salts inherently or from contamination	1. Walls; roofs; floors subject to wetting and drying cycles 2. Structures in contact with contaminants – groundwater, sea water, effluents, acid rain

Table T2.1/1   *(Continued)*

Cause	Effect	Duration, frequency	Examples of materials or components affected	Notes
6. *Movements in soils*				
(a) Loading	Settlement	1. Soils: varies with season; 2. Fill: loading/compaction of fill dependent	1. Soils: silts and peaty ground (clays) particularly susceptible; 2. Fill: ground-floor slabs	1. Soils: proximity of trees or effects of felled trees need consideration 2. Fill: need to consider analysis (load-bearing and sulphate content)
(b) Mining subsidence, swallow holes, landslips, soil creep, earth tremors	Settlement			

Based on *Materials for Building*, Vol. 1, Table 2.03/1 and BRE Digest 223, *Wall cladding: designing to minimise inaccuracies and movements*, March 1979, Table 1, p. 3 and BRE Digest 361, *Why do buildings crack*, May 1991, Table 1.

other influencing factors include how a material is used, where it is used and the way it is exposed to water, heat and so on. The mechanism of cracking and evaluating the risk of it occurring in individual cases is complex.

Therefore what is required initially for the general practitioner is what engineers sometimes refer to as a first approximation, a broad-brush approach that assists in identifying the most likely problem areas and their solutions in principle. Such an approach is used in this study by applying elementary considerations of strength and understanding how directions of movement in a material together with modes of their restraint can be translated reasonably simply into forces and then into stresses.

This approach is in contrast to that used in *Materials for Building*, Vol. 1 (*2.03 Cracking in buildings*, pp. 110–134), where principles were explained and much use was made of case studies. In this book the latter have been taken into account, absorbed and assessed for presentation in a different way. Similarly, account has been taken of the results of further research and knowledge gained from further case studies reported by BRE and others.

2 Significance

Cracking in building materials, including particularly the elements of which they form a part, is probably one of the commonest and, in some senses, most tiresome failures that occur in practice. Slight cracks are usually only unsightly and unacceptable, if not worrying, to occupants or owners. Excessive cracking may lead to serious structural weakening.

Some cracking is probably inevitable. *As a principle*, good practice prefers that every reasonable attempt is made to reduce cracking to the absolute minimum. Small or apparently insignificant cracks may provide the initial paths for water to penetrate and cause dampness where it was not intended. The dampness may cause further cracking through chemical and/or physical changes. In this way the cracking becomes progressively worse, as may the decay and deterioration that accompanies it. Alternatively, cracking may reduce the performance of the building fabric through air infiltration, increased heat loss and reduced sound insulation.

The occurrence of cracks initially or during the use of a building usually means additional maintenance costs. Where cracking results in serious structural weakening, the cost of rectifying the cause of the failure could be considerable.

3 Structural weakening

For reasons that are explained later, a material generally fails because it has not been able to resist the tensile stresses induced in it by one or other of the movements given in Table T2.1/1. When a material cracks it has been weakened structurally to some extent. The reduction in strength which the cracking causes will depend on the size, depth and distribution of the cracking that has taken place. Consequently, slight cracks (the term 'slight' in this context being relative to circumstances) are normally not considered to be serious, as they do not necessarily cause a significant loss of stability. As guides, Tables T2.1/2 and T2.1/3 classify the degree of visible damage for walls caused by settlement.

However, it is important that the determination as to whether cracks in a particular circumstance are serious or otherwise should be based initially on the relationship between the actual strength of the material in question and its size and shape relative to the induced stresses. On the other hand, when determining the seriousness of cracks, account should be taken of any loss of efficiency of the building and its fabric in other ways (see '2 Significance' above).

4 Occurrence

(a) Principal causes

The movements that are principally responsible for cracking are given in Table T2.1/1. The main characteristics of each are:

1. *Externally applied loads*: These relate to direct loading of the structural and other elements. The main effect is to cause the elements to deflect – either sagging or hogging.
2. *Restraints of internal movements*: These are caused by changes either in temperature or moisture in materials. The movements (either expansion or contraction) and the restraint result in indirect loading. Most cracking is caused by one or both of these movements. It is normally not possible to prevent either cause of the stresses entirely. The stresses involved are always associated with the material or element 'wanting to move'. Consequently, the stresses induced may be reduced by allowing the material or element to move or to deform to some extent: in other words, providing some space for movement to take place – some 'elbow room'.
3. *Vibrations*: These are a form of momentary loading caused by traffic, machinery or the wind. As far as is currently known their main effect is the possible loosening of fixings or the disturbance of glazing areas rather than cracking. The position may change.
4. *Chemical changes*: These include corrosion of ferrous metals (expansion), sulphate attack of Portland cement and the products made with it (expansion) and carbonation of porous Portland cement products (contraction). With any of these there is no question of providing space for the movements to take place. The change has to be prevented from occurring, almost always by controlling dampness.
5. *Physical changes*: These include ice or crystalline salt formation (expansion) and loss of volatiles (contraction or loss of plasticity). As

Table T2.1/2   *Classification of visible damage to walls with particular reference to ease of repair of plaster and brickwork or masonry*

Category of damage	Degree of damage[a]	Description of typical damage (Ease of repair in italic type)	Approximate crack width (mm)
0	Negligible	Hairline cracks of less than about 0.1 mm width are classed as negligible	Up to 0.1[b]
1	Very slight	*Fine cracks which can easily be treated during normal decoration.* Perhaps isolated slight fracturing in building. Cracks rarely visible in external brickwork	Up to 1[b]
2	Slight	*Cracks easily filled. Redecoration probably required. Recurrent cracks can be masked by suitable linings.* Cracks not necessarily visible externally; *some external repointing may be required to ensure weathertightness.* Doors and windows may stick slightly	Up to 5[b]
3	Moderate	*The cracks require some opening up and can be patched by a mason. Repointing of external brickwork and possibly a small amount of brickwork to be replaced.* Doors and windows sticking. Service pipes may fracture. Weathertightness often impaired	5 to 15[b] (or a number of cracks up to 3)
4	Severe	*Extensive repair work involving breaking-out and replacing sections of walls, especially over doors and windows.* Window and door frames distorted, floor sloping noticeably.[c] Walls leaning[c] or bulging noticeably, some loss of bearing in beams. Service pipes disrupted	15 to 25[b] but also depends on number of cracks
5	Very severe	*This requires a major repair job involving partial or complete rebuilding.* Beams lose bearing, walls lean badly and require shoring. Windows broken with distortion. Danger of instability	Usually greater than 25[b] but depends on number of cracks

[a] It must be emphasized that in assessing the degree of damage account must be taken of the location in the building or structure where it occurs, and also of the function of the building or structure.

[b] Crack width is one factor in assessing category of damage and should not be used on its own as a direct measure of it.

[c] Local deviation of slope, from the horizontal or vertical, of more than 1/100 will normally be clearly visible. Overall deviations in excess of 1/150 are undesirable.

From BRE Digest 251, *Assessment of damage in low-rise buildings with particular reference to progressive foundation movement*, July 1981, Table 1, p.5, Building Research Establishment, Crown Copyright.

with chemical changes, there is no question of providing space for the movements to take place. Unlike chemical changes, it may not be possible to prevent the change. In that case an alternative material that is more likely to resist the change should be selected. In the case of sealants, replacement as required may be considered appropriate.

6. *Movements in soils*: This is a different type of change from the others. It is a rather special cause but can be related to both externally applied loads ((1) above) and movements caused by temperature and moisture ((2) above). However, other factors are also involved.

(b) Materials commonly affected

Cracking is more common with all materials used in the wet-trade processes (e.g. concrete, screeds, plasters, renders, bricks and ceramics). The cement included to form the product (concrete, screed,

Table T2.1/3 *Classification of visible damage caused by ground-floor slab settlement*

Category of damage	Degree of damage	Description of typical damage	Approximate (a) crack width (b) 'gap'[a] (mm)
0	Negligible	Hairline cracks between floor and skirtings	(a) NA (b) Up to 1
1	Very slight	Settlement of the floor slab, either at a corner or along a short wall, or possibly uniformly, such that a gap opens up below skirting boards which can be masked by resetting skirting boards. No cracks in walls. No cracks in floor slab, although there may be negligible cracks in floor screed and finish. Slab reasonably level	(a) NA (b) Up to 6
2	Slight	Larger gaps below skirting boards, some obvious but limited local settlement leading to slight slope of floor slab; gaps can be masked by resetting skirting boards and some local rescreeding may be necessary. Fine cracks appear in internal partition walls which need some redecoration; slight distortion in door frames so some 'jamming' may occur necessitating adjustment of doors. No cracks in floor slab although there may be very slight cracks in floor screed and finish. Slab reasonably level	(a) Up to 1 (b) Up to 13
3	Moderate	Significant gaps below skirting boards with areas of floor, especially at corners or ends, where local settlements may have caused slight cracking of floor slab. Sloping of floor in these areas is clearly visible (slope approximately 1 in 150). Some disruption to drain, plumbing or heating pipes may occur. Damage to internal walls is more widespread with some crack filling or replastering of partitions being necessary. Doors may have to be refitted. Inspection reveals some voids below slab with poor or loosely compacted fill	(a) Up to 5 (b) Up to 19
4	Severe	Large, localized gaps below skirting boards: possibly some cracks in floor slab with sharp fall to edge of slab (slope approximately 1 in 100 or more). Inspection reveals voids exceeding 50 mm below slab and/or poor or loose fill likely to settle further. Local breaking-out, part refilling and relaying of floor slab or grouting of fill may be necessary; damage to internal partitions may require replacement of some bricks or blocks or relining of stud partitions	(a) 5 to 15 but may also depend on number of cracks (b) Up to 25
5	Very severe	Either very large, overall floor settlement with large movement of walls and damage at junctions extending up into first-floor area, with possible damage to exterior walls, or large differential settlements across floor slab. Voids exceeding 75 mm below slab and/or very poor or very loose fill likely to settle further. Risk of instability. Most or all of floor slab requires breaking-out and relaying or grouting of fill; internal partitions need replacement	(a) Usually greater than 15 but depends on number of cracks (b) Greater than 25

[a] 'Gap' refers to the space – usually between the skirting and finished floor – caused by settlement after making appropriate allowance for discrepancy in building, shrinkage, normal bedding down, etc.

From BRE Digest 251, Table 2, p.7, Building Research Establishment, Crown Copyright.

plasters or renders) or to provide adhesion (brickwork, plastering, rendering, mosaic or tiling) is normally the initiator, so to speak, of the cracking, in the early days of the relevant materials in particular. The cracking is usually attributable to lack of proper precautions for mixing, drying or adhesion. Importantly, all the materials involved are brittle. This accounts for the comparative ease with which they crack.

In composite elements, such as brickwork or brickwork/concrete clad with plaster, rendering, mosaic or tiling, the position is slightly more complicated. The maximum tensile strength will be determined by either the strength of the individual materials (in brickwork, for example) or the adhesion between the cladding and its background.

(c) Thermal effects

There is now a greater possibility of cracking occurring in some recent constructions due mainly to significant increases in expansion and contraction from changes in temperature. The reasons include:

- The use of larger units with fewer joints. This results in greater potential movement at any one point.
- The use of newer materials some of which, such as plastics, may have thermal coefficients of expansion up to 10 times greater than most traditional materials.
- The increased use of thermal insulation. This results in greater movements of other materials employed with the insulants in layered or composite constructions.
- The use of thin sections whose low thermal capacity increases the range of temperature experienced by a material when subjected to extremes of temperature.

Classification and recognition

1 Movement/restraint and force relationship

In order to understand more easily the mechanism of cracking and to recognize the various forms it can take, it is first necessary to convert the movement and restraint of a material to forces. This is a relatively simple matter because arrows used to show the direction of movement in a material or element become the direction of the force. The force related to the direction of the movement can be considered as the counteracting force; the restraint as the acting force. This enables the nature of the force within the material or element to be determined. Common types of movement and the forces they create within elements are illustrated in Diagram D2.1/1. These are explained below. It should be noted that, in all cases, cracking occurs *normal* to the line of the force. This is a useful *aide-mémoire*.

(a) Expansion – uniform

The expansion is assumed to take place uniformly in the material or element. The desired movement outwards is restrained. The effect of the restraint is to induce compression.

(b) Contraction – uniform

The contraction is assumed to take place uniformly in the material or element. The desired movement inward is restrained. The effect of the restraint is to induce tension.

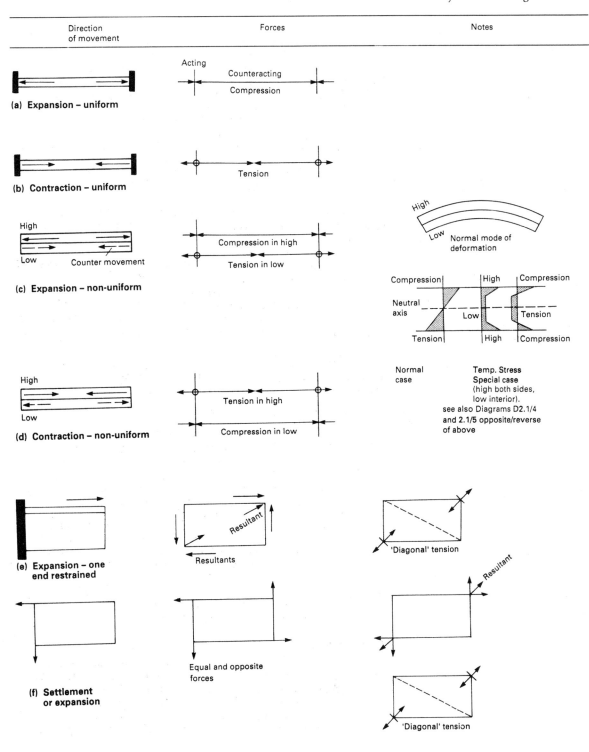

| Direction of movement | Forces | Notes |

(a) Expansion – uniform

Acting   Counteracting

Compression

(b) Contraction – uniform

Tension

(c) Expansion – non-uniform

High

Low   Counter movement

Compression in high

Tension in low

High

Low

Normal mode of deformation

Compression   High   Compression

Neutral axis   Low   Tension

Tension   High   Compression

(d) Contraction – non-uniform

High

Low

Tension in high

Compression in low

Normal case

Temp. Stress Special case (high both sides, low interior).
see also Diagrams D2.1/4 and 2.1/5 opposite/reverse of above

(e) Expansion – one end restrained

Resultant

Resultants

'Diagonal' tension

(f) Settlement or expansion

Equal and opposite forces

Resultant

'Diagonal' tension

Diagram D2.1/1 *Forces induced by movements*

(c) Expansion – non-uniform

The material or element may be homogeneous but the expansion is not uniform throughout the section. This may occur when there is either a temperature or moisture gradient across the section (see 'Gradients', p. 55). The same applies if the temperature or moisture content of the surface layer alone is increased. Alternatively, two materials with different responses to changes in temperature or moisture may be bonded together to form a composite. If the gradients are considered as separate layers then all the examples behave in the same way. Accordingly, because the outer layers are at a higher temperature or moisture content, these will expand more than the inner layers. Relatively, the latter can be considered to be contracting, thereby providing the restraint and so the acting force. Thus compression and tensile forces will be induced as illustrated. However, except in the case of actual surface layers, the deformation will tend to be in the form of bowing, as noted in the diagram, because of the relative weakness of the restraint.

In addition to a temperature or moisture gradient, there will also be a stress gradient. The outer layers will be in tension, the inner layers in compression. Each is at its maximum at the exposed surface. In the exceptional case of actual surfaces there will be tension at or near the exposed surface and compression in the interior, as noted in the diagram. The distribution of these stresses help to explain the phenomenon of surface crazing in many cement-based products.

(d) Contraction – non-uniform

This case is the opposite of non-uniform expansion described in (c) above.

(e) Expansion – one end restrained

A commonly occurring example is the expansion of a roof slab that is restrained at one end. This is a complex case that can be simplified by using the resultant of the forces. For the present purpose (and simplicity) the stresses within the roof slab itself can be ignored. What is important is the effect that its uni-directional expansion has on the wall or other structure supporting the roof slab. It is assumed that the latter cannot move at its free end (i.e. it is restrained by its supporting structure). This restraint results in the basic opposing forces that are the characteristic of shear (see Diagram D2.1/1(e)) and should not be confused with the restraint at the other end of the roof slab that is responsible for its uni-directional movement. To achieve equilibrium, each of the forces is accompanied by another force acting at right angles. The resultant of these two forces is a force acting roughly along the diagonal. The relationship between the resultant forces at each end is to induce tension on either side of the diagonal.

(f) Settlement or expansion

This is a form of shear except that the movement takes place in two directions, outwards and downwards, and is known as rotation. It may be caused by settlement as in the case of foundations or ground-floor slabs or by expansion, as in the case of walls supporting roofs or having returns. Where walls support roof slabs, rotation applies to the support and the shear described in (d) above to walls at right angles to the support.

The outward and downward forces are each balanced by a force acting in the opposite direction. When combined at the opposite corners, the resultant forces pull outwards to be counteracted by forces

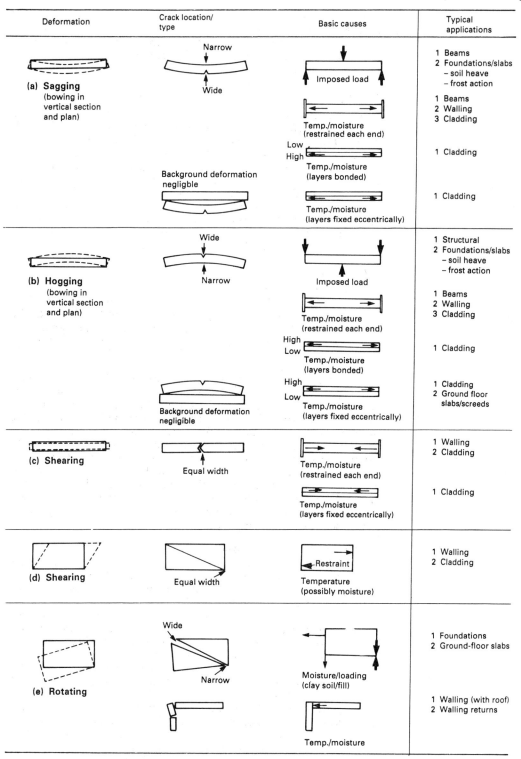

Deformation	Crack location/type	Basic causes	Typical applications
(a) **Sagging** (bowing in vertical section and plan)	Narrow / Wide	Imposed load	1 Beams 2 Foundations/slabs – soil heave – frost action
		Temp./moisture (restrained each end)	1 Beams 2 Walling 3 Cladding
		Low / High — Temp./moisture (layers bonded)	1 Cladding
	Background deformation negligble	Temp./moisture (layers fixed eccentrically)	1 Cladding
(b) **Hogging** (bowing in vertical section and plan)	Wide / Narrow	Imposed load	1 Structural 2 Foundations/slabs – soil heave – frost action
		Temp./moisture (restrained each end)	1 Beams 2 Walling 3 Cladding
		High / Low Temp./moisture (layers bonded)	1 Cladding
	Background deformation negligible	High / Low Temp./moisture (layers fixed eccentrically)	1 Cladding 2 Ground floor slabs/screeds
(c) **Shearing**	Equal width	Temp./moisture (restrained each end)	1 Walling 2 Cladding
		Temp./moisture (layers fixed eccentrically)	1 Cladding
(d) **Shearing**	Equal width	Restraint / Temperature (possibly moisture)	1 Walling 2 Cladding
(e) **Rotating**	Wide / Narrow	Moisture/loading (clay soil/fill)	1 Foundations 2 Ground-floor slabs
		Temp./moisture	1 Walling (with roof) 2 Walling returns

Diagram D2.1/2 *Classification of crack initiation*

inwards across the diagonal. As in the case of shear, this results in induced tension. However, as rotation normally involves hingeing at the corner away from the movements the tension along the diagnonal is unequal, being greatest at the end away from the hinge. As will be seen later, this influences the shape of the crack.

2 Crack initiation

The examples described above ('1 Movement/restraint and force relationship') can be classified into five types of deformation. For each type there is a recognizable location and type of crack; the basic causes related to each type and typical locations in which they apply may be identified. All these are illustrated in Diagram D2.1/2, in which deformation is indicated by dashed lines and loading and direction of movements by different types of arrows.

The classification is intended to provide the basis for the broad-brush approach mentioned earlier under 'General considerations', (p. 37). Explanatory comments on each class are given below. It should be noted that in each class the extreme case is illustrated.

(a) Sagging

(1) Deformation
This is bending, in which there are tensile stresses in the bottom half of the deformed section and compressive stresses in the top half. The sagging may also occur in vertical section or in the plan of a construction, in which cases it is more commonly referred to as bowing.

(2) Crack location/type
In the extreme case illustrated, the crack occurs in the zone of greatest tension, i.e. in the centre of the sag. It is initiated at the outer edge, forming a wedge or triangular shape across the width of the material or element. The shape of the crack arises because the tension across the section reduces (i.e. the crack stops at the tension/compression interface or the neutral axis, to use the structural description). In practice, the crack seldom extends to the neutral axis. It is usually restricted to a zone near the outer leaf. Equally important, there may be more than one crack. These usually occur on either side of the centre line and diminish in depth and width as they move away from that line. The size, location and distribution of cracks will therefore depend on circumstances in individual cases.

(3) Basic causes and typical applications
- *Imposed load*: The imposed loads are associated with elastic deformation included in Table T2.1/1 and relate mainly to beams. In foundations or ground-floor slabs soil heave or frost action would tend to produce the same type of loading.
- *Temperature or moisture expansion – 1: with end restraint*: The condition is as illustrated in Diagram D2.1/2(a) and applies to beams, walling and cladding. The bowing, the deformation most likely to occur, could, of course, take place in either way or it could be restrained (by the background to which claddings may be attached, for example). Materials or elements of thin section are more susceptible to potential cracking.
- *Temperature or moisture expansion – 2: bonded layers*: The basic cause relates to either homogeneous materials with a **high>low** (temperature or moisture) gradient across them or to composite materials or elements in which two layers are bonded together (see Diagram D2.1/1 and related text). In both cases there is a difference in the relative movements of the two zones or layers in which the lower

moves more than the upper. Deformation of the whole thickness occurs but cracking is usually confined to the lower zone or layer. The size and distribution of the cracking would be the same as in the case of imposed loading described above. Claddings are most commonly affected.

- *Temperature or moisture expansion – 3: layers restrained by fixings*: The cause is more likely to relate to vertical sections. The direction of the bow will depend on circumstances in particular cases. However, this condition is characterized, in contrast to (2) above, to fixing of one layer eccentrically (i.e. mainly at or near the perimeter of the material or element) to another. The two layers can then move independently. In practice, it is the bottom layer that bows; the other layer (i.e. the background) normally does not deform much. The consequences in terms of cracking of the bottom layer are the same as described for (2) above. In some cases the cause may relate to plastered soffits of reinforced concrete slabs, where unintentionally, the perimeter of the whole area of plaster, or selective parts of it, are edge-fixed by better adhesion.

(b) Hogging

(1) Deformation
This is exactly the same as sagging except mirrored. The comments for sagging apply.

(2) Crack location/type
The comments given for sagging apply, bearing in mind that everything is mirrored.

(3) Basic causes and typical applications
Again, comments given for all the examples are the same as for sagging, bearing in mind that everything is mirrored. There is one exception that should be noted. This relates to the example where the background deformation is negligible. Screeds may tend to hog during drying out; ground-floor slabs will tend to hog if there is sulphate attack of the concrete (see Table T2.1/1, 4(b)).

(c) Stretching

(1) Deformation
The deformation is characteristic of tensile stresses.

(2) Crack location/type
The material or element is torn apart at or near the centre line. The crack is of equal width on both sides.

(3) Basic causes and typical applications
The basic causes are associated with contraction – see 1(b) and (d), under '1 Movement/restraint and force relationship' earlier. There is a strong resemblance to the causes of sagging or hogging, except that the direction of the movements are reversed. Walling and cladding are commonly affected.

(d) Shearing

(1) Deformation
The deformation follows conventional shear. The mechanics and resolution of the complexities are explained in '1 Movement/restraint and force relationship', (e).

(2) Crack location/type
The cracking is initiated along or near to the diagonal between the corners at opposite ends of the movement/restraint. The crack or cracks are of uniform width or nearly so. The cracking may not, of course, extend from end to end of the diagonal but be localized, depending on the location of the highest stress and the resistance the material can provide to the tension involved. This is for much the same reasons as given for '(a) Sagging' earlier.

(3) Basic causes and typical applications
Expansion by temperature is the most usual basic cause; moisture changes may also be causative. Walls and claddings are most likely to be affected.

(e) Rotating

(1) Deformation
The deformation results from either settlement or expansion, as explained in '1 Movement/restraint and force relationship', (f) earlier.

(2) Crack location/type
There are two cases to consider: those associated with foundations and ground-floor slabs and those with walling (roofs and returns). In both cases the crack is wide at one end and narrow at the other. With settlement, the affected walls will tend to crack along the diagonal as illustrated – similar to shearing except that the crack has an elongated wedge shape. In the case of ground-floor slabs the wedge shape would be across the thickness of the slab. This is what happens with walling, the rotation occurring along a weak plane (e.g. bed joint in masonry).

(3) Basic causes and typical applications
Where foundations and ground-floor slabs are involved, moisture changes (contraction) and loading, acting separately, are responsible. With walling, temperature changes are usually the predominant cause; moisture changes can sometimes be significant.

Effects of variations

1 Overview

In practice, whether or not cracking does take place, the way the cracks form and how they are distributed depends on a number of variable if not interacting factors. For example, where cracking is due to changes in temperature or moisture an alteration in one of the primary influencing factors is often accompanied by compensating changes in others. It is now possible to assess quantitatively the stresses that may be induced and thereby the cracking that may result from them.* The theoretical considerations given earlier in this part still retain their value for first approximations.

In addition to the theory and basic mechanisms explained earlier, it is necessary to keep in mind those factors that are likely to change the extent or the pattern of cracking that may occur. In some cases the effect of some of the factors may make the cracking that actually takes place less severe than the theory would suggest; in others the reverse could happen.

Quite separately, but nevertheless related, is the effect that the nature of a construction, the distribution or the shape of a building may have on the pattern of cracking. In this chapter the relevant factors are explained, illustrated or listed.

*In, for example, BRE Digests 227, 228 and 229, *Estimation of thermal and moisture movements and stresses*: Parts 1, 2 and 3, the examples given in Part 3 are of a simple kind but the estimation, as to be expected, requires the application of mathematics and structural principles.

Material		Value	Modulus of elasticity [E] – N//mm² × 10³
Metals	Aluminium	70	
	Brass	100	
	Bronze	100	
	Copper	95–135	
	Iron — Cast	80–120	
	Iron — Wrought	193	
	Lead	14	
	Steel (all types)	200–210	
	Zinc	140–220	
Timber	Softwoods (mean all grades)	5.5–12.5	
	Hardwoods (mean all grades)	7–21	
	Laminated — Softwoods	6.9–11.7	
	Laminated — Hardwood	8.6–18.6	
	Plywood	6–11	
Stones	Granite	20–60	
	Limestone	10–80	
	Sandstone	3–80	
	Marble	35	
	Slate	10–35	
Concrete	Dense	15–36	
	Lightweight	8	
	Aerated	1.4–3.2	
Brickwood	Concrete – Dense	10–25	
	Lightweight	4–16	
	Aerated	3–8	
	Clay	14–18	
	Calcium silicate	4–26	
Plastics	Thermosetting	5.5–10	
	Thermoplastic	0.1–3.5	
	Glass	70	
	Particleboard	2.0–2.8	

(Scale: 0, 50, 100, 150, 200)

2 Influencing factors

The factors that may influence the initiation or pattern of cracking are all likely to vary to some extent. The relevant factors, excluding those responsible for stress concentrations that are dealt with separately, include the following.

(a) Properties of materials

Strength (and modulus of elasticity in particular) and other properties, such as response to temperature and moisture movements, are relevant. The variation in the moduli of elasticity of a range of materials

Chart C2.1/1 *Moduli of elasticity (Young's modulus, E) of various materials. (Data from BRE Digest 228, Estimation of thermal and moisture movements and stresses: Part 2, August 1979, Table 1, pp. 2 and 3)*

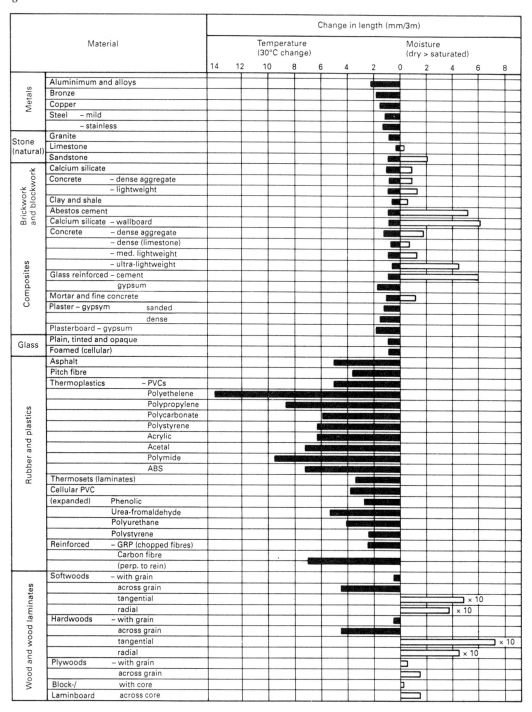

	Material	Change in length (mm/3m)													
		Temperature (30°C change)							Moisture (dry > saturated)					
		14	12	10	8	6	4	2	0	2	4	6	8
Metals	Aluminimum and alloys												
	Bronze												
	Copper												
	Steel – mild												
	– stainless												
Stone (natural)	Granite												
	Limestone												
	Sandstone												
Brickwork and blockwork	Calcium silicate												
	Concrete – dense aggregate												
	– lightweight												
	Clay and shale												
Composites	Abestos cement												
	Calcium silicate – wallboard												
	Concrete – dense aggregate												
	– dense (limestone)												
	– med. lightweight												
	– ultra-lightweight												
	Glass reinforced – cement												
	gypsum												
	Mortar and fine concrete												
	Plaster – gypsym sanded												
	dense												
	Plasterboard – gypsum												
Glass	Plain, tinted and opaque												
	Foamed (cellular)												
Rubber and plastics	Asphalt												
	Pitch fibre												
	Thermoplastics – PVCs												
	Polyethelene												
	Polypropylene												
	Polycarbonate												
	Polystyrene												
	Acrylic												
	Acetal												
	Polymide												
	ABS												
	Thermosets (laminates)												
	Cellular PVC												
	(expanded) Phenolic												
	Urea-fromaldehyde												
	Polyurethane												
	Polystyrene												
	Reinforced – GRP (chopped fibres)												
	Carbon fibre												
	(perp. to rein)												
Wood and wood laminates	Softwoods – with grain												
	across grain												
	tangential												× 10
	radial												× 10
	Hardwoods – with grain												
	across grain												
	tangential												× 10
	radial												× 10
	Plywoods – with grain												
	across grain												
	Block-/ with core												
	Laminboard across core												

Chart C2.1/2 *Movements in selected materials 3 m long and subjected to a range of thermal moisture changes. (Based on values in BRE Digest 228, Table 1, pp. 2 and 3)*

is compared in Chart C2.1/1, comparative moisture movements are given in Chart C3.3/6 and the coefficients of linear expansion are compared in *Materials for Building*, Vol. 4, Chart C4.03/1, p. 57. For convenience, thermal and movements for a given length and for a given **high>low** are compared in Chart C2.1/2 for a limited range of materials.

The differences in properties between different materials is obviously of fundamental importance in cracking. However, variations that may tend (and do) occur inherently in different 'samples' of the same material are significant in the way they may influence the initiation and pattern of cracking. In practice, it is these variations that give rise to cracking occurring often in unsuspected places or patterns.

(b) Degree of restraint

There are basically two types of restraint:

- *Inherent restraints,* which relate to corners of constructions (in plan or in section), gravity and the support given by such elements as foundations; and
- *Built-in restraints,* which relate to forms of construction and workmanship. It is the variations that may occur in the built-in restraints that need special consideration.

The degree of the built-in restraint actually obtained is likely to be different from that expected or intended. What may or may not be achieved on a building site is very dependent on the standard of workmanship and accuracy applied in particular cases. Often what was expected is not achieved – **creative pessimism** again. Depending on circumstances, this may be found to reduce or increase the cracking that does take place. Consider a simple example of the strength of a mortar mix. If a weaker mix than intended is actually obtained, cracking may not occur at all; the achievement of a stronger mix than intended may produce cracking where none was expected.

(c) Stress relaxation

(1) Generally
The magnitude of the stresses induced by the restraint of movements due to temperature or moisture will depend on four factors:

1. The magnitude of the movement in the material if unrestrained;
2. The modulus of elasticity of the material;
3. The capacity of the material to creep or flow under load; and
4. The degree of restraint to the movement of the material by its connection to the other elements in the structure.

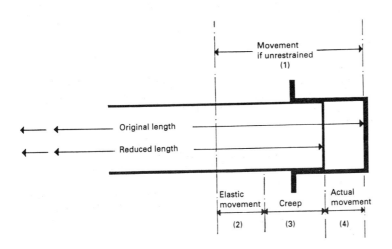

Diagram D2.1/3 *Factors influencing stresses due to temperature or moisture changes. (From* Principles of Modern Building, *Vol. 1, 3rd edn, HMSO, London, 1959, Fig. 2, p. 22, Building Research Establishment, Crown Copyright)*

The interdependence of these four factors is illustrated in Diagram D2.1/3. This shows that the unrestrained movement is equal to the sum of the actual movement, the elastic movement and the creep. Of these for it is only the actual movement that takes place physically. This movement is inversely proportional to the restraint. Elastic movements and creep are movements that do not take place physically but are so called because they are movements that would normally occur if the member was subjected to an externally applied load and was free to move. Under conditions of restraint therefore, and because this movement (i.e. in this context, strain) does not take place, stresses are either created (for elastic movements) or relaxed (for creep). The elastic movement is proportional to the induced stresses and inversely proportional to the modulus of elasticity. The creep, on the other hand, increases with the induced stresses and depends on the time during which movement takes place.

(2) Effects of compensating changes
The theoretical consideration described above cannot, unfortunately, be applied easily in practice to make quantitative assessments of the likelihood of a building component to crack due to induced stresses. This is because a change in any of the first three listed above is often accompanied by *compensating changes* in the others and by alterations in the strength of the material.

As an example, a concrete mix with a high water/cement ratio has greater unrestrained shrinkage movement than a mix with a low one. The additional water in the mix has the effect of reducing the modulus of elasticity, which allows a larger elastic movement for a given stress. At the same time, there is an increase in creep which, in turn, reduces the degree of elastic movement necessary. Thus, for a particular degree of restraint, the induced stresses due to shrinkage are less than with concrete mixes having lower water/cement ratios. However, there is still a tendency for greater shrinkage cracking with wet mixes (i.e. with high water/cement ratios) because they have reduced tensile strength.

(3) Effect of reinforcement
In the wet-trade processes generally, further complications arise in connection with shrinkage cracking. Reinforcement of a material (e.g. in mortars or renders) serves to act as a restraint and so stresses are induced. However, the reinforcement does help to distribute the shrinkage cracks so that these become *much finer* than they would have been in the unrestrained material. In this the extent of the reinforcement is important.

(4) Reducing serious cracking
In view of the difficulties, there are no precise recommendations for avoiding shrinkage cracking due to induced stresses. There appears to be no better advice* than that the likelihood of *serious* cracking can be *reduced* by:

- *Avoiding* materials that, when restrained, deform considerably as a result of moisture or temperature changes;
- *Avoiding* unnecessary restraint to shrinkage or expansion of the material; and
- *Using* materials or combinations of materials in a building component in such a way that the extensibility (i.e. the total elastic and creep deformations before cracking occurs) is as large as possible.

*Principles of Modern Building, 3rd edn, Vol. 1, HMSO, London, 1959, p. 22.

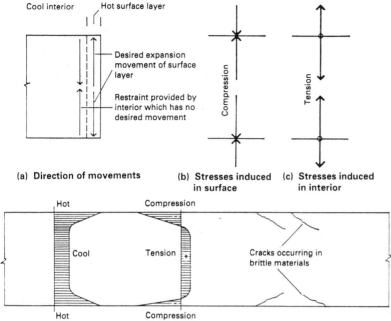

Diagram D2.1/4 *The effects of heating the surface of a material*

(d) Gradients

There is another aspect of stress inducement due to restraint that relates to differential movements between the surface and the interior of a material caused by a gradient between the two. The gradient may arise because of differences in temperature (e.g. during heating or cooling of the material) or moisture (e.g. drying out or wetting). The inducement of the stresses are best explained using the heating and cooling of a material* and then noting the effects on materials generally, including differential drying shrinkage.

(1) Heating the surface

A temperature gradient occurs across the section of a material when its surfaces are heated (see Diagram D2.1/4(d)). For simplicity, consider one surface. As this is hotter than the interior it will want to expand, but this expansion is restrained by the interior, as shown in Diagram D2.1/4(a). The result of this restraint is to induce a compressive stress (b) in the surface and tensile stress in the interior (c) and shown graphically in (e). If the material is ductile and the stress high enough, it will yield, and in so doing the stress will be relieved. Under conditions of high temperature the stress distribution will be zero, as yielding will take place at lower stresses than usual. On the other hand, a brittle material may fail by compression in the surface, in which case there will be a shearing failure on planes of 45°, known as *spalling* and shown in (f).

(2) Cooling the surface

Cooling of the surface will have the reverse effect (i.e. tensile stresses in the surface and compressive stresses in the interior), as illustrated in Diagram D2.1/5(a) and (b)). As in the case of heating, a ductile material will relieve the stresses on yielding but a brittle material may crack.

*Based on Richards, Cedric W., *Engineering Materials Science*, 1961, Chapter 12.5, 'Thermal stresses'.

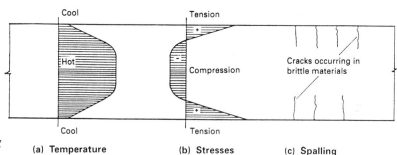

Diagram D2.1/5 *The effects of cooling the surface of a material*

Because there is tension in the surface, the cracks will occur at right angles to the surface instead of at 45°.

(3) Exfoliation

From these two examples, as a general rule the hot region is in compression while the cool region is in tension. Materials such as the *sedimentary stones* are classed as brittle and consist of layers of material. They may therefore fail due to the shearing action that takes place at the junction of the surface and the interior when there is differential movement due to heating or cooling. In such cases the surface virtually 'peels off', the process being known as *exfoliation*.

(4) Differential drying shrinkage

The principles applicable to heating and cooling of materials can, within certain limits, also apply to differential drying shrinkage that occurs in the wet-trade processes such as plastering, rendering, screeding or concreting.

If the surface layers dry out before the interior, then the surface will act as a restraint to the shrinkage that must still take place in the interior. The directions of movement will be the same as in Diagram D2.1/4(a), except that the desired movement will occur in the interior. Nevertheless, the stress situation will be the same and the surface will be subjected to compression and the interior to tension. As in the case of heating, cracking of the surface will occur as the materials are brittle. This is in part why it is so important that the drying shrinkage is carefully controlled by ensuring that it takes place relatively slowly and, as importantly, evenly.

It should be noted that there are other important aspects such as the amount of water used (i.e. the water/cement ratio) and the compaction of the mix that would include the effects of trowelling the surface, which also influence the degree of shrinkage cracking. It is for this reason that careful drying out alone is not sufficient to control shrinkage cracking.

When a restraint is provided to drying shrinkage of the surface layer, then the stress situation is similar to that illustrated in Diagram D2.1/5. For this analogy to be complete and to be related to practice, the surface layer must also include in its meaning a whole material, as in the case of plastering, where restraint is provided by the background. It may be noted that crazing of precast concrete cladding (or other) panels is thought to be due to differential moisture shrinkage between the surface and adjacent 'interior' layer. Surfaces that require fine aggregates and much trowelling appear to be more susceptible to the fine surface crazing.

(e) Environmental conditions

The environmental conditions, and those related to temperature and/or moisture in particular, to which materials may be exposed before, during and after construction are not necessarily uniform and, in the UK, can fluctuate between extremes over a short period. The rate at which changes may occur has a marked effect on the risk of cracking taking place. Generally, the faster the rate, the greater the risk.

When considering movements due to changes in temperature it is important to take into account variations that may occur due to:

- Thermal capacity (i.e. the slower response to temperature fluctuations of materials with a high thermal capacity and the quicker response of those with a low thermal capacity);*
- The position of a layer (or layers) or thermal insulation in a construction, element or component† together with ventilated cavities that may provide a cooling effect; and
- The colour of the exposed layer (i.e. where lighter colours tend to reflect more heat than darker ones).‡ Generally, insulation located immediately behind an exposed layer will cause that layer to heat up rapidly, subject again to the effects of thermal capacity and surface colour of the exposed layer.

The effects of these on the service temperature ranges of (selected) materials can be deduced from Table 2.1./4.

When applying the values given in Table 2.1/4 (or data applicable to moisture movements) in the design of joints it is important that estimates of likely changes are related to the temperature (or moisture content) at the time of construction. In some cases it may be more appropriate to estimate this temperature from local climatic data related to the construction programme. Normally it is sufficient to assume a single mean value of 10°C.

In practice, the temperature at the time of construction is unlikely to be at either extreme, the minimum in particular. The total movement of a material will be that governed by the range of temperature but the amount of contraction and expansion it undergoes will be influenced by the temperature at the time of construction (i.e. from the temperature at the time of construction to the minimum value for the contraction and the temperature at the time of construction to the maximum for the expansion). If, for example, the minimum temperature is −25°C, the maximum is 80°C (i.e. for an external cladding that is lightweight, dark in colour and over insulation in Table T2.1/4) and the temperature at the time of construction is 10°C, then the contraction will be governed by a temperature difference of −15°C (i.e. −25°C + 10°C) and the expansion by a temperature difference of 70°C (80°C − 10°C). It should be noted that in some cases the temperature at the time of construction may be governed by a process of construction (e.g. sealants cannot generally be applied satisfactorily below an ambient temperature of about 5°C).

3 Stress concentrations – 1: General considerations

(a) Line of least resistance

Cracking will always follow the line of least resistance (i.e. the weakest part of a construction). That path will also be dependent on the extent of the stresses involved. Parts of a construction that might not normally provide such paths might do so if they are subjected to a concentration of stress.

*Materials for Building, Vol. 4, 4.05 Thermal response, p. 97.

†Materials for Building, Vol. 4, 4.04 Thermal insulation, under 'Temperature distribution', p. 80.

‡Materials for Building, Vol. 4, 4.05 Thermal response, pp. 97 and 99–101.

Table T2.1/4   *Examples of service temperature ranges (°C) of materials in the UK*

	Min.	Max.	Range
External			
Cladding, walling, roofing			
Heavyweight			
Light colour	−20	50	70
Dark colour	−20	65	85
Lightweight, over insulation			
Light colour	−25	60	85
Dark colour	−25	80	105
Glass			
Coloured or solar control	−25	90	115
Clear	−25	40	65
Free-standing structures or fully exposed structural members			
Concrete			
Light colour	−20	45	65
Dark colour	−20	60	80
Metal			
Light colour	−25	50	75
Dark colour	−25	65	90
Internal			
Normal use	10	30	20
Empty/out of use	−5	35	40

The following situations are not included in the above examples, and may give rise to temperature extremes more severe than those listed:

Dark surfaces under glass (e.g. solar collectors).
Materials used in cold rooms or refrigerated stores.
Materials used for, or in proximity to, heating, cooking and washing appliances, or flues and heat-distribution networks.

From BRE Digest 228, *Estimation of thermal and moisture movements and stresses: Part 2*, August 1979, Table 2, p. 5, Building Research Establishment, Crown Copyright.

(b) Main causes of stress concentration

Changes of various kinds cause stress concentrations. Restraint is also associated. As explained earlier, tensile stresses are responsible for cracking. The way these stresses may be created or induced was discussed earlier under 'Classification and recognition' (see, in particular, '1 Movement and force relationship' and Diagram D2.1/1). The most common alterations in constructions that may cause cracking include changes in:

- Line due to loading;
- Line due to physical changes;
- Dimensions or stiffness;
- Direction;
- Area;
- Exposure;
- Material or form of construction;
- Adhesion/restraint; and
- Chemical composition.

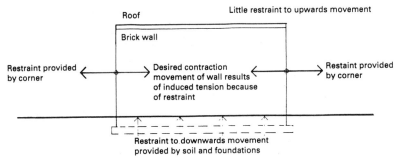

(a) **Elevation of rectangular building showing induced
tension in external wall due to restraint of shrinkage**

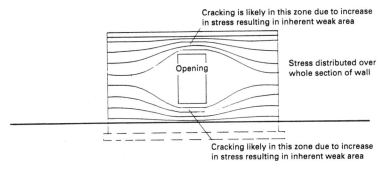

(b) **Stress concentration due to opening in wall
creating inherent weak areas**

Diagram D2.1/6 *Stress concentrations
due to an opening in a wall*

(c) Example of change in area

Typical locations in which the relevant stress concentrations occur are
illustrated in Diagram D2.1/8 and discussed under '4 Stress concentrations – Typical examples' below. A simple, yet frequently occurring,
example of stress concentration is the effect of induced stresses due to a
change in area of a wall because of an opening in it (see Diagram
D2.1/6). Without an opening, the stresses in the wall will be equally
distributed throughout its height ((a) in the diagram) in accordance
with the rule that stress is equal to force divided by area. With an
opening, the stresses will be concentrated in the (reduced) area above
and below the opening ((b) of the diagram). Cracking is likely to occur
in either or (more usually) in both of these areas. By contrast, if the wall
is long but without openings, then it is likely to be divided into equal
areas. A more common example of this is a masonry partition in a
framed building (see Diagram 2.1/7).

4 Stress concentrations – 2: Typical locations

In Diagram 2.1/8 the main causes of stress concentrations (see 3(b)
above) are related to typical locations in which they occur. A brief
explanation of each category follows with the sub-paragraphs related to
each cause given in the diagram. Generally in each case, a form of
movement control (e.g. movement joints or strengthening of the
construction to resist the stresses) is required at the point(s) of stress
concentration, but see 'Precautions' later.

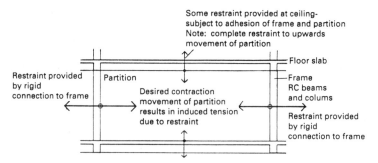

Elevation of basic conditions

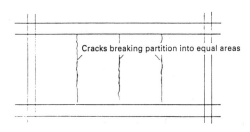

Diagram D2.1/7 *Conditions likely to cause cracking of partitions in a framed building*

Elevation showing one possible result of induced tension

(a) Changes in line due to loading

All horizontal structural elements such as beams and slabs deflect to some extent when loaded. For the present purposes deflection is regarded as a change in the horizontal line of the beam or slab. The change may cause stress concentrations in the element or other elements supported by or abutting it. Concrete beams and slabs are used for illustrative purposes; the principles apply equally to other structural materials such as steel or timber.

1. The stress concentration occurs in the lower part of the simply supported concrete beam.
2. In a continuously supported reinforced concrete beam the deflection and stress concentration now occurs in two places: between the beam 'supports' (i.e. similar to (1) above) and over them.
3. The deflection of the reinforced concrete beam (or slab) will cause stress to be concentrated in the wall (or partition) abutting the underside of a reinforced concrete slab. It should be noted that in some cases the deflection has been accentuated by the inclusion of shrinkable aggregates in the concrete.
4. The deflection of the reinforced concrete slab (beam) induces stresses in the wall (or partition).

(b) Changes in line due to physical changes

Apart from heave due to tree roots (see p. 78 ff), freezing of certain hardcores (see *3.8 Frost action*, p. 549) may cause bowing with resulting stress concentrations and cracking.

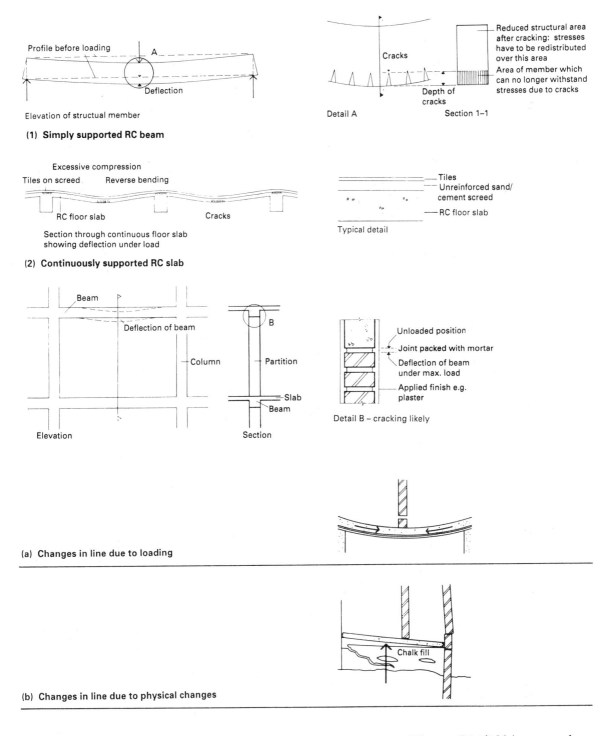

Profile before loading

A

Deflection

Elevation of structual member

(1) Simply supported RC beam

Cracks

Depth of cracks

Reduced structural area after cracking: stresses have to be redistributed over this area

Area of member which can no longer withstand stresses due to cracks

Detail A

Section 1–1

Excessive compression

Tiles on screed

Reverse bending

RC floor slab

Cracks

Section through continuous floor slab showing deflection under load

(2) Continuously supported RC slab

Tiles

Unreinforced sand/cement screed

RC floor slab

Typical detail

Beam

Deflection of beam

Column

Partition

B

Slab

Beam

Elevation

Section

Unloaded position

Joint packed with mortar

Deflection of beam under max. load

Applied finish e.g. plaster

Detail B – cracking likely

(a) Changes in line due to loading

Chalk fill

(b) Changes in line due to physical changes

Diagram D2.1/8 *Main causes and typical locations of stress concentrations or restraint*

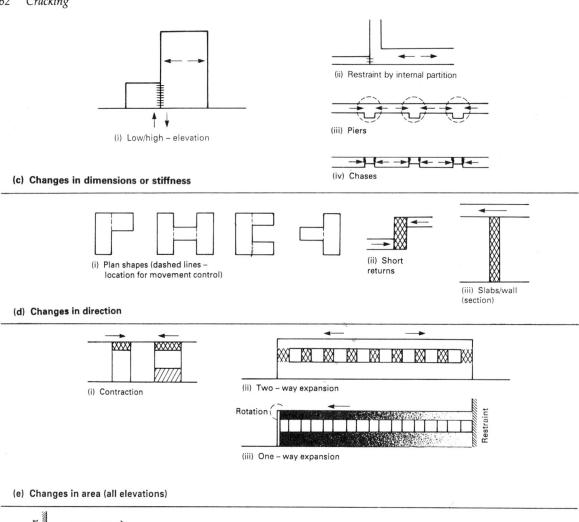

(i) Low/high – elevation

(ii) Restraint by internal partition

(iii) Piers

(iv) Chases

(c) Changes in dimensions or stiffness

(i) Plan shapes (dashed lines – location for movement control)

(ii) Short returns

(iii) Slabs/wall (section)

(d) Changes in direction

(i) Contraction

(ii) Two – way expansion

Rotation

Restraint

(iii) One – way expansion

(e) Changes in area (all elevations)

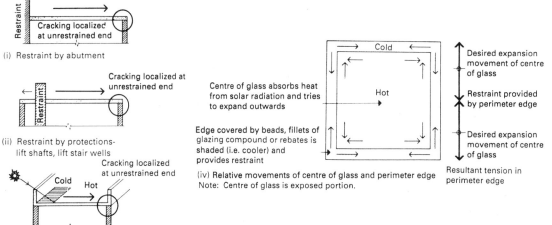

Restraint

Cracking localized at unrestrained end

(i) Restraint by abutment

Cracking localized at unrestrained end

Restraint

(ii) Restraint by protections- lift shafts, lift stair wells

Cracking localized at unrestrained end

Cold   Hot

(iii) Restraint by areas shaded for long periods

Centre of glass absorbs heat from solar radiation and tries to expand outwards

Edge covered by beads, fillets of glazing compound or rebates is shaded (i.e. cooler) and provides restraint

Cold

Hot

(iv) Relative movements of centre of glass and perimeter edge
Note: Centre of glass is exposed portion.

Desired expansion movement of centre of glass

Restraint provided by perimeter edge

Desired expansion movement of centre of glass

Resultant tension in perimeter edge

(f) Changes in exposure

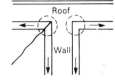

(i) Vertical/horizontal abutment

(ii) Column wall abutment

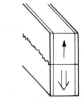

(iii) Abutment within a wall

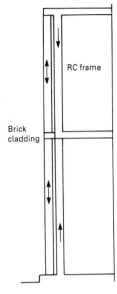

(iv) Expansion/shrinkage

(g) Changes in materials construction

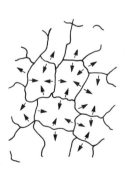

(h) Changes in adhesion/restraint

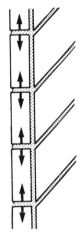

(i) Sulphate attack of mortar

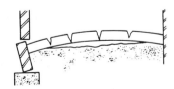

(ii) Sulphate attack of ground-floor

(i) Changes in chemical composition

(c) Changes in dimensions or stiffness

1. Differential movements are chiefly responsible for the stress concentration at the junction of a low and high wall.
2. The intersecting wall provides restraint.
3. The piers provide a buttressing effect with the stress concentration occurring at them.
4. The stresses are concentrated at the thinner part of the wall. The movements in the thicker walls on either side may be in either direction depending on whether expansion or contraction is involved.

(d) Changes in direction

1. L-, H-, U- and T-shaped plans are inherently susceptible to stress concentration at the junctions where a change in direction occurs. The intensity of the stresses are liable to be increased if one part is higher than the other (see (b)(1) above).
2. Short returns in masonry walls (in cavity walls notably) are likely to be subjected to the effects of movements (usually expansion) of the longer walls on either side. In new work, the moisture expansion of the bricks may be sufficient to cause stress concentrations (see *3.3 Moisture content*, p. 235).
3. This is a special case in which there is a change in direction of two elements (e.g. a roof and a wall or partition abutting its underside), one of which, the roof slab, is subjected to thermal movement. As a result of the latter, stresses are concentrated in the wall (or partition), causing it to move sideways.

(e) Changes in area

1. The reasons for the stress concentration above and/or below openings in a wall due to contraction of the wall have been explained above (3(c)).
2. The weakness of a wall with a run of window openings and subject to expansion is at the piers between the openings. These will be subjected to a shearing action.
3. In this case there is a run of window openings as in (2) above but there is restraint to the expansion at one end. The shearing action resulting from the expansion of the roof in particular occurs in the areas of walling above and below the openings. Furthermore, the greatest concentration of stresses occurs at the unrestrained end. The wall at that end is also subjected to a rotating action by the movement of the roof.

(f) Changes in exposure

1. Shading of part of a roof from the sun 'converts' the shaded area into a 'zone of restraint' in much the same way as the restraint provided by a wall or a projection through the roof, as illustrated in the accompanying diagrams. In all these cases the stresses are concentrated at the unrestrained end (see also (c)(3)).
2. The shading effect of edge cover to window glass, particularly those glasses with large inherent movements (e.g. coloured, heat absorbing or clear glass with applied colour or a dark background), is to induce stresses as illustrated. The glass behind the edge cover is shaded from direct solar radiation. This gives rise to a temperature

differential (usually considerable, as glass is a poor conductor of heat) between the centre of the glass and the perimeter edge. Tensile stresses occur in the edge as a result and these are proportional to the difference in the temperature between the edge and the centre of the glass. It should be noted that the tensile stresses occur at the edge that is in any case weakened by minute 'flaws' caused when the glass is cut to size. Knocks and abrasions and 'nipping off' cause more serious flaws. The flaws determine the limiting tensile stress of the glass while the failure occurs by the enlargement of an existing flaw into a crack.

(g) Changes in material/construction

1. The abutment of a wall (or partition) with the underside of a concrete floor or roof slab with both surfaces finished with plaster. The shrinkage of the latter, together possibly with the shrinkage of the walling, results in cracking at the junction.
2., 3. The two examples are simple cases where masonry abuts a concrete column (2) and brick and block masonry abut (3). Stress concentrations occur at the abutment and these lead to cracking of finishes such as plaster.
4. This is an example of the combined effects of moisture expansion in the brick cladding (see *3.3 Moisture content*, p. 271) and the drying shrinkage and creep of *in-situ* concrete columns. The stresses are concentrated where the brickwork is supported/restrained by a concrete (usually reinforced) nib. The latter is fractured; the brickwork cracks. The principles embodied in this example are also applicable to other cases where the brickwork or stone cladding (natural or pre-cast) may be restrained. In some cases, the drying shrinkage of the concrete alone may be sufficient to cause stress concentrations.

(h) Changes in adhesion/restraint

Applied finishes such as renders and plasters shrink as they dry out. The drying out and adhesion to the background may not be uniform. Changes such as these tend to encourage the finish to dry out in small areas, causing stress concentrations at their interfaces. This results in what is normally known as 'map' or 'crocodile' cracking. It differs from the mode of cracking normally associated with the effects of sulphate attack (see (i) below).

(i) Changes in chemical composition

Expansion may be caused by a number of chemical reactions such as those resulting from reactive aggregates in concrete (see *3.6 Chemical attack*, p. 396 ff), sulphate attack (see *3.6 Chemical attack*, p. 405 ff) and corrosion of ferrous metals embedded in concrete and other materials (see *3.8 Corrosion*, p. 479). The examples show the effects of sulphate attack.

1. The expansion of the mortar due to sulphate attack causes stress concentrations in the joint that are 'transferred' to the rendering, producing more or less horizontal cracks and not the map cracking normally associated with shrinkage cracking.
2. In a concrete ground-floor slab subjected to sulphate attack the expansive forces result in hogging of the slab that, in turn, concentrates stresses at the abutting wall.

Precautions

1 Generally

(a) Aims

Some cracking is almost inevitable. Fortunately, not all cracks cause problems; those that do not are more of a nuisance than anything else. The primary aim of any precautions should therefore be to reduce cracking to its absolute minimum. However, it should be clear from the preceding sections that many factors are involved and most of them are interactive to some extent. The whole business of cracking is complex. There is therefore a need, initially at any rate, to try to simplify things by determining the key issues. This may be done by applying most of the principles for building described in Chapter 1, p. 17 ff. Once the key issues have been identified, then it is time to consider which of the factors are likely to dominate and to devise appropriate solutions accordingly.

(b) Design and constructional achievement

Design is the starting point for all precautions. Importantly, intentions must be capable of achievement on-site and if they are then contractors need to apply all the appropriate checks, balances and supervision to ensure that all aspects are complied with, **continuity** no less.

(c) Applying the principles for building

(1) **Separate lives**
In general, almost all cracking is associated with the principle of **separate lives** (e.g. materials with different movement characteristics used together in a construction or elements of construction with different movement characteristics abutting, adjoining or otherwise closely related to each other). Where the differential movements are sufficiently large, the two materials or elements of construction have to be given sufficient elbow room in which to move harmlessly. The separation (i.e. the discontinuity) between the two has to be absolute.

(2) **Continuity**
For the separation to be complete means applying the principle of **continuity**. Where there is conflict with other performance requirements such as water exclusion, compensation has to be applied through the use of discontinuous continuity, for which there are many workable solutions (see *3.4 Exclusion*, p. 290 and p. 295).

(3) **Balance**
The principle of **balance** is important from two points of view. First, it is the principle governing the inherent 'desire' of all materials to be in equilibrium with their environment. They move to some extent as they change from one condition of imbalance to another. Second, in composite units or elements the overall balance that the various layers can achieve will influence the way the unit or element deforms. This is important whether thermal or moisture movements are involved.

(4) **Creative pessimism**
In the light of complexity generally and the many interactive factors and (in some cases) unknowns involved, **creative pessimism** would caution that *more rather than less* should be the controlling philosophy in determining precautions in individual cases.

2 Movement joints

(a) Locations

There are certain locations in all buildings where movement joints may be needed inherently. The locations where movement control is required may be determined from Diagram D2.1/8. The form that this control should take (i.e. by using movement joints or otherwise dealing with the induced stresses) will depend on the circumstances in particular cases.

(b) Classes of joints

To be effective, a movement joint has to be completely continuous. However, to meet other requirements of performance (e.g. weather, dirt and sound exclusion), the joint has to be filled. The discontinuous continuity has to be achieved in one of two ways, depending on the direction of the movement and the relationship of the parts of the construction that have to be separated (see Diagram D2.1/9). As an *aide-mémoire*, the first type may be thought of using the principle of springs ((a) in the diagram) and the second the principle of rollers ((b)).

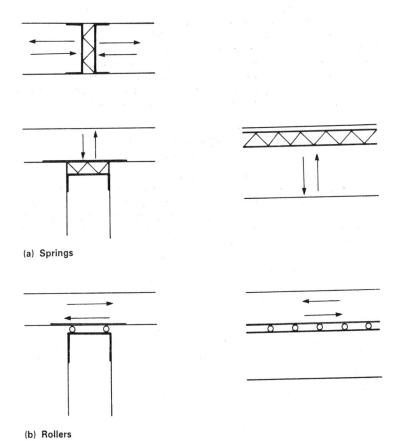

(a) Springs

(b) Rollers

Diagram D2.1/9 *Classification of movement joints. Arrows indicate direction of movement*

(1) Springs

The function of the spring type is to allow the movements to take place by 'absorbing' a sufficient amount without creating stresses. The material must, of course, be capable of springing back to keep the joint closed. This function is achieved with what are known as elastomeric materials (certain types of sealant) or resilient materials (specially formulated foamed strips and boards).

(2) Rollers

The function of the roller type is to enable the element that is moving to slide harmlessly backwards and forwards without inducing stresses in the element below it. In practice, it is, of course, quite impracticable to use real rollers. Traditionally, felt dpcs have been used in walls, but to be effective they should be bonded to a mortar bed.* Lead is another material that is used where heavy loads are involved.

(c) Forms

Movement joints may take many different forms, depending on particular circumstances. Most of the commonly used forms are described and/or illustrated in *3.4 Exclusion*, p. 292 ff; the functions and treatment of jointing materials are explained in *2.2 Strength and the use of materials*, p. 117 ff.

(d) Rules for masonry

(1) Demise of the 100 ft rule

At one time there used to be what may be described as the '100 foot' rule for the location of movement joints. This rule was simple to apply. Unless other methods of accommodating induced stresses (e.g. using reinforcement) were used, a movement joint should be located whenever a structure exceeded 100 ft (about 30 m) in length. Experience has taught that movement joints are required more frequently and that the frequency changes according to the unit (and other factors).

(2) BS 5628

For the majority of situations, BS 5628: Part 3: 1985 gives the following empirical recommendations for the spacing of vertical movement joints:

- *Fired-clay masonry*: not to exceed 12 m;
- *Calcium silicate masonry*: between 7.5 m and 9 m – never exceeding 10 m. The ratio of length to height of the panels should not exceed 3:1. This type of masonry also needs special consideration for movement joints or other ways of restraining movement at openings;
- *Concrete masonry*: 6 m, unless the manufacturer recommends otherwise in particular cases.

(e) Panel walls in frame structures

Horizontal movement joints, usually at or near floor level, are now essential where masonry panel walls are used in steel or concrete frame buildings. Provision for the movement is complicated by the need for also providing restraint for the panel. More or less standard solutions are available.

3 Seeing the problem as a whole

*In some cases the felt dpc has acted unintentionally as a roller layer because it was laid dry on its mortar bed. Unexpected cracking has normally been the result.

Important as precautions aimed specially to reduce the risk of cracking are, there is still a need to consider these in conjunction with those that should be taken in respect of other factors covered later. Consideration of the problems associated with cracking as part of building generally cannot be overemphasized.

The need to control moisture content is one important and relevant case; the control of thermal aspects is another. As to moisture content, there are a number of precautions that should be taken when materials are being delivered to or stored and handled on-site, and these are discussed in *3.3 Moisture content*, under specific materials. Keeping materials as dry as possible at all times during the life of a building is equally important. How this may be achieved in principle is explained in *3.4 Exclusion*. Dryness during the life of a building also relates to problems associated with (3.6) chemical attack, (3.7) corrosion and (3.8) frost action. The deleterious results of any one of these may, of course, result in cracking.

Sources

The following are included here for convenience but relate to all the preceding sections.

Addleson, Lyall, *Materials for Building*, Vols 1–3, Iliffe Books, London, 1972 (o/p)

Addleson, Lyall, *Materials for Building*, Vol. 4, Newnes-Butterworths, London, 1976

Billington, N. S., *Thermal Properties of Buildings*, Cleaver-Hume, London, 1952 (o/p)

Lenczner, D., *Movements in Buildings*, 2nd edn, Pergamon Press, Oxford, 1981

Rainger, Phillip, *Mitchell's Movement Control in the Fabric of Buildings*, Batsford, London, 1983

BRE

Current papers
CP 94/74, *The rippling of thin flooring over discontinuities in screeds*, by J. Warton and P. W. Pye, 1974

Defect action sheets
18, *External masonry walls: vertical joints for thermal and moisture movements*, new edition, February 1985
75, *External walls: brick cladding to timber frame – the need to design for differential movement*, April 1986
76, *External walls: brick cladding to timber frame – how to allow for movement*, April 1986

Digests
75, *Cracking in buildings*, reprinted 1975 (superseded by 361)
79, *Clay tile flooring*, revised 1976
104, *Floor screeds*, new edition 1973, revised 1979
130, (First series) *Asbestos sheets cracking*
157, *Calcium silicate (sandlime flintlime) brickwork*, new edition, 1981
223, *Wall cladding: designing to minimise defects due to inacurracies and movements*, March, 1979
227, *Estimation of thermal and moisture movements and stresses: Part 1*, July 1979
228, *Estimation of thermal and moisture movements and stresses: Part 2*, August, 1979
229, *Estimation of thermal and moisture movements and stresses: Part 3*, September 1979
361, *Why do buildings crack*, May 1991 (supersedes 75)

Information papers
IP 6/84, *The movement of foam plastics insulants in warm deck flat roofs*, by J. C. Beech and G. K. Saunders, April 1984

BRE BRAS
TIL 5: 1971, *Devices for detecting changes in width of cracks in buildings* (covers simple mechanical devices: more sophisticated versions now available commercially)
Building Technical File, *Cracking and bulging of plaster*, BRAS, 4, pp. 45 and 46, January 1984

BSI
BS 5262: 1976, *Code of practice for external rendered finishes*, September 1976
BS 5628: Part 3: 1985, *British Standard Code of practice for use of masonry Part 3. Materials and components design and workmanship* (formerly CP 121: Part 1), 29 March 1985
BS 6262: 1982, *British Standard Code of practice for glazing for buildings* (replaces CP 152), 30 June 1982

BDA
BDA Technical Note 3, *Further observations on the design of brickwork cladding to multi-storey rc frame buildings*, by Donald Foster (Editor J. R. Miller), April 1975

Soils

1 Scope

This section includes a general review of soils on which most buildings are founded. The construction of buildings on solid rock is excluded. Particular attention is given to the movements associated with clay soils. Low-rise buildings founded on these soils are inherently at risk from cracking. Fills are considered separately in the next section.

2 General considerations

(a) The effects of movements in general

Soils of various kinds support the entire load (i.e. both dead and imposed) of buildings. Consequently, movements that may occur in the soil are likely to have an effect on the stress distribution of both the foundation and the superstructure. There may be some distortion of either the foundation or the superstructure or both. However, the movements will also cause stress redistribution. Depending on the relationship between this and the type of foundation and superstructure, the distortions may, at best, result in cracking of walls, partitions, slabs or their finishes and, at worst, complete structural collapse.

(b) Modes of distortion and crack patterns

Soil movements result in either a sagging or a hogging distortion. The latter may, however, not be complete in that the hogging takes place on one side only (i.e. a form of cantilevered hogging). This type of hogging is more commonly associated with tree roots. There are therefore three basic modes of distortion. Each are characterized by particular crack patterns. Diagram D2.1/10 illustrates both, drawing on a practical example for the initiation of cracking due to sagging and hogging described earlier in this chapter (see 'Crack initiation', p. 46 ff).

Diagram D2.1/10 *Crack patterns associated with different modes of distortion – relate to Diagram D2.1/2 for sagging, hogging and rotating. (From BRE Digest 251, Assessment of damage in low-rise buildings with particular reference to progressive foundation movement, July 1981, Fig. 7, p. 6, Building Research Establishment, Crown Copyright)*

(c) Principles for dealing with movements

Although the interaction of soil, foundation and superstucture present a complex system, there are two alternative basic principles which apply when dealing with this:

- To ensure that there is as little movement as possible in the soil once it has been loaded;
- To ensure that the building as a whole, i.e. foundation and superstructure, is specifically designed to accommodate movements which it is known are likely to occur.

The solution to both these problems requires specialized knowledge first, of the soil (a subject usually covered under soil mechanics) and, second, of structural mechanics. It is not proposed to deal with either of these subjects in detail, but rather to consider and note the main points in so far as they are relevant to the likelihood of cracking taking place.

(d) Soil types and variations

In the context of being the main supporting element of buildings, soils constitute some of the most fundamental, and in many ways, most important materials to be considered in practice. Most soils consist of solid particles of varying shapes and sizes, with water and, to a lesser extent, air filling the spaces between them. Their characteristics, behaviour and general properties are, in many ways, quite different from those of normal building materials. However, an important similarity between soils and naturally occurring materials used in buildings does exist. Like these materials, soils of the same type may differ quite considerably in their properties. Furthermore, there are a number of different types of soils each with their own properties. For convenience, soils are divided into two broad groups:

- Cohesive soils which consist chiefly of the silts and clays and whose strength is largely dependent on the amount of water they contain (clay notably); and
- Granular soils, which consist chiefly of the gravels and sands and whose strength is dependent on the relationship of the closeness of the packing of their particles to the size of the external force.

The relative properties of these two groups are given in Table T2.1/5, while Table T2.1/6 indicates the grading of particle size in each group. Chart C2.1/3 illustrates the approximate load-bearing capacities of the various types.

In common with the differences in properties which can be expected when dealing with soils, it is also important to note that on any building site it is quite likely that *variations* in the properties of the soil will occur. In addition, there may be a number of different types of soil, especially at different levels below the surface. Because these and other variations may exist, it is always important that a thorough site exploration is undertaken prior to the design of foundations and superstructure.

(e) Angular distortion

Damage due to differential movement is dependent not only on the amount of movement that takes place but also on what is known as 'angular distortion' relative to the type of construction. 'Angular distortion' is a measure of the differential movement, defined as h/AB

Table T2.1/5   *Properties of cohesive and granular soils*

Property	Cohesive soils	Granular soils
Strength	Strength derived from power of cohesion of the particles. Strength does not necessarily increase with increase in depth below surface	Strength derived from internal friction between particles which increases under load. Strength increases with increases in depth below surface
Voids	Low proportion of voids	High proportion of voids
Cohesion	Marked cohesion	Negligible when clean
Compressibility	Very compressible	Only slightly compressible
Reaction to compression	Compression takes place slowly over a long period	Compression takes place almost immediately
Permeability	Practically impermeable	Permeable

Note: Account should be taken of the ranges which exist between the two extremes quoted above.

Table T2.1/6   *Particle size of different soils (mm)*

Gravel	Particles larger than 2.00
Sand	Particles between 2.00 and 0.06
Silt	Particles between 0.06 and 0.02
Clay	Particles smaller than 0.002

Note: there are many systems for grading particle size. The above grading is the MIT system adopted in BSS 1377: 1948, p. 40.

Diagram D2.1/11 *Angular distortion used in connection with soil movements. (From BRE Digest 63, Soils and foundations, October 1965. Building Research Establishment, Crown Copyright)*

(Diagram D2.1/11), where h is the differential vertical movement and AB the distance between two points (the span). Brickwork and plaster show the effects of differential movement quickly, and from the limited data available the onset of cracking can be associated, according to BRS (now BRE), with an angular distortion of about 1/300. This is equivalent to a differential movement of 9.5 mm over a span of 3.0 m. Warehouse and factory buildings of framed construction can usually tolerate larger angular distortions but at about 1/150 structural damage may be expected unless the joints have been specially designed to tolerate them.

3 Settlement

(a) Primary function of foundations

One of the primary functions of a foundation is to transfer the total load of a building onto the soil. This load transfer should be as even as possible, while it should be spread over sufficient area of the soil for safe bearing. Thus the area required for the distribution of the load will be dependent on the safe bearing capacity of the soil. This deals with one basic problem. The next problem is how the soil behaves once it has been loaded.

(b) Effects of load, soil type and time

In general, all soils will compact or *consolidate*, i.e. there will be a downwards movement, commonly known as settlement, on being loaded. The load on the soil, as applied through the foundations, increases both water and soil pressures. Water is squeezed from between the solid particles and driven to areas where the water pressure is less *(high>low)*. The soil particles, on the other hand, are

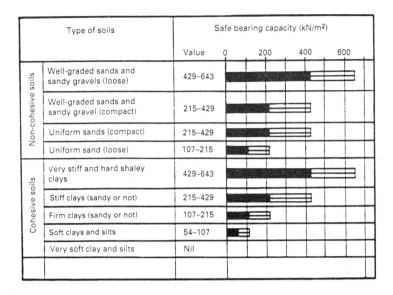

Type of soils		Safe bearing capacity (kN/m²)				
		Value	0	200	400	600
Non-cohesive soils	Well-graded sands and sandy gravels (loose)	429–643				
	Well-graded sands and sandy gravel (compact)	215–429				
	Uniform sands (compact)	215–429				
	Uniform sand (loose)	107–215				
Cohesive soils	Very stiff and hard shaley clays	429–643				
	Stiff clays (sandy or not)	215–429				
	Firm clays (sandy or not)	107–215				
	Soft clays and silts	54–107				
	Very soft clay and silts	Nil				

Chart C2.1/3 *Comparison of the safe bearing capacities of various types of soil. Note: For illustration of relative differences. The values should be regarded as approximate guides only, as similar soils vary from one to another with no marked definition while there may be disagreement on the meaning of terms such as 'hard', 'stiff' and 'soft' as applied to clays or 'compact', 'loose' and 'uniform' as applied to sands. (Values from Faber, John and Mead, Frank,* Foundation Design Simply Explained, *Oxford University Press, Oxford, 1961)*

forced closer together. Consolidation continues until the water pressure has fallen to its original value and the forces between the particles have increased by an amount equal to the newly applied load. The total settlement will be dependent on the type of soil and the imposed load. In addition, there is the *time* aspect of settlement. The imposed load on a soil is generally increased as construction proceeds. There is, however, a marked difference in behaviour between sandy and clay soils, as is illustrated in Diagram D2.1/12.

(c) Differences between granular and cohesive soils

From this diagram it will be seen that the duration of movement in sandy soils relative to the imposed load is short, i.e. each increment of load causes the total settlement of the soil for that load. Thus there is no further settlement once the total load has been imposed. Clay soils, on the other hand, do not react immediately to the loading increments and thus settlement proceeds for a considerable time after the final load has

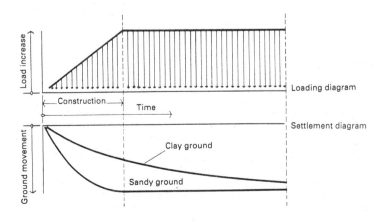

Diagram D2.1/12 *Time and load-dependent building settlements liable to occur on sandy and clay soils. (Based on 'Some principles of foundation behaviour', A. W. Skempton, RIBA Journal, 1942)*

been imposed. Foundations on sandy soils settle quickly once the load is applied because the soil particles, and thus the pore spaces, are large which, in turn, allows rapid water movements. Clay soils, on the other hand, offer considerable resistance to the expulsion of water and thus settlement caused by consolidation can continue for years after construction. It may be noted here that if the process is reversed (as might happen if the load on the soil is reduced by excavations), swelling of the soil occurs as water tends to move towards the unloaded areas. Shrinkage and swelling may often act simultaneously.

Because of the natural variations which occur in all soils, it is likely that there will not be the same total settlement everywhere. This is then one cause for differential settlement occurring, and it can take place even when the building load has been evenly distributed. If, on the other hand, the building load is not evenly distributed then the amount of differential settlement will increase. Although it is usually impossible to eliminate differential settlement completely, due to the nature of soils, it is always wise to ensure that at least the building load is distributed as evenly as possible.

(d) Type of foundation required in principle

The *type of foundation* required to ensure even distribution of the building load will depend not only on the type of soil but also on the disposition of load-bearing elements. Some common forms of foundations are:

1. *Strip foundations*, where the loads are transmitted to the foundation by walls and where special soil conditions do not exist. The depth of foundation will depend on the properties of the soil at any depth. Thus where soil conditions are good these may be both narrow and shallow; on poor soil they have to be wider and sometimes deeper, with transverse reinforcement used to spread the load over sufficient area.
2. *Isolated pad foundations* often support framed buildings where the loads are transferred by columns. As there is no connection between the pads special account has to be taken of differential settlement.
3. *Raft foundations* are used with all types of load transmissions on poor sites with weak soils in order to spread the load over a large area.
4. *Piled foundations* may be required on extremely weak soils when it may be necessary to transfer the load to a lower stratum.

The time aspect of settlement already noted does not influence eventual differential settlement nor the precautions which should be generally adopted. However, knowledge of this aspect is useful when any assessments of the possibility of the worsening of any trouble in a new building have to be made.

4 Clay soils

(a) Occurrence of cracking

Most of the problems of cracking of buildings founded on clay soils are due to moisture content changes in such soils. Cracking occurs most frequently in small-scale buildings, generally with light loads. Housing in particular falls into this category. The reason for these failures is often associated with the lack of knowledge of the behaviour of clay which, because of the comparatively light loads involved, leads to the use of shallow strip foundations. For certain types of clay, the latter are unsuitable; piled foundations are needed instead.

(b) Clay and sandy soils compared

As changes in moisture content play an important part in the movement of cohesive soils, particularly clay, it is necessary to draw a basic distinction between clay and non-cohesive soils. The latter, i.e. sandy and gravel soils, consist of non-absorbent and indestructible particles of rock, chiefly quartz. Thus once these particles have been completely compacted there cannot be any movement due to changes in moisture content which may occur in the voids between the particles.

Clay, on the other hand, consists of very much finer particles of matter. As a result, there is much more absorption of water by clay (the material is colloidal in character and much water is adsorbed) which results in a considerable change in size according to the moisture content present. Like other materials, there is expansion on absorption and contraction on desorption.

(c) Nature of clay soils

Clays are soils that contain a large proportion of very small mineral particles (i.e. having a diameter of less than 0.002 mm and invisible to the naked eye). Characteristically, clays are plastic, and smooth and greasy to the touch. These characteristics are more pronounced as the amount of clay present in the soil increases relative to any silt or coarse-grained material present. As with all naturally occurring materials, clay is seldom found in its 'pure' state. This means that the properties of clays vary, sometimes considerably.

In the wet state, clays are soft and sticky. When dried out, they shrink and crack. Intact lumps become hard to break. Importantly, firm shrinkable clay softens only slowly when immersed in water without disintegrating. If silt or coarse-grained material is present, disintegration occurs quickly. In the field, firm shrinkable clays can be identified by:

- Their highly fissured nature;
- The high polish left by digging tools; and
- The extensive crazing that occurs as the clay dries out on the sunny side of a trench.

The behaviour of clays is determined by the type and size of minerals present. For example, the very 'clayey' minerals hold water within their molecular structure much as jelly does. 'Fat' clays contain the very 'clayey' minerals and in nature always hold more water than the 'lean' clays, which contain fewer of the 'clayey' minerals.

(d) Determination of clay-nature and plasticity

The clay-nature and plasticity of soils is usually determined by simple, standardized mechanical index tests as described in BS 1377. From these tests the *liquid limit* and *plastic limit* are determined:

- *The liquid limit* defines the amount of water (expressed as a percentage of dry weight) required to bring the clay to a very weak, plastic consistency.
- *The plastic limit* is at the other extreme, and is the amount of water in a clay soil below which it is no longer plastic and pasty but breaks up when worked in the hands.

Below the plastic limit there is insufficient water to fill the spaces between the solid particles, and the voids become increasingly filled with air as the water content reduces. The effect of air entering the voids is to reduce the amount of shrinkage. Soils behave plastically between the plastic and liquid limits when the soil is saturated with water that lubricates and separates the solid particles. (Typical values for a clay with potentially very high shrinkage are just under 30% for the plastic limit and about 75% for the liquid limit.)

When saturated, the soil shrinks as it dries out. Importantly, the volume change is in direct proportion to the amount of water removed. As explained earlier, air enters the soil when it is dried out having moisture contents near the plastic limit. This causes the volume shrinkage to become less than the amount of water removed.

(e) Behaviour of clay soils – 1: Background

A clay soil will shrink as its water content is reduced below its 'natural state'. The residual soil exerts more and more suction as it dries and shrinks. The soil will swell subsequently by absorbing water brought into contact with it. The concept of soil moisture suction in relation to moisture content is important when considering the way trees and other large vegetation influence the behaviour of shrinkable clays.

A change in clay volume can occur as a result of load change and through a change in its moisture content. (For the present purposes, the former is excluded.) A change in moisture content can be caused through:

- Moisture movement near the ground surface where moisture evaporates in dry weather and is replenished by rainfall and by upward migration from the water table;
- The action of the roots of vegetation which cause the transpiration of soil moisture from greater depths than in (1). In general, the larger the vegetation, the greater will be its demand for moisture in dry weather. Hence the importance of trees that grow near to buildings founded on clay soils.

According to BRE, a distinction can be made between the depth of volume changes of clay soils due to:

- The presence of vegetation generally; and
- Large trees.

The seasonal volume changes due to the presence of vegetation generally extend to about 1–1.5 m into the clay ground, with significant clay shrinkage and swelling confined to the top 1 m or so. In contrast, large trees have a greater influence in deep zones of dried and shrunken clay beneath them. Because the rate of penetration of water into the more impermeable dry clay is slow, the zones extended during a dry summer will not be replenished with water when the wet weather returns. This means that the larger dry zones become permanent.

Serious drying (and shrinkage) of clay soils below a depth of 1.5 m is, for practical purposes, almost always associated with the removal of moisture by the roots of trees. It is not easy to determine the amount of shrinkage a particular clay (in a particular site, for example) will undergo. Samples need to be tested either according to BS 1377 or by comparing measuring blocks of the soil. As a guide, Table T2.1/7 gives what BRE call the clay shrinkage potential. It will be seen that the higher the clay fraction, the greater the plasticity index and hence the higher the shrinkage potential. The values in the table need to be used with care. BRE have indicated (Table 2.1/8) how they may be applied to particular clays.

Table T2.1/7 *Clay potential shrinkage*

Plasticity index (%)	Clay fraction (%)	Shrinkage potential
>35	>95	Very high
22–48	60–95	High
12–32	30–60	Medium
<18	<30	Low

From BRE Digest 240, *Low-rise buildings on shrinkable clay soils: Part 1*, August 1980, Table 1, p.3, Building Research Establishment, Crown Copyright.

Table T2.1/8 *Shrinkage potential of some common clays*

Clay type	Plasticity index (%)	Clay fraction (%)	Shrinkage potential
London	28	65	Medium/high
London	52	60	High
Weald	43	62	High
Kimmeridge	53	67	High/very high
Boulder	32	–	Medium
Oxford	41	56	High
Reading	72	–	Very high
Gault	60	59	Very high
Gault	68	69	Very high
Lower Lias	31	–	Medium
Clay silt	11	19	Low

From BRE Digest 240, Table 2, p.3, Building Research Establishment, Crown Copyright.

(f) Behaviour of clay soils – 2: Examples

In the UK the seasonal nature of the moisture content changes in clay soils is significant. Changes in moisture content are also influenced by other aspects, as explained earlier in (e) above. Examples include the following.

(1) Shelter
The amount and depth of shrinkage of the clay will depend on the shelter the ground may receive from the sun by the building itself or from other neighbouring obstructions (e.g. buildings or trees). In the absence of such shelter differential settlement may occur between those parts exposed to the weather and those covered by the building itself. It should be noted that the shrinkage of the clay occurs both vertically and horizontally, as illustrated in Diagram D2.1/13. The diagram also shows the stepped pattern of the cracking that results from these combined movements. It will be seen that the cracking has resulted from the movement of the clay at the periphery of the building.

(2) Grasses
Generally, moisture content changes of the clay under grasses extend to about 1 m. The greatest amount of settlement is usually about 25 mm but is confined to the surface only. Larger settlements have been known to occur between April and September.

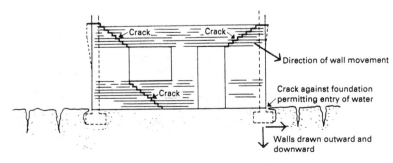

Diagram D2.1/13 *Cracking associated with shallow foundations on shrinkable clay. (From BRE Digest 63, Building Research Establishment, Crown Copyright)*

(3) Growing trees

The roots of trees and other vegetation are capable of penetrating considerable distances sideways and downwards. Permanent drying to depths up to 4.6 m with a shrinkage of 75 mm or more have been measured in the UK. Trees can cause particularly severe damage, as illustrated in Diagram D2.1/14.

Trees and shrubs do vary in the penetration or extension of their roots. Generally, the higher the tree, the greater the penetration. However, simple rules as to the distance that trees should be from a building for safety are not reliable. Advice should be sought. Importantly, the root-penetration characteristics of new trees to be used should be checked beforehand.

(4) Uprooted trees

Where established trees have been cut down or removed, soil heave (i.e. a reverse action) occurs. On removal of the roots or their primary moisture-absorbing function, clay soils re-absorb, albeit usually slowly, the moisture previously removed by the roots. The clay swells and lifts foundations, usually unevenly, as the swelling is most marked close to the site of the pre-existing trees. The resulting upwards movement (heave) is illustrated in Diagram D2.1/15. It can continue for several years.*

Diagram D2.1/14 *Particularly severe damage to the external brick wall in the form of stepped diagonal cracking (see Diagram D2.1/2, Rotating) due to changes in moisture content of the clay soil caused by tree roots. (Based on BRE Digest 75,* Cracking in buildings, *reprinted 1975, Fig. 6, p. 5, Building Research Establishment, Crown Copyright)*

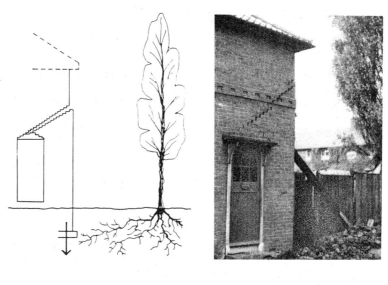

Diagram D2.1/15 *Consequences of tree felling: clay soil previously dried by the tree roots reabsorbs water. The pressures developed are often greater than those applied by shallow foundations. The resulting upward movement can continue for several years – relate to Diagram D2.1/2, Hogging. (From BRE Digest 63, Building Research Establishment, Crown Copyright)*

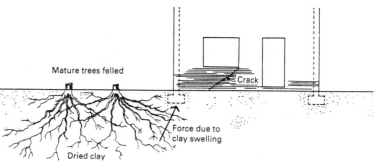

Mature trees felled

Crack

Force due to clay swelling

Dried clay

*In an office building built on a site cleared of trees, BRE observed during a five-year period that the heave was about 6.4 mm per year. This movement, BRE estimated, could continue for up to ten years. If it did, the total heave would be about 64 mm.

(g) Categories of clays in the UK

As a guide, there are basically three classes of clays in the UK, the distribution and characteristics of which are:

1. So-called *'overconsolidated'* clays, many of which are found in south-east England. These are firm, shrinkable clays, capable of supporting buildings of up to three or four storeys on shallow foundations. Examples of these clays are London, Gault, Weald, Kimmeridge, Oxford, Woolwich, Reading, Lias, Barton and glacial drift clays (in East Anglia, for example) (Diagram D2.1/16). The normal state of overconsolidated clays is a fully saturated condition. The moisture contents are close to the plastic limit (i.e. typically 25–30% for London Clay). However, close to the ground surface (i.e. in the top 1–1.5 m) fluctuations in moisture content from as little as 15% in dry summer weather to 40% in wet winters occur frequently.

2. What may be called *sandy clays* are found in the area further north from that shown in the map in Diagram D2.1/16. The potential shrinkage of these clays are smaller than those that occur in south-east England.

3. *Soft alluvial clays* are to be found in, and adjacent to, estuaries, lakes and river courses. Examples of these are the Fens, the Somerset Levels, the Kent and Essex marshes alongside the Thames and the clays of the Firths of Forth and Clyde. All these clays have two 'layers': the top has a firm, shrunken crust that is drier than the underlying body of clay. This means that there are two foundation problems: (i) of clay shrinkage in the top layer, and (ii) of the avoidance of excessive settlement due to loading the softer clay and peat beneath the top layer.

Diagram D2.1/16 *General distribution of shrinkable clays in the UK, based on BRE records and Institute of Geological Sciences maps. (From BRE Digest 240, Low-rise buildings on shrinkable clay soils: Part 1, August 1980, Fig. 2, p. 3, Building Research Establishment, Crown Copyright)*

(h) Precautions

1. The likelihood of cracking from the movement of foundations can be reduced by using foundations that are so deep that soil movements are unimportant. An alternative, and one which is sometimes found to be more economical, is to support the building on bored pile foundations, with the piles designed with adequate strength and sufficient resistance to settlement so that relative movements of piles are unimportant.
2. Where it is proposed to erect buildings close to *single trees*, shallow foundations should *not* be used if the buildings are to be nearer than the mature height of the tree. Where there are roots of *groups or rows of trees* competing for water over a limited area, the limiting distance should be one and half times the mature height of the trees. These distances are only *rough guides* and while existing trees have been specifically mentioned it is important to note that young trees should not be planted closer than the distances suggested.
3. When trees are felled to clear a site for building, considerable time should be allowed for the clay, previously dried by the tree roots, to regain water.

5 Special conditions

(a) Mining subsidence

Serious settlement may occur in those areas that suffer from mining subsidence. Local authorities and/or British Coal usually have records of areas where large settlements can be expected. To minimize the damage that could be caused to buildings, one of three different methods may be used:

1. Allowing the building to act as a single unit so that it may tilt without cracking even though its support by the ground may be reduced in a few localized areas.
2. Having a raft foundation or one using beams and slabs with three-point support from the ground. If there are movements the foundations and building would tilt together.
3. Having flexible joints in the building so that relative movements are accommodated in the structure in a predetermined way. The Clasp system employs this principle.

(b) Freezing of the soil

Expansion of soils can take place in any soil that is capable of holding water either within its particles or in voids between the particles. Soils most commonly associated with what is termed 'frost heave' are fine sands, silts and chalks. Generally, it is not normal to anticipate expansion from freezing at a depth below 0.46 m, although during severe winters in the UK frost may penetrate to a depth of 0.61 m or so.

The comparatively small amount of damage to buildings as a result of frost heave is usually confined to severely exposed parts of buildings founded on predominantly chalky and fine sandy soils. In some of these, there may be a build-up of ice layers that cause the foundations to lift, but then only in exceptionally severe winters. Diagram D2.1/17 illustrates some typical effects, from a house that was newly built on

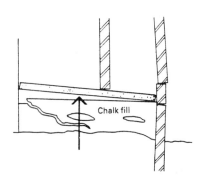

Diagram D2.1/17 *The results of frost heave on foundations. Chalk used as hardcore under oversite concrete froze while waterlogged and lifted the oversite concrete and brickwork over it. (Based on BRE Digest 75, Fig. 5a, p. 5, Building Research Establishment, Crown Copyright)*

chalky soil with chalk fill under the floor slabs. As it was unoccupied and therefore unheated, the parts most affected were those most exposed. Although heated buildings are unlikely to be damaged in this way, it should be noted that garages and other outbuildings which are often built on concrete slabs and footings shallower than those of adjoining houses may be affected. As a precaution, fine sands, silts and chalks should not be used as fillings beneath ground-floor slabs even in buildings that are eventually to be heated on completion.

There are, of course, special types of buildings, such as cold storage rooms, where problems of water freezing in the soil beneath them, especially in the early life of the building, can be expected. These are exceptional cases and accordingly require specialized treatment. (It has been known for a cold storage room to be completely lifted off the ground due to water freezing in the soil beneath the floor slab. This was due to inadequate insulation.)

(c) Loss of ground

Although dense beds of sand are normally excellent foundation soils, the finer particles may occasionally be washed away by water, thus leaving the coarser material in a less stable condition. Under such conditions the soil can lose much of its bearing capacity, particularly if an excavation is made.

(d) Lateral movement

Natural or geological phenomena, artificial agencies or a combination of both can cause some foundation movements. Slopes and landslips (clay soils notable) and swallow holes (chalk and limestone areas significant) are the chief natural or geological phenomena. Mining subsidence referred to earlier is a good example of an artificial agency.

Sources

Addleson, Lyall, *Materials for Building*, Vol. 1, Iliffe Books, London, 1972 (o/p).
Cutler, D. F. and Richardson, I. B. K., *Tree Roots and Buildings*, Construction Press, London, 1981. A delightful, authoritative little book from the Royal Botanic Gardens, Kew, giving, among other valuable advice on the subject, the related characteristics and damage potential of a variety of trees common in the UK.
Lenczner, D., *Movements in Buildings*, 2nd edn, Pergamon Press, Oxford, 1981
Rainger, Phillip, *Mitchell's Movement Control in the Fabric of Buildings*, Batsford, London, 1983.

BRE BRAS
TIL 5: 1971, *Devices for detecting changes in width of cracks in buildings* (covers simple mechanical devices: more sophisticated versions now available commercially).

BRE Digests
240, *Low-rise buildings on shrinkable clay soils: Part 1*, August 1980.
241, *Low-rise building on shrinkable clay soils: Part 2*, September 1980.
242, *Low-rise buildings on shrinkable clay soils: Part 3*, October 1983.
298, *The influence of trees on house foundations in clay soils*, June 1985. Read with Digests 240, 241, 242, and 251.

Fill and hardcore

1 General considerations

Fill and hardcore are used to make up ground (i.e. to raise its level). Fill is used to raise the level of the ground over a considerable area of site; hardcore for limited areas of a building as infill within foundations or beneath an oversite concrete slab. Both settle due to loads imposed on them, but the nature and extent of the settlement is different from that associated with soils. Importantly, the materials used as fill and hardcore may contain chemicals or other deleterious substances that may crack or degrade materials in close contact with or near to the fill or hardcore. Some materials, in fills in particular, may be hazardous to health or harmful to the environment or building. Emphasis is given here on aspects related to cracking; degradation is dealt with in *3.6 Chemical attack*. Aspects associated with health or harm to people and the environment are not covered in detail.

2 Fill

(a) Types of fill

A wide variety of materials may be used as fill. Table T2.1/9 includes the main types. For the present purposes fill is classified as existing or new:

1. *Existing fill*: This type of fill usually contains a substantial amount of industrial, mining or domestic wastes. The constituents of the fill may vary considerably, as might the extent to which they have been compacted. In most cases the fill is deep. In urban areas, relatively shallow fills occur where marshy land has been reclaimed by raising

Table T2.1/9   *Compressibility of fills*

Fill type	Compressibility	Typical value of constrained modulus (kN/m^2)
Dense well-graded sand and gravel	Very low	40 000
Dense well-graded sandstone rockfill	Low	15 000
Loose well-graded sand and gravel	Medium	4 000
Old urban fill	Medium	4 000
Uncompacted stiff clay fill above water table	Medium	4 000
Loose well-graded sandstone rockfill	High	2 000
Poorly compacted colliery spoil	High	2 000
Old domestic refuse	High	1 000–2 000
Recent domestic refuse	Very high	

From BRE Digest 274, *Fill, Part 1: Classification and load carrying characteristics*, Table 2, p.3, Building Research Establishment, Crown Copyright.

the ground level and sites have been infilled with the rubble of demolished building. Nowadays, many existing docks are being filled.

Fill that has large voids or materials liable to decay is especially hazardous. For example, refuse dumps may contain metal containers that rust and leave large voids and plastics containers which compress but do not decompose.* Large settlements may occur when organic matter breaks down. The decomposition of such matter may also result in potentially hazardous concentrations of methane or carbon dioxide. Underground fires may occur (with no sign of them on the surface) if the fill contains combustible material.

2. *New fill*: Generally, soils and what may be termed 'hard rubble' are used for new fill. Unlike existing fill, new fill is used for a particular site on which it is intended to build in the short term. The material(s) to be used for the fill may therefore be selected for the specific purpose.

Ideally, granular soils such as rockfill, gravels and coarse sands should be used, as these materials drain readily and consolidate quickly. The time required for consolidation is dependent on the grain size. Smaller grains take longer to consolidate. When clay soils are used the moisture content of the clay should be controlled. There should be no filling with clay during wet weather.

All fill should be placed in layers and compacted.† When various types of materials are used they should be placed in horizontal layers across the site. With granular and cohesive soils it is better that the granular layers are sandwiched between the cohesive layers, as this will assist in drainage and reduce the time for the cohesive material to consolidate.

Appreciable and uneven settlement of the surface of fill occurs if the natural ground on which it rests is soft and compressible (e.g. layers of peat or soft clay). This settlement is additional to that resulting from the compression of the fill. Filling may cause landslip if the fill is laid on unstable sloping sites.

(b) Compression of fill

Fill may be compressed by one or more of three causes:

1. *Self weight*: This is usually the main cause of long-term settlement where the fill is deep. Granular fill and poorly compacted unsaturated fills of all types compress almost immediately a load is applied. (The weight of the fill produces a load sufficient to cause compression.) Most of the settlement due to self-weight therefore occurs as the fill is being placed. However, subsequently there may be what is termed 'creep' settlement, and this can be significant.‡ The rate of creep compression varies with the type of fill (see Diagram 2.1/18). BRE have described a creep compression rate parameter – α. This is defined as the percentage vertical compression of the fill that occurs during $\log_{10}$ cycle of time, say between one year and ten years after the fill was placed. Table T2.1/10 gives some typical values of α, the magnitude of which can depend on the depth of the deposit as well as on the nature and degree of compaction of the fill. Using α, the conditions in the fill must remain unaltered for predictions of the settlement of a fill to be valid. Much greater movements could be caused if there is an increase in stress due to a change in the applied load or in moisture content.

*Some plastics are biodegradable. These are being used increasingly nowadays.

†The Department of Transport's *Specification for Road and Bridge Works* gives useful guidance on layer thickness and compaction for different types of soil. All types of soil listed in that publication are not necessarily suitable for supporting buildings.

‡This type of settlement is caused under conditions of constant effective stress and moisture content.

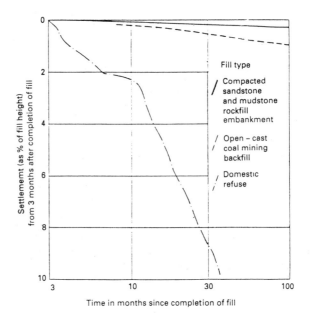

Diagram D2.1/18 *Settlement rates of different types of fill (vertical compression plotted against $\log_{10}$ time). (From BRE Digest 274,* Fill, Part 1: Classification and load carrying characteristics, *June 1983, Fig. 1, p. 2, Building Research Establishment, Crown Copyright)*

2. *Loads of the building constructed on the fill*: There is a wide variation in the compressibility of fill materials (see Table T2.1/10). The variations depend on the nature of the fill; its particle size distribution; compactness; the existing stress level; the stress increment; and the moisture content.

   The imposition of a structural load on fill invariably causes most of the compression, as in the case of self-weight. Fine-graded saturated fills do not compress in this way. Generally, movements that occur after a building has been constructed can be more of a problem than those that take place while it is being built. This makes the long-term creep component particularly significant. The values of creep compression rate parameter given earlier for compression due to self-weight (see Table T2.1/9) are also applicable to settlement produced by loads applied by the building. With the latter, zero time corresponds to the application of the load rather than the placement of the fill.

3. *Inundation*: A loose unsaturated fill is liable to collapse if it is inundated with water. Consequently there may be a serious settlement problem if inundation occurs after construction. Apparently, this is often a major cause of settlement problems with buildings on restored open-mining sites.

Table T2.1/10   *Creep compression rate parameter*

Fill type	Typical values of α (%)
Well-compacted sandstone rockfill	0.2
Uncompacted opencast mining backfill	0.5–1
Domestic refuse	2–10

From BRE Digest 274, Table 1, p.3, Building Research Establishment, Crown Copyright.

(c) Consolidation

The causes of settlement described above are related mainly to loose uncompacted fills formed by tipping. Settlement due to both self-weight and applied loads is, however, controlled by a consolidation process if fine material is placed under water, as in a tailings lagoon. The fine material forms a soft cohesive fill. When loaded, the excess porewater pressures dissipate slowly as water is squeezed out of the voids of the fill. This type of fill could be susceptible to liquefaction. If a firm crust forms over the surface of the lagoon deposit this is likely to be thin and overlie very soft material.

(d) Differential settlement

As explained under 'Soils' earlier in this chapter, cracking is due to differential settlement. Poorer types of fill, such as those containing organic matter or which were placed without control, are variable in their properties and therefore are liable to large differential settlements. (They are also prone to large total settlements.) Differential settlements are also likely to occur at the perimeter edges of filled areas and in places where the depth of fill changes rapidly.

(e) Settlement after construction

Estimation of the magnitude of movement of a fill needs to be undertaken by an expert in the field, and field loading tests are usually advisable. The characteristics of the fill need to be classified in terms of the nature of the material, particle size distribution, degree of compaction, depth, age and the level of the water table.

As a guide, movements are likely to vary between small and very large in three categories:

1. *Small movements*: Vertical compression of the fill subsequent to construction is everywhere smaller than 0.5%. This amount of movement is likely with a granular fill that has been placed under controlled conditions and has received adequate compaction.
2. *Significant movements*: Vertical compression of the fill subsequent to construction is estimated to have a maximum value between 0.5% and 2%. This amount of movement is likely with a granular fill placed without compaction, has a little organic material within it and has been in place for some years. Special attention needs to be given to foundation design. Usually, piling is advisable. Alternatively, a ground treatment to improve the load-bearing capacity of the fill so that it conforms to the first category may be used. Very substantial foundations may be required where large differential settlements can occur. In these cases the building units should be kept small and simple in plan.
3. *Very large movements*: Vertical compression of the fill subsequent to building is estimated to exceed 2%. Fills in this category are likely to include recently placed domestic refuse with a high organic content liable to decay and decomposition and fine-grained materials which have been transported in suspension and discharged into lagoons, where they form highly compressible cohesive fill which may be subject to liquefaction. Problems due to both total and differential settlement will be severe and ground-improvement techniques are unlikely to be successful. Methane emission from fills of recently placed domestic refuse may cause a major problem.

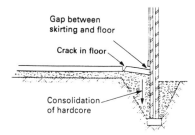

Diagram D2.1/19 *Filling below a ground-floor slab on sloping ground and the cracking resulting from the natural consolidation of the hardcore – see text. (From BRE Digest 276, Hardcore,* August 1983, Fig. 2, p. 3, *Building Research Establishment, Crown Copyright)*

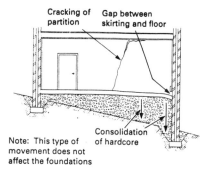

Diagram D2.1/20 *Filling of deep foundations and the cracking resulting from the 'natural' consolidation of the hardcore – see text. (From BRE Digest 276, Fig. 3, p. 3, Building Research Establishment, Crown Copyright)*

3 Hardcore

(a) Consolidation

Like fill, hardcore needs to be laid in relatively thin layers and each layer properly compacted. Suitably graded material is preferable to coarse ungraded. With graded material most of the voids are filled during compaction. This means virtually no consolidation of the hardcore after building operations are complete (i.e. less settlement). The hardcore also needs to be free of materials that can degrade when wet if consolidation is to be minimal. Looked at in another way, significant consolidation (i.e. sufficient to cause cracking) is likely to occur if the hardcore is ungraded or contains material that will be degraded by water. Deep filling with hardcore (usually depths in excess of 600 mm) can also result in significant consolidation.

To prevent cracking from occurring, the ground-floor slab needs to be supported evenly everywhere. This support may be lost in places where consolidation of the hardcore occurs. The slab and any partitions built on it can crack.

Typically, two conditions of the hardcore may cause such cracking: either sloping hardcore (i.e. progressive increase in its depth), as illustrated in Diagram D2.1/19, or filling of deep trenches at the perimeter of the slab (i.e. local increase in the depth of the hardcore), as shown in Diagram D2.1/20.

(b) Materials used

1. *Gravel and crushed hard rock*: Ideally, these should be suitably graded. If not, then they should at least form the base layer.
2. *Quarry waste*: This material is clean, hard and safe to use. It is usually unevenly graded and therefore difficult to consolidate to give a good working surface.
3. *Chalk*: This is susceptible to frost heave in floors of buildings when a prolonged frost has occurred during construction. To reduce the risk of damage during winter building the hardcore should be protected (by heat, if necessary) from frost during long, cold spells. Usually, a few days of continuously freezing conditions are needed before difficulties are likely to be encountered. Special consideration should be given where buildings are unheated or cold stores are involved.
4. *Concrete rubble*: Good hardcore results if the material is clean and suitably graded. Rubble from general building demolition can contain materials that lead to problems under wet conditions. Among these, gypsum is notable, as it can cause sulphate attack if it is placed close to concrete or brickwork (see *3.6 Chemical attack*, 'Sulphates', p. 411).
5. *Brick or tile rubble*: The material should contain fine material (from bricks that are soft and crumble easily) so as to facilitate compaction. Care is needed to establish whether the bricks used contain sulphates and in using old bricks to which gypsum plaster is adhering (see *3.6 Chemical attack*, 'Sulphates', p. 411). In particular, care is needed to ensure that there are no pieces of timber, more so if the rubble is from an old building where there has been an outbreak of *dry rot* (see *3.6 Chemical attack*, 'Wood-rots', p. 432). Refractory bricks such as those used in chimneys and furnaces should be checked to ensure that they are not of the type that expand when exposed to moisture.
6. *Blastfurnace slag*: Slag from making iron is a hard, strong material. It is suitable for hardcore. (Slags from steelmaking are not recom-

mended as they may expand on wetting.) It contains sulphates but these are normally predominantly calcium sulphate, which has limited solubility in water. However, the slag to be used should be analysed to ensure that it does not have a significant sulphate content (i.e. the extract contains less than $1\,g/l$ of SO_3 (see *3.6 Chemical attack*, 'Sulphates', p. 412). One advantage of slag is that it is free draining and so will not retain water in contact with concrete or brickwork.

However, it is not advisable to place large volumes of slag in wet stagnant conditions. This may cause water pollution by sulphur species leached from the slag. Unless sampled and tested to ensure that they do not contain wastes other than from blast furnaces, slags from old 'slag banks' should not be used. Importantly, the slag should not be old or partially vitrified, as swelling may take place from sulphate solutions arising from either groundwater or other components of the fill.

7. *Colliery spoil*: In general, unburnt colliery shale is to be preferred. This is likely to vary in its quality, although it has good free-draining characteristics. However, it has contributed to much sulphate attack of concrete ground-floor slabs (see *3.6 Chemical attack*, 'Sulphates', p. 415). There is no convincing evidence that the swelling of colliery spoils has caused failures.

8. *Oil shale residue*: Spent oil shale results from the extraction of oil from oil shale and is available in the Lothians area of Scotland. It is tough and resists disintegration by weather. It may be used as hardcore provided its soluble sulphate content is low – this content can be variable and appreciable. Care is needed on wet sites to provide suitable protection against sulphate attack on concrete or mortar (see *3.6 Chemical attack*, 'Sulphates', p. 416).

9. *Pulverized fuel ash*: PFA or fly ash is the fine material from the precipitators of coal-burning power stations. It is suitable for use as hardcore. Unless mixed with furnace bottom ash – the other major waste from the burning process – the total sulphate content is relatively low, and this content should always be checked. When conditioned with water, the ash produces a lightweight material which has self-hardening properties.

Sources

Addleson, Lyall, *Materials for Building*, Vol. 1, Iliffe Books, London, 1972 (o/p)
Rainger, Phillip, *Mitchell's Movement Control in the Fabric of Buildings*, Batsford, London, 1983

BRE Digests
272, *Fill. Part 1: Classification and load carrying characteristics*, June 1983
273, *Fill. Part 2: Site investigation, ground improvement and foundation design*, July 1983
276, *Hardcore*, August 1983

Strength and the use of materials

Introduction

This chapter considers certain practical applications of the strength of materials, both in construction and in use. One aspect that is not dealt with here is structural design on the larger scale. From the point of view of design and construction what is of concern here are the properties of materials which affect their *assembly*,* from shaping materials to their intended form, joining materials and components together, and sealing the joints between them.

The aspects of the strength of materials in use dealt with in this chapter concern not so much how buildings physically stand up over time (which would include, for example, the effect of chemical attack on the concrete of foundations) but rather the mechanical *weathering* of materials. In particular, this includes the resistance of finishes to abrasion and impact and therefore applies predominantly to the interior of buildings. With the latter it is usually impracticable to employ calculations to predict performance in terms of wear and tear. Resort has to be made to the lessons of experience, and such experience might well include accelerated tests on the material in question. Despite the fact that it is not always possible to have a quantitative basis on which to work, it is nevertheless important that some form of checklist of performance criteria should be developed so that strength, as it affects the use of materials, may be given its due consideration.

Strength and the use of materials in construction

This section is divided into three main areas. The first concerns the *shaping and forming* of materials, in other words, the preparatory processes which are applied to a particular material to shape its form, either into modular pieces of material for use directly in construction (such as bricks and blocks) or as an intermediate stage for further use in the fabrication of components (such as sawn timber sections). The second topic is *jointing theory*, which it is essential to grasp before considering *jointing, fixing and sealing*. These are the practical processes of joining materials and components together in the assembly of a building.

Shaping and forming

1 General principles and objectives

The degree to which civilizations have modified the environment, taking raw materials and transforming them, is a measure of the sophistication of their technology. There are two key ideas here: first, the objective of achieving something new from a raw material, such as

*An aspect of the principle of **separate lives**.

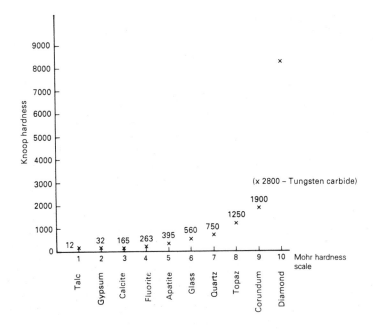

Chart C2.2/1 *Comparative hardness of materials*

taking clay and transforming it into brick, and second, the use of tools to achieve this. Technologically primitive cultures assemble structures by taking materials as found with little modification, whereas in more advanced societies techniques have been developed which enable complex transformations to be made. Significant changes took place in the Industrial Revolution when hand tools began to be superseded by machinery. Tools which made tools were invented, and jigs developed which enabled repetitious manufacture of similar elements. With mechanization came the need for standardization, which led to such familiar standards as the Standard Wire Gauge and the British Whitworth Thread. From this period and since come many of our modern materials, formed by technologically advanced processes – concrete*, steel,† glass,‡ and plastics.¶ Older materials began to be made in modern ways – notably bricks and glass.

2 Shaping by cutting

Cutting is a method of shaping *solid* materials in which there is a loss of material. The separated parts may all be of use, or there may be waste involved. Cutting relies on the relative hardness of the cutting tools and the material being cut. (See Chart C2.2/1 for comparative hardness of materials.) Where there is a large hardness differential between the two materials, a single edge is sufficient as a cutting tool. A stainless steel knife is ideal for cutting putty but not timber. An intermediate stage is the use of scoring – repeated running of a sharp edge across the softer material, making a deepening groove until the material is penetrated. Pipe-cutting tools work on this principle.

Where the hardness differential is smaller, tools are used which cut away small fragments of the material and remove these fragments as an integral part of the cutting process. A handsaw for cutting timber has

*Portland cement was invented in 1824 and reinforced concrete developed in the second half of the nineteenth century by Coignet, Hennebique, de Baudot and, after 1900, August Perret.

†Cast iron was first used on a large scale at Coalbrookdale in 1779. Mild steel was made possible by the Bessemer process after 1855, becoming widespread in the 1890s.

‡The most significant development in glass technology was fairly late: float glass was invented by Alastair Pilkington in 1951 and came into production in 1959.

¶For example, PVC began in commercial production in the 1930s, polythene has been made since the 1940s and 'Perspex' (methylmethacrylate) since 1940.

individual teeth which cut fragments of timber ('sawdust') and, by means of the 'set' of the teeth, in which they are splayed alternately to one side and the other, the dust is pushed out with the downward stroke of the saw through the cutting slot, which is wider than the individual teeth. A morse thread drill bit, used for cutting cylindrical holes in, for example, timber or steel, has two helical grooves up which the cut fragments ('swarf') are passed.

Cutting is, of course, one of the oldest methods of shaping materials of all types in their solid state. Under this broad heading there are many different techniques for different materials and applications, each requiring tools of different character. Shaping may be done by sawing, routing, guillotining, planing, drilling, chiselling, reaming and filing.

Historically, development of cutting methods has been towards techniques which allow better control over the shaping process. In particular, this has meant the use of *jigs*, which are secondary tools used for guiding or controlling the use of the main one. Many cutting tools are *self-jigging* to varying degrees, such as the plane where the blade is held in a fixed relationship to the work. This is an advance over the medieval adze, used for shaping large timbers to the desired profile, which required much greater manual dexterity in its use. Similarly, a pair of scissors can be seen as a self-jigging hand-held guillotine – offering greater control over accurate cutting than a freely held knife. Separate jigs commonly used include a mitre block used in controlling the angle at which a saw is held or a vertical drill stand.*
Further development of cutting has been towards increasingly comprehensive mechanization, culminating in robotic workstations driven by computer-generated instructions.

Cutting also has a role to play in demolition and site clearance. Table T2.2/1 lists the development of cutting technology of this type.

Table T2.2/1 *The development of cutting technology*

Historical precedent	Present method
Battering ram	Demolition ball
Hammer and chisel	Concrete breaker Rock drill Tungsten carbide flail
Hammer and wedge	Hydraulic burster Gas-expansion burster
Gunpowder	Explosives
Quarrying	Diamond drill/saw
Metal working	Thermic lance Powder cutting
Mining and stone working	Rocket jet burner
Industrial/manufacturing technology	Future developments: Microwave cutting Laser Eddy-current heating

*For a thorough discussion of the relationship between hand, tool and jig see *The Nature and Art of Workmanship* by David Pye, Studio Vista, London, 1968.

From Lazenby, D. and Phillips, P., *Cutting for Construction: A handbook of methods and applications of hard cutting and breaking on site*, Architectural Press, London, 1978.

3 Shaping by bending

Bending is a method of shaping materials in their *solid* state in which there is no loss of material. The strength properties of a material before shaping by bending are of importance, particularly the relationship between stress and strain, which gives rise to the properties of *ductility* and *malleability*.

Ductility applies to shaping a material by bending in two ways. First, the material must be sufficiently plastic, when the loads are applied to form it, to be able to bend without fracturing. Second, once bent, it must be able to keep its form without either springing or creeping back to its original shape; in other words, it must be neither too elastic nor viscous nor prone to creep. These properties are best illustrated on the stress/strain curve in Diagram D2.2/1. Up to a certain stress the material obeys Hooke's Law and is elastic. Beyond that it will elongate without much further stress but also without breaking. The explanation for this kind of behaviour lies in the crystalline structure of the material, and it so happens that the appropriate crystalline structures for ductility occurs almost exclusively in metals. Materials which fail these requirements by being too stiff (such as glass), too elastic (such as rubber) or too viscous (such as asphalt) can only be formed by moulding, described below.

There is something of a grey area between forming materials by bending and by moulding. Materials other than metals are formed by bending, particularly timber, but they have to be processed first by moistening or heating to allow bending to be done initially and to achieve a 'set' in the new form on drying or cooling.

Bending may be used to shape materials into highly complex forms. Some typical uses include handrails for staircases (in metal or timber), pressed metal frames for windows, doors, and cladding, and flashings for roofs and dpcs in walls. The actual process of bending may take place either under factory-controlled conditions or on the building site.

Materials shaped by bending have in common the fact that their *arrises* can never be absolutely sharp but are always rounded. In general terms, the thicker the material, the more rounded the arris. The radius of the arris is primarily governed by the minimum radius through which the material is capable of being bent. However, for specific materials the minimum radius for a given thickness will be governed by its strength in tension, compression and shear. Smaller radii are possible when materials are bent over a support or *former*. In the bending process the outer face of the materials will be in tension while the inner one will be in compression. For a given radius, which here is measured from the inner face, the outer face will have to stretch much further as the thickness of the material increases. At the same time, there will be greater compression of the inner face. Failure of the material may occur if the stress in either face is excessive: on the other hand, the difference between the tensile and compressive stresses may lead to shear failure. The relative differences, in terms of dimensions of the inner and outer faces, for various radii and thicknesses of material are illustrated in Diagram D2.2/2 and Chart C2.2/2.

In practice, it is uncommon for elaborate calculations to be undertaken to determine the minimum bending radius of a material. Reliance is placed on experience and, where there is none, on prototypes. The diagram and the chart, therefore, are only intended to show in principle the mechanical aspects associated with bending, and also to explain the absence of sharp arrises obtained with other shaping processes.

As tensile strength plays such an important part in the bending process it follows that materials which are weak in tension are

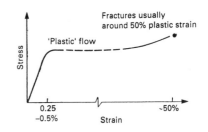

Diagram D2.2/1 *Stress/strain curve for a ductile metal. Note that 50% strain means an extension of the original length by 50%. (From Gordon, J. E.,* The New Science of Strong Materials*)*

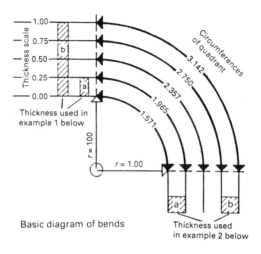

Basic diagram of bends

Examples of relative difference
between circumference of outer
and inner face (based on outer/inner face ratio)

1 Change in thickness

(a) $r = 1.00$ $t = 0.25$ (1:0.25)

$$\text{ratio} = \frac{\text{outer face}}{\text{inner face}} = \frac{1.965}{1.571} = 1.25$$

∴ outer face is <u>0.25</u> longer than inner face

(b) $r = 1.00$ $t = 1.00$ (1:1)

$$\text{ratio} : \frac{\text{outer face}}{\text{inner face}} = \frac{3.242}{1.571} = 2.00$$

outer face is <u>1.00</u> longer than inner face

2 Change in radius

(a) $t = 0.25$ $r = \underline{1.00}$ (1:4)

as 1(a) above
outer face is <u>0.25</u> longer than inner face

(b) $t = 0.25$ $r = 1.75$ (1:7)

$$\text{ratio} : \frac{\text{outer face}}{\text{inner face}} = \frac{3.142}{2.750} = 1.14$$

outer face is <u>0.14</u> longer than inner face

Diagram D2.2/2 *A composite diagram illustrating the effect of bending (through 90°) on the dimensions of the inner and outer faces of a material with changes in radii and thickness of material, with examples of the relative differences between the two faces*

unsuitable for shaping by bending. Thus brick, concrete and stone (which, in addition to being relatively weak in tension, are also liable to brittle fracture) are incapable of being bent to form them. Materials which are more commonly shaped by bending are timber, metals and plastics, and are described separately below.

(a) Timber

The practice of bending timber, mainly for staircase handrails, is fairly traditional. A more recent use of bending has been in laminated timber sections, particularly for structural applications.

For *dry* timber the maximum stretch possible without breaking is approximately 2% of the original length. However, in its dry state timber is too elastic to remain deformed. Treatment of timber by

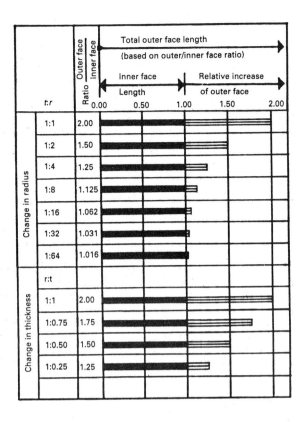

t:r	Outer face / Inner face Ratio	Inner face Length	Relative increase of outer face
Change in radius			
1:1	2.00		
1:2	1.50		
1:4	1.25		
1:8	1.125		
1:16	1.062		
1:32	1.031		
1:64	1.016		
Change in thickness r:t			
1:1	2.00		
1:0.75	1.75		
1:0.50	1.50		
1:0.25	1.25		

Chart C2.2/2 *Comparative graph showing the relationship of lengths of inner and outer faces for different thickness/radii (t:r) and radii/thickness (r:t) ratios*

steaming, boiling or heating in wet sand alters the characteristics of the material in two ways, namely, (1) a compressibility of up to 30% of the original length is possible and (2) the fibres of the timber are altered such that it becomes more plastic. After drying out the timber remains in the bent shape.

(b) Metals

The bending of metals falls into two categories, first, that done in a factory with elaborate machines (breakpresses, rolling machines, panel-forming presses, etc.) and whose primary use is the production of actual building components, and second, that carried out on the building site in connections with flashings, dpcs, roofing, pipework, etc. In each case metals have their special bending characteristics which derive from their crystalline structure. When the metal is bent the crystals take up new positions.

Bending is a form of work done on the metal, and if the work is carried out while the metal is in its cold state *work hardening* occurs, making the metal harder and stronger at the expense of its ductility. Metals and their alloys vary in their susceptibility to work hardening. *Lead* is really unique in the way that it can be bent and rebent many times more than other metals, making it useful for situations such as roof flashings, where continual bending is necessary. In all metals work hardening may be removed by annealing, which is a heat treatment that rearranges the crystals to their original, softer, state.

(c) Plastics

Shaping of plastics is an essential part of their manufacturing process. Thermosetting plastics are, on the whole, extremely difficult to shape by bending in the solid state. Special grades of laminated plastics are made for what is known as post-forming.

The thermoplastics, on the other hand, are comparatively simple to bend after moderate heat treatment. Examples of such forming include, in particular, floor finishes, where it is possible to bend the material at the junction with walls to form a skirting, and dpc work.

4 Shaping by moulding

In this technique solid materials are formed to the desired shape by preparing them in a liquid or plastic state which is then either poured into moulds or extruded through or drawn over a die. Strength development of materials formed in this way takes place during the cooling, setting or hardening process after shaping. The technique is used with all inorganic materials which have solid and liquid phases (i.e. metals, glass, polymers (including both thermosetting plastics and thermoplastics) and rubbers), ceramic materials in which the liquid or plastic state is achieved by the addition of water, cementitious materials, and organic materials such as wood particleboards, which are moulded by compression. As mould costs are usually high, this is a technique for forming which is associated with mass production. One major exception is reinforced concrete which, when formed *in situ*, often only uses the mould (called formwork) once.

(a) Metal casting

This is a moulding process applicable to both ferrous and non-ferrous metals. Castings are made by pouring molten metal into moulds and allowing it to solidify. Moulds may be permanent or non-permanent. When permanent moulds are used the process is called *die casting*.

Non-permanent moulds may be made from moulding sand, a synthetic sand/resin mixture or other special materials. The moulding medium is formed around a *pattern* within a *moulding box*. The pattern is removed and the molten metal poured into the mould. Castings formed in this way usually require a further process of cleaning up: this is called *fettling*.

(b) Plastics

There are five principal methods for moulding plastics: compression moulding, extrusion forming, injection moulding, vacuum forming and blow moulding. Thermosets are always moulded in compression moulds whereas thermoplastics can be formed by all five methods.

In compression moulding, split dies or moulds are used in hydraulic presses which incorporate heated plattens. The material, in granular or pellet form, is placed in the lower half of the mould. Pressure and heat are applied, fusing the material and forcing it to the shape of the mould. There are various types of moulds and methods. For example, in transfer compression moulds the material is forced into the mould by an auxiliary ram before full compression occurs.

Extrusion is a moulding process common to plastics, ceramics and metals. Thermoplastic powder or compound is delivered from a feed hopper to an Archimedean type of screw, where it is heated and then forced through a die. As the plastic cools, it retains the shape of the die.

This process is used for many building products, including edge trims and weatherboard claddings. In injection moulding, which again is used exclusively with thermoplastics, the material is heated and then forced into a split mould. In vacuum forming a single-sided mould is used. The material is laid in sheet form over it and heated. A vacuum produced from the mould side forces the material to follow the lines of the mould. Blow moulding is similar, but here a positive air pressure is used to force a heated blank into conformity with a mould. This process is particularly used for bottle-shaped products.

(c) Bricks and blocks

Clay bricks are moulded by two methods. *Wire-cut* bricks are formed by extruding clay through a die (which may incorporate perforations) onto a table where it is cut with taut parallel wires. *Pressed* bricks are made by forcing a slab of clay into a mould by either mechanical means or by hand. Sandlime bricks and concrete blocks and bricks are moulded in a way similar to pressed bricks.

5 Finishing

The final process of shaping metals, stones and timber is often done by grinding and polishing. The principle here is to use an abrasive material which cuts on a very fine scale.

Grinding is, in effect, a method of machining by means of abrasive particles rather than cutting tools. A grinding wheel comprises a large number of abrasive particles held together by a bonding material. In a well-designed wheel the bonds are strong enough to hold the particles together while they are sharp but allow them to be torn out when they become blunted. Aluminium oxide (Al_2O_3, which forms 90% of corundum and 70% of emery) and silicon carbide (SiC, more familiar as carborundum) are abrasives commonly used. Abrasive wheels and sheets are graded by particle size, bond strength and openness of structure.

Polishing with abrasives (not to be confused with polishing as an applied finish rather than a forming method) is similar to grinding except that the abrasive is carried in a liquid medium rather than bound together. As the finishing proceeds, increasingly finer grades of abrasive powder (in the form of 'flours') are used.

Jointing, fixing and sealing

The aspects of fixing and assembling considered here are those associated with the joining together of two or more preformed elements with the aid of some intermediate material or component in order to ensure structural or constructional stability. In addition to an introduction to jointing theory, three key groups of materials associated with the process of jointing are covered: mechanical fixings, specific adhesives and sealants.

1 Jointing theory

(a) Introduction

The process of assembling the components and materials in the construction of a building includes as a fundamental part the making of joints. An understanding of basic jointing theory underlies both fixing

Diagram D2.2/3 *The terms used in describing joints.* Notes: (1) *Jointing material and jointing section are both jointing products. Jointing material is a formless jointing product. Jointing section is formed to a definite cross section.* (2) *Joint face is the part of the joint profile used in the calculation of the component's work size (see Diagram D2.2/9). (Source: BS 4643 : 1970)*

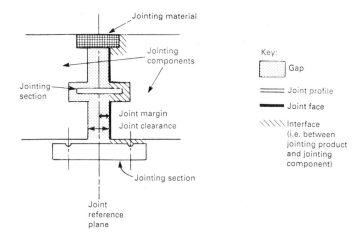

and assembling, the subject of this section, and the performance of the building fabric, particularly the problem of exclusion, which is discussed in *3.4 Exclusion.**

Since the first edition of this book was published there has been much research into the theory of jointing from all aspects, particularly from the point of view of achieving a good fit on-site (for example, see BS 5606: 1978, *Code of practice for accuracy in building*). The early British Standard on dimensional coordination in building (BS 3626: 1963, 1986, *Specification for modular coordination in building*) introduced a basic conceptual framework of grids and spaces and reflected the increasing use of components that could not be cut and fitted in traditional ways. The problem of dealing with tolerance continues to present a major challenge to all parties involved in building.

The principles for building which are relevant to jointing are **continuity**, **separate lives** and **balance**.

BS 6093 states that 'since joints are breaks in the physical continuity of construction they are potential breaks in the continuity of performance'. The objective of continuity of performance, such as thermal insulation or the exclusion of moisture, must be achieved in spite of the physical discontinuity. **Continuity** therefore is the fundamental principle.

However, the fact of **separate lives** is unavoidably present: at the construction stage as 'the process of assembly', with all the problems of deviation and fit that that implies, and then throughout the building's life as the parts which are joined are subjected to differential movements and have differential durability.

Balance as an objective is important, because it underlies the appropriate choice of jointing material. An example of imbalance would be the use of a heavy-duty expansion anchor to fix back a lightweight cladding material.

(b) Definitions

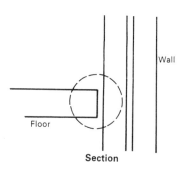

Diagram D2.2/4 *A joint as a meeting of elements of different functions*

The primary reference for jointing theory is BS 6093: 1981, Code of practice for the design of joints and jointing in building construction.

'Joint' is a broad term, and includes both the placing or connecting together of units of the same material (such as mortar joints in brickwork or roll joints in lead sheet roofing) and the junction between different materials and building elements (for example, the connection between a door frame and a wall). Diagram D2.2/3 illustrates the various terms used in describing joints.

(c) The need for joints

Joints are necessary for a number of reasons:

1. As a meeting of adjoining elements which perform different functions (Diagram D2.2/4);
2. As a barrier or separator required by a change in conditions (Diagram D2.2/5);
3. As breaks in the continuity of construction required by the limitations of working methods (Diagram D2.2/6);
4. As connections between components whose size is limited by manufacture, handling, storage, transport, or assembly;
5. As a means of allowing for the effects on both assembly and service of inherent and induced deviations (Diagram D2.2/7).

However, because of the problems of achieving continuity of performance, as a general rule (which must frequently be broken for the practical reasons listed above), unnecessary use of joints should be avoided.

(d) The function of joints

As stated above, the overriding function of all joints is to achieve continuity of performance. ISO 3447 has a comprehensive list of these performance functions, given in Table T2.2/2. Of course, not all joints will be required to perform all these: in particular, some joints will not be required to accommodate movements. However, this is a key aspect from the point of view of *sealing*, and is discussed in some detail below.

(e) Allowing for deviations

When an architect preparing a detail draws and dimensions an element, such as a window frame or part of a wall of brickwork, allowance must be made for the fact that in reality the sizes as built will not be the same as the precise dimensions shown. The design dimensions are only *target* ones and the true dimensions will deviate from this target.

There are two basic kinds of these dimensional deviations. *Induced* deviations are those caused by work done, and which therefore arise from inaccuracies in operations such as manufacture, setting out, and erection (see Diagram D2.2/8). *Inherent* deviations, on the other hand, are those caused by an inherent property of the material, such as its response to changes in humidity, temperature or stress. The first kind of deviation is static. That is, it occurs as a result of assembling materials with a specific manufactured size resulting in a joint of particular dimensions on-site. The second kind is dynamic, in that these kinds of deviations change constantly in response to environmental conditions.

BS 5606 and BS 6954 set out a method of predicting induced deviations, in terms of the predicted variability in the size of clearance between work size and coordinating size (see Diagram D2.2/9).

Inherent deviations can be calculated from information on coefficients of thermal and moisture movement published in BRE Digests 227, 228 and 229, *Estimation of thermal and moisture movements and stresses*. Thermal movement per unit length of a material can be calculated from the material's coefficient of thermal expansion, the temperature range in service (bearing in mind the exposure of the

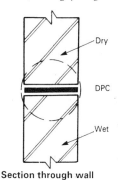

Section through wall

Diagram D2.2/5 *A joint as a separator between different conditions*

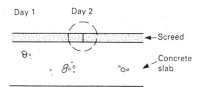

Section through floor

Diagram D2.2/6 *A joint as a break in the continuity of construction*

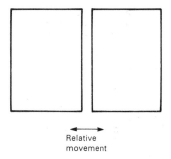

Diagram D2.2/7 *A joint as a means of allowing for deviation in the size of elements*

Table T2.2/2   *General list of joint functions, grouped under design aspects*

1. *Environmental factors*

1.1 To control passage of
  (a) Insects and vermin;
  (b) Plants, leaves, roots, seeds and pollen;
  (c) Dust and inorganic particles;
  (d) Heat;
  (e) Sound;
  (f) Light;
  (g) Radiation;
  (h) Air and other gases;
  (j) Odours;
  (k) Water, snow and ice;
  (l) Water vapour.
1.2 To control condensation.
1.3 To control generation of
  (a) Sound;
  (b) Odours.

2. *Capacity to wisthand stress either during or after assembly)*

To resist stress in one or more directions due to:
  (a) Compression;
  (b) Tension;
  (c) Bending;
  (d) Shear;
  (e) Torsion;
  (f) Vibrations (or any other type of stress which may induce fatigue);
  (g) Impact;
  (h) Abrasion (indicate, for each particular case, the type of wear);
  (j) Shrinkage or expansion;
  (k) Creep;
  (l) Dilation or contraction due to temperature variations.

3. *Safety*

3.1 To control passage of fire, smoke, gases, radiation and radioactive materials.
3.2 To control sudden positive or negative pressures due to explosion or atmospheric factors.
3.3 To avoid generation of toxic gases and fumes in case of fire.
3.4 To avoid harbouring or proliferation of dangerous microorganisms.

4. *Accommodation of dimensional deviations*

4.1 To accommodate variations in the sizes of the joint at assembly due to deviations in the sizes and positions of the joined components (induced deviations).
4.2 To accommodate continuing changes in the sizes of the joint due to thermal, moisture and structural movement, vibration and creep (inherent deviations).

5. *Fixing of components*

5.1 To support joined components in one or more directions.
5.2 To resist differential deformation of joined components.
5.3 To permit operation of movable components.

6. *Appearance*

6.1 To have acceptable appearance.
6.2 To avoid
  (a) Promotion of plant growth;
  (b) Discoloration due to biological, physical or chemical action;
  (c) All or part of the internal structure showing;
  (d) Dust collection.

7. *Economics*

7.1 To have a known first cost.
7.2 To have a known depreciation.
7.3 To have known maintenance and/or replacement costs.

8. *Durability*

8.1 To have specified minimum life, taking into account cyclic factors.
8.2 To resist damage or unauthorized dismantling by man.
8.3 To resist abrasive action.
8.4 To resist action of
  (a) Animals and insects;
  (b) Plants and microorganisms;
  (c) Water, water vapour or aqueous solutions or suspensions;
  (d) Polluted air;
  (e) Light;
  (f) Radiation (other than light);
  (g) Freezing of water;
  (h) Extremes of temperatures;
  (j) Airborne or structure-borne vibrations, shock waves or high-intensity sound;
  (k) Acids, alkalis, oils, fats and solvents.

9. *Maintenance*

9.1 To permit partial or complete dismantling and reassembly.
9.2 To permit replacement of decayed jointing products.

10. *Ambient conditions*

10.1 To perform desired functions over a specified range of
  (a) Temperature;
  (b) Atmospheric humidity;
  (c) Air or liquid pressure differential;
  (d) Joint clearance variation;
  (e) Driving rain volume.
10.2 To exclude from the joint if performance would be impaired:
  (a) Insects;
  (b) Plants;
  (c) Microorganisms;
  (d) Water;
  (e) Ice;
  (f) Snow;
  (g) Polluted air;
  (h) Solid matter.

Note: Any one joint will have to satisfy a selection of functions only. This list cannot be comprehensive and the designer may have to identify additional functions applying in a specific situation.

From BS 6093: 1981.

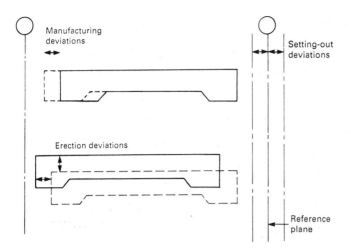

Diagram D2.2/8 *Illustration of induced deviations*

material and its colour. See Table T2.1/4, p. 58) and the restraint of the movement. The formula is

$$TM = l \times C \times t$$

where TM = total unrestrained thermal movement, l = original length, C = coefficient of thermal expansion and t = temperature range (usually ranging from both sides of the installation temperature).

Both inherent and induced deviations occur in three dimensions, which must be considered in the design of the joints. Three-dimensional deviations include tapering in elevation, bowing and twisting, misalignment of component faces and surface irregularities.

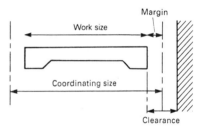

Diagram D2.2/9 *Definition of terms used in dimensional coordination*

(f) Joint types

Joints can be categorized into types in a number of different ways, according to the purpose of the classification:

1. Classification by geometry: Joints can be classed as butt, lap or mated (see Diagram D2.2/10).
2. Classification from the point of view of assembly: Rostron* distinguishes between integral and accessory joints (see Diagram D2.2/10).
3. Classification by allowance for deviations: joint types can be classified by the way they deal with the two types of deviation. BS 6093 lists four types:
   Type 1: no allowance is made for deviations (Diagram D2.2/11);
   Type 2: allowance is made for both types of deviation (Diagram D2.2/12);
   Type 3: allowance is made for inherent deviations only (Diagram D2.2/13);
   Type 4: allowance is made for induced deviations only (Diagram D2.2/14).

It is useful to consider how the various deviations occur throughout the assembly in deciding what type of joint to use in which location.

*Rostron, R. M., *Light Cladding of Buildings*, Architectural Press, London, 1964.

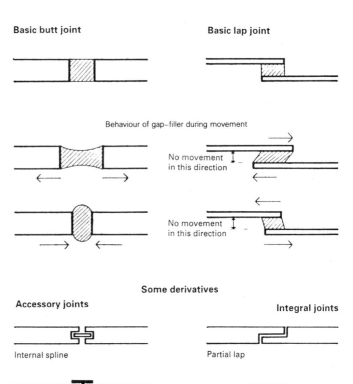

Diagram D2.2/10 *Classification of joint types by geometry. Accessory and integral joints are derivatives which reflect different methods of assembly. (After Rostron)*

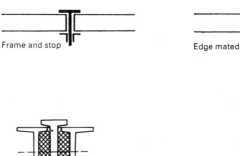

(a) Plain bolted joint. Holes sized for bolt clearance

(b) Coupling bar between two steel window frames. Mechanical fixing allows negligible movement

(c) Base of partition panel fixed with non-adjustable cleat

Diagram D2.2/11 *Examples of type 1 joints: no allowance made for any deviations. (Source of this and the next three diagrams: BS 6093: 1981)*

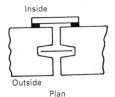

(a) **Two stage joint with rain baffle and air seal**

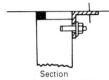

(b) **Joint at head of partition panel with restraining cleat.** This is designed to allow movement. Prefixing of cleat to soffit is a type 1 joint.

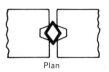

(c) **Joint between panels with gasket seal**

Diagram D2.2/12 *Examples of type 2 joints: allowance made for both induced and inherent deviations*

(a) **Tongue jointed floor boarding.** Close butted initially. Joints subsequently subject to movement.

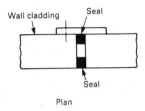

(b) **Joint between wall panels.** Joint clearance at assembly made with fixed (constant) width gauge or stop.

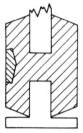

(c) **'Zipper' structural gasket in rubber/plastic**

Diagram D2.2/13 *Examples of type 3 joints: allowance made for inherent deviations only*

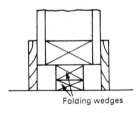

(a) **Base of partitioning component with folding wedge packings**

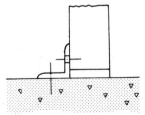

(b) **Base of suspended panel fixed with adjustable cleat**

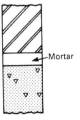

(c) **Horizontal mortar joint**

Diagram D2.2/14 *Examples of type 4 joints: allowance made for induced deviations only*

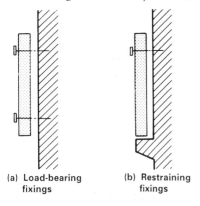

(a) Load-bearing
fixings

(b) Restraining
fixings

Diagram D2.2/15

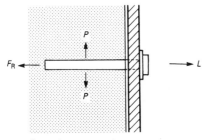

For anchor to succeed:

$L \leq \Sigma F_R$ where $F_R = P.A. \mu$

L = Applied load
P = Expansion pressure
F_R = Frictional force
A = Expansion area
μ = Coefficient of friction

Diagram D2.2/16 *Frictional grip fixings. (This diagram and D2.2/17, 19–27 are based on Hilti,* Handbook of Anchor Technology)

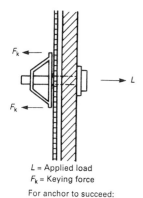

L = Applied load
F_k = Keying force

For anchor to succeed:

$L \leq \Sigma F_k$

Diagram D2.2/17 *Keying hold fixings.*

2 Mechanical fixings and fasteners

(a) Principles

1. *Function*: fixings have essentially two functions. One is to act in a *load-bearing* capacity, the other is to *restrain* (see Diagram D2.2/15). Fixings are often required to perform both functions.
2. *History*: fixings as a way of joining and fastening materials have been used since early times. The Greeks used U-shaped cramps to hold masonry together. In medieval timber-framed construction wooden dowels were employed at joints. Both of these fixings are primarily of the restraint type.
3. *Types*: basically, there are three ways in which fixings work, and for each of these categories there is an extended family of sub-types. The first method is to use the fixing to make an *anchor* in one part so that the other part can be gripped to it. The second is to squeeze the two parts together by means of a *through-fixing*. Adhesion is the third method, and here the bonding forces are distributed evenly over the interface between the two parts.

   Types of anchor and through-fixings and procedures for their use are discussed below in (b): adhesion is considered separately in *3 Adhesion*, p. 108.

   These types can be related to the difference between the two parts in terms of their size, relative structural importance or their position within the constructional sequence. Thus through-fixings are generally used where the two parts are of roughly equal size or importance, such as the nuts and bolts in a steel frame. Where there is a clear distinction between the *base* and the component fixed to it, anchor-type fixings or adhesives are generally used. Examples include fixing battens to brickwork or bonding ceramic tiles to walls.

4. *Fixing theory: how fixings hold*:

- Frictional grip (see Diagram D2.2/16): This is a rather loose but descriptive term used to denote the compressive forces that are permanently created between the surfaces of two elements which will prevent one of them sliding past the other. These forces may act in conjunction with the *keying hold* described below. Frictional grip may be achieved at one of two stages: (1) as the fixing device is *passed through* the elements, or (2) subsequent to the passing-through process. To be effective, the expansion pressure produced when the fixing is made must cause a frictional force between the fixing and the hole wall that is greater than the applied load. Nails and pins create the expansion pressure purely as they are driven in. Expansion anchors achieve frictional grip between the two surfaces during the 'tightening-up' process. They do, however, usually work in combination with a keying hold.

   Fixings of the anchor type rely for their success on the frictional grip that can be obtained, which, in turn, depends on the mechanical strength properties of the materials involved. These are discussed later.

- Keying hold (see Diagram D2.2/17): Some fixings are designed so that the applied load is transmitted to the base material by means of a mechanical interlock or key. This principle is used in its most pure form by through-fixings, such as nuts and bolts and cavity fasteners. The latter includes many proprietary fixings – bulb rivets and toggle fasteners to name just two (see Diagram D2.2/18). To be successful, the applied load must not exceed the total keying force capacity of the material.

Screws work on a combination of this mechanical keying and frictional grip, but, due to the pitch of the threads, the majority of the force which resists pull-out comes from the interlock.

- Adhesion: Shear forces between the fixing and the base material can be carried if there is adhesion between the two (see Diagram D2.2/19). The applied load must always be smaller than the shear forces at the glue line. Note that the adhesion may be purely *specific* adhesion or, as is more usual, combined with *mechanical* adhesion – in other words, a keying hold.

Resin anchors used in concrete are a modern equivalent of the traditional rag bolt where the fixing is locked into place using an adhesive – resin as opposed to cement grout. In both cases some mechanical keying plays a part in the fixing's load-bearing capacity.

5. *Fixing theory: modes of loading*:

- Direction and position of load: Fixings may be loaded axially or laterally (in shear) or a combination of the two (see Diagram D2.2/20). With stand-off fastenings used for cladding and fixing of services, bending forces are set up in the fastening which must be resisted (see Diagram D2.2/21).
- *Types of loading*: The graph in Diagram D2.2/22 illustrates the kinds of load to which fasteners may be subjected. Wind on buildings may cause all three types of dynamic loading. This is a particular problem and is discussed at some length in (c) p. 107.

6. *Fixing theory: modes of failure*: Fixings may fail in a number of different ways:

- Snapping or shearing of the fixing (see Diagram D2.2/23);
- Rupture or spalling of the base material around the fixing, which originates from a plane at right angles to the axis of the fixing (see Diagram D2.2/24);
- Pull-out of the fixing from the base material (see Diagram D2.2/25);
- Cracking of the base material in the same plane as the fixing, typically when it is set too close to edge of the material (see Diagram D2.2/26).

7. *Factors of safety*: If a hundred fixings were made in the same base material and then tested to failure, they would not fail at identical loads. In the nature of things the results would vary, but if they were plotted out as a graph they would lie on a bell-shaped curve, known by statisticians as a 'normal distribution'. The majority of results lie in the middle of the distribution, or 'mean' (average). A small proportion lie a long way from the average, those which are either unusually weak or unusually strong.

A statistical term for describing the proportion of results which lie away from the mean is the 'standard deviation'.* 32% of all readings lie within the band of the mean plus or minus one standard deviation; 95% lie within the band of the mean plus or minus two standard deviations. Thus (and this is the crucial point from the safety point of view) the great majority of fixings (97.5%) reach a value of mean-2s.d. This value is termed the 'fractile'.

Safe working loads are defined in relation to this fractile point, i.e. safe working load is less than or equal to (mean-2s.d./factor of safety). This is shown graphically in Diagram D2.2/27. A typical value of the factor of safety is three.

In some cases the safety factor would need to be larger, for example, where:
(a) The results of failure would be particularly devastating;
(b) The assumed loading conditions are particularly uncertain;

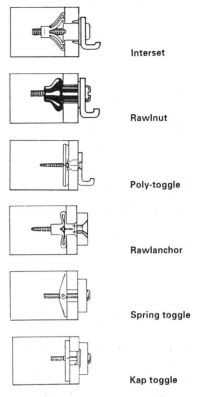

Interset

Rawlnut

Poly-toggle

Rawlanchor

Spring toggle

Kap toggle

Diagram D2.2/18 *Examples of proprietary cavity fixing devices. (Courtesy: Rawlplug Co. Ltd)*

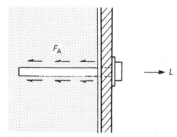

For fixing to succeed:
$L \leq \Sigma F_A$ where $\Sigma F_A = A \cdot \tau$.

L	= Applied load
F_A	= Adhesive shear forces
A	= Adhesive area
τ	= Shear strength of adhesive

Diagram D2.2/19 *Adhesion fixings.*

*Standard deviation is defined as s = square root of sum of $(X_i - \bar{X})^2/n$.

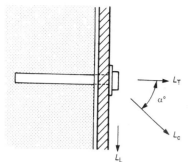

L_T = Tensile load (α = 0–10°)
L_L = Lateral load (α = 80–90°)
L_c = Combined load (α = 10–80°)

Diagram D2.2/20 *Direction of load.*

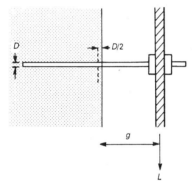

$M_b = L\,(g + (D/2))$

Where:
M_b = Bending moment in fixing
L = Applied lateral load
g = Stand-off distance
D = Diameter of fixing

$$\sigma_b = \frac{M_b}{W_x}$$

Where:
σ_b = Actual bending stress
W_x = Axial moment of resistance

e.g. $\dfrac{D^3}{10\mu}$ for round section

Diagram D2.2/21 *Bending forces in stand-off fixings.*

(c) The performance of the base material is dubious;
(d) There is a risk of dynamic loading whose effects cannot be tested.

(b) Fixing types, processes and procedures

This section is concerned primarily with the principles applicable to the different types of fixing described above, and the detailed *rules* relating to the great range of products available is outside its scope. Reference should be made to manufacturers' information in respect of particular products and for guidance on selecting the appropriate fixing for a specific job.

(1) Anchor fixings – 1: nails

Nails and similar sharp-pointed devices with smooth surfaces are driven into materials by impact or a series of impacts. This may be made by a hammer or a mechanical means such as a power hammer or shot-firing. The strength of the fixing device must be related to the compressive strength of the materials into which it will be driven and the type of impact. The fixing device must also be strong enough to resist being deformed during the whole of the process. In this respect it is interesting to note the development of nails made of specially tempered steel which are capable of being driven into materials of relatively high compressive strength such as concrete, brick, stone and steel. The comparative brittleness of these special fixings makes it extremely important from a safety point of view that they are installed in the correct way.

As far as the elements are concerned, the ease with which they can be pierced depends primarily on their hardness and their resistance to crushing. However, although materials of low compressive strength are more easily pierced, they will form a weaker frictional grip per unit length of the fixing device than materials with a high compressive strength. In addition, the ability of materials to accommodate the expansion which takes place as the fixing device is driven in will be related to their tensile strength. The stress that is created locally as this happens, and hence the material's resistance to fracture, depends on the cross-sectional area of material near the fixing. All materials are

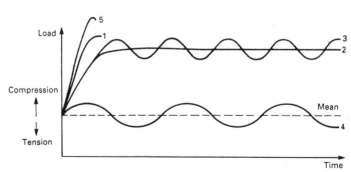

1 – Short-term loading
2 – Long-term loading
3 – Dynamic pulsating loading
4 – Dynamic alternating loading
5 – Shock loading

Diagram D2.2/22 *Types of loading.*

more vulnerable to cracking or splitting if the fixing device is driven in near to an edge than if they are in the centre of the element. This is shown in Diagram D2.2/28. In *isotropic* materials such as brick, stone or concrete the cracking may occur in any direction. Initial failure of materials such as these is often seen at the surface of the material which receives the first impact. In *anisotropic* materials (i.e. those with a grain such as timber), cracking will occur along the plane which is weakest (see Diagram D2.2/29).

Once the material has been pierced and is able to resist the tensile forces, it must then be elastic enough to provide permanent compressive forces at the surface of the fixing device to achieve the required frictional grip. In all these cases the success of the frictional grip will depend (in addition to those mechanical properties already described) on maximum surface contact in order to obtain the maximum compressive force between the surfaces.

(2) Anchor fixings – 2: screws

Screws and similar threaded devices are capable of tapping and sometimes boring through materials. These all require a shank which is tapered and threaded over most of its length. Since the boring takes place as the device is being turned, resistance to torsion is one of the main requirements of the fixing. In addition, it must be harder than the material through which it passes for the threads to indent or tap it. In most cases the strength requirements of the fixing device may be considerably reduced if the base material is pre-drilled with a hole which is slightly smaller in diameter. This also has the effect of reducing the stresses which the base material must sustain since there is less material to be displaced.

(3) Anchor fixings – 3: expansion bolts and plugs

There are a variety of fixings of this type available. However, in each case frictional grip is obtained by the expansion of the device which is inserted into a pre-drilled hole. The hole is normally made just fractionally larger than the insert. In the simplest case a thin-walled cylinder or plastic plug is inserted into the pre-drilled hole. A screw is then driven into the plug causing it to compress against the walls of the hole. A soft plastic is used so that it is forced into the pores of the base material to obtain a mechanical key to supplement the friction forces created.

Expansion bolts consist of an expandable sheath of metal over a bolt. After insertion of the device and as it is tightened up, the sheath is made to expand and compress against the walls of the hole. These are all proprietary devices and are the subject of continuing research and

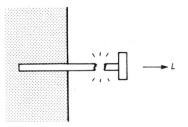

Diagram D2.2/23 *Failure by fracture of the fixing.*

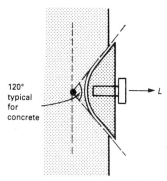

120° typical for concrete

Diagram D2.2/24 *Failure by spalling of the fixed material.*

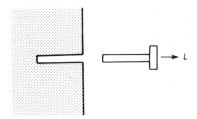

Diagram D2.2/25 *Failure by pull-out of the fixing.*

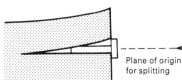

Plane of origin for splitting

Diagram D2.2/26 *Failure by splitting or cracking of the element.*

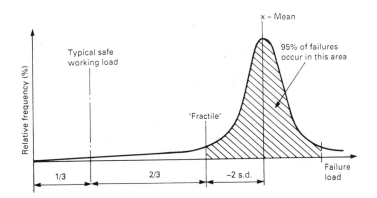

Diagram D2.2/27 *Statistical explanation of safe working loads.*

development to improve their performance. They are predominantly designed for use in concrete.

(4) Anchor fixings – 4: purpose-made sockets
These work on the principle of a keying hold in which the key is prefabricated as a purpose-made slot or threaded socket which can be cast into concrete, welded to steelwork or screwed onto or into timber.

(5) Through fixings
This group includes all those fixing devices which pass through the assembly to form a keying hold at each side and incorporate a means of 'tightening up' as part of the fixing process. The traditional nut and bolt has no frictional grip between the surface of the hole and the shank of the bolt. The diameter of the pre-drilled hole must be slightly smaller than the shank to avoid damage to the threads. Frictional grip is created between the surfaces of the two elements in contact as the nut is tightened up. In this the compressive strength of the materials of the

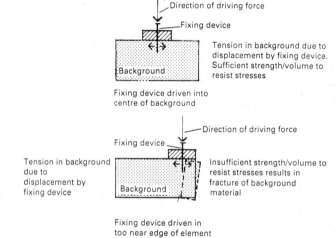

Diagram D2.2/28 *Diagrammatic representation of the relative weakness of the edges of an isotropic material to resist the tensile stresses due to displacement by a nail or screw-type fixing device. (Note: the inherent weakness of the edge is greatly increased as the impact driving force is increased. Special care is therefore always required when firing techniques are used)*

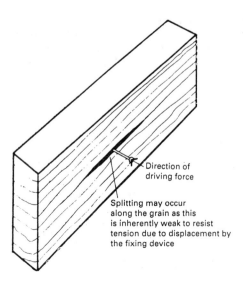

Diagram D2.2/29 *Representation of the inherent weakness along the grain of timber relative to the displacement caused by a nail or screw-type fixing device*

two elements is important. It is usually necessary to increase the spread of pressure by the use of washers.

With ordinary mild steel black bolts this frictional grip between the surfaces is not of great significance. With high-strength friction-grip (HSFG) bolts the two parts are held tightly enough together for the friction to be an important factor in the structural behaviour of the joint.

Rivets are now rarely used in structural steel work but are commonly employed in light metalwork and other sheet materials such as plastics and plywood.

Cavity fixings, of which there are many proprietary kinds, work on the basis of forming a keying hold blind, against which a nut or screw can be tightened. (See Diagram D2.2/18 for some examples.)

(c) Strength of fixings in use

Once the fixing has been made, the strength of all the materials involved, both the fixing itself and the elements joined, should be sufficient to support the loads imposed. In the case of restraint fixings the load may consist of only a small component of the dead weight of one of the elements, whereas with load-bearing fixings the total load is likely to comprise the dead weight of one of the elements combined with a live load.

Externally, a key component of the live load is the force exerted by wind, which affects both pitched and flat roofs and facades. Subject to the form and shape of the building and its surrounding obstructions, considerable pressures and suctions may be exerted on the roofing or cladding. Such suctions are capable (as shown by many examples) of lifting individual components or the entire roof structure. Where the dead weight of the roofing or cladding component is small in relation to its area, the part played by the fixings in resisting these forces becomes critical. British Standard recommendations for additional fixings at the perimeter of tiled and slated roofs must be followed.

In addition to maximum loads, light cladding and roofing is susceptible to the effects of vibration. Vibrations can be caused either by the dynamic nature of the wind (particularly vortices at parapets and other abrupt changes in the building's geometry) or by the resonant response of cladding units fixed at intervals to a steady force. On this aspect there is at present no authoritative published guidance on problems that may arise. However, there has been practical experience of problems of fixings loosening as a result of wind-induced vibrations. This has been particularly marked with self-tapping screws fixed through insulation giving a long lever arm, as shown in Diagram D2.2/30. This type of fixing should be used with caution.

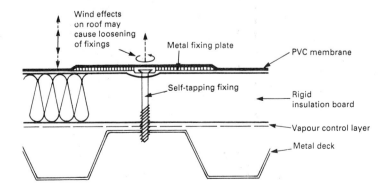

Diagram D2.2/30 *An example of a stand-off fixing vulnerable to the effects of movement. In this example wind effects on the waterproof membrane caused the fixings to unscrew*

3 Adhesion

(a) General principles

Adhesion is defined as the attachment of two surfaces by interfacial forces consisting of molecular forces, chemical bonding forces, interlocking action or a combination of these. Of the two principal types of adhesion, *mechanical* and *specific*, the latter generally uses an additional material to form the adhesive bond between the two surfaces of material to be joined.

In this section the focus is on practical applications of the physical laws, including an outline survey of the different types of adhesives and the rules governing their selection and use.

Definitions
- *Adhesive*: a third material used to promote the intimate interfacial contact between the surfaces of the adherends being joined that is necessary to achieve adhesion.
- *Adherend*: each of the two materials or components joined by the adhesive.
- *Glue line*: the adhesive at the interface between the two adherends.
- *Tack*: (1) the resistance offered by an adhesive film to detachment from the adherend surface, or (2) the stickiness of an adhesive film as it forms an instant bond when brought into low-pressure contact with an adherend.

(b) Effects of movement

The principle of **separate lives** suggests that the two materials being joined by an adhesive bond are likely to be subjected to differential movements. Forces set up as a result of such movement will place a stress on the adhesion bond: to be effective, the bond must be sufficient to resist these stresses. At the same time, movements of all kinds must be controlled or limited, as dictated by circumstances. One common type of failure occurs with plasters or renders, particularly those with a cement base, applied to a background of wet construction. This is used here as an example to explain the effects of movement (see Diagram D2.2/31).

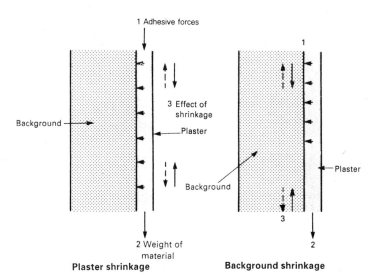

Diagram D2.2/31 *The effects of movement on an adhesive bond. When 1 > [2 + 3] then adhesion is maintained. When 1 < [2 + 3] then adhesion is lost. Note that the effect of expansion of the plaster is the same in principle – the direction of shrinkage forces is shown by dotted lines. A combination of both forces can occur at the same time*

Both the wall and the plaster are subject to moisture movements. Such movement persists until the materials have reached moisture content equilibrium. Backgrounds which appear dry may still be damp enough for some moisture movement to occur. On the other hand, the plaster, after application, must complete its drying shrinkage. During the drying of both the wall and the plaster it is possible for the shrinkage which occurs to be sufficient to 'shear through' the bond between them. Excessive amounts of water in the plaster tend to increase its drying shrinkage, in addition to making the material weaker. Some **balance** must therefore be created. Greater dependence on a mechanical bond will assist in maintaining adhesion, but care must be taken not to depend entirely on this and in so doing disregard basic precautions to ensure satisfactory specific adhesion.

Any movements, from whatever source, which occur after application of the plaster will, if they are large enough, cause a similar loss of adhesion.

(c) Practical precautions

Securing and maintaining good adhesion requires observance of a few simple precautions. When dealing with any form or type of adhesive it is essential to remember two basic rules: (1) that adhesion is concerned with the bonding of the *surfaces* of two materials (see Diagram D2.2/32) but that (2) no adhesive can effect a bond which is stronger than the surface to which it is applied. Thus the fact that an adhesive may be 'stronger than the wood itself' is of no benefit.

As outlined earlier, adhesion is, in general, the result of the combined effects of both mechanical and specific adhesion. In many cases, as in the use of traditional adhesives such as mortars, plasters and renders, it is difficult to lay down precise rules as to what proportion of each should contribute to the bond. Consequently, as a general rule it is safer to ensure that provision is made for the *maximum amount* of each to take place. This rule may not necessarily apply to some of the special adhesives which are specifically formulated to bond smooth surfaces together and have poor gap-filling properties. The manufacturers' recommendations for their products should always be ascertained and followed.

In order to achieve good 'wetting' of the adherends it is normal to use adhesives in a liquid form, rather than to try to effect the even and true matching of the solids that would otherwise be necessary for specific adhesion to occur. Therefore the liquid content of both the adhesive and the adherend is often critical. Where the adhesive is insufficiently liquid, the 'wetting' process cannot take place to the full. Again, a typical example is the application of plaster to a brick wall. However, if the wet material is applied to a dry and absorbent background, much of the water will be absorbed, leaving insufficient water either for the 'wetting' to be completed or for the plaster to set properly. Similar problems occur if paint is applied to a dry absorbent background.

The precautions which should be taken to secure and maintain good adhesion may be summarized as follows:

1. *Cleanliness*: All surfaces to receive the adhesive should be clean and there should be no intervening or foreign substances between the adhesive and the background which will reduce the 'wetting' properties of the liquid content of the adhesive. Typical intervening substances are grease, oil, dirt, dust, flaking paint, chemicals (particularly salts), scum, laitenance, mill scale and rust.
2. *Dryness*: This, before and after application, is particularly necessary for those surfaces which are to receive adhesives with non-water

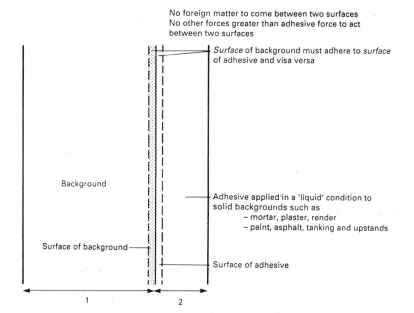

No foreign matter to come between two surfaces
No other forces greater than adhesive force to act
between two surfaces

Surface of background must adhere to *surface*
of adhesive and visa versa

Background

Adhesive applied in a 'liquid' condition to
solid backgrounds such as
    – mortar, plaster, render
    – paint, asphalt, tanking and upstands

Surface of background

Surface of adhesive

1

2

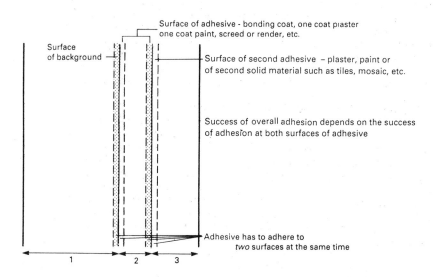

Surface of adhesive - bonding coat, one coat plaster
one coat paint, screed or render, etc.

Surface
of background

Surface of second adhesive – plaster, paint or
of second solid material such as tiles, mosaic, etc.

Success of overall adhesion depends on the success
of adhesion at both surfaces of adhesive

Adhesive has to adhere to
two surfaces at the same time

1

2

3

Diagram D2.2/32 *Illustration of the*
importance of surfaces in adhesion.
Above: *two materials and two*
surfaces; below: *three materials and*
four surfaces

bases. In the case of water-based adhesives a certain amount of dampness may be required to avoid excessive suction of the water content of the adhesive.

3. *Key*: Some provision should be made to ensure sufficient key in the background material, unless recommendations for a given adhesive do not specifically stipulate the necessity of a key. Materials which do not normally have 'keyed' surfaces, such as metals, should have their surfaces roughened. Porous materials may need additional keying. On dense concrete, the surface of which is not keyed during construction and which is to receive plaster or render, liquid-applied bonding agents are sometimes used. Such bonding agents may be imperative on highly absorbent lightweight concrete surfaces to secure adhesion of calcium sulphate plasters in particular.

4. *Avoiding suction*: Any absorbent materials to receive adhesive should be sufficiently sealed to prevent excessive absorption of the liquid content of the adhesive. This requirement may not necessarily apply to specially manufactured adhesives.

5. *Chemical decomposition*: Any agencies which are likely to lead to the chemical decomposition of either the adhesive or background after application must be avoided.

6. *Avoiding movement*: Any excessive movement, including vibrations, of the background, adhesive or of elements and units which may be joined by an adhesive should be avoided in order to maintain permanence of bond.

(d) Special adhesives

(1) Background
Given the number of possible combinations of materials that might be joined using adhesives, it is not surprising that the range of special adhesives is so large and continually developing. About 60 years ago the only adhesives of major importance were animal glues and other natural products which had not changed in centuries. New adhesives were developed to meet the demand of new technologies. For example, casein adhesives (based on milk) were employed first in World War I for wooden aircraft constructions. They suffered from poor moisture resistance. The main source of more recent development has been the introduction of synthetic resins. The first of these was phenol formaldehyde. Epoxy resins and modified phenolics were used structurally in aircraft in World War II.

In the building industry today, performance criteria for particular applications, such as water resistance of adhesives used to bond wall tiles in swimming pools, are pushing the refinement of adhesives away from 'general-purpose' types towards a proliferation of products for specific applications. The use of structural adhesives to replace conventional fasteners is expanding. These products have been aimed at mechanical engineering applications hitherto but they are likely to be adopted increasingly for structural ones.

(2) Advantages and disadvantages of adhesive bonding
The benefits and limitations of adhesives compared with mechanical fixings are set out below:

Benefits
- Ability to join a variety of materials of differing composition, thickness and coefficient of thermal expansion where mechanical fasteners would cause distortion or damage, particularly foils and thin sheets;
- Elimination of protrusions;

- Fabrication of complex shapes;
- More uniform distribution of stresses over the entire bonded area, thus saving weight and cost;
- The high damping capacity of adhesive bonds and the elongation properties of many adhesives allows stresses to be absorbed or transferred and gives good resistance to damage from vibration;
- Weight reduction and, since no holes are required, the maintenance of structural integrity;
- The ability to join heat-sensitive materials which would be damaged by welding or brazing.

Limitations
- Residual stress in bonded joints, arising from differential expansion of the adhesive and component surfaces, where cure temperatures are different to the ambient temperature of the adherends;
- Optimum bond strength usually takes time to develop, unlike welding;
- Low peel strength and tendency to creep under load are common problems;
- The effects of severe environments on the long-term performance of adhesive bonds is often unknown;
- Inflammability and toxicity are common characteristics, particularly of solvent-based adhesives;
- Bonded structures are not easily dismantled;
- Rigorous quality control is needed to maintain reliable joints.

(3) Adhesive materials and their properties
Adhesive materials are classified under a number of different heads. Shields* describes the following:

- End use (e.g. metal to metal, paper, timber, etc.). This is not a satisfactory basis for classification because of the great variation in bonding characteristics found within each class.
- Setting action. There are three main types:
  Solvent-based adhesives, which solidify on loss of solvent. They include natural product glues as well as synthetic resins such as PVA;
  Temperature-setting adhesives, which liquefy on warming and set on cooling;
  Chemical-setting adhesives are generally employed in structural applications subjected to high loads or adverse conditions. They are generally insoluble and infusible and may be formulated to enhance a particular property such as peel strength.
- Physical form. Adhesives come in a variety of forms, including:
  Liquids of high or low viscosity;
  Paste or mastic. These are used where void-filling and non-slump characteristics are required;
  Film or tape, supported or unsupported (e.g. by paper backing). These are suitable for smooth adherends and guarantee a uniform thin glue line;
  Powder. These require heat or water to activate;
  Granules, rods, etc. These are hot-melt materials, activated by heat.
- Method of application.
- Chemical composition. British Standards for adhesives are broken down according to chemical types, which include (see Table T2.2/3):
  Natural products, comprising materials of vegetable or animal origin;
  Thermoplastics, comprising natural and synthetic resins. These include hot-melt adhesives, PVA and cyanoacrylates (the 'super-glues');

*Shields, J., Engineering Design Guides, *Adhesives Bonding*.

Thermosets, based on synthetic polymers such as epoxy resins (e.g. Araldite), polyesters and phenol formaldehyde;

Elastomers, comprising rubber-like materials of natural or synthetic origin. SBR rubber, neoprene and silicone come under this heading;

Two-polymer compositions, consisting of composite materials derived from categories 2, 3 and 4.

- Suitability of the adhesive for the environment.
- Origin (e.g. natural/synthetic).
- Method of bonding (e.g. hot melt, solvent release, pressure sensitive).

Table T2.2/3 *Classification of adhesives according to chemical type*

Category	Representative types	Physical forms	Joint properties	Joint materials	Applications
Natural	Starch, dextrin, animal and fish glues, casein, rosin, shellac, Canada balsam, and gums	Solvent dispersions, emulsions, pastes, solids, and powder	Low strength with poor moisture resistance, but fair resistance to heat and chemicals	Paper, cork, packaging materials, textiles, wood (furniture), some plastics, and metals	General domestic purposes, industrial packaging, and other rapid set assemblies
Thermo-plastics	Cellulose derivatives, polyvinyl acetate, polyvinyl alcohol, polyacrylates, polyethers, and oleoresins	Solvent dispersions or water emulsions, films, and solids	Fair peel strength but low shear strength and subject to creep under high loads; maximum service temperature approximately 90°C	Non-metallic materials: wood, plastics, leather, textiles, and paper	Generally in assemblies subject to low loads in service
Thermosets	Urea and melamine formaldehydes, polyesters, polyimides, furanes, epoxides, and phenolics	Liquids, films, pastes, and powder	High shear strength and resistant to creep, but low peel strength; maximum service temperature is between 200°C and 250°C	Metals, wood, ceramics, and glass	Structural assemblies involving stressed joints based on metals and wood
Elastomers	Natural and reclaim rubbers, acrylonitrile-butadiene, polyurethane, polychloroprene silicone, butyl, butadiene-styrene, and polysulphide	Films, solvent dispersions, and water emulsions	High peel strength and flexibility, but low shear strength and poor creep resistance; maximum service temperature is between 80°C and 100°C, with silicones exceeding 200°C	Plastics, rubbers, fabrics, and leather	Unstressed joints of light materials and flexible bonds
Two-polymer types	Phenolic–nitrile, phenolic–neoprene, phenolic–polyvinyl acetal, and epoxide (modified)	Liquids, films, and pastes	Strength properties dependent on formulation, but generally higher strength over a broader temperature range than other types; good resistance to chemicals	Metals, ceramics, glass, and thermosets	Structures subject to high stresses or adverse service conditions such as heat and moisture

Source: Shields, J. Engineering Design Guides, Adhesive bonding.

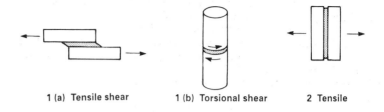

1 (a) Tensile shear   1 (b) Torsional shear   2 Tensile

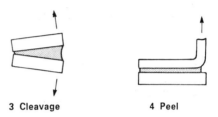

3 Cleavage   4 Peel

Diagram D2.2/33 *Types of stress in adhesive joints. Diagrams D2.2/33–35 are based on Shields*

- Structural/non-structural.
- Durability: BS 1204 uses the following classification:
  INT – interior use
  MR – moisture resistant
  BR – boil resistant
  WBP – weather- and boil-proof (only achieved by phenolic resins).

The properties of the different products are too diverse to discuss here and a guide such as Shield's *Adhesives Handbook** together with manufacturers' information should be consulted.

(4) Design of bonded joints
To form satisfactory joints using adhesives the following rules should be observed: first, as much of the bonded area as possible should contribute towards the joint strength and, second, the joint design should use a favourable geometry from the point of view of the stresses to which it will be subjected.

Four main types of stress are recognized as occurring in adhesive bonded joints, as shown in Diagram D2.2/33.

In *shear* loading, stress is distributed throughout the joint, although non-uniformly, so that all the adhesive is at work simultaneously. Where the stress is parallel to the plane of the glue line the joint area is used to the best advantage, giving an economical bond that is most resistant to failure. This is the favoured geometry for an adhesive joint, and the one most frequently used.

In *tensile* loading the forces are perpendicular to the plane of the glue line and are uniformly distributed over the entire area. Therefore no part of the joint carries more or less than its share of the load, which follows the first of the above rules. In practice, tensile stress may not be the only stress present, and if the applied loads are offset the stress distribution becomes uneven.

In *cleavage* loading, which occurs as a result of an offset tensile force, the stress is concentrated to one side of the joint. This means that a larger bonded area is required than with tensile loading, and should be avoided.

*Shields, J., *Adhesives Handbook*, 3rd edn, Butterworths, London, 1985.

In *peel* loading the stress is concentrated at a fine line on the joint edge and a high stress is thus placed on the extremity of the glue line. This is the condition which places most stress on the joint, and takes least advantage of the total adhesive area, and should be avoided where possible.

Lap joints are the most commonly used form of adhesive joint. If a simple lap joint is subjected to a tensile force the non-uniform stress distribution curve shown in Diagram D2.2/34 is found. This shows that the greatest stress on the adhesive occurs at both ends of the overlap. The peaks are greatest for rigid adhesives and smallest for flexible, extendable ones. Failure load for lap joints is proportional to the *width* of the joint, but the benefit of increasing the *length* decreases to a limiting value. However, the load-carrying performance in relation to the overlap length also depends on the thickness of the glue line. For a given overlap length, as the glue line becomes thinner (and therefore the ratio of overlap length to glue line thickness becomes greater) the shear strength decreases. Lap joint performance is improved if the geometry is modified to reduce stress distortions on the glue line. Examples are shown in Diagram D2.2/35.

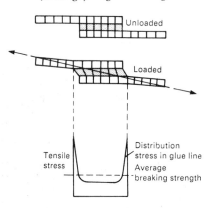

Diagram D2.2/34 *Distribution of stress in a lap joint subjected to tensile force.*

(5) Adhesive selection
Adhesives have to be chosen to function as an integral part of the joint in question. Designers are generally non-specialists and should therefore seek guidance from the specialist manufacturers.

As a guide, Chart C2.2/3 shows compatible adhesives and adherends. Note that, in general, any two adherends may be bonded together if the chart shows that each is compatible with the same adhesive.

(6) Bonding techniques
Adhesives may be applied in various ways, including brushing, flowing, roll coating and spraying.

Having applied the adhesive, there are usually three steps to complete the adhesive bond:

● Liquefaction of the adhesive to ensure 'wetting' of the adherend.

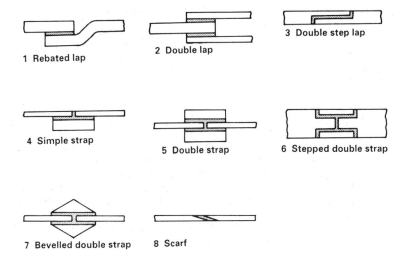

Diagram D2.2/35 *Representative types of adhesive bonded lap joints.*

Chart C2.2/3 *Complementary adhesives and adherends.*

Adherends \ Adhesives	Natural	Animal glues	Starch	Dextrine	Casein	Elastomers	Acrylonitrile butadiene	Polychloroprene	Polyurethane	Silicone rubber	Polybutadiene	Natural rubber	Butyl	Thermoplastics	Cellulose nitrate	Polyvinyl alcohol	Polyvinyl acetate	Polyacrylate	Silicone resin	Cyanoacrylate	Thermosets	Phenolic formaldehyde	Urea formaldehyde	Resorcinol formaldehyde	Melamine formaldehyde	Polyesters (unsaturated)	Epoxy resins	Polyimides	Phenolic-vinyl formal	Phenolic-polyvinylacetal	Phenolic nitrile	Phenolic epoxy	Inorganic	Sodium silicate
Metals							●	●				●								●							●	●	●		●	●		
Glass, ceramics		●					●								●		●										●			●		●		●
Wood					●							●			●		●					●	●	●	●		●							
Paper		●	●	●	●										●	●	●																	●
Leather		●					●	●				●			●																			
Textiles, felt		●					●					●			●		●																	
Elastomers																																		
Polychloroprene (neoprene)								●																										
Nitrile							●													●														
Natural							●					●								●										●				
Silicone										●																								
Butyl							●						●																					
Polyurethane							●	●	●																									
Thermoplastics																																		
Polyvinyl chloride (flexible)							●	●	●																									
Polyvinyl chloride (rigid)							●	●	●																		●							
Cellulose acetate								●							●					●														
Cellulose nitrate															●					●														
Ethyl cellulose															●					●							●				●			
Polyethylene (film)								●			●							●																
Polyethylene (rigid)																											●				●			
Polypropylene (film)								●			●							●																
Polypropylene (rigid)																											●				●			
Polycarbonate								●																			●							
Fluorocarbons												●							●			●					●							
Polystyrene								●												●							●							
Polyamides (nylon)								●														●			●		●				●			
Polyformaldehyde (acetals)								●												●										●	●			
Methyl pentene								●												●														
Thermosets																																		
Epoxy																				●		●		●			●							
Phenolic								●												●					●		●				●			
Polyester																										●	●							
Melamine								●	●																		●							
Polyethylene terephthalate							●	●										●									●							
Diallyl phthalate							●																			●	●							
Polyimide																											●	●						

(*From Shields.*) Note: In general, any two adherends may be bonded together if the chart shows that they are compatible with the same adhesive

- Setting or curing of the adhesive. Depending on the setting type, this may be achieved by removal of redundant components of the adhesive such as solvents and carriers, or by chemical curing. Since, with the former, bond formation takes place with the loss of volatiles, they must be able to evaporate or be absorbed by the adherend. With the latter, bond formation occurs by a chemical change within the adhesive. Chemical cures are often temperature sensitive.
- Application of pressure to the joint during the curing process to maintain the integrity of the assembly.

4 Sealing

(a) General principles

BS 6093 defines a seal as a 'physical barrier that is notionally impenetrable and is in contact with the components forming a joint', noting that the term does not presuppose the use of any particular material or mechanism but implies effectiveness in sealing against whatever agent is relevant, such as water, air, fire or sound. Sealing is therefore an integral part of the formation of joints and may be achieved by the physical connection of the components or by sealant as a further material.

Where such a further material is used, sealing may be achieved by *sealants, gaskets* or *baffles* (known as 'jointing products', see Diagram D2.2/3). A sealant is a material which, 'when applied to a joint in an unformed state, seals it by adhering to appropriate surfaces within the joint'. A gasket is a flexible preformed material that constitutes a seal when compressed. A baffle is a component of a two-stage joint (see *3.4 Exclusion*, p. 284) that forms a seal against rain but not air.

The particular problem of sealants is how they cope with the actual dimensional deviations that occur in the process of assembly. Jointing theory was discussed above on p. 95: this section is concerned not so much with the principles of joint design but rather with jointing products and their properties from the mechanical point of view. It is important, though, to be aware of the dimensional deviations which are an integral part of the process of sealing.

As discussed above, dimensional deviations are of two kinds: *induced* and *inherent*. Sealing materials are used in joints which are designed to make allowance for both induced and inherent deviations. There is, however, a considerable difference between the kind of sealing material that is used to seal joints which only allow for induced deviations and those which allow for inherent deviations as well, which is the need to accommodate movements over time. An example of the former is an ordinary mortar joint in brickwork, where the wet mortar accommodates the variations in the size and profile of the bricks which occur naturally as a result of manufacture. The latter must have an appropriate degree of *elasticity*, and examples of typical applications include the joints between storey-height GRP cladding panels (in which thermal movements predominate) or expansion joints in a blockwork wall (in which moisture movements predominate).

Seals play a crucial part in the success of *one-stage* joints (see *3.4 Exclusion*, p. 284). This kind of joint depends entirely on the effectiveness of the jointing products to seal against water and air penetration simultaneously. Jointing products have to maintain contact with the joint faces and keep their own cohesive integrity in use. Their durability is therefore important. Since most sealants have a durability of less than the life of the building, allowance has to be made for their maintenance and replacement. Their life is extended if they can be protected from weathering or abuse.

Seals for one-stage joints are available in two forms: those that depend for the transmission of movement forces on adhesion (such as sealants) and compression (such as gaskets). Gaskets require more precisely dimensioned joints than sealants. Another difference between sealants and gaskets is that the assembly sequence for sealants is usually component/component/seal, and for gaskets component/seal/component.

Variability in the size of components and their setting out, together with variations in joint width due to movement, are the crucial factors in the performance of seals. The performance limitations of jointing products will determine the permissible variations in joint width, which, in turn, will affect the frequency of joints and the standards of accuracy of manufacture and setting out required.

The *movement-accommodation factor* is the full range of movement between maximum compression and maximum extension that the sealant can accommodate. It is usually expressed as a percentage of the minimum design joint clearance. Typical low, medium and high factors are 5%, 15% and 25%.

Minimum design joint clearance (C_d) can be calculated from the following information:

1. The movement accommodation factor (MAF) of the sealant in question;
2. The estimated total movement (TM) of the components of the joint.

The formula to use is:

$$C_d = (TM \times 100/MAF + TM)$$

This only takes account of inherent deviations, the dynamic movements. Allowance also needs to be made for induced deviations, the variations in joint size that arise from manufacture and setting out.

The profile of the sealant is important as well as its actual width. The aim is to produce a joint cross section such that there is low internal stress and a satisfactory area of adhesion. As a result, the optimum profile for seals generally is a depth-to-width ratio of 1:2 (see Diagram D2.2/36). This profile is also satisfactory from the point of view of other performance criteria such as resistance to wind pressure.

Sealant joints are affected by the mode, frequency and rate as movement, as well as its magnitude. These factors will influence the choice of sealant.

(b) Types of sealant

Sealants are classified as elastic, elastoplastic, plasto-elastic or plastic according to their response to movement. Table T2.2/4 summarizes the various types of sealant and their properties.

(c) Practical precautions for sealants

Sealants must be applied against a firm backing so that the material is forced against the sides of the joint to achieve good adhesion. The back-up material ensures that the correct depth and proportions of the seal are achieved. Typical back-up materials are closed-cell foamed plastics such as polyethylene or rubber, cork or fibre boards.

It is important that adhesion only takes place between the sides of the joint and not between the sealant and the back-up material. To avoid this adhesion, a bond-breaking tape is used at the back of the sealant: some back-up materials such as polyethylene foam perform both functions.

Although components can be erected in various conditions of temperature and moisture content, the environment at the time of installation must be taken into account when the joints are sized. For example, if components are assembled in hot weather the joint width will be at its minimum. As the components contract with falling temperatures the joint will open up, putting tensile stress on the sealant.

High moisture contents of components make it difficult to achieve good adhesion between the sealant and the components. Good quality control on-site is needed to ensure that sealants are not applied at such times.

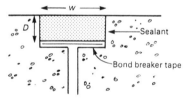

Recommended
depth-to-width ratio:

Sealant type:			
Plastic	3:1	to	1:1
Plastic-elastic	2:1	to	1:1
Elasto-plastic	1:1	to	1:2
Elastic	1:2		
Minimum depth for non-porous materials	6 mm		
Minimum depth for porous materials	10 mm		
Minimum width	5 mm		

Diagram D2.2/36 *Examples of sealant joints and recommended profiles*

Table 2.2/4 Types of joint sealant and their properties.

Sealant type	Application type	Max. joint width (mm)	Movement accommodation (as percentage of joint width)	Character[a] (after cure)	Typical uses	Comments	Expected service life (years)
Bituminous and rubber/bitumen	Hot poured	50	10	Plastic	In contact with bituminous materials	Poor durability in external movement joints. Available to BS 2499 in four types	5
	Hot applied, non-sag	25	10–15	Elasto-plastic	–	–	5–10
Oleo-resinous ('oil-based mastics')	Gun applied, non-curing	20	5–10	Plastic	Pointing around window and door frames	Forms surface skin. Regular maintenance necessary	Up to 10
Butyl rubber	Gun applied, non-curing	20	5	Plastic	Pointing, bedding	Not recommended for exposed joints. Properties vary with formulation	
Acrylic (solvent)	Gun applied, non-curing	20	15	Plasto-elastic	Pointing, e.g. around timber frames treated with exterior wood stains, etc.	Good adhesion. May need warming before aplication	Up to 15
Acrylic (emulsion)	Gun applied, non-curing	20	10–20	Plasto-elastic	Internal pointing	Low health hazard. Durability uncertain	
1-part polysulphide	Gun applied	20	10–25	Elasto-plastic	Movement joints in heavy structures	Slow curing; vulnerable to damage by movement until fully cured	
1-part polyurethane	Gun applied	20	15–30	Elastic	Movement joints with light (e.g. metal) components		
2-part polysulphide	Gun applied	25	30	Elasto-plastic	Both fast-moving joints in lightweight structures and slow-moving joints in large heavy structures	Mixed on-site so must be used within 'application life'. Maintenance costs low	Up to 20
2-part polyurethane	Gun applied	25	30	Elastic			
1-part silicone (low modulus)	Gun applied	20	50	Elastic	Joints between plastics and metal components	High initial cost. Careful surface preparation essential	
1-part silicone (high modulus)	Gun applied	20	20	Elastic	Sanitary ware. Fast-moving joints	Unsatisfactory on porous surfaces	
Polyepoxide/polyurethane (2 part)	Gun applied	50	50	Elasto-plastic			20+

Notes:

[a] Plastic materials exhibit plastic flow and have little or no recovery after deformation. Elastic materials (elastomeric) have physical properties similar to rubber and return to the original shape after deformation. Elasto-plastic and plasto-elastic exhibit partial elastic and partial plastic properties.

1. Movement accommodation is expressed as the total reciprocating movement occurring at the joint. Manufacturers may quote either this or ± figure based on the median joint width.
2. Multi-component sealants require mixing prior to application and must be used within their pot life.
3. Elastic and elasto-plastic sealants generally require primers on porous substrates.
4. This table is not an exclusive list of sealant materials available.

Based on BRE IP 25/81, *The selection and performance of sealants*, and 'Sealants, mastics and gaskets', *DOE Construction*, **18**, June 1976.

(d) Gaskets

Unlike sealants, for gaskets to work it is essential that they remain under compression at all times. They are thus made from a family of materials with similar properties: natural and synthetic rubbers or plastics.

Gasket joints are best designed by specialists: the bulk of their use is in components such as windows, where design development using mock-ups is possible and factory fabrication provides a better guarantee of the necessary dimensional accuracy.

Strength and materials in use

1 Abrasion

Abrasion is defined as 'the wear or removal of the surface of a solid material as a result of relative movement of other solid bodies in contact with it'.* It may occur whenever the surfaces of two materials are rubbed against one another. The factors which influence its extent are:

1. The period of time during which the rubbing takes place;
2. The surface characteristics of the two materials in contact;
3. The resistance to rubbing (i.e. friction) provided by the two surfaces.

The last is directly related to the properties of the surface of the materials and the pressure exerted during rubbing.

(a) Significance

During some manufacturing processes positive use is made of abrasion either to shape or to provide particular surface finishes to materials. In some cases it may be necessary to roughen an otherwise smooth surface to provide a key for the application of an adhesive.

However, during the construction and life of a building, abrasion is more normally associated with actual damage to the surface of materials. Such damage may be confined to comparatively small areas. At best, it may only affect the appearance of the material or, at worst, cause permanent damage. Further, the abrasion of the surface of a material may lead to more rapid deterioration of the material as a whole, either by an increase in the rate of abrasion or by reducing the resistance to damage by other agencies such as water.

It has been difficult to make positive quantitative assessments of the susceptibility of materials to abrasion in use. These difficulties arise mainly because of the complexities of the mechanisms taking place and the unpredictability of local factors which cause abrasion. Thus the whole question of abrasion has to be treated primarily on the basis of experience. Some conditions within buildings are known to be more conducive to damage than others, while at the same time some materials are known to be more resistant to abrasion. Once the severity of conditions has been assessed, it is then possible to consider three courses of action:

1. To protect the surface of materials with some form of applied finish, usually thin and often temporary, and to ensure that frequent renewal is possible;
2. To select finishes which, although they may be liable to damage in the severest conditions anticipated, are used in such a way that affected areas may be easily renewed; or
3. To select materials which are known to resist abrasion in the severest conditions.

*BS CP 3: 1950: Chapter IX, *Durability*.

(b) Finishes

As abrasion is essentially concerned with the wearing away of surfaces of materials the problems associated with it are normally confined to finishes. These may be applied or integral to the material, in which case, replacement is difficult. Surfaces subject to abrasion should have finishes chosen which are either highly resistant or are easily replaced.

(c) Factors influencing abrasion

Although the time period during which abrasion takes place clearly has a direct effect on its extent, the more fundamental factors are the properties of the surfaces and the abrasive medium, if there is one.

The key properties of a surface affecting abrasion are its hardness and surface texture.* It is convenient to regard as the abrasive the material which is harder and worn away least. Abrasion works by the cumulative scratching by the harder material of the softer one. The harder material may actually be one of the surfaces, such as hob-nailed boots on lino, or it may be in the form of a medium between two surfaces of approximately equal hardness, such as ordinary grit and dust brought in by footwear from the street onto a polished floor. The rate of abrasion increases with the magnitude of the pressure with which the abrading surface is applied to the abraded one.

(d) Agencies causing abrasion in buildings

In order to determine the severity of conditions likely to lead to damage by abrasion it is necessary to consider the agencies which cause the abrasion to take place. In most practical building examples damage occurs where the abrasive acts as a medium between two surfaces being rubbed together. The abrasion is carried out to greater effect if one of the materials is soft enough for the abrasive medium to be embedded in it.

Dust, dirt and grit are the most common abrasives encountered in buildings. This makes floors particularly vulnerable, since the action of walking or the use of wheeled traffic constitute the necessary rubbing action for abrasion to occur.

(e) Floors and abrasion

As dirt and grit, acting as the abrasive materials, account for a sizeable proportion of the damage to floors, it follows that regular cleaning is likely to reduce the amount of damage. Doormats have an important function in removing some of the street grit from footwear. Nevertheless, at the entrance to buildings, lifts, and on stairs where there is a high intensity of foot traffic floor finishes should be specifically selected for their resistance to abrasion. All sheet flooring materials will have their durability reduced if they are laid on an uneven base, since abrasion is concentrated on local areas which are proud.

Cement-based floors such as granolithic are liable to what is known as 'dusting', caused by the abrasive effects of foot and wheeled traffic. When this happens the dust itself acts as an abrasive medium, accelerating the process. The liability to dusting can be controlled by special surface treatments (sodium silicate, magnesium or zinc silico-fluoride or paints may be used).

*For a comparative chart of the hardness of materials see Chart C2.2/1. Materials used deliberately as abrasives in finishing processes are generally selected from the upper end of the hardness scale, such as corundum.

(f) Maintenance and cleaning

Damage by abrasion may also occur as a result of the maintenance and cleaning of various surfaces in buildings. Such maintenance may and should be carried out regularly. However, this does not mitigate the care which many finishes require during these processes. The appearance and surfaces of many materials may be ruined when the incorrect cleansing agent is used.

(g) Handling of materials

It is only in special circumstances during the life of a building that occasional scratching by hard, sharp objects may be reasonably anticipated. However, during the construction stage the liability of such damage to the surfaces of materials is far greater.

Damage generally occurs during the handling of materials, but may also be caused by plant and equipment. Surface finishes which are easily prone to scratching or which are not easily touched up on-site (and this would include many self-finished materials) require consideration to be given to protection of the surfaces. Temporary transparent plastics protective coatings which are peeled off once the risk of damage is removed are effective.

2 Impact

(a) Context

By definition, impact is concerned with the sudden application of a load on a material. It is apparent that there are innumerable instances, both during construction and the life of the building, when materials may be subjected to this form of loading. Whenever it occurs, impact results in stresses that are momentarily higher than those from the same static load. In simple terms, impact takes place when *two* objects, at least one of which is in motion, collide with one another. Depending on the *relative* strengths of the materials and the intensity of the forces resulting on impact, there may be damage to the object in motion or to the static object, or in some cases to both. The strength of the materials and the intensity of the force are obviously closely related, but it is convenient to consider these separately.

Impact is not of great importance in buildings from the structural point of view. However, it is very relevant as one of the causes of physical 'weathering' of materials in the building in use.

(b) Strength characteristics

The strength characteristics which affect the kind and extent of damage which occurs as a result of impact are rather complex. Damage results when a material is unable to absorb completely the kinetic energy lost during impact. The kind of damage that may occur as a result of impact is either permanent deformation (in other words, denting or buckling) or fracture, which may be localized as chipping or cracking, or may affect the element or component as a whole.

Resilience is the property of materials to absorb energy without permanent deformation: hence resilient materials are the most resistant to impact damage. Animals have adapted to an environment where impacts are common: skin, muscle and feathers are very much more resilient than most building materials. The lessons from nature are applied more for temporary protection against impact damage, such as wrapping the vulnerable arrisses of prefabricated units with foam plastic, than for permanent materials.

As far as impact is concerned, failure can generally be said to have occurred if the materials are permanently deformed, whether by denting, chipping or fracturing. *Toughness* of materials (in other words, resistance to fracture) is therefore not the same as resistance to impact. Ductility and malleability are properties common to some metals which make them tough because it enables them to absorb large amounts of energy without fracturing. However, these same properties make them vulnerable to indentation on impact. Brittle materials are less vulnerable to being dented but are more prone to fracture. As long as the service impact energies are less than the energy needed to cause fracture, brittle materials may be more resistant to impact than tough, ductile materials. Brickwork is a case in point.

In considering the relationship between impact and strength, the manner in which the material is actually used is important, particularly its size, shape and method of fixing. Thin panels of material, whether brittle or tough, are more vulnerable to moving objects; corners are particularly prone to impact damage and both hard and soft materials may require protection.

An aspect of the method of fixing is the amount of support the material receives from a background. Hard plaster will serve as an example.

Proper adhesion between the plaster and the background implies that there are no gaps between them. Any impact will therefore be resisted by the plaster and the background together. If, on the other hand, there has been poor adhesion which results in gaps being formed (in other words, if there are planes of weakness) and impact occurs in the zone of the poor adhesion, then the plaster must resist the impact alone. Hard tiles, particularly those used as wall or floor finishes, may suffer in the same way if they are not bedded properly.

(c) Intensity of forces

Although it is possible to make quantitative measurements of the forces which may be involved during impact, it is not usually practical to rationalize these measurements to assist in the problems associated with local damage. Some of the difficulties may be apparent when it is realized that the intensity of forces involved during an impact are dependent on a number of factors. First, there is the energy of the moving object, and second, the area of the object which comes into physical contact with the static obstruction.

(d) Significance of damage

The nature of damage which may result from impact varies widely. It may lead to corners of elements being chipped: in other cases the element may be deformed by dents or impressions, while in the worst case the element may be severely cracked, leaving it structurally weakened, if not unusable. It is thus possible that such damage as may occur may not necessarily adversely affect the main properties for which the material was selected, such as its basic strength. Nevertheless, other factors, such as appearance or the susceptibility of the damaged material to further deterioration, need to be taken into account.

In anticipating possible damage from impact, two factors should be considered:

1. *Ease of replacement*: Vulnerable materials, or materials in a high-risk location, should be able to be replaced without extensive disassembly of adjoining construction. This is a further example of allowing for the principle of **separate lives**.
2. *Ease of making good*: In areas where impact is likely, self-finished permanent materials which are vulnerable to impact damage should be avoided. The ease with which materials or finishes can be made good to the standard of adjoining areas should be considered.

(e) Choice of materials

The susceptibility of materials to impact damage during the life of the building is generally less in comparison to site conditions. Local areas where activities are of a kind to cause possible impact damage warrant special attention, such as the service areas of buildings where items are unloaded and refuse is handled. Vandalism or wilful damage, whatever its root cause, can inflict more damage on a building in

Table T2.2/5   *Impact performance categories for walls*

Category	Description	Examples
A	Readily accessible to public and others with little incentive to exercise care. Prone to vandalism and abnormally rough use.	External walls of housing and public buildings in vandal-prone areas
B	Readily accessible to public and others with little incentive to exercise care. Chances of accidents occurring and of misuse.	Walls adjacent to pedestrian thoroughfares or playing fields when not in category A.
C	Accessible primarily to those with some incentive to exercise care. Some chance of accidents occurring and misuse.	Walls adjacent to private open gardens. Back walls of balconies.
D	Only accessible, but not near a common route, to those with high incentive to exercise care. Small chance of accident occurring or of misuse.	Walls adjacent to small fenced decorative gardens with no through-paths.
		A, B, C, D all applicable to zone of wall up to 1.5 m above pedestrian or floor level.
E	Above zone of normal impacts from people but liable to impacts from thrown or kicked objects.	1.5–6 m above pedestrian or floor level at locations A and B.
F	Above zone of normal impacts from people and not liable to impacts from thrown or kicked objects.	Wall surfaces at higher positions than those defined in E.

From BS 8200: 1985.

minutes than would occur in normal use in decades. As a result, where vandalism is a risk, ease of replacement and repair are probably more important than sheer toughness.

(1) Arrisses
Corners of walls and columns in vulnerable areas may be protected by reinforcing angles or beads.

(2) Internal walls
In buildings with much wheeled traffic such as hospitals or kitchens, and in waiting rooms where the backs of chairs repeatedly strike the wall, buffer rails should be considered.

(3) External cladding
Research has been carried out* to try to provide a common basis for measuring the resistance of claddings to impact damage, and thereby to set performance levels for claddings in typical locations. Guidance on these locations, on the desirable impact resistance, and the measured impact resistance of a range of cladding materials are set out in Tables T2.2/5–7.

(4) Floors
Floors may be subjected to foot traffic alone, or to wheeled traffic, or to major impacts from objects such as beer kegs or specific industrial processes. In buildings such as shopping centres or airports the cumulative effect of millions of feet is a major impact problem.

Table T2.2/6 *Test impact requirements for retention of performance of exterior wall finishes*

Wall category[a]	Test impact energy for impactor shown[b]		
	H1	H2	S1
A		See note 1	
B		10	120
C	6		
D		See note 2	
E	6		
F	3		

[a] See Table 2.2/5.
[b] See Table 2.2/7.

Notes:
1. In each case the type and severity of vandalism needs to be carefully assessed and appropriate impact values determined.
2. With category D walls the risk of impact is minimal and impact test values are therefore not appropriate.
3. For test types see Table 2.2/7.

From BS 8200, 1985.

Table T2.2/7 *Hard and soft body test parameters*

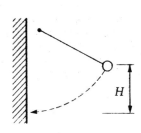 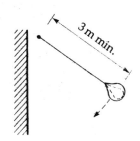

Hard body test   Soft body test

$H = e/9.8m$
e = Test impact energy (Nm)
m = Mass of impactor

	Impactor	Dia. (mm)	Mass (kg)
H1	Hard body	Steel ball	50	0.5
H2	Hard body	Steel ball	62.5	1.0
S1	Soft body	Canvas spherical bag filled with 3 mm dia. glass spheres	400	50

From BS 8200: 1985.

*BRE CP 6/81, *Assessment of external walls: hard body impact resistance*, by R. P. Thorogood.

(5) Glazing

In certain locations the impact resistance of ordinary float glass is inadequate. Such locations include:

- Glazing in doors;
- Windows below a height of 800 mm above floor level;
- Shop fronts;
- Security glazing in banks and, for example, jewellers;
- Rooflights.

Float glass can have its impact resistance increased by two methods. *Toughened* glass is made by a heat treatment. *Laminated* glass is made tougher by building up plies of ordinary float glass and a special vinyl sheet. In terms of impact these work rather differently. The effect of laminating glass is to increase its resistance to penetration (important in security locations), because the plastic holds the glass layers together: the actual resistance of the glass to chipping is unchanged. Toughened glass is altogether tougher in that its work of fracture is increased.

In areas where very high impact resistance is needed polycarbonate is used as a glazing material. This is an extraordinarily tough material compared with glass. However, it performs less well than glass in terms of resistance to abrasion, ultraviolet light and cost.

(f) Precautions

Materials and components are particularly vulnerable to impact damage during the construction process, for obvious reasons, namely, the amount of moving around of both components designed primarily to be fixed in a static location, the exposure to moving plant and other materials, and the generally lower standards of care and protection than apply to the completed building.

Although on most building sites there is plenty of scope for improving standards of protection in storage and handling, there are realistic limits to the care with which materials may be physically handled. For this reason, it is necessary to devise special measures for reducing damage caused to components and materials by impact. Progress continues to be made, for example, in the packaging of bricks into pallets to reduce waste. The use of expanded foam and corrugated plastic sheet and taping of prefinished metal components are ways of reducing impact damage.

Sources

Hilti, *Handbook of anchor technology*
Laundberry, Bill, *AJ Handbook of Fixings and Fasteners*, Architectural Press, London, 1971
Lazenby, D., *AJ Handbook of fixing and fasteners*, Architectural Press, London 1971
Martin, B., *Joints in Buildings*, Longman, Harlow, 1977
Pye, D., *The nature and art of workmanship*, Studio Vista, London, 1968
Shields, J., *Adhesives Handbook*, (3rd ed.) Butterworths, London, 1985
Shields, J., *Adhesive Bonding*, Engineering Design Guides
Steeds, W., *Engineering Materials, Machine Tools and Processes*, 4th edn, Longman, London, 1964

BRE
CP 6/81, Assessment of external walls: hard body impact resistance, R. P. Thorogood
IP 25/81, The selection and performance of sealants

Digests
137, *Principles of joint design*
199, *Getting good fit*
227, 228, 229, *Estimation of thermal and moisture movements and stresses,*
Parts 1, 2 and 3, 1979

BSI
BS CP3, Chapter IX, 'Durability', 1950
BS 1202: 1979, *Specification for nails, parts 1–3*
BS 1204: 1979, *Synthetic resin adhesives (phenolic and aminoplastic) for wood,*
Parts 1 and 2
BS 1210: 1963, *Specification for wood screws*
BS 4174: 1972, *Specification for self-tapping screws and metallic drive screws*
BS 4190: 1967, *Specification for ISO metric black hexagon bolts, screws and*
nuts
BS 4395: Parts 1 and 2, 1969, part 3, 1973, *Specification for high strength*
friction grip bolts and associated nuts and washers for structural engineering
BS 5442, *Classification of adhesives for construction*
BS 5606: 1978, *Code of Practice for accuracy in building*
BS 6093: 1981, *Code of practice for design of joints and jointing in building*
construction
BS 6100, *Glossary of building and civil engineering terms:* subsection 1.5.2
1987 *Jointing,* subsection 1.5.1 1984 *coordination of dimensions, tolerances*
and accuracy
BS 6104, *Mechanical properties of fasteners,* Parts 1–3
BS 6213: 1982, *Guide to selection of constructional sealants*
BS 6262: 1982, *Code of practice for glazing of buildings*
BS 6750: 1986, *Specification for modular coordination in building*
BS 6954, Parts 1–3: 1988, *Tolerances for buildings*
BS 8200: 1985, *Design of non-loadbearing vertical enclosures of buildings*
DD 69:1980, *Method of classifying the movement capacity of joint sealants*

Journal
DOE Construction, **18**, June 1976, 'Sealants, mastics and gaskets'

Part 3

Water and its effects

Water is the most common substance on earth, having probably the best known of all chemical formulae: H_2O. It is one of only two liquids occurring naturally in appreciable quantities, the other being petroleum. Water covers more than three quarters of the earth's surface. Without it there would be no weather. Despite its abundance and apparent simplicity, water is an unusual compound* because it has special qualities that enable it to play its vital role in both organic and inorganic fields. Water is an almost universal solvent, and when it provides the medium of acid-base reactions it is also an active participant. In short, besides being essential to life, water has a unique combination of solvent, thermal capacity, chemical stability, permittivity† and abundance.

Depending on circumstances, the effects that water may have in any of its three states of aggregation (i.e. solid, gas and liquid) can be considered as either constructive or destructive. In the natural environment, the net effect of these would normally be equated to what is sometimes known as the 'balance of nature'. In building the properties of water have a positive effect in that it is needed in many chemical reactions during the manufacture or use of materials and in the servicing of buildings to provide conditions fitted to human convenience and comfort. More commonly, water is associated with the destructive effects it may have on materials in constructions. The durability of almost all materials used in these is largely dependent on the resistance that materials have to the effects of water.

The physical and/or chemical states of materials in use depend on the relationship between the precise quantities of water present in, on or between them and the degree of pollution of the water, the rate at which it is either lost or gained and the length of time over which it is retained. Water may change either of these states by physical and/or chemical alterations in the materials or their surroundings. Whether the changes can be deleterious depends on particular cases. As important is the extent to which any change affects the performance of function deleteriously. In these contexts, water can be considered as a 'material' against which it is necessary in varying degrees to design or to guard throughout the construction and life of a building. So that a better understanding may be obtained of the significance of the effects of water in respect of the use, selection and performance of materials in constructions, emphasis is given in this part to the deleterious effects of water.

Apart from its harmful effects on materials, water also needs to be considered in the broad context of dampness. This may affect certain performance functions (such as thermal insulation that is reduced as the moisture content of related porous materials increases) or health. In the past, dampness was more commonly related to its influence on or aggravation of disease or infirmity. Nowadays it has also to be related to comfort. The problems of dampness concerning both durability and health therefore need to be considered as part of the overall effects of water.

Water obviously needs to be excluded from materials wherever they may occur in a construction if its deleterious effects are to be prevented or minimized. In some cases the exclusion of water must be total; in others it may only be necessary to exclude it to some extent. Whenever considering the possible effects of water it is important that cognisance is taken of the fact that water seldom, if ever, acts in its pure state in almost every aspect of building, in the performance of materials in particular. The way that water is a carrier of other substances, the derivation and possible effects of 'carried' substances on materials, their performance or the performance of function and the means by which water is transferred from places (i.e. in the environment generally and within or through materials) needs to be understood. This part aims to provide that understanding. Accordingly, it includes consideration, mostly in detail, of

*Compared with other liquids, Cotterill describes water as exotic (Cotterill, Rodney, *The Cambridge Guide to the Material World*, Cambridge University Press, Cambridge, 1985, p. 88).

†A property used in electrical engineering.

the properties of water, the sources and pollution of water (i.e. the exposure generally of materials to water), the modes by which water may be transferred and the mechanisms associated with and the possible effects of moisture content. Description of these aspects precedes consideration of the exclusion of water. The latter is then followed by details of particular physical and chemical effects such as efflorescence, chemical attack, corrosion of metals, frost action and changes in appearance. The extent of repetition and/or cross referencing in each of the separate sections into which the subject matter has been divided should help to emphasize the need in practice to consider water and its effects as a whole rather than in separate compartments.

General considerations

Scope

This chapter considers the relationship between the occurrence and behaviour of water in the natural and the built environments and the general factors that are related to materials used in building constructions. These considerations provide the background and basis for the subsequent chapters. Accordingly, consideration is given here to

- The basic properties of water;
- The transfer of water and its effects generally in the natural environment;
- The definition, basic causes and measurement of dampness in buildings;
- The basic flow of water over the exposed surfaces of materials; and
- General guidance on the assessment of the severity of conditions of exposure of materials in constructions.

Basic properties of water

The use of the term 'basic' is intentional. Only those properties related specifically to the needs of this study are considered in detail. For convenience, the properties of water are included and compared with other substances in Table T3.1/1. Notes are included separately on those aspects which need amplification in the context of materials and their performance. These notes do not necessarily follow the table in chronological order. A comparison of the chemical composition of fresh and sea water has been added.

The ease with which water can change state, physically or chemically, is referred to later as the 'instability of water'. This is demonstrated clearly by the fact that in the now superseded Imperial and MKS systems the precise state of water was defined when it was used as a standard against which other substances could be compared. Examples included relative density (originally specific gravity), specific heat capacity and viscosity. In addition, definitions of quantities of heat such as the British Thermal Unit (BTU) or the calorie used water as one of the defining elements.

1 Chemical formula

Chemically, one molecule of water has the formula H_2O, that is, two atoms of hydrogen to one of oxygen. There are variations in the manner in which these molecules may be arranged or in the kinds of hydrogen and oxygen atoms making up the molecules, which give rise to 'special' types, such as 'heavy water'. Because of the minute quantities in which these occur they are generally considered of little importance in the natural environment.

Table T3.1/1 *Properties of water*

Property	Value	Comparison with other substances/Notes
1. Quantity in the outermost 5 km of the earth[a]	–	Three times that of all other substances together Six times that of felspar, the next in abundance
2. State of aggregation[a]	–	Almost the only inorganic liquid present in nature and the only substance occurring in all three states of aggregation
3. Solvent power[a]	–	More general than any other fluid
4. Density at 4°C 0°C	$1000\,\text{g/mm}^3$ $999.9\,\text{g/mm}^3$	Relative density of sea water (salinity of 35) at 0°C = 1.025

5. Viscosity[b] at 20°C
          0°C
          10°C
          25°C
          500°C
          100°C
	$1.0000 \times 10^{-3}\,\text{Ns/m}^3$ (approx.)
	$1.7910 \times 10^{-3}\,\text{Ns/m}^3$ (approx.)
	$1.3077 \times 10^{-3}\,\text{Ns/m}^3$ (approx.)
	$0.8950 \times 10^{-3}\,\text{Ns/m}^3$ (approx.)
	$0.5490 \times 10^{-3}\,\text{Ns/m}^3$ (approx.)
	$0.2338 \times 10^{-3}\,\text{Ns/m}^3$ (approx.)
		Used as standard of viscosity
		
		At 25°C: benzene = $0.649 \times 10^{-3}\,\text{Ns/m}^3$
		          kerosene = $2.375 \times 10^{-3}\,\text{Ns/m}^3$
		          spindle oil = $92.000 \times 10^{-3}\,\text{Ns/m}^3$
		          castor oil = $620.000 \times 10^{-3}\,\text{Ns/m}^3$
		
		Affected by temperature and pressure: values given for normal atmospheric pressure

6. Specific latent heat[b]	$4.187\,\text{kJ/kg}°\text{C}$	(*Note*: for comparative purposes, the specific latent heat of the materials listed below have been compared to that of water (i.e. water = 1.0) and are therefore *ratios*. To find the actual value of any material in kJ/kg°C multiply by 4.187)

Most rocks vary from	0.21 to 0.19
Sea water	0.940
Ice	0.502
Steam (ave.)	0.488
Common metals vary from	0.22 to 0.0395
Alcohol	0.60
Benzene	0.41
Helium	1.25
Hydrogen	3.402
Oxygen	0.242
Nitrogen	0.235

2 Instability of water

Water, as generally found in the natural environment, can be distinguished from nearly all other substances and especially fluids in two important ways. First, it is the only substance which occurs *in all three states of aggregation*, that is, solid, liquid and gas. As important, it can occur in these states under normal conditions of temperature and pressure, and in this it departs from the normal when compared with

Table T3.1/1   *(Continued)*

Property	Value	Comparison with other substances/Notes	
7. Boiling point[b]	100°C	Sea water (normal)	104°C
		Mercury	356.7°C
		Olive oil (about)	300°C
		Turpentine	159°C
		Alcohol	78.3°C
		Ether	34.6°C
		Sulphur melts	115°C
		Tin melts	232°C
		Lead melts	327°C
8. Freezing point[b]	0°C	Expansion on freezing is very exceptional (about 10% by volume)	
		Sea water (normal)	−9°C
		Benzene	5.4°C
		Turpentine	−10°C
		Mercury	−38.8°C
		Alcohol	−115°C
9. Latent heat of vaporization[a, b]	2253 kJ/kg°C (under normal atmospheric pressure)	Greatest of all substances	
10. Latent heat of fusion of ice[b]	334 kJ/kg°C (under normal atmospheric pressure)	To convert one kilogram of ice into steam requires:	
		Latent heat of fusion	334 kJ/kg°C
		Heating from 0° to 100°C	419 kJ/kg°C
		Latent heat of vaporization	2253 kJ/kg°C
			3006 kJ/kg°C
11. Thermal capacity[a]	–	After ammonia, thermal capacity and heat of fusion, the greatest of all fluids and solids	
12. Thermal conductivity[a]	–	Greatest of all fluids except mercury	
13. Surface tension at 20°C[a]	73.5×10^{-3} N/m	Highest of all fluids except mercury	
14. Transparency[a]	–	Very considerable and almost equal for all visible rays	
15. Electrolytic dissociation[a]	–	Very small	
16. Dielectric constant[a, b]	80	Highest of all fluids	

[a] Kuenen, P. H., *The Realms of Water*, Cleaver-Hume Press Ltd, London, 1955.
[b] Fox, Sir Cyril S., *Water*, The Technical Press Ltd, London, 1951.

other substances. Second, *the solvent power* of water is more than that of any other fluid. Consequently, it is seldom found in its pure state, when it would be odourless and have an insipid taste. (In this state it is said to be colourless, transparent, tasteless and scentless.)

Both these factors are highly significant. They do, in some measure, account for the difficulty of defining water in concise terms.* Thus it is suitably descriptive to refer to the instability of water. In fact, this instability underlines the importance of always considering water in all

The Concise Oxford Dictionary defines water as 'Colourless transparent tasteless scentless compound of oxygen and hydrogen in liquid state convertible by heat into steam and by cold into ice, kinds of liquid consisting chiefly of this seen in sea, lake, stream, spring, rain, tears, sweat, saliva, urine, serum. etc.'. This should illustrate the difficulties involved.

three states of aggregation and as being 'polluted'. For almost obvious reasons, water in its visual perceptible states, that is, as liquid or solid, is readily appreciated. However, in the invisible state, as a gas, this may not be as apparent. Nevertheless, even in this state it can be active, chemically and physically, while in addition its occurrence, especially in the UK, is extensive and ubiquitous. Finally, in many situations water in the gas state may readily and profusely be converted into the liquid state, without changes of either temperature or pressure in the general atmosphere but by contact with colder surfaces, as commonly occurs, for example, in the formation of dew.

3 Transportation by water

The ability of water to transport not only small particles of matter but also large land masses must again, in addition to its greater relative density and internal friction, be associated with its three states of aggregation, although more usually with the solid and the liquid states.

Although water may carry smaller fragments of matter, such as mud, or roll small grains and pebbles, it also has a *lubricating effect* which allows broken particles to slide over one another more easily. The breakdown of large masses of rocks, for instance, into smaller particles is facilitated by the solvent action of water. In this it acts as the vehicle for the actual corrosive agents, such as acids, which may themselves be transported relatively great distances. In this context the corrosive agents are in their dissolved form, although this does not necessarily reduce their potential destructive powers. Breakdown may also occur when water, in the liquid state, trapped or otherwise held in cracks and fissures, subsequently freezes. The 'abnormal' property of water to expand on freezing results in extremely great pressure being exerted. Thus, in addition to actually assisting the physical and chemical breakdown of particles, water also has the ability of subsequently transporting these. Kuenen has aptly said that 'water takes the lion's share of all transport on earth'.

4 Derivation of energy

The power of movement apparently inherent in water is dependent on kinetic energy (except that due to tides) derived from the action and movements of the atmosphere. The primary source of these displacements is energy derived from solar radiation.

5 Surface tension and capillarity

The surface tension of water is also greater than any other liquid. This has a marked effect on drop formation (important in precipitation), wave formation, flow and capillarity. The latter, while also associated with the 'wetting' power of water, plays an important part in the transfer of water, mainly in the liquid state, in and about the earth (rocks and soils).

Capillarity is also of significance in the transfer of water in and about building materials, wherever capillary paths may be formed. These may be *either* within materials, as in porous materials, *or* at the junction (overlapping or abutting joints) of two or more materials. The latter could be a combination of either porous or non-porous materials.

6 Thermal properties

The thermal properties of water, namely, its great capacity for absorbing, transmitting and storing heat, together with heat of vaporization, heat of fusion of ice and specific heat, all of which are significantly different from other substances, play an important part in regulating thermal conditions everywhere. The effects of all these properties are closely interrelated. Nevertheless, in general terms water does act as a moderator of temperature fluctuations. The amount of water vapour in the atmosphere is important for the part it plays in what has been described as the 'hot-house effect' of the atmosphere. (This should not be confused with the 'greenhouse' effect.) Control of the amount of vapour evaporated from the earth, which affects the amount of vapour present in the atmosphere, is partially influenced by the great heat of vaporization of water.

7 Density and freezing

The expansion which takes place when water freezes is another significant departure from the normal. Because of this expansion, which is roughly 10% by volume, ice has a lower density than liquid water. (This accounts for floating ice and also for the preservation of much underwater life during winter conditions.) The maximum density of water is reached at 4°C, at which contraction ceases. Thereafter, further reductions in temperature result in expansion taking place. As already mentioned, water which freezes in entrapped situations induces great pressures on the 'walls of the container' in which it is held, and this accounts for some considerable breaking up of rocks, etc. in the natural environment. As will be seen later in *3.8 Frost action*, this process is equally important in the built environment.

It is relevant to note that the properties of sea water are completely different from those of fresh water. Contraction on cooling, as with other substances, proceeds steadily until freezing takes place. In addition, the actual temperature of freezing, namely −9°C, is lower than that of fresh water.

8 Electrolytic dissociation and dielectric constant

The extremely small degree of electrolytic dissociation of water together with the large dielectric constant, when compared to other liquids, are important factors which affect all solutions in water.

9 Compressibility

The compressibility of water is extremely small.

10 Chemical composition of fresh water and sea water

Under '2 Instability of water' reference is made to the fact that for most general purposes it is desirable to consider water as being 'polluted'. This has been done in order to emphasize that water usually contains other substances and is therefore seldom found in its pure form.

River water is a useful representative of fresh water. On average, sea water contains nearly 200 times as much dissolved solids as river water. A comparison of the salt content of river water and sea water is given in Table T3.1/2. The analyses given in the table are average values.

The degree of salinity, of the total salt content of water, is usually expressed in parts per thousand. Although the saltiness of sea water is due mainly to the presence of the chloride of sodium, other salts are also present. On average, the salinity of sea water is said to be 35, that is, 35 parts of dissolved solids per thousand grams of sea water, and these solids are constituted as shown in Table T3.1/3.

Table T3.1/2   *Comparison of salts in river and sea water*

	River water		Sea water	
Salt	*%*	*%*	*Salt*	
Calcium carbonate	42.90	0.30	Calcium carbonate	
Magnesium carbonate	14.80		Magnesium carbonate	
Sodium chloride	2.20	77.70	Sodium chloride	
	–	10.80	Magnesium chloride	
Calcium sulphate	4.50	3.60	Calcium sulphate	
	–	4.70	Magnesium sulphate	
Sodium sulphate	4.20	–		
Potassium sulphate	2.70	2.50	Potassium sulphate	
Sodium nitrate	3.50	–		
Aluminium and iron oxide	3.60	–		
	–	0.20	Magnesium bromide	
Silica	9.90	–		
Other salts	1.30	0.20	Other salts	
Organic substances	10.40	–		
	100.00	100.00		

Values taken from *Water* by Sir Cyril Fox, p. 12.
Note: Percentages are of total anhydrous solids.

Table T3.1/3   *Analysis of dissolved solids per 1000 g of sea water*

		g	*%*
Sodium chloride	(NaCl)	27.213	(2.7213)
Magnesium chloride	(MgCl$_2$)	3.807	(0.3807)
Magnesium sulphate	(MgSO$_4$)	1.658	(0.1658)
Calcium sulphate	(CaSO$_4$)	1.260	(0.1260)
Potassium sulphate	(K$_2$SO$_4$)	0.863	(0.0863)
Calcium carbonate	(CaCO$_3$)	0.123	(0.0123)
Magnesium bromide	(MgBr$_2$)	0.076	(0.0076)
		35.000	(3.5000)

Values from *Physical Geography*, by P. Lake, Table 2, p. 143.
Note: Percentage values have been added as they illustrate the relative concentration of salts.

It should be emphasized that the degree of salinity is not the same everywhere, although the proportion of the constituents does not change. Density at a standard temperature (0° or 15°C) is a guide to the degree of salinity.

Thus at a temperature of 15°C:

Density (g/mm³)	Salinity
1000.0	0.0
1013.8	20.0
1026.0	35.0

(From *Water* by Sir Cyril S. Fox)

Sources

Kuenen, P. H., *The Realms of Water*, Cleaver-Hume Press, London, 1955.

Fox, Sir C. S., *Water*, The Technical Press, London, 1951.

Monkhouse, F. J., *Principles of Physical Geography*, 5th edn, University of London Press, 1963.

Lake, Philip, *Physical Geography*, 4th edn, Cambridge University Press, Cambridge, 1958.

Meteorological Office, *A Course in Elementary Meteorology*, HMSO, London, 1962.

Holmes, Arthur Nelson, *Principles of Physical Geography*, new and fully revised edition, 1965. Profusely illustrated with diagrams and photographs.

The water cycle

1 Background

The association of water in the three states of aggregation in which it may be found in the natural environment can be extended to include a related, albeit broad, classification of its distribution. Thus, in its *liquid state* it occurs in the oceans, lakes, rivers, springs and generally underground, that is, groundwater; in its *vapour state* it is more easily recognized as cloud or mist, although it is present in the air in an invisible form, in which case humidity terms are relevant; and finally in its *solid state* the most common form of occurrence is as ice (particularly in the polar regions, where it is to be found in abundance), or as glaciers. Although useful for the general picture, such a classification does not convey any idea of the *constant transfer* of water which takes place and during which, among other things, water is *changing from one state to another*. The emphasis is usually on the change of state from vapour to liquid and vice versa, but the significance of the solid state should not be overlooked.

This continual transfer of water, known as the 'Hydrolic' or *water cycle*, results from the interaction of extremely complex factors in which the thermal and other properties of water play a part in addition to solar radiation, wind and pressure. All these are the primary factors which may in turn be affected by local topographical and climatic conditions. The actual method by which the transportation of water takes place is in itself fairly complex. However, in essence, the basic process consists of evaporation of the liquid water, subsequently to be condensed and then, finally, to be precipitated more commonly as liquid in the form of rain, although snow as a means of precipitation should be included.* The actual amounts of evaporation, condensation and precipitation vary in general not only from one part of the world to another, but from one season to another (daily, monthly or local variations may also be significant), depending on the interrelated factors already mentioned.

*Detailed studies of these meteorological factors, together with changes due to local conditions, are not included. The emphasis is being placed on the transfer of water, but this should not undermine the importance of the agencies which cause the transfer to take place.

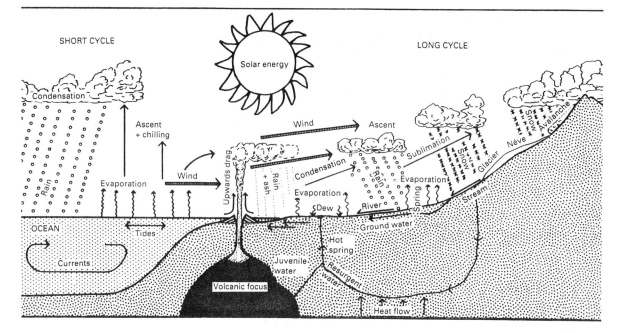

Diagram D3.1/1 *Principal aspects of the water cycle. (Based on a drawing in Kuenen)*

2 Types of cycle

As it is the oceans which act as the main and largest reservoir of all water on earth, they serve as the primary source from which evaporation can take place. However, the amount lost by evaporation is, by and large, eventually returned. (It may not necessarily be exactly the same water returned to exactly the same place of origin, but this is immaterial.) The 'route' by which it is returned is, on the whole, devious and tortuous. Thus, in general terms, the oceans are at once the points of departure and the points of return of water. Again in general terms, the length of time that it takes for the evaporated water to return depends on whether a short or long cycle is involved. Diagram D3.1/1 illustrates the principal aspects of the water cycle, including the short and long cycle.

(a) Short cycle

During the short cycle water evaporated from the ocean follows the basic pattern of condensation and subsequent precipitation in the form of rain. There is no transfer of the water vapour away from the environs of the ocean, that is, over land, and thus there is evaporation from and the immediate return of water to the ocean. (A shorter cycle occurs when condensation in the atmosphere evaporates without reaching the earth's surface.)

(b) Long cycle

The long cycle includes evaporation from the sea except that precipitation takes place on 'dry' land. The main route for the return of water to the sea is via rivers. These, in turn, may be fed by springs, which assist in removing groundwater (this, in turn, is partly replenished by precipitation, melting ice and by lakes, etc.) or, as occurs in the longest cycle of all, by glaciers which involves a change of state. The complexity of the long cycle may be appreciated if it is realized that within it shorter cycles also occur.

3 Equilibrium and distribution

The interchange and transfer of water is virtually a continuous process, during which there are likely to be both losses and gains of water. The effect of these is to establish, more or less, a **balance** of the quantities of water distributed over the earth as a whole. Thus it is estimated that 97% of all the water on earth is in the oceans; of the remaining 3% about one-third occurs as snow and glaciers, while about two-thirds is to be found in rivers, lakes and groundwater in soils, etc. Finally, approximately 1/2000% occurs in the atmosphere.

4 Microcosm – land areas

(a) Local distribution

The overall distribution does not take into account the extremely wide variations in the relative distribution which occurs in the microcosm, whatever area this is deemed to include. Here it refers to land areas. In some ways, therefore, the almost disproportionate quantity of water in the oceans may tend to reduce the significance of the relative quantities of water which exist on 'dry' land. As far as the latter is concerned, another point which is important is the fact that wide variations occur in the precise quantities of water present, not only between one area and another but also between one season and another (daily variations may also be important).

(b) Wet and dry times

Thus there may be dry times and wet times in which the general condition of the ground in terms of dryness and wetness can be associated. These variations are of particular significance when considered in relation to the short cycle which occurs within the long cycle. This short cycle, here termed the 'local cycle' in order to avoid confusion with that which occurs over the oceans, is important in the natural environment and virtually fundamental to the problems of the built environment. In both the effect of local variations must always be taken into account. However, an examination, in general terms, of this local cycle is useful.

5 'Local cycle' on land

(a) Sources of evaporation

The sources of water which may be evaporated on dry land include the relatively large and obvious masses, as in lakes, reservoirs, etc., and the surface of the ground itself. Lakes are likely to produce variations not only in the micro-climate but also in the amount of water in the surrounding atmosphere and ground. They also help to feed underground watercourses which may eventually discharge far from the feed point. The possible increases of water in the atmosphere are also likely to modify some aspects of the cycle as, for example, by the formation of local condensation, mist, etc.

Subject to the amount of water present in the surface layers of the ground, the latter can provide an important source for evaporation. Static water in the ground is largely, though not entirely, held in capillaries formed between the grains and particles of the soil. Such capillaries are not necessarily all interconnected at all levels below the surface. Thus zones or patches of water may occur at various levels. These zones may, when dry, be filled from above during rain, for example, or from below if they occur immediately above the water table. The latter is almost an underground reservoir, although its top surface may not be level. However, it is the water which occurs at or near the surface which is mainly evaporated and which leads to dry conditions. The quantities of water at lower levels may be reduced as a result of this evaporation, although, in general, this is not considered the primary agency. Water from lower levels may be brought to the surface by thermal and other conditions. This water would discharge into watercourses, but inevitably some would return to the ground.

(b) Absorption and flow patterns

However, after dry periods, it is the surface layers which first absorb water during any subsequent falls, while any excess percolates to lower levels. This excess would either replenish water previously lost or flow to some other place via underground watercourses. The latter have more or less the same flow patterns as those which occur on the surface. In all this the precise quantities of water and rate of absorption or percolation are dependent, detailed weather conditions apart, on the nature of the ground which, among other things, may be extremely heterogeneous in its constitution and thus variable. Damp surface layers may also, under certain conditions, encourage local condensation in the form of dew, thus maintaining damp conditions without precipitation. The dew may, of course, change to frost with resulting expansion problems.

(c) Effect of plant life

Plant life, especially trees, may also make their own contributions, which in certain situations may be highly significant, to the transfer process. In the ground some transfer of water takes place by 'root action'. Roots may, in fact, encourage transfer from depths lower than those at which ordinary surface evaporation is effective. During dry periods this may be extremely important. However, through 'respiration' the water is returned, if sometimes only in part, to the atmosphere in the form of vapour, subsequently to take its place in the condensation and precipitation processes. The local concentrations of vapour in the atmosphere may result in almost permanent wet conditions prevailing at ground level, as often occurs, for example, in forests.

6 Prevailing wet and dry conditions

(a) Variations

Mention has already been made of the wide variations that may occur from one place to another and from one time of the year to another. These variations cannot be attributed to one cause only, but are the net result of the action and interplay of a number of factors, among which 'weather' conditions are important. Knowledge, in quantitative terms, of the variations which are likely, between wet and dry extremes of an area is apparently useful in agricultural undertakings. On the other hand, this knowledge also assists in gauging the probable length of time during which objects are likely to be wet or dry. If this is then related to such other factors as the degree of pollution of water, *a guide*, albeit extremely crude, may be obtained of the likely amount of deterioration which may result. The dampness of the air and the ground may be attributed, in loose terms, to high humidity, low temperatures, lack of intensive solar radiation, occurrence and speed of winds and high rainfall.

In the UK it is normal to refer to damp conditions in the winter period and to associate this with little drying out, while drier conditions are normally expected in the summer period.

(b) Analysis

An indication of the effect of micro-climatic conditions in terms of the state of dryness of the soil, for instance, can be obtained by comparing average monthly rainfall and evaporation figures. Chart C3.1/1 compares these figures for six stations in various parts of Britain. Before highlighting some of the relevant facts to be derived from the charts, it is necessary to make a few comments on the figures presented and which were supplied by the Meterological Office, on request.*

(c) Units of measurement

The amount of both rainfall and evaporation is gauged in terms of millimetres. A millimetre of rain, for example, is that layer of water that would cover a level surface. Quantitatively, it would be equal, over an area of $1\,km^2$ to approximately $1000\,mg$ or $10^6\,l$ of water and exerting a force of $9.807\,MN$. On a smaller scale it would be equivalent to approximately $1\,l/m^2$.

(d) Evaporation measurements

These are obtained with a standard Meteorological Office type of evaporation tank, and are therefore not taken from the ground as such. The number of stations for evaporation measurements throughout the UK are much sparser than those for rainfall, while the averages are based on comparatively few years (the Cawood figures, for instance, refer to the five years 1960–1964 only); those for rainfall are the average for the years 1916–1950.

The values for the winter tend to be less reliable than those for the summer. This is not considered to be of much importance as the values are, in any case, extremely small.

(e) Key points

Returning to the chart and accepting the limitations referred to above, the following points are of note:

1. Despite detailed variations, the disparity, during the winter period, between rainfall and evaporation substantiates the generally wet conditions that prevail and offers some support of the slow drying-out concept. The situation is reversed almost completely during the summer period. Subsequent comparison with other meteorological data such as temperature, sunshine, wind, pressure and humidity for the two periods would generally show that these, too, are, for the purposes required here, reversed, although variable.

*The values used are 'as supplied' while some of the comments are based on those intimated by the Meteorological Office. In the absence of any authoritative basis for establishing standards of dryness or wetness and drying out times, etc., the chart is intended to assist in a general understanding of the problems being outlined.

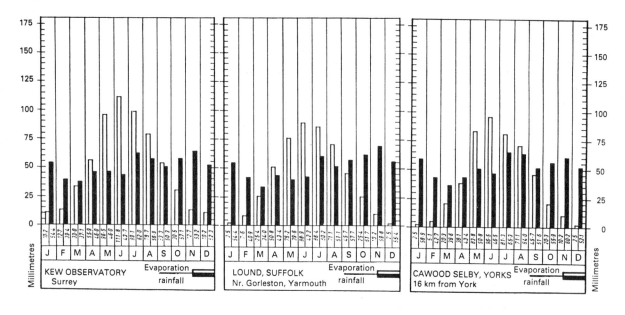

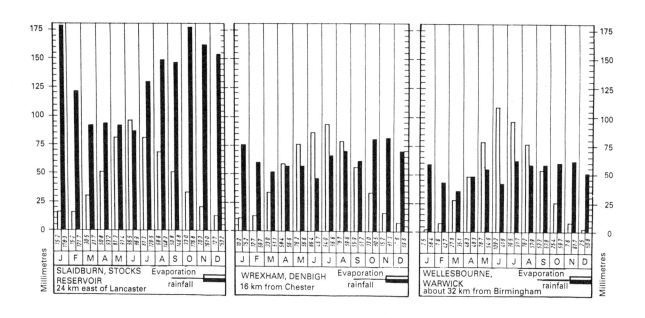

Chart C3.1/1 *Comparison of average monthly rainfall and evaporation in millimetres for six stations in the UK. Note: Evaporation averages are based on data collected over a few years and are only guides. Rainful averages based on data for period 1916–1950. (Courtesy: Meterological Office on request – Crown Copyright)*

2. Most of the annual total of evaporation is accounted for in the half-year from April to September, with the peak rate occurring at about midsummer at every station.
3. The average annual evaporation 'rhythm' is generally considered to be roughly the same throughout the country. It is worth noting that in winter there is a tendency for the windier northern and western districts to evaporate a little more than the southern districts. Thus the wind plays as important a role as the sun.
4. Despite general similarities in the annual pattern, there are notable differences in both rainfall and evaporation figures for each month when the stations are compared. At the Slaidburn Stocks Reservoir there is almost a dramatic increase in the average annual rainfall, where it is 1580 mm compared to, say, Kew Observatory at 608 mm (the lowest of the six stations quoted). The comparable evaporation figures (Slaidburn 554 mm and Kew 607 mm) do not, however, follow the pattern of the rainfall figures.
5. In practical building terms the wetness of the winter and dryness of the summer or the balance achieved between the two should not be confused with problems relating to weathering of building surfaces or water penetration. Guides on both of these are available from driving rain indices (see *3.2 Exposure*, '2 Rain', p. 180 ff).

Significance of the cycle

(a) Lessons

The water cycle has two main lessons to teach. First, it demonstrates the principle of **balance**. This, in varying degrees, is applicable to the behaviour of water in and around constructions. The processes that occur are also applicable on a small scale to building constructions. Second is the corrosive or destructive effect of water. This is obviously important in considerations of the durability of materials. Aspects related to the latter are considered below.

(b) Constant changing of transfer and state

Clearly, water behaves in a dynamic way. The significance of this lies mainly in the fact that, because of the constant transfer which is taking place and including the changes of state involved, the aggressive 'nature' of water is, in a sense, greatly enhanced, particularly when it is considered that during the transfer processes water has ample 'opportunities', due to its great powers of solvency, to 'pick up' corrosive agencies. By continual action and interaction, the actual composition of the water along any 'given course' is also likely to change.

(c) Transfer processes

The extent to which the action of water, with or without corrosive agencies, will cause materials to deteriorate is also dependent on the transfer processes, in that these influence the moisture content, that is, wetness of dryness, on or within the materials. The moisture transfer in and content of the ground is also significant, especially when it is considered that the most important elements of buildings, the foundations, are permanently embedded in it and in some instances at considerable depths. Furthermore, there are a number of other building elements which may be in close contact with or otherwise associated with the ground. Even in the natural environment conditions in the ground are significant.

(d) Interrelation of influencing factors

The inclusion of the whole problem should also serve to emphasize the importance of the interrelation of a great many complex factors, many of which are subject to wide variation. In general, it is of fundamental importance to be able to assess, for any given situation, the severity of conditions which are likely to prevail. In this context the immediate presence of obvious masses of water (oceans, lakes, etc.), for instance, are likely to present severe conditions. As a generalization, this is sufficient. But at the same time it is as well to be aware of the fact that, subject to other factors, the micro-climatic conditions may, among other things, be such as to lead to conditions as severe as those used in the generalization. The effect of such variations as may occur may not necessarily be as significant in the natural environment as they generally are in the built one, where conditions above and below ground are always relevant. In this, the analysis and subsequent synthesis of local conditions cannot be overemphasized. Intuitive assumptions based on the basic average pattern can very often be misleading.

Sources

All as for 'Basic properties of water'.

Weathering

1 Background

(a) Agencies involved

The agencies which are involved in what is known as the sculpturing of the earth's surface are often covered by the single term *denudation*. Among the agencies are included *weathering* (the disintegration and decay of solid rock* by the elements *in situ*); erosion (the attack and subsequent removal, i.e. transportation, of the 'debris' by the same agent, as, for example, running water, ice, wind, waves, tides and current); *corrosion* or *abrasion* (the combined work of the transporting agent and the load which acts as the abrasive in the wearing away process); and, finally, *attrition* (the wearing away of the abrasive medium or load). Although less conspicuous, these agencies also act in sculpturing the underground world.

(b) Geological sculpturing

In the geological meaning of sculpturing both wearing away and the depositing together with reforming elsewhere of the 'waste' material are normally implied. In the practical terms associated with sculpturing this could be taken as being analogous to shaping processes involving cutting, chiselling, etching, etc., on the one hand, and moulding or casting, on the other. (In the natural environment the casting may not necessarily involve a 'mould'.) In building terms the natural wearing away processes are usually taken to be deleterious.

*The use of the term 'solid rock' is not as restrictive as it may appear at first, especially when it is considered that size is not usually stipulated. On the other hand, solid rock does form an overwhelming proportion of the earth's crust, occurring in all shapes and sizes and in an almost inexhaustible range of chemical combinations. However, some reference is also made later to the part played by decayed organic matter, which can also be said to have disintegrated.

(c) Emphasis

In order to draw attention to the part played by water, particularly the chemical and physical changes involved, the main emphasis is placed here on weathering aspects. However, it is necessary to include the other agencies while some references to those aspects of weathering, in which water plays no part at all, are also made to complete the picture.

Under the influence of 'the elements' weathering may take place basically by either *chemical* or *physical* means. Although the manner by which disintegration and decay take place is quite different (in chemical weathering new substances are formed, but no new substances are formed during physical weathering), the effect of either means is essentially to produce a loose layer of material which can be easily removed by the various transportation agents. If the complementary action of the two means are taken into account (a chemical change may facilitate a physical change subsequently to take place and vice versa), then it is true to say that any substance, whatever its constituents, may, in time, be deteriorated. Fortunately, such deterioration requires for each substance particular circumstances which may not always be 'conveniently' available. Thus in the geological time scale deterioration is an extremely slow process. The rate of disintegration, including decomposition of materials used in buildings, may not necessarily be slow when related to the time scale of buildings.

2 Chemical weathering

(a) Importance of water

All substances may be defined with a definite chemical composition. In the case of rocks, which are aggregates of mineral particles, the latter may also be chemically defined. Under dry conditions the constituent minerals remain unchanged.* However, under damp conditions chemical decomposition may take place, although some minerals, such as quartz, are virtually unaffected, while others, such as rock salt, are removed in solution. These are the extremes. In chemical weathering water is the most influential agent, whether directly or indirectly. As mentioned earlier, the potency of water is primarily influenced by the corrosive agents which it may carry at any given time. The degree of concentration of these agents is also important. However, another interrelated factor is the actual length of time during which water may be present and hence be able to 'act'. The greater part of weathering is often attributed to the effects of standing water, with horizontal surfaces usually suffering far greater than vertical ones. (This consideration is particularly pertinent in buildings.) For this reason, any factor which is likely to maintain damp conditions will, in general, increase the amount of weathering which will take place. Nevertheless, this should not detract from the part played by the corrosive agents, the derivation of which is of some importance.

(b) Corrosive agents

In the natural environment the main corrosive agents commonly in existence are humic acid, carbon dioxide (only active in the dissolved state – carbonic acid) and oxygen. The action, particularly on rocks, is important. *Humic acid,* a highly active chemical agent, is derived mainly from decayed vegetable matter and occurs in groundwater flowing

*Chemical weathering in desert areas is almost non-existent.

through such matter.* (The decay of the vegetable matter in itself also requires the presence of water, although other agencies may also be at work.) *Carbon dioxide* and *oxygen* are 'freely' available in the atmosphere and are found in both rainwater and groundwater. Often humic acid and carbon dioxide may be present in the same water, thus increasing the corrosive potential of the water.

(c) Decomposition of rocks

The decomposition of some of the more common minerals in rocks is given below:

1. *Felspar*. More than half the volume of the earth's crust is composed of felspar, which is a common constituent of granite. Decomposition by rainwater (carbon dioxide plays a part) results in the felspar being converted into clay, with calcium and sodium compounds particularly going into solution. Colloidal silica is also removed. The chemical equation is complex. The process is known as 'hydrolysis'.
2. *Limestone*. Calcite, the crystal constituent of limestone (the rock consists in any case mainly of lime – $CaCO_3$), is the only rock-forming mineral which dissolves readily in water. The solvent action of water is considerably increased when carbon dioxide and humic acid are present. The process is known as 'carbonation'. The results of the reaction is to form, among other things, calcium bicarbonate, which, when associated with water supply, is a cause of temporary hardness. During the transfer processes the lime solution may, after evaporation, fill in cracks in rocks to form calcium carbonate, or, more impressively, stalactites (downward growth) and stalagmites (upward growth), i.e. columns of 'limestone' in underground caves. The two may join to form one continuous column. The whole process is a long one.

   Lime also forms the cementing agent in some sandstones, the decomposition of which can often be attributed to the weakening of the 'cement'. Where the grains are of quartz, as is often the case, these remain unaffected chemically, although physically they will be dislodged from the main body of material. Thus the rock crumbles and in this sense disintegrates. The removal of the cementing agents in any rock has a similar effect, i.e. a tendency for the rock to crumble. This may occur at an increasing rate.
3. *Iron*. Iron is a very common element occurring in many rocks. (Next to aluminium, iron is the most widely distributed metal – each tonne of earth contains about 50 kg.) The action of oxygen on the iron, during oxidation, is to produce a crust of ferric oxide which readily crumbles.
4. *Miscellaneous*. Certain minerals which have the property of being able to take up water expand, during a process known as hydration, and thus initiate the break-up of the rock containing them. Another effect of rocks such as basalt which are capable of expanding on taking up water is similar to exfoliation, i.e. peeling of the surface layer. The process is known as 'spheroidal weathering' and occurs when the outer surface is affected by penetrating water. The process may continue as fresh layers are exposed. This usually results in the rocks being rounded.

(d) Effects of other processes

Although the greatest amount of chemical weathering takes place as a result of standing water, in which could be included contact with moist air, considerable weathering also occurs due to the action of running

*The influence of sulphur dioxide (eventually formed into either sulphurous or sulphuric acid) may also be quite considerable in weathering. The emission of sulphur dioxide (and other corrosive agents) into the atmosphere as a result of man-made activities such as fuel burning is relatively recent, and in relation to the geological time scale has been of short duration. However, the presence of sulphur dioxide and other corrosive agents has given rise to what has become known as 'acid rain' that is now known to cause much damage and deterioration in the natural environment. In this chapter the influence and significance of naturally occurring agents is considered. However, the effects of sulphur dioxide and a general discussion on 'acid rain' is included in *3.2 Exposure*, 'Atmosphere', p. 174.

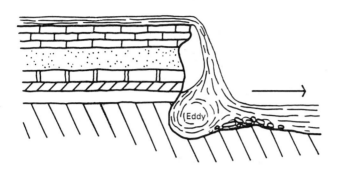

Diagram D3.1/2 *The formation of waterfalls. The effect of horizontal beds of resistant rock: undercutting produces a steep overhanging face while less resistant rocks are cut back, thus causing a rapid upstream recession of the fall. The effect of the eddy has parallels in certain projecting features on the external face of a building.* (*Courtesy: Monkhouse, F. J.,* Principles of Physical Geography, *5th edn, University of London Press, 1963*)

water, as is found in the erosive processes. The combined effect of waves, currents and tides on the sculpturing of the sea wall and floor are often extremely impressive. In these processes the abrasive action of the 'debris' removed elsewhere and carried by the water account for some accelerated weathering.

In many cases the combined action of erosion and abrasion are extremely significant, particularly as they are, in apart at any rate, responsible for the carving of the earth's crust to form watercourses such as rivers, streams, etc. The precise route of these may not always be readily predictable but, once commenced, the weathering processes contribute to their gradual widening and deepening. (The relative resistances of the materials of which the surface is composed are important. Thus, loosely adhering particles as would be commonly found in soils are more readily removed than the constituents of rock.) This, then, virtually establishes the watercourse along which most subsequent water fed into it will flow. Some significant results may occur when less resistant rocks are worn away and undercutting takes place, as in waterfalls (Diagram D3.1/2). The continual flow of water down certain parts, particularly of the exposed faces of buildings, may, in miniature, initiate similar watercourses. 'Waterfall effects' may, under some conditions, also occur on the external faces of buildings, particularly those with suitable projections.

(e) Contribution of pollution

So far, attention has been drawn to the main elements affected by weathering. However, it is to be noted that many other elements are involved, all of which contribute not only to the weathering processes but also to the pollution of the water. An indication of some of the other substances, many of them corrosive, has been given in the analyses of river and sea water (Tables T3.1/2 and T3.1/3). Among these, the sulphates (including others which do not occur in the tables) are of particular relevance in building practice, in connection with the problem known as sulphate attack. As, mentioned under 'The water cycle', any of the corrosive agents may be present during the various water transfer processes, not least those which occur underground in groundwater. Some of the corrosive agents require special consideration in water supply practice, for their effects not only on materials but also on health. Others such as the carbonates (a general term usually implied to include bicarbonates) of calcium and magnesium are associated with temporary hardness problems, while the sulphates and chlorides of calcium and magnesium with permanent hardness. (See *3.2 Exposure*, 'Water supply', p. 207.)

3 Physical weathering

1. *Frost action* is the one physical weathering process which requires the presence of water. As mentioned earlier, water expands roughly 10% by volume on freezing, with the result that rocks may be broken up quite considerably when water in cracks and crevices subsequently freezes.
2. *Temperature fluctuations*, especially where there is a marked diurnal range, are responsible for the development of cracks (known usually as joints) in rocks as a result of the expansion and contraction which occurs. Ultimately, large blocks will break away into smaller units. In those cases where the shell or surface layers of rocks become excessively heated, i.e. in relation to the interior, *exfoliation* or 'peeling off' of the shell takes place.
3. *Wind*, like water, is more commonly considered as an agent which transports abrasive materials and, in this way, is regarded as being capable of assisting in physical weathering. The effect of wind action can be quite pronounced on loosely adhering or relatively soft materials.
4. *Rain*, one of the most important of the 'elements', is also one of the more obvious sources of water, and in this way can be said to assist in physical weathering such as frost action, erosion, or where it acts as a lubricant (whole slopes may become unstable after heavy rain). In some processes changes in moisture content may also result in a physical breakdown. As in the case of wind, the action of rainwater on loosely adhering and soft materials (soils are again obvious examples) can be considerable.

(a) Pressure releases

Finally, a process of physical disintegration can take place during what is known as 'unloading' or *'pressure release'*. If the weight-caused pressure of overlying rocks is removed as a result of denudation, the newly exposed rocks may then be able to expand, thus causing joints to form. This facilitates 'frost shattering'.

4 Effects of biological agencies

The chief way in which plants particularly may assist in chemical weathering is to maintain damp conditions on the surface of rocks, although they may also 'exude' acids. On the other hand, plants may assist in preventing rapid deterioration by binding the surface layers together and thus increasing the difficulty of removal of the debris. Soil erosion is more often than not accelerated by the removal of trees and other vegetation. Other things being equal, the rate of deterioration can be increased by exposure of fresh surfaces.

Mechanical weathering can be influenced to some degree by the root system of plants. Roots do have extraordinary penetrating and expanding powers, thus being able to break open or widen existing cracks and crevices. Such opening up then allows access for water and air, and subsequent chemical weathering.

5 Significance

(a) Exposure to aggressive agencies

As there may, at first, seem to be such a marked difference between the materials of the natural environment when compared with those used in the built environment, it may not be readily apparent what the significance of the weathering, already outlined in general terms, has for the built environment. However, despite recognizable differences in materials, both environments have one thing in common. Subject to variations which, in the present context, are not really significant, both environments are, in general, exposed to the same aggressive agencies, which would also include 'the elements'. In the built environment, moreover, some of these agencies may occur internally, i.e. within buildings, while the effects of other corrosive agents, such as sulphuric acid, 'byproducts' from fuel burning and manufacturing processes, have to be considered. In some cases, albeit local, these may be pertinent to weathering.

(b) Importance of similarity

Similarity between the materials is, therefore, more important than any apparent dissimilarity, especially as chemical weathering must be related to chemical composition and the 'laws' implied by this, rather than to what can best be termed colloquial description, whether this is technical or not.

As an example, consider iron. This substance occurs in abundance in the natural environment but in forms not easily recognizable. It is converted into iron oxide during oxidation and may, among other things, stain other materials. The fact that iron, as commonly used in buildings, is not recognizable as the same substance* which occurs naturally does not detract from the fact that, given the right conditions, the 'man-made' material will also be changed into iron oxide, more commonly known as rust (i.e. a case of *reversion*). Therefore similar analogies could be drawn of many other substances which occur in the natural environment and subsequently in materials used for buildings, albeit differently arranged and composed during manufacture. Physical weathering of materials is not restricted to those of a specific composition, although the latter will influence the likelihood of disintegration. Thus, even here there is a parallel with the problems of the built environment.

(c) Interaction of variations

Weathering in the natural environment also illustrates, sometimes extremely dramatically, the effects and interaction of variations in both the agencies which cause disintegration and the resistance of materials to these agencies. The results of the 'free for all' are usually quite marked.

(d) Change and control

The fundamental and perhaps most obvious difference when dealing with the built environment is that choice of materials to resist specific conditions is possible. Such *choice* may very often be limited, but nevertheless to this can be added the vital element of direct *control* which is possible. Thus, for example, the life of materials may be prolonged by protection, i.e. by the use of additional materials ('coatings') or by shelter from the aggressive agents, on the one hand, and by maintenance, particularly cleaning, on the other.

Choice and control are themselves dependent on the potential of aggressive agents likely to be present in any given situation. Accurate assessment of these is often extremely difficult. Resort has, therefore,

*It is recognized that this is an oversimplification. In both the natural and built environments iron is associated with other substances. However, this does not materially alter the fact that in both cases iron can be oxidized, nor the principle being outlined.

to be made to the most accurate guide available. If necessary, a suitable factor of safety, as far as the materials are concerned also has to be incorporated – **creative pessimism**.

Sources

All as for 'Basic properties of water'.

Dampness

1 Background

(a) Scope

The meaning and use of the term 'dampness' is discussed and a brief outline of the principal causes of dampness is given as an *aide-mémoire*, as all causes are described in detail elsewhere in this book. The way porous materials absorb, give off and retain moisture is covered in detail in *3.3 Moisture content*. Guidance on the measurement of dampness is given and the methods of measurement generally available are summarized.

(b) The problem

Everyone appears to know when dampness occurs in buildings and what it is. In the UK about two and a half million buildings suffer from it (1985 estimate) and about one half of the problems in buildings are caused by it. However, in building work, the dryness of a material or of its surface is important before finishes such as screeds, plasters, renders, paints and other protective coatings or finishes that depend on adhesion are applied. Consequently, there is a need to define both, but for most practical purposes it should be sufficient to define one, dampness. Dampness and dryness are more commonly associated with porous materials, although in some cases the dampness or dryness of the surfaces of impervious materials may also be important where adhesion is required.

2 Meaning and use

(a) Criteria

Dampness cannot be defined absolutely for the simple reason that porous materials such as timber, brick, tile, mortar, plasters, renders, screeds and concrete contain a certain amount of moisture in them in their normally dry state. Data on this state of materials are not readily available. The normally dry moisture contents of five materials should illustrate the range to be expected:

- Timber: 10–20% with about 8–14% normal for timber in a centrally heated building;
- Lightweight concrete: 1–5%;
- Dense concrete: 0.5–1.75%;
- Plaster: 0.2–1%;
- Clay brick: 0.25–0.75%.

   In general, the normally dry state actually attained by a material will depend on its structure and the prevailing moisture content of the air surrounding it. The latter will govern the moisture equilibrium or **balance** achieved by the material.

Questions of health apart, a criterion that is almost overriding in defining dampness is the quantity of moisture needed to cause deterioration of a particular material during its life in a construction. So far, there are no definitive, if any, data on this aspect.

The definition of dampness has therefore to be related to individual materials and circumstances. Clearly, most porous materials contain some moisture in their normally dry state, and for convenience this state can be described as being normal. Therefore it should be sufficient as a general guide to consider a material to be damp if its moisture content is above normal. In time, data may be made available to enable 'normal' to be defined more positively than has been possible in the discussion above.

(b) Surfaces and interiors

During the construction of most buildings, considerable quantities of water are used (see *3.2 Exposure*, 'During construction', p. 200). That water has eventually to evaporate until the materials reach their normal state of dryness. Before completion, dryness of surfaces may be required before finishes are applied. Damp surfaces usually, but not always, signify that the interior is also damp (i.e. the whole body of the material is damp); on the other hand, dry surfaces do not necessarily signify that the interior is also dry. An understanding of the way porous bodies dry out is important, particularly if the results of moisture measurement are to be interpreted properly.

It is normal to expect the exposed surface(s) of a material to dry out before its interior. During the process, moisture has to be drawn to the surface. At any given time during the process, drawing of moisture from the interior may have stopped, perhaps temporarily. In that case the surface will appear to be dry visually and physically, the latter from measurements. In certain cases it may therefore be necessary to check the dampness of the interior.

It should be noted that the same principles apply when buildings or parts of buildings have to be dried out because of water penetration or flooding. Whatever the need for drying out may be, it is important that attempts are not made to accelerate the process. This normally results in cracking or other deformation – see *2.1 Cracking*, p. 56.

3 Causes of dampness

Dampness, whether on the surface or within the interior of porous materials, may result from any one of the causes set out below. The order in which these are arranged is of no significance. At the same time, dampness may result from one or more of the causes. Looked at in another way, these causes could be regarded as sources of water and are, in fact, considered in this way later under *3.2 Exposure*.

1. *Construction water*: Residual moisture from water introduced during the construction of the building. (Concerns both internal and external elements.)
2. *Rising damp*: Moisture rising by capillarity in those elements such as walls and floors in close contact with the ground.
3. *Atmosphere and condensation*: Moisture in the atmosphere and by formation of condensation or absorption by hygroscopic materials, as, for example, deliquescent salts. (Almost universally applicable.)

4. *Rain penetration*: Rain penetration of building units. (Primarily concerned with external elements.)
5. *Leaks from services*: Leakage of water from faulty plumbing (including rainwater), drainage, heating and other waterborne services. (Universally applicable.)
6. *Cleansing and maintenance*: Spilling of water during cleaning, etc., (indeterminate) or during actual usage (determinate in most cases) as, for example, in kitchens, laboratories, industrial processes, etc.

4 Measurement

(a) Background

In some instances, particularly in existing buildings, it may be sufficient to rely on sight or touch (feelings of undue coldness may also be indicative) for dampness 'measurements' and thus the possibilities of impending deterioration. Often, of course, actual deterioration of materials draws attention to dampness. However, as mentioned earlier, during building a need has arisen for more accurate methods of determining dampness where adhesive processes to surfaces are involved. Such a need is obviously met better by quantitative rather than subjective measurements, and this has, therefore, led to the development of a variety of physical methods. The unit most suited for making such quantitative measurements is moisture content, although meters, for simplicity, are gauged in coloured zones, such as green (dry enough), amber (drying, i.e. not dry enough) and red (still wet) and related to specific materials or a range of materials.

The basic requirements for 'measuring devices' appears quite simple at first. However, difficulties have been encountered relative to normal building, rather than laboratory, conditions.

Some of the difficulties are concerned with the necessity for checking, albeit occasionally; length of time often required before useful readings can be taken; sensitivity to damage and perhaps also cost relative to accuracy obtained, i.e. greater accuracy usually implies greater cost in capital and labour charges.

(b) Methods

(1) Hygrometer – air dampness
Often the humidity of the air (i.e. terms of moisture content or vapour pressure) can provide useful guidance on the potential for drying out or the risk of condensation. To obtain humidity, the temperature and the relative humidity of the air has to be measured and the results 'processed' with the aid of a psychrometric chart. Temperature and humidity may be measured with a whirling hygrometer (Diagram D3.1/3 – rather slow and tedious) or with one of a number of electronic thermohygrometers. Some moisture meters have a thermometer that can also be used as a wet bulb.

(2) Oven drying
This is the most accurate method for the determination of moisture content of materials (*3.3 Moisture content*, p. 224). Although it is possible to measure large samples, as in the case of timber, for which this method was largely developed, it is more common to extract samples of material by carefully drilling. The amount of moisture in the samples is then measured by direct weighing, first damp and then dry. This method provides the opportunity of determining the moisture gradient

across a material or construction. It is time consuming. The results are not affected by the presence of salts in the sample but the salts are not detected either. Salt detection requires another method.

(3) Carbide meter

Developed by the BRE, this is similar to oven drying of drilled samples. Instead of heating, the measurements are made with a calcium carbide meter on-site. Like oven drying, the results are not affected by the presence of salts nor are these identified.

(4) Hygrometer – surface dampness

A hygrometer determines the equilibrium humidity produced in an air space in contact with the surface to be tested (Diagram D3.1/4). The equilibrium humidity results may then relate to specific requirements (e.g. suitability for painting).

(5) Conductivity meter

This type of meter has probes that are pressed onto the surface to be tested (Diagram D3.1/5). If the surface is relatively soft (e.g. timber) the probes can penetrate to the interior to some extent; if not, the condition of the surface only is measured. Specially insulated probes are available for measuring the dampness of the interior. Holes have to be drilled for this purpose.

The electrical resistance between the probes is measured. The results may be shown by a needle on an absolute or relative scale or bands of different colours (or both). Electronic versions use lights instead of a needle. As the electrical resistance is further reduced by soluble salts, care is needed in interpreting the results if the presence of such salts are suspected. Comparative measurements from timber nearby can be a useful check. Alternatively, salt-detecting devices are available.

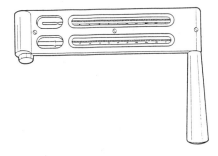

Diagram D3.1/3 *Sling or whirling hygrometer for measuring dry and wet bulb air temperature for interpretation into humidity. Slow and tedious in use. Electronic versions quicker and more convenient*

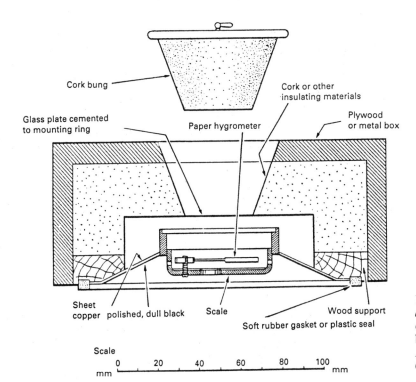

Cork bung

Cork or other insulating materials

Glass plate cemented to mounting ring

Paper hygrometer

Plywood or metal box

Sheet copper polished, dull black

Scale

Wood support

Soft rubber gasket or plastic seal

Scale

0   20   40   60   80   100
mm   mm

Diagram D3.1/4 *Hygrometer – apparatus for measuring dampness. (From BRE Digest 163,* Drying out buildings, *March 1974, Building Research Establishment, Crown Copyright)*

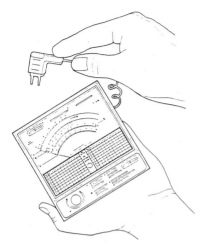

Diagram D3.1/5 *An example of a conductivity meter. Note probes. Readings are taken from a scale; some versions have coloured bands instead of numbers; electronic versions use flashing lights with numbers and/or coloured bands*

(6) Capacity or capacitance meter
The probe with this type of meter has a flat face. This is pressed against the surface and registers the moisture content at or near the surface. A rough surface may cause inaccuracies in the reading. Soluble salts may also introduce errors.

Sources

BRS Digests
33 (1st Series), *Causes of dampness in buildings*, August 1951. This digest, old though it may be, still retains a remarkable freshness about the causes of dampness. Some aspects have been overtaken by time and these have to be recognised

BRE Digests
18, *Design of timber floors to prevent dry rot*, January 1962. Explains the method of use of an hygrometer
152, *Repair and renovation of flood-damaged buildings*, 1984
163, *Drying out buildings*, March 1974. Describes meters for wood and explains the method of use of an hygrometer, the latter as in Digest 18
198, *Painting walls – Part 2: Failures and remedies*, 1984. Example of the increasing use of quantitative data in the evaluation of dryness in finishing processes
245, *Rising damp in walls, diagnosis and treatment*, 1986. Has detailed explanation of analysis of samples obtained by drilling for soluble salts and moisture content, the latter either by oven drying or by carbide meter

Flow

1 Definition and scope

The term 'flow' as used here relates to the basic paths that water, in its liquid state, takes over vertical and along soffits of horizontal surfaces of materials. The flow of water influences the actual degree of exposure of certain parts of a wall construction. The patterns that the water paths may take are affected basically by the characteristics and geometry of the surfaces of materials. Wind action and other factors influence the flow patterns, and the effects of these are covered in *3.2 Exposure*.

Flow is important when considering the exclusion of water and changes in the external appearance of materials. The basic paths of water described here and the effects they may have on either are considered in more detail in *3.4 Exclusion* and *3.9 Flow and changes of appearance**.

2 Basic exposure

(a) Generally

In general, rainwater strikes the exposed faces of a wall at an angle. A measurement of 10° from the vertical has been used as a guide, but this is not always the case. When there is little or no wind during rainfall, water from horizontal surfaces such as projecting balconies and string courses will flow over the wall surfaces below.

*For convenience, relevant diagrams in 3.9 Flow and changes of appearance are repeated in this chapter.

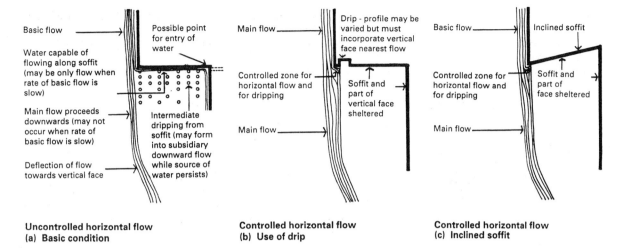

Uncontrolled horizontal flow
(a) Basic condition

Controlled horizontal flow
(b) Use of drip

Controlled horizontal flow
(c) Inclined soffit

Diagram D3.1/6 *Controlling the horizontal flow of water along a soffit. The basic condition (i.e. the uncontrolled flow) is illustrated in (a) without regard to the effects of surface characteristics; (b) and (c) show two ways of controlling the flow so that it is contained at or near the leading edge*

(b) Quantity of water

The quantity of water that may flow down a vertical surface will be influenced by the extent to which materials that form that surface absorb water. Although the flow of water is generally downwards, certain conditions of exposure to wind can cause the water to flow upwards while the wind is blowing. It is convenient to imagine the downward or horizontal flow of water as a film of variable thickness and area. Importantly, there is a progressive increase in the quantity of water with height, so the greatest amounts pass at or near the base of a wall.

(c) Pattern of flow

The pattern of flow, namely, those areas over which water actually passes, will be influenced partly by the obstructions in its path and partly by the rate of flow. The rate of the flow of water is, of course, less during a short shower than during a heavy downpour or even a continuous drizzle. However, for any given intensity of rainfall the rate of flow will be less at the commencement and the end of the fall than during its duration. Importantly, the flow usually continues for a while after the rain has stopped. With slow rates of flow, obstructions tend to cause the flow to become fragmented and, by analogy, broken up into rivers, streams and lakes.

3 Soffits

(a) Basic mechanism and control

Water will flow along the soffit of a horizontal surface due to surface tension effects. This flow may be controlled either by interrupting the surface or by defeating the surface tension effects. Diagram D3.1/6 illustrates in (a) the basic condition that gives rise to the horizontal flow; in (b) the interruption of the surface by the use of a drip; and in (c) defeating surface tension effects by sloping the soffit upwards.

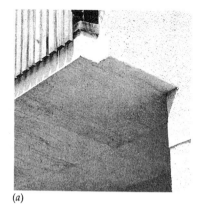

(a)

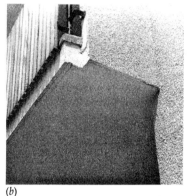

(b)

Photograph P3.1/1 *Comparative illustrations of the use of a drip to control the flow of water along soffits. Without a drip as in (a), water travels a considerable distance. The drip controls that flow as in (b). (Compare with Diagram D3.1/6)*

The horizontal flow of water is more commonly associated with projections from the walls. Photograph P3.1/1 (taken from the same building) illustrates in (a) the considerable distance that water can travel along a soffit and in (b) how a drip at the leading edge has stopped the flow. In both cases the slight downstand to the rough rendered surface to the returns has provided a drip.

(b) Within constructions

Within constructions, horizontal flow may occur along soffits of projections through walls. Reinforced concrete slabs or beams bridging across a cavity are examples. In these, water may penetrate at the bed joint and then flows along the soffit of the slab or beam. A drip near the inside face of the outer leaf should be incorporated so as to prevent water bridging to the inner leaf (see Diagram D3.1/7).

(c) Importance of continuity

An important point about drips is that they must be continuous, otherwise bridging will occur. Small discontinuities are likely to lead to localized bridging. Care needs to be taken to ensure that drips are not filled in or blocked in places by, for example, mortar during bricklaying or slurry in the forming of concrete.

4 Profile of front edges

The extent to which water will be thrown off the face of a building will be influenced basically by the profile of the front edge of the projection (Diagram D3.1/10). Where the front edge is rounded, as in (a), the water will tend to follow the rounded profile; where the front edge has a sharp arris, as in (b), the water will tend to be thrown clear of the projection. Thus a projection with a sharp arris to its front edge results in a greater throw-off than one with a rounded edge.

The extent to which water may be thrown off depends on two factors: (a) the head of water that is dependent on the height above the sill, and

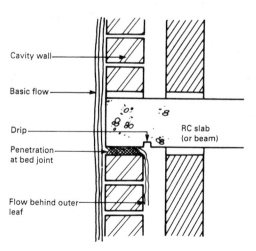

Diagram D3.1/7 *Use of a drip on the soffit of an RC slab bridging a cavity*

(b) the intensity of the flow above the sill. The greater the head and the intensity of flow above the sill, the greater the throw-off.

5 Eddies

Water thrown off by a projection will be deflected towards the vertical surface below it. If the surface below the projection is a recess, the surface of the recess may receive the thrown-off water at an angle. This may result in the creation of an eddy, as shown in Diagram D3.1/9. An eddy imposes additional exposure to joints at or near the junction of the vertical face and the projection. The traditional *stooling* of sills and other projections ensured that the vulnerable joint was lifted out of harm's way.

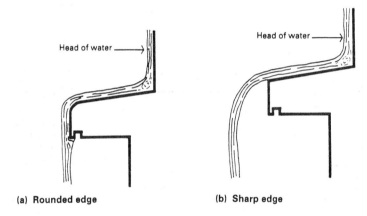

(a) Rounded edge

(b) Sharp edge

Diagram D3.1/8 *The effect of the profile of the top front edge of a projection on the flow of water*

6 Returns of projections

Most of the water flowing over a projection will discharge over the front edge. Some will also flow over the return (Diagram D3.1/10). The amount so flowing will be dependent on the slope of the projection. The greater its slope, the greater the amount likely to flow over the return edge. A form of tray will ensure that water does not flow over the return edge.

Assessment of severity of conditions

1 Scope

If materials are to be used with as little waste as possible, the *known* properties of any one material must be closely related to the conditions to which it is *expected* to be exposed during its working life. The extent to which waste may be reduced at any given time in the development of

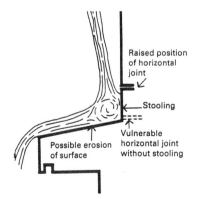

Diagram D3.1/9 *Creation of an eddy at the junction of the top of a projection and the vertical face of the wall above due to angular flow of water. Similar effects may be obtained when the flow is perpendicular*

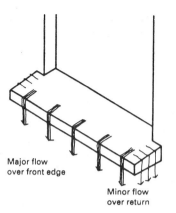
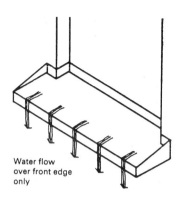

Major flow
over front edge

Minor flow
over return

Water flow
over front edge
only

(a) Basic condition and problem   **(b) Possible solution-tray principle**

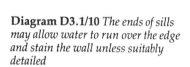

Diagram D3.1/10 *The ends of sills may allow water to run over the edge and stain the wall unless suitably detailed*

building is primarily dependent on the *degree* of accuracy which can be practically achieved in the determination of the variables which are involved. The need for any degree of accuracy in connection with problems associated with water and its effects (as in other fields) will be dependent on a number of factors, chief among which are social/ economic relationships. Greater advantage in the use of any material also implies the use of smaller factors of safety. However, as in the case of the mechanical properties of materials, three criteria, namely, knowledge, quality and variations, will determine the actual factor of safety applicable in any given circumstance.

With changing requirements and new ways of satisfying them, most empirical approaches have been found wanting. A need arose for greater accuracy and reliability in assessing the severity of exposure. Assessments of such conditions based on 'scientific' methods and resulting in a quantitative approach aim to provide more reliable *guides* than empirical, rule-of-thumb methods have done. Consequently, attempts have been, and continue to be, made to increase the number already available. Compared to the structural mechanical field, the development of these methods is more difficult. The purpose of this section is mainly aimed at drawing attention to some of the problems involved and to some of those methods which are now available. By so doing it is hoped that some understanding might be gained of the reasons for many of the 'gaps' in knowledge, which perforce leads, in many instances, to the use of somewhat generalized statements often strictly qualified. Detailed values are not generally given here, as these are included in those sections to which they essentially belong.

2 Precedent

Classifications which have been used in the past included such comparative and descriptive terms as '*severe*', '*moderate*' or '*sheltered*'. In general, these are of little practical value, particularly for comparative purposes, unless accompanied by wide experience of the use of a particular material or of a particular location, when it is then possible to take into account variations such as might exist between, for example, macro- and micro-conditions. The number of qualifications necessary, whether by way of general comparison or other means, in order to allow for the widest use of a particular classification often become misleading. In many cases, however, the key to the difficulty would appear to rest with the non-measurable basis used, which generally leads to hazardous guesswork.

All classifications are useful only as guides, and should always be regarded as such. Accordingly, they require intelligent interpretation in all applications. However, in providing guidance they should aim to reduce the amount of guesswork needed to the absolute minimum. Thus classifications become of far greater practical value and are much more reliable as guides when they can be related to some quantitative method of measurement. This then allows measurements to be made for a specific building problem on a basis similar to that used for the classification. On the other hand, some of the information necessary to enable a suitable assessment to be made may very well be obtainable from other published sources.

The use of quantitative methods of assessment, particularly those based on calculation, are more common (and have been for a considerable time) in structural mechanical work. In the newer sciences associated, for instance, with the use of thermal and acoustic properties of materials, great reliance is also placed on the use of methods of calculation. Although the methods used in these fields help to reduce the hazards of guesswork, they are not necessarily practical or feasible when considering problems associated with water and its effects. Revaluation leading to certain modifications of known methods and, in a sense, innovations are usually necessary.

3 Difficulties

A different approach is necessary when dealing with water and its effects, but chief among these would appear to be the difficulty of resolving the many factors involved in terms of methods of calculation. On the other hand, to be of any significant value, a method of assessment based on quantitative methods of measurement must also be capable of being related to methods of measuring the related properties of materials and other factors. Among other things, the units of measurement used are generally required to be the same in all cases or, if not the same, at least capable of being related to each other simply.* In many cases methods of assessment are, or could be, easily made available, but these may not have been adequately related to the properties of materials or, for that matter, to buildings as a whole. Finally, the requisite properties of all types of materials may not be available – in the past there may not have been the need for such properties.

4 Examples

Among the examples included, the one on 'Rain penetration', developed by the Meteorological Office and the BRS, is described in detail in order to illustrate the difficulties which are normally encountered in devising quantitative methods of assessment and the manner in which these are being overcome. The remaining examples

*Simplicity would appear to be an essential requirement of any methods used if such methods are to find wide acceptance among designers and many others associated with practical building problems.

are dealt with in general terms and, where necessary, 'gaps' in knowledge are highlighted.

(a) Rain penetration

(1) Application of climatic data

There is a great deal of climatic data available on the climatic factors involved such as rainfall, wind, sunshine, etc., covering many stations throughout the UK and in some cases measured over a long period of time. (The time element is important if more accurate forecasts of probable future conditions are to be made.) All these factors are significant when considering, for example, moisture movements, the exclusion of rain from a building, efflorescence, chemical attack, corrosion, frost action and weathering.

However, the methods of measurement, made separately for each element, are seldom, if ever, made by meteorological offices so that they may be related directly to the way in which they influence the properties of materials or, as important, the way in which they 'act' on buildings. As an example, consider driving rain, that is, wind-driven rain, usually carried at an angle so that it impinges on vertical surfaces. This is of far greater importance in making an assessment of the severity of exposure conditions than the amount of rain falling on level ground, as obtained from normal rainfall measurements. In order to make an assessment of the measure of the severity of exposure of a wall, it is necessary to know the direction and speed of the wind during any given intensity of rain together with the amounts 'collected' from the vertical faces of buildings. Other factors such as raindrop size and the angle of incidence which vary from one rainstorm to another are also significant.

(2) Combining wind and rainfall

Following investigations into the amount of rain collected from the vertical faces of buildings and other related data, it has been shown that the amount of rain driven onto a wall is directly proportional to the product of the rainfall recorded on level ground and the wind speed during the rain. Thus, once the values of the two factors are known, it is then possible to determine the severity of conditions of exposure which might obtain in a given location on a quantitative basis, by the use of a suitable index – a driving rain index (DRI).

(3) Mapping of DRI

The pioneering work on the mapping of driving rain indices was done by the Norwegian Building Research Station during the early 1950s. By 1965, four DRI maps had been prepared – for the UK, Norway, Canada and Denmark. The map prepared by BRS (Lacy) together with the Meteorological Office (Shellard) showed omni-directional average annual indices. This was achieved by relating information then available on rainfall and wind speeds. That information did not include, except for a few sites, the mean wind speed during rain. Consequently, by making certain assumptions to compensate for the absence of the relevant data, the DRI was calculated by multiplying the *average* annual rainfall (in millimetres) and *average* (all hours) windspeed (m/s), expressed as m^2/s.

The first map published of the UK showed the variations of the indices (as contours with 'Sheltered', 'Moderately exposed' and 'Severely exposed' gradings). Another accompanying map showed the annual *relative* DRI for each of eight wind directions. This map contains what are termed 'driving rain roses'. The petal of each rose indicates the proportion of the total index attributable to winds from different directions. The rain roses were intended to make it possible to assess which faces of buildings were likely to be more heavily exposed than others (see *Materials for Building*, Vol. 2, pp. 28, 29, 35 and 36).

Later, more detailed maps for the whole of the UK were published by BRE. However, even these did not help to overcome the inherent coarseness of the index. The indices dealt with the macro-climatic condition. Assessment of the changes likely to occur in the micro-climate of a particular site was rather crude. Relating an index to a particular form of construction for a particular site could not be done with absolute conviction.

Despite their limitations, the maps with the indices and rain roses on them did, for the first time, make it possible to compare directly the exposure to driving rain of different sites in the UK. The indices also provided the basis for work by BSI and BBA to aid assessment of the suitability of cavity walls in different parts of the UK and of different heights to receive cavity fill. Similarly BSI tests for window leakage have been made available.

(4) Increasing accuracy

Experience has shown the accuracy of the indices and the rain roses to be wanting as reliable guides in the assessment of exposure to driving rain. The annual indices, now seen as guides on weathering rather than rain penetration, are not sufficiently accurate for that purpose. It is now also recognized that the rain roses should provide guidance on the risk of rain penetration, but they do not. In the meantime, more relevant meteorological information is available; computers can model reality more closely than hitherto. More recent work by the Meteorological Office (Caton) and BRE (Prior) has been directed at fine tuning of the original BRS DRI or making the DRI of more practical use, for masonry constructions especially. As the original rain roses were based on annual averages, it was not possible to assess the amount of driving rain received by a wall over a period of hours or days. The short-term exposure of a wall is the governing factor. This has given rise to what is now termed 'the driving rain spell index'. More accurate maps and ways of allowing for local topographical and other conditions have been prepared. It is now possible to determine with greater accuracy than hitherto the exposure to driving rain of walls with different orientations in different parts of the country. The methods for assessing such exposure have been published by BSI (see *3.2 Exposure* for an explanation). More work has yet to be done to make the annual driving indices more relevant for the weathering of materials.

(b) Dampness and adhesion

The measurement of moisture content of materials and the significance of such measurements in the adhesive processes in buildings have already been outlined in this chapter under the heading of 'Dampness', p. 150. At this point it may be noted that even such general methods of assessment which are available would not be of very much practical value unless the degree of dampness which any of the adhesive processes can 'tolerate' without failure was established. The practicality of the available methods is still, to a certain extent, limited by the fact that not all the adhesive processes have been fully investigated.

(c) Sulphate attack

Among the chemical attacks on materials which have received particular attention is that related to sulphate attack of cement products used below ground level. In this a classification of soil conditions* has been developed in which there are five classes, each based on a stipulated concentration of sulphur trioxide. Precautionary measures which are advisable relative to each of the five basic concentrations and to the type of product concerned have been put forward. It is therefore important that an analysis of the sulphate content of any soil is undertaken so that the necessary measures may be taken to avoid attack. The classification is a useful and simply presented guide. (Sulphate attack is discussed in detail in *3.6 Chemical attack* p. 405 ff.)

(d) Atmospheric pollution

Atmospheric pollution plays an important part in chemical attack on a variety of materials. Classifications intended to act as guides for the severity of conditions are still general in character and in descending order of pollution are usually given as: (1) industrial and heavily polluted urban areas; (2) normal urban areas; (3) coastal areas; and (4) rural areas. None of these classes takes into account the increase in the severity of conditions often caused by micro-climatic conditions or the proximity of the source of pollution. In part this is due to lack of sufficient quantitative measurements of the degree of atmospheric pollution throughout the country which could then possibly be related to the resistance to deterioration by chemical means which may be expected from various materials. Some guidance is available on the rate of corrosion of mild steel and zinc coatings on steel relative to the four classes of atmospheres quoted above. The rates of corrosion used in conjunction with tables of thickness of steel plate or of zinc coatings enable an *estimation* to be made of the thickness required for a given life. The rates of corrosion may also be used to estimate the life of a given thickness of metal, with or without a coating.

(e) Dimensional fluctuations

An assessment of the extent of dimensional fluctuations in porous materials due to changes in their moisture content is necessary if the basic shape and size of any member or element, usually determined by other factors, is to be suitably modified to limit distortion or prevent the likelihood of cracking taking place. In the main, modifications, in addition to changes in the basic dimensions, including area, usually involve the use of movement joints. In this the detailed relationship of two or more elements of different materials may be important.

Among the factors relative to the properties of individual materials which may influence the extent of change likely in any given situation are relative humidity, thermal conditions and air movement of the environment. Most of the information (and there is a considerable amount available) on the dimensional changes to be expected in various materials is generally related to relative humidity. It is, therefore, necessary for some estimation to be made of the effect of the other factors involved. Because of these and other difficulties in making the

*Table 1 in Concrete in *Sulphate-bearing soils and groundwater*, BRS Digest (2nd series) No. 90, HMSO, February 1968.

necessary calculations, rule-of-thumb methods, particularly for the 'traditional' materials, are more often used and usually found to be fairly reliable as guides. However, this may not necessarily be true of the newer materials, in which case results based on calculations, despite their own limitations, generally provide sufficient guidance to enable an assessment to be made of the necessity or otherwise of taking precautions. Where it is desired to limit the actual dimensions of any movement joint it is important to take into account the limitations of the calculations. Accordingly, it is always advisable to incorporate a suitable factor of safety. In most cases this means adding an appropriate amount to the calculated size – **creative pessimism**.

Exposure

Introduction

All materials may be exposed to water in varying degrees during construction and the subsequent life of a building. An assessment of the severity of conditions of exposure to water under which materials are likely to perform is essential before selecting, using or maintaining them. The same applies to selecting materials for use in particular forms of constructions or methods of assembly. To assess the severity of conditions, an understanding of the relationship between sources of water and particular elements or parts of a building is needed. In addition, the degree to which each source may be polluted should be understood.

This chapter examines those factors which affect the *basic exposure* of materials to water. Detailed consideration is excluded of those factors that are likely to alter this exposure, particularly within a material or construction. These, together with the principles that may be adopted to exclude water, are covered in *3.4 Exclusion*. Similarly, detailed consideration of the resistances that various materials may offer to water are included under the relevant headings of later chapters (*3.5 Efflorescence, 3.6 Chemical attack, 3.7 Corrosion* and *3.8 Frost action*).

Eight sources of water have been identified for inclusion:

Atmosphere
Rain
Ground
During construction
Water supply
Faulty services
Maintenance and general usage
Entrapped moisture.

It should be recognized that one particular source of water may not affect all parts of a building at all times during its life. Some may only be relevant during the construction of a building or for a short time after completion (during drying out, for example). Parts of a building may only be exposed to particular sources (for example, foundations with groundwater or flat roofs with entrapped moisture).

Where relevant, the extent to which each source may become polluted is considered. Water may 'pick up' pollutants throughout the whole water cycle (see *3.1 General considerations*, 'Water cycle', p. 137). In buildings this cycle is complex. Some building materials may provide a source of pollutants. It is therefore difficult here to take into account fully the effects of any 'chain reactions' that do occur as water follows its cycle, in either its liquid or vapour state. (In building, the solid state plays little part in pollution.) Nevertheless, it is important when making an analysis in practice of the severity of conditions in particular cases that account is taken of the whole of the water cycle in buildings.

Atmosphere

1 Generally

Water vapour is always present in the atmosphere. The comparative ease with which gases are transferred from place to place or even through materials means that virtually all parts of a building, whether these occur internally, externally or within the thickness of a construction, are perpetually exposed to water to some degree.

The exposure of materials to vapour in the atmosphere needs to be considered from two different points of view:

1. The effects of the water vapour in (better, moisture content of) the air on moisture-absorbing materials. Timber, timber products, cement-based materials and plasters are among materials whose moisture content is dependent basically on the prevailing moisture in the atmosphere with which they are in contact. Changes in their moisture content influence their state of dryness (*3.1 General considerations*, 'Dampness') or the moisture movements they may undergo (*3.3 Moisture content*). Deliquescent salts on the surfaces of materials can absorb moisture from the atmosphere, thereby causing dampness. Air with much moisture in it for long periods can help to sustain mould growth (*3.6 Chemical attack*).
2. The conversion of the water vapour into liquid during condensation. The latter is considered in detail in *Materials for Building*, Vol. 4, *4.06 Condensation*. For convenience, general points related to exposure are discussed later.

2 Expression of humidity

An important characteristic of air is the change in its ability to hold water vapour with variations in the temperature of the air holding it. The higher the temperature, the greater the amount of water vapour that the air can hold. Air that contains the maximum amount of water vapour is said to be saturated.

The humidity may be expressed as:

- *A vapour pressure*, that is, the partial gaseous pressure exerted by the water vapour and is normally measured in pascals or millibars;
- *The moisture content* of the air in kilograms of water vapour per kilogram of *dry air*. (This is equivalent to the *mixing ratio*, i.e. the ratio of the mass of water vapour to the mass of dry air with which it is associated.)

It is unusual for the air to be saturated. Its degree of saturation is stated in terms of its *relative humidity* (or wet-bulb temperature). This is the actual vapour pressure expressed as the percentage of the saturation vapour pressure at the *same* temperature. Importantly, relative humidity, or wet-bulb temperature, on its own gives no information about the absolute moisture content of the air. To obtain the latter it is necessary to have the dry-bulb temperature recorded simultaneously with it. However, relative humidity on its own is important in some moisture problems, as mentioned above (for example, in the equilibrium moisture contents of organic materials such as timber and timber products or in cement-based products, as referred to earlier). The same applies to the sustenance of mould growth.

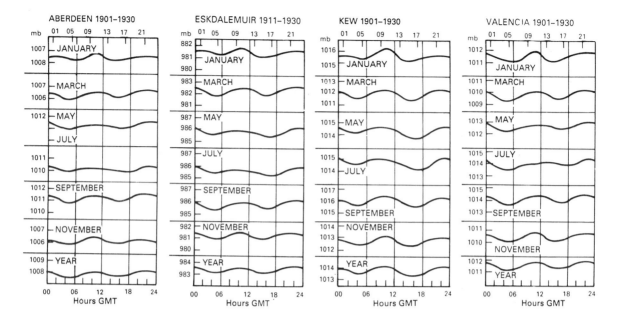

Diagram D3.2/1 *Average hourly pressure at four sites in the UK. An example of information on humidity (available from the Meteorological Office). (Data from* Climatological atlas of the British Isles, *Meteorological Office, HMSO, London, 1952, p. 14, Crown Copyright)*

3 Variations

(a) Information

The Meteorological Office publishes information on water vapour in the atmosphere in terms of vapour pressure or relative humidity for many sites in the UK (Diagram D3.2/1). Information on related aspects derived from the basic information on vapour pressure and/or relative humidity are also available.

(b) Location, season and time

Variations in vapour pressure and relative humidity across the UK are generally as follows: the values are greatest in the south-west and least in the north; they decrease with distance from the sea.

As to season, there is twice as much water vapour in the air in July as there is in January. Generally, the diurnal variation is, on average, small, being about 5% higher during the day because of surface evaporation. At all times of the year the relative humidity is lowest in the early afternoon and drying can be expected to be most rapid at this time.

(c) Specific example

An example of the average diurnal and seasonal variations of observations made at Kew Observatory over a number of years are given in Chart C3.2/1. From this it will be seen that the diurnal range is approximately 78–87% in January (a difference of 9%), rising to approximately 58.5–88.5% in July (a difference of 30%). The annual range is approximately 67.5–88.5% (a difference of 22%).

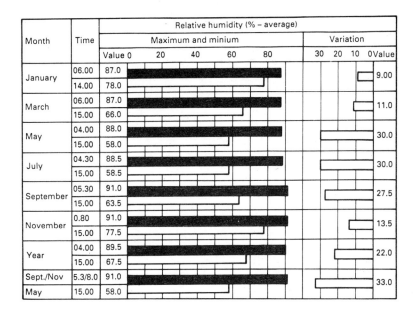

Month	Time	Relative humidity (% – average) Maximum and minium (Value)	Variation (Value)
January	06.00	87.0	9.00
January	14.00	78.0	
March	06.00	87.0	11.0
March	15.00	66.0	
May	04.00	88.0	30.0
May	15.00	58.0	
July	04.30	88.5	30.0
July	15.00	58.5	
September	05.30	91.0	27.5
September	15.00	63.5	
November	0.80	91.0	13.5
November	15.00	77.5	
Year	04.00	89.5	22.0
Year	15.00	67.5	
Sept./Nov	5.3/8.0	91.0	33.0
May	15.00	58.0	

Chart C3.2/1 *Range of average maximum and minimum relative humidity at various months of the year observed at Kew, London, 1901–1930. (Data from* Climatological atlas of the British Isles, *Meteorological Office, HMSO, London, 1952, p. 14, graphs of average relative humidity, Crown Copyright)*

(d) Rapid changes

In contrast to many other countries, the UK may 'suffer' from rapid changes in the moisture content of the air over short periods of time. This is because the country may be successively covered by air streams of different origins. Depending on the location and season, each of these air streams has its own special characteristics of temperature, humidity, etc. Cold Continental air from central Europe usually brings low absolute humidities; air coming from the Atlantic relatively high absolute humidities.

(e) Special cases inside buildings

Internally, the relative humidity will approximate to that prevailing outside, except when central heating is used, special humidity control is exercised or excessive quantities of vapour are produced, such as may occur in kitchens, bathrooms, laundries and certain industrial processes. All these present special cases which must be considered on their merits and are, therefore, outside the scope of this book. However, whatever the condition, the range of relative humidities likely to be encountered in practice is important.

4 Condensation

Today, condensation as a source of dampness in buildings is almost as significant as rain. Before the results of years of research and experience into the more traditional forms of construction showed ways of successfully avoiding exclusion of rainwater the increased use of new forms of rapid heating in new forms of construction brought widespread problems of surface condensation and mould growth. Later, other developments, in insulated roofing particularly, highlighted the occurrence of condensation within a construction (i.e. interstitial condensation). Condensation (better, the risk of it occurring) has become an important aspect in exposure to water.

Both forms of condensation, either surface or interstitial, are attributable to a number of factors, among which higher internal air temperatures and better thermal insulation are, paradoxically, important. Local concentrations of surface condensation are usually due to the lack of **continuity** in the thermal insulation of the enclosure, resulting in 'cold bridges' being formed, on which moisture contained in the warm air may condense. (The term 'cold' is used relative to the temperature and the thermodynamic properties of air in which 'dewpoint' is significant.) Uninsulated cold water pipes also provide 'suitable' cold surfaces for the formation of condensation – such condensation is very often mistaken for a 'leaky' pipe.

Interstitial condensation, resulting mainly from differential vapour pressures (temperature differentials are also involved), must be carefully considered whenever 'sandwiched' forms of constructions, including within their thickness materials of high thermal insulation, are proposed. The actual position of the insulation is critical. The use of vapour control layers (vcl)* also requires careful consideration. When such layers are to be incorporated account should be taken of the probable necessity for the adequate drainage of water which may condense on the vcl itself. An example of the 'unwitting' use of a vcl may be noted here. Impervious claddings used on the exterior of buildings automatically form efficient vcls and water vapour permeating from the inside air can condense behind such claddings. Drainage of the condensed vapour is important.

The quantity of water likely to result from condensation may be predicted for given design conditions. This information may be useful in some analytical cases. However, the main purpose of predictions of the risk of condensation is to provide guides as to what best to do by way of precautions. Alternatively, they may indicate a need for a change in the conditions or in the form of construction, including materials. Certain patterns of use (kitchens, bathrooms, swimming baths and many industrial processes) imply, without the need for calculation, severe conditions of exposure. (See *Materials for Building*, Vol. 4, *4.06 Condensation*, pp. 126 and 127.)

Although greater emphasis has been given to condensation occurring internally, it should be noted that condensation, often referred to as dew, does occur frequently externally. Thus external elements are not entirely exempted from due consideration. In general, it is the metals which require special care in that suitable precautions should be taken to ensure that condensed vapour is not allowed to remain in cracks, crevices, etc.

Finally, condensed flue gases provide an important source of water in chimneys. Furthermore, such condensed vapour may be highly polluted. Special precautions are, therefore, necessary in the selection and use of materials making up the flue (see *3.6 Chemical attack*, 'Sulphates', '7 Domestic chimneys' p. 421).

5 Pollution

(a) The effects of changes

The causes of changes in the amount of pollution in the atmosphere are illustrated in Diagram D3.2/2. The basic causes are divided between human activity and the weather. The control of smoke by the Clean Air Act 1956 is a human activity aimed at reducing the amount of air

*The use of the term *vapour control layer* replaces the misused and misunderstood terms 'vapour barrier' and 'vapour check', neither of which indicated the degree of vapour resistance being provided in given cases. When vcl is used, it is expected to be defined by its vapour resistance.

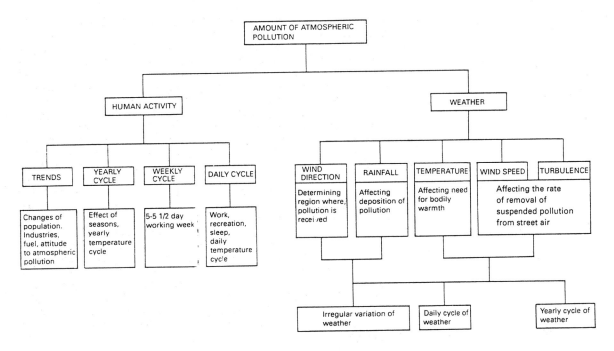

Diagram D3.2/2 *The causes of changes in atmospheric pollution in chart form. Note: This diagram should be read downwards with the words 'depends on' inserted at every step. (Based on Meetham, A. R., Atmospheric Pollution, its Origins and Prevention, 3rd revision, Pergamon Press, Oxford, 1964, Table 333)*

pollution. Thirty years or so on from the enforcement of the Smoke Control Order, the amount of smoke in the atmosphere has been reduced considerably. As forecast at the time, there has been a decrease in the amount of sulphur dioxide emitted. However, with smoke banished from the atmosphere almost completely, more of the sun's heat is allowed through the air. The effect is to set up convection, thereby increasing the dispersion of the sulphur dioxide to the upper atmosphere. The wind helps to disperse the sulphur dioxide widely but it also brings it down to ground level. Today 'acid rain', which includes sulphur dioxide, is of major concern even in countries where air pollution has not been a problem in the past. (Acid rain or, better, acid deposition, is discussed under 'Rain' later.)

The historical outline and the significance of smoke and other pollutants relative to materials is covered in detail in *Materials for Building*, Vol. 2, pp. 50–62, under *3.2 Exposure*, 'Pollution, 1 Atmospheric pollution'. In this book emphasis is given to the present position together with a brief explanation of the source and possible effects of the relevant pollutants.

(b) Significance

Atmospheric or air pollution plays an important part in causing water to contain corrosive agencies more potent than would be the case in the absence of the pollution. Additional abrasive particles may also be present in the water as well as the chemical substances. In general, it is perhaps more usual to consider the contamination of rainwater, either in the chemical or physical sense, caused by impurities present in a polluted atmosphere. However, in the context of building materials, and taking into account the comparative ease with which gases can be transferred, it is more realistic to include the effects of air pollution on water formed during condensation, both surface and interstitial. In this the action of flue gases in chimneys requires special consideration. The fact that polluted water vapour may often be in contact with building materials, if not transferred through them, makes it necessary to pay some attention to this aspect too. In short, the destructive effects of atmospheric pollution when associated with water can be widespread and are, therefore, applicable to all parts of a building, whether these are inside or outside. This is important when considering exposure.

Although quite a substantial quantity of water becomes corrosive as the impurities present in a polluted atmosphere are 'picked up' and dissolved during rainfall, the corrosive agencies may also be formed as a result of water wetting the surfaces of buildings and subsequently absorbing impurities which are then oxidized. The importance of this method of formation lies chiefly in the fact that, on evaporation, the corrosive agencies are liable to become highly concentrated. The destructive effects of such concentrations are accentuated if they are held by materials for any length of time.

A further related point may also be noted here. The corrosive agencies of polluted rainwater may be changed when there is a reaction between the contaminated water and certain other substances. As an example, consider any building material containing carbonates. In addition to the fact that the material itself will suffer damage, soluble sulphates and chlorides are formed as a result of the action of polluted rainwater on the insoluble carbonates. The 'new' substances may, of course, be transferred to other materials which, in turn, may suffer damage.

(c) Basic constituents

The air normally contains many basic constituents, the major of which, expressed as a percentage by weight, are: nitrogen (75%), oxygen (23%), argon (1.26%), water vapour (0.70%) and carbon dioxide (0.04%). Neon, krypton, helium, xenon and hydrogen are also present but in a comparatively minor way. Some of these, when combined with water (free carbon dioxide forming a weak acid – carbonic acid – is a notable example), may be corrosive.

However, within the present context, none of the major or minor constituents of air is usually considered to be a cause of atmospheric pollution.* Other substances, such as the oxides of sulphur, the chlorides, chlorine and fluorine compounds, resulting essentially from fuel burning, fuel handling and chemical processes (in order of importance) or, on the other hand, the salts from a natural source such as the sea and found in areas at or near the coast, form the main constituents of a polluted atmosphere. All these are capable of causing water to be corrosive and occur in a wide range of concentrations.

When compared to the major constituents of air, the quantities of the pollutants which may be present even in a highly polluted atmosphere are small. In some cases they are also small when compared with the quantities of the minor constituents of air. An indication of the scale of the difference may be obtained by considering the fact that the major constituents of air may be measured in g/m^3, and the minor constituents and the atmospheric impurities in mg/m^3. However, despite the comparatively small quantities involved, the potential (and, in fact, actual) aggressiveness of the impurities, particularly when associated with water, should not be underestimated.

The additional or direct *annual* costs of the maintenance of buildings in polluted atmospheres for painting and decorating, cleaning and depreciation of buildings and the corrosion of metals are considerable, and rising.

*The pollution of water by the 'natural' constituents of air is extremely important relative to decay and deterioration of materials. However, this aspect has been purposely omitted from this section so that detailed consideration may be given to those pollutants which are essentially 'man-made' and therefore potentially controllable. The effects of the natural constituents of air on materials are noted as relevant in the other sections dealing mainly with decay and deterioration in detail.

(d) Problems of assessment

In the past the greatest attention has, naturally, been given to those materials which were commonly found in practice to be most affected by atmospheric pollution in the presence of water. Accordingly, stone and metals (particularly the ferrous metals) have received considerable study. However, others such as sacrificial coatings (paint, galvanizing, etc.), bricks and mortar may also be included. From the point of view of exposure, classifications were made according to the type of atmosphere. In descending order of severity these are given as Industrial, Urban, Coastal and Rural. On the basis of these classifications an indication of the probable performance of a material could be given. Few attempts have been made to classify atmospheres relative to building performance, that is, in terms of the precise degree of pollution which may be expected in any area. The reasons for this include:

1. The wide variation in the degree of pollution, including the substances present, which may occur in any locality. These variations are due to the interaction of many variable factors, including the effects of human activity and weather (Diagram D3.2/2).
2. The lack of systematic measurements over the whole country, that is, polluted as well as non-polluted areas. The great number of measurements in the UK made since 1933 with a variety of measuring devices have mostly been confined to polluted areas and essentially for local needs.
3. The lack of any systematic correlation between the measurements of airborne pollution and the performance of materials, particularly when water is present.
4. There has been no apparent need on a large scale for any greater degree of accuracy.

The gaps are being filled, albeit slowly, since the publication of the Beaver Report on Air Pollution, 1954. Measurement and collection of data was first undertaken by Warren Spring Laboratories. The Scientific Adviser to the former LCC and GLC has reported on work carried out by his department. In 1985 BRE commenced a site and laboratory study of the effects of acid deposition on building materials. The Department of the Environment publishes annually a Digest of environmental protection and water statistics.

(e) Impurities from fuel burning

In general, the greatest amount of air pollution in the UK results from the burning of fuel of various kinds. The impurities or pollutants, which include solids, liquids and gases, are chiefly smoke, grit and dust, sulphur dioxide, carbon monoxide, nitrogen oxides, chlorine and fluorine. These are dealt with separately below, but before doing so it is convenient to note the amounts of the pollutants which are discharged annually over the whole country. Smoke and sulphur dioxide are, in the first instance, the most important (the amounts of these are usually much higher than the others, while they are the major destructive agencies) and are accordingly the ones most commonly measured. They also form the basis of assessment of the degree of pollution in an area.

The graphs in Diagrams D3.2/3 and D3.2/4 illustrate how the emissions of smoke and sulphur dioxide have fallen in the UK since the early 1960s. The changes that have occurred over ten years in the

Diagram D3.2/3 *Fall since 1960 in the emission of smoke from coal combustion and average urban concentrations in the UK. (From DOE,* Digest of environmental protection and water services, *No. 9, HMSO, London, 1986, Fig. 1.1, p. 1, Crown Copyright)*

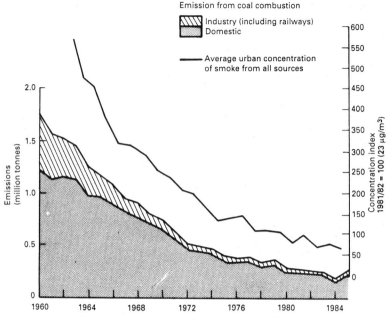

Diagram D3.2/4 *Change since 1960 in the emission of sulphur dioxide from fuel combustion and average urban concentrations. (From DOE,* Digest of environmental protection and water services, *No. 9, Fig. 1.2, p. 2, Crown Copyright)*

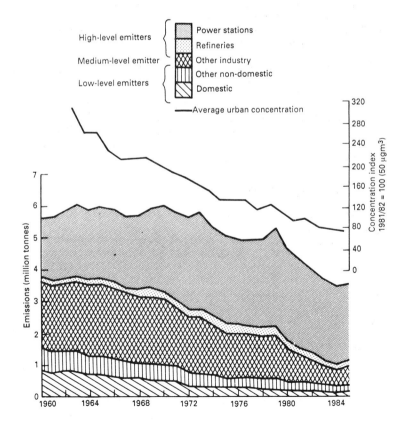

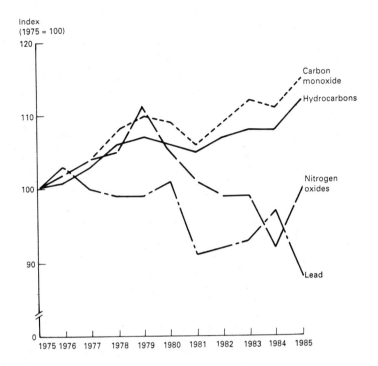

Diagram D3.2/5 *Estimated emission from petrol-engined road vehicles of hydrocarbons, carbon monoxide, nitrogen oxides and lead between 1975 and 1985. The graph is based on constant emission factors and does not take into account possible reductions in emission factors as a result of EC Directives. (From DOE*, Digest of environmental protection and water services, *No. 9, Fig. 1.3, p. 3, Crown Copyright)*

emissions of carbon monoxide and nitrogen oxides are illustrated in Diagram D3.2/5.

It is important to note that the amounts quoted in the graphs are only indicative of the *potential* of air pollution. The quantities are those *discharged* rather than those which are deposited at or near ground level. The concentrations of deposited matter are, in general, far more significant. In the context of the contamination of water, and taking into account tall buildings, this may not necessarily be the case. Conclusive evidence on this point is, however, still not generally available.

(1) Smoke
The term 'smoke' is applied to the visible products of imperfect combustion, and commonly seen in smoke trails, which may be short or long, from chimneys. The particles in smoke are extremely small (not to be confused with grit and dust, which consist of much larger particles) and behave like a gas. Thus smoke ultimately becomes so well mixed with air that it is invisible except as a bluish haze which obscures distant objects. Despite the mixing which takes place, smoke in this 'diluted' form still remains a potential cause of dirt and damage.

Coal contains a high proportion of carbon. Thus the finely divided particles resulting from its incomplete combustion consist of carbon or carboniferous matter, which, when viewed in bulk, are nearly black. In addition, coal also contains tarry hydrocarbons which add to the adhesive power of the particles and a tendency for the formation of *sooty deposits*. An important aspect of the finely divided nature of the particles is that they are supported by air and, unlike the heavier particles, do not fall readily to the ground. Thus they may be carried in currents of air, sometimes great distances from the source of emission, and eventually enter buildings via windows, cracks, vents, etc. with the incoming air. Whereas ordinary gas molecules will 'bounce off' the enclosing surfaces of an interior which they are continually bombarding, the smoke particles will, because of their adhesive qualities, remain behind. This then accounts for the progressive accumulation of dirt.

A similar procedure takes place on the external surfaces of buildings where, perforce, greater quantities of smoke are available. The resultant sooty deposits are not, unfortunately, readily removed by rainwater unless the material on which they have been deposited is soluble, as in the case of limestones. In addition to spoiling appearance, the sooty deposits are generally *hygroscopic* and are thus capable of holding fairly large amounts of water which, because of the chemical nature of the deposits, becomes corrosive. In addition to the tarry matter and carbon which causes the blackening effect, soot also carries with it free acids and soluble salts. (The action of acids and the effects of solubility of stones on decay and appearance are discussed in detail in *3.6 Chemical attack*, under 'Acids'.)

(2) Grit and dust

Grit and dust are emitted from furnaces whether or not there is visible smoke. Generally, the particles are large enough to fall under the influence of gravity. Ash, which is the unburnable solid material that is set free when fuel is burnt, mainly escapes with the flue gases, although some does remain behind. Some of the ash particles, it should be noted, may be so finely divided as to remain suspended in air. Unlike smoke particles, grit and dust tend to be deposited near the source of emission.

(3) Sulphur dioxide

Sulphur dioxide (SO_2) is released in considerable quantities as a result of the burning of fuels such as coal, coke and certain fuel oils which contain sulphur. Other sulphur compounds are also formed but in far smaller quantities. The percentage of sulphur, from which the sulphur dioxide is formed when combined with oxygen during combustion, in various types of fuels in common use is given in Chart C3.2/2. Only 80–90% of the sulphur in coal and coke is emitted from a chimney as sulphur dioxide, the remainder being retained in the ashes. In the case of oil, all the sulphur is emitted as sulphur dioxide.

The corrosive action of sulphur dioxide, itself soluble in water, is due to the formation of sulphurous or sulphuric acid. Some of the sulphur dioxide is removed from the air by solution in cloud droplets falling as

Chart C3.2/2 *Percentages (by weight) of sulphur in fuels in common use (1963 – average values for the UK). (Data from Notes on Air Pollution, No. 4, Pollution by sulphur dioxide, Warren Spring Laboratory (DOE), Crown Copyright)*

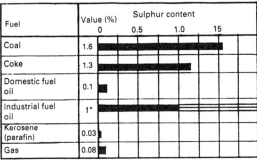

Fuel	Value (%)	Sulphur content			
		0	0.5	1.0	15
Coal	1.6	████████████████			
Coke	1.3	████████████			
Domestic fuel oil	0.1	█			
Industrial fuel oil	1*	════════════			
Kerosene (parafin)	0.03	▌			
Gas	0.08	█			

*Note: The percentages rises up to 4% in residual oil.

rain and in surface water. It is estimated that only one-fifth of the sulphur dioxide emitted into the atmosphere is brought down with rain; a further one-fifth is blown into the sea, while the remainder is mostly dissolved in water on buildings, soil and vegetation, after rain and at other times. (It is estimated that the average time for sulphur dioxide to remain in air is less than 12 hours.)

The formation of sulphuric acid as a result of rainwater wetting surfaces of buildings is of particular interest. (The process involved is, of course, also applicable to surfaces other than those of buildings.) Rainwater wetting the surfaces of buildings when combined with the sulphur dioxide present in the atmosphere forms a weak solution of sulphur dioxide. On evaporation a concentrated solution of sulphur dioxide is formed and when this is oxidized sulphuric acid results.

As already mentioned, the sooty deposits on surfaces are hygroscopic. Accordingly, when they are moist or wet they absorb sulphur dioxide from the polluted atmosphere. The absorbed sulphur dioxide is converted into sulphurous acid and then to sulphuric acid, which slowly becomes more concentrated on drying. The interaction of the sulphuric acid which will eventually flow down surfaces or be absorbed into materials may result in the formation of sulphates, chlorides and other substances which may cause destruction of many materials. These compounds also form powerful electrolytes which accelerate the corrosion of metals in general and the ferrous metals in particular.

(4) Carbon monoxide

Carbon monoxide is mainly associated with the burning of petroleum products, particularly in motor vehicles. The concentrations are usually high in busy streets. The black smoke from diesel engines is of an oily nature and dirt deposited on buildings is often difficult to remove. In general, carbon monoxide is more harmful to human health than to building materials.

(5) Nitrogen oxides

These are also released during fuel burning. The amounts emitted are small compared to sulphur dioxide. Work currently in progress may show that nitrogen oxides cannot necessarily be ignored when considering acid deposition.*

(6) Chlorine and fluorine compounds

These compounds are formed from the combustion of coal, the amounts involved being small (the concentrations are usually too small for chemical measurement) when compared with sulphur dioxide. However, when associated with water the compounds form hydrochloric acid (HCl), hydrogen fluoride (HF) and silicon tetrafluoride (SiF_4), all of which can harm materials.

(f) Dispersion – 1: Height of emission

In addition to the type of pollutants involved during emission and the velocity of the emission, the height above ground level from which the pollutants are emitted is of some importance when considering dispersion. In general, the higher the chimney (a source of emission), the greater the dispersion. However, other influences should be noted. Domestic chimneys, being low, do not disperse sulphur dioxide as effectively as the taller industrial ones. On the other hand, the quantities of fuel burnt industrially are so large that their contribution to ground-level concentrations of sulphur dioxide becomes extremely significant.

*The causes of the so-called 'greenhouse effect' due to the depletion of the ozone layer by carbon dioxide and gases such as CFCs are also being investigated by BRE. These investigations will be mostly concerned with thermal effects rather than with the degradation of materials by acid rain.

(1) Ash

(2) Combustible matter

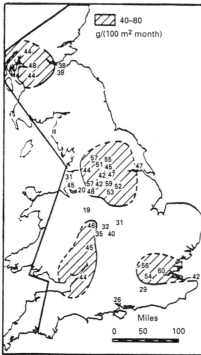

(3) Sulphates

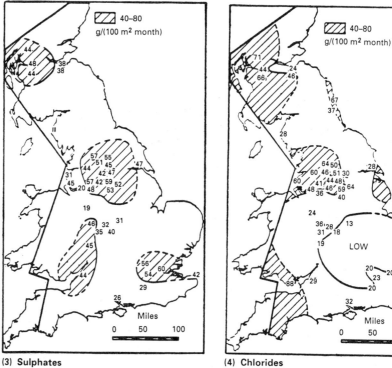

(4) Chlorides

Diagram D3.2/6
Distribution of insoluble and dissolved matter in country areas well away from large towns. Intended to illustrate the general trend. Some country areas may have higher concentrations than those suggested in the maps due to special local conditions (e.g. proximity of chemical plants). (From Meetham, Figs 87–90, p. 190)

(g) Dispersion – 2: Thermal and wind conditions

(1) Generally

Of the meteorological conditions which influence the dispersion of pollution by the atmosphere, thermal and wind conditions are probably the most important. The action of rain, for example, generally causes the pollutants and especially sulphur dioxide to be brought down to ground level in solution. Thermal and wind conditions are inevitably interrelated.

Although wind under normal conditions acts as an efficient mixer with consequential increases in pollution, the worse pollution of all arises when the air temperature increases in height and there is no wind. Under conditions such as these an *inversion* occurs, in which case the pollution remains near the level of emission. (When the air temperature of the surroundings decreases with height, dispersal is usually rapid, as there is no tendency for the pollution to accumulate.) Although inversions accumulated overnight may be destroyed the following day as a result of solar heat gain, the resultant mixing which takes place due to compensating up-and-down currents may cause the pollutions to be brought down to ground level in high concentrations. This is known as *fumigation*. Valleys, it may be noted, are usually particularly susceptible to inversions. In these areas fumigation is also a serious problem.

In general, wind is usually regarded as the agency by which pollution may be dispersed away from the source of emission. Although this may well decrease the concentrations of pollution near the source of emission, there is still the possibility of the pollution being carried distances up to 80 km away. This aspect is of particular importance when considering areas which may otherwise be regarded as comparatively free from pollution because they do not have immediate major sources of pollution. Topography or buildings around a chimney may influence significantly the degree of localized pollution during a wind. The relationship of the chimney to its surroundings may be such as to cause the pollution to be brought to the ground rapidly. The complex nature of the performance of a plume of smoke from a chimney and particularly relative to tall buildings requires that each case should be considered on its own merits. Help may be obtained from local authorities and others.

(2) Country areas

Despite the lack of measurements of the main pollutants in all areas, an attempt has been made to show the distribution of pollution, both insoluble and dissolved matter, in country areas well away from large towns or some miles to the west or south-west of the nearest source of pollution. The results of the attempts are shown in maps (1) to (4) in Diagram D3.2/6. The diagram should be read with the accompanying Chart C3.2/3 which gives average values for various types of district. It should be noted that the maps have been prepared without information

Chart C3.2/3 *Average values in g/100 m^3 per month of pollution deposited in the UK – compare with maps in Diagram D3.2/6. (Data from* Investigation of atmospheric pollution, *Warren Spring Laboratory, 31st Report (DSIR, 1959), Table 30, Crown Copyright)*

Type of district	Insoluble matter (g/100 m²/month)					Dissolved matter (g/100 mw/month)			
	Values (Combustible)	Values (Ash)	Value (Sulphates)	Values (Chlorides)
Country	35.28	50.96	50.96	35.28
Residential suburb	156.8	196.0	78.40	78.40
Central park	156.8	235.2	117.60	78.40
Industrial	274.4	470.4	117.60	78.40
Extreme industrial	1,1760 ×10	4312.0 ×10	431.20	156.80

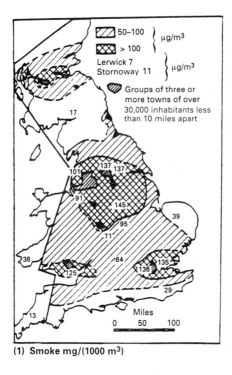

(1) Smoke mg/(1000 m³)

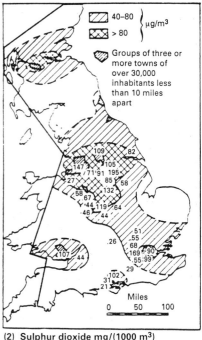

(2) Sulphur dioxide mg/(1000 m³)

Diagram D3.2/7 *Winter mean (1) smoke and (2) sulphur dioxide in country air (units: mg/1000 m³) – see text. (From Meetham, Figs 91 and 92, p. 190)*

for Wales, South-west England, Sussex, Kent, East Anglia and the Scottish Lowlands and Highlands. The distribution of chlorides are due more to sea spray than to chimneys. Despite the inherent limitations of the information contained in the maps, they do illustrate, in association with maps in Diagram D3.2/7, in which maps (1) and (2) show the winter mean smoke and sulphur dioxide in surface country air, the fact that pollution may be carried away from the source of emission to other areas, particularly the dissolved matter, as most of the insoluble matter is usually deposited near the point of emission.

(h) Distribution of pollution

(1) Generally
As already noted, the total amounts of the pollutants discharged over the country as a whole do not give any indication of the concentrations of the pollutants at ground level in any particular locality. Furthermore, the totals do not give any idea of the variations due to season or the fact that the dispersion of the pollutants, especially those which are finely divided or gaseous, from the point of emission is materially affected by many factors, including chimney height (or any other point of emission), the velocity and amount of emission, the velocity and direction of the wind and other meteorological and topographical conditions. It is not intended to discuss all these factors in detail, but to deal with some of these aspects. An indication of the variation in distribution in London during winter is given in Diagram D3.2/8.

(2) Seasonal variations
It is perhaps axiomatic that there will be a greater amount of pollution during the winter than during the summer in the UK. The range of variation of seasonal average pollution from place to place is given in

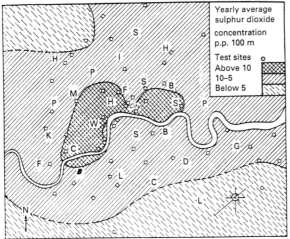

(1) Sulphur dioxide

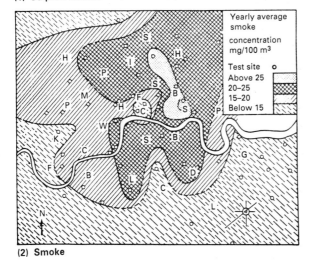

(2) Smoke

Diagram D3.2/8 *Yearly distribution of (1) sulphur dioxide and (2) smoke in London, 1957–1958. Although the amount of smoke now emitted is far less, the relative distribution of sulphur dioxide is similar to that shown. (From: Meetham, Figs 94 and 95, p. 190)*

Chart C3.2/3. From the point of view of the damage to building materials, the cumulative effect of pollution is, in general, of more significance than the seasonal variations or, for that matter, the daily variations, which can be considerable. However, the latter cannot be necessarily overlooked as they are indicative of the severity of conditions which prevail during a period when it is wetter than in the summer. An increase in the degree of severity of conditions is, in turn, indicative of possible accelerated destruction of materials.

(i) Impurities from other sources

Although fuel burning is always regarded as the main source of pollutants in general, other sources, essentially industrial, contribute significantly to what can best be called 'localized pollution'. In terms of the effects on buildings *near* the source of emission they cannot be discounted. In some cases the proximity may be of extreme importance in the selection and use of materials. The concentrations of any of the pollutants will be dependent on the same variables as those concerned with fuel burning. However, each case should be treated separately.

The main impurities from *industrial sources* are:

1. Paint particles and solvent odours as would be emitted, for example, from car factories.
2. Gases from chemical works resulting in oxides of sulphur; sulphuretted hydrogen; carbon bisulphide; nitrogen oxides; chlorine and hydrogen fluorides. Various other substances may also be involved. Industrial processes other than chemical works also produce corrosive substances. Thus, for instance, fluorine compounds are emitted from aluminium works and brick and pottery kilns. In the case of lime works, carbonates of calcium or magnesium are burnt through the agency of coal.
3. Grit and dust may also be emitted in large quantities from certain industrial processes. Cement works, among others, are notorious emitters of grit and dust. Demolition of buildings may also cause a local temporary nuisance.

Smoke control has only recently been possible since the passing of the Clean Air Act 1956. However, the Alkali, etc. Works Regulation Act 1906 has for some considerable time controlled the amounts of pollutants allowed to be emitted from industrial processes. This Act resulted from the uncontrolled escape into the atmosphere of hydrochloric acid, which caused damage to vegetation and property in the early days of the alkali industry, when sodium carbonate was made by the strong heating of sodium chloride and sulphuric acid (the Leblanc process). It may also be noted here that many corrosive substances may be discharged from industrial processes into rivers or underground watercourses*.

Sources

Lacy, R. E., *Climate and building in Britain*, BRE Report, HMSO, London, 1977

Meetham, A. R., *Atmospheric Pollution: its origins, sources and prevention*, 3rd ed., Pergamon Press, Oxford, 1964

Notes and reports of Warren Spring Laboratory (formerly in DSIR, now in DOE) on air pollution

BRE News of Construction research, April 1989, 'Building and the greenhouse effect'

Climatological Atlas of the British Isles, Metereological Office, HMSO, London, 1952

DOE, *Digest of Environmental protection and water services*, 9, HMSO, London, 1986

Addleson, Lyall, *Materials for Building*, Vol. 2, 3.02 *Exposure*, pp. 46–48, Iliffe Press, London, 1972 (o/p)

Addleson, Lyall, *Materials for Building*, Vol. 4, 4.02 *Exposure*, pp. 22–65 and 4.06 *Condensation*, pp. 124–142, Newnes-Butterworths, London, 1976

Rain

Walls above ground and roofs are the elements of the external fabric of the envelope of a building usually associated with exposure to rain and these are considered in this section. Of the many functional requirements of these elements, preventing the penetration of rainwater is one; changing the appearance of the external surfaces of which the elements are composed is another. Importantly, changes in appearance, implying alterations of either a physical or chemical nature (or both), may result in decay or deterioration.

*Lethal substances may also be released into water courses from disused mines or other underground workings. The disused Cornish tin mines are a case in point.

The effects of the failure to prevent the penetration of rainwater are inherently short term in that they manifest themselves soon after the completion of a building. Leaks and/or dampness of internal surfaces are the immediate effects; decay and/or deterioration of materials of which the element is composed usually takes longer to occur. In contrast, changes in appearance are inherently long term in that they usually take some time after the completion of a building to be seen. In both cases a number of variable factors need to be considered, and these are interrelated to some extent. However, as to determining their basic exposure, one approach appropriate to the short term is required for preventing penetration of rainwater and another appropriate to the long term for changing appearance. All the relevant factors and approaches are considered in this section.

1 The meaning of 'rain penetration'

It is normal to consider the problems associated with the exposure to rain as being related in a general way to the prevention of rainwater totally from the external fabric. However, in practice, and taking into account the variability of the properties of the various materials which are available to make up the respective elements of which the external fabric is composed, it is better and more realistic to consider the extent to which the penetration of rainwater should be *controlled*. The exposed surfaces, that is, those facing onto the rain, provide the lines of demarcation, so to speak, beyond which the penetration of water should be controlled. This does not in any way conflict with the reasons behind the necessity for the prevention of rain penetration, which are essentially aimed to maintain dry conditions, (1) for human health and (2) to control the deterioration, physical or chemical, of materials, surface and/or interior, used to make up the external fabric, or others which may be closely associated with it. (With reference to the last reason, it may also be noted here that dampness in certain materials often results in the undesirable reduction in their thermal properties.) None of these reasons requires that the whole thickness of the external fabric should necessarily remain dry at all times.

The extent to which water may be allowed to penetrate past the exposed surface of materials will be dependent on a number of factors, but one of the most important of these is concerned with the 'absorption/evaporation' performance of the materials. (The time element, that is, the period during which dampness may persist, is of primary importance.) Accordingly, a distinction should be drawn between (1) *the porous materials* which allow absorption and evaporation, although the quantities involved in each type of material are subject to wide variations, and (2) *the impervious materials* which, in the present context, do not absorb water. Thus the control of the penetration of rainwater into porous materials is necessary.

The necessity for the *use of joints* in virtually all materials, whether porous or impervious, means that special attention should be given to this aspect of an element as a whole. It is extremely common to find that control of penetration of rainwater has been defeated by 'leakage' through the joints (see *3.4 Exclusion* for coverage of joints and junctions).

2 The significance of the wind

The significance of the wind in connection with the formulation of the Driving Rain Index is explained generally in *3.1 General considerations,* Assessment of severity of conditions, 4 Examples (a) Rain penetration, p. 159 ff). This section includes a wider and more detailed coverage of

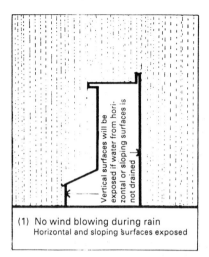
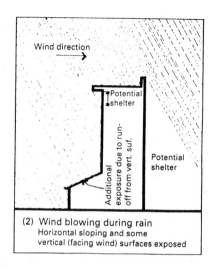

Vertical surfaces will be exposed if water from horizontal or sloping surfaces is not drained

(1) No wind blowing during rain
Horizontal and sloping surfaces exposed

Wind direction

Potential shelter

Additional exposure due to run-off from vert. suf.

Potential shelter

(2) Wind blowing during rain
Horizontal sloping and some vertical (facing wind) surfaces exposed

Diagram D3.2/9 *Representation of the basic effect of wind during rain relative to the exposure of horizontal, sloping and vertical surfaces using the section through a hypothetical building*

the relevant aspects. A differentiation is made between (1) fundamental effects and (2) special effects relative to walls and roofs. The consideration of special effects relates to the exposure to rain only. (Other effects on such aspects as structural loading, ventilation or smoke from chimneys are excluded – those related to fixings of roofs are covered in *2.2 Strength and the use of materials.*) Particular consideration is given to the way other buildings tend to change the direction and/or speed of the wind and how corners of buildings (in plan and section) create vortices. (A vortex produces a swirling action similar to that to be seen when water is running out of the outlet of a bath.)

(a) Fundamental effects

One of the fundamental effects of wind during rain is to change the direction of the fall of water. In building terms this implies the deflection of water so that it impinges on vertical surfaces. In the absence of any wind, rain would fall completely perpendicularly. Thus water would only impinge on horizontal or sloping surfaces, while vertical surfaces would be sheltered. A representation of this relative to a section through a hypothetical building is shown in (1) of Diagram D3.2/9. It should be noted that the vertical surfaces only remain sheltered if the water collected by the horizontal or sloping surfaces is adequately drained and therefore prevented from flowing over the vertical surfaces. The effect of wind blowing during rain is shown in the comparative diagram (2) from which it will be seen that in addition to the horizontal and sloping surfaces, the vertical surfaces facing the wind now receive rainwater. The diagram also serves to illustrate the fact that the *run-off* from vertical surfaces will increase the load, that is, the exposure, of any abutting horizontal or sloping surfaces, unless, of course, the run-off load is drained away separately.

(b) Special effects – generally

The diagrams used, it should be appreciated, are an oversimplification, but they do illustrate the principle involved. No account has been taken of wind speed and precise direction, nor the effect of the profile of the building on the 'shape' of the wind flow around the building. Windspeed is important in determining the basic exposure, as is

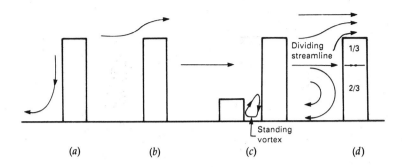

Diagram D3.2/10 *Wind and the effects of pressure changes on the wind flow patterns – see text. (From 'Wind flow over walls', by Tom Lawson, AJ Handbook,* Building environment, Technical study, Air movement 2, AJ, 27 November 1968, Fig. 1, *p. 1293)*

discussed later in '3 The basic exposure of a site'. The direction of the wind influences the faces of the building which will be exposed. In diagram (2) and assuming a rectangular building, two faces are likely to be exposed. Both wind speed and direction are likely to change either during a particular fall of rain or, more importantly, during rainfalls occurring at different times. In this, seasonal variations are important. Another factor which is likely to cause those surfaces which may be potentially sheltered (as shown in (2), for instance) to become exposed are the effects of the shape of the building on wind flow (see '4 Effects of building design' later). Such flow patterns may also be important when the rain has ceased (water may still be flowing down surfaces) but the wind is still blowing. The effects of the profile and other features of the external envelope of a building on the flow patterns over surfaces is considered in *3.9 Flow and changes in appearance*.

(c) Special effects – walls: Changes in direction/speed and vortices

When the wind acts normally on the face of an isolated building a pressure gradient down the exposed face of the building causes the air to flow down that face (Diagram D3.2/10(a)). For air near the top of the building it is easier for it to flow over the building than go around it (Diagram D3.2/10(b)). (It should be noted that it is this flow that increases the exposure of parapets not only to the effects of the wind but also to rainwater.) If there is a building in the way the air flows *up* the face of this building until it meets the main airstream and a standing vortex is created (Diagram D3.2/10(c)). In either case there is a dividing streamline which separates the flow going upwards and over the building and that going downwards to ground level (Diagram D3.2/10(d)).

As a general rule, the position of the dividing streamline is about two-thirds of the height of the building from the ground – details of the building and its surroundings will determine the precise position. In general, the upwards flow of the air (and rain with it) is less important with low buildings. Among other things, the speed of the upward flow is relatively slow. Low buildings may, of course, be sheltered by surrounding buildings and screens (such as boundary walls). However, the reverse is true of tall buildings, more so those with low buildings or other forms of shelter nearby. The detailing of joints and junctions needs special care.

The corners are particularly vulnerable to the scouring effect of the wind, the scouring being accentuated by grit in the atmosphere. (In simple terms the wind may be said to be 'turbulent at corners'.) The lower two-thirds of the wall are affected by a vortex at the corner (Diagram D3.2/11(a)). A similar action takes place at the top of the wall (i.e. at parapet level). Re-entrants in plan result in a vortex that travels up the wall (Diagram D3.2/11(b)).

Diagram D3.2/11 *Wind vortices at corners (re-entrants of walls) causing scouring – see text. (From 'Typical wind patterns', by Tom Lawson, AJ Handbook,* Building environment, Information sheet, *Air movement 4, AJ, 11 December 1968, Fig. 8, p. 142)*

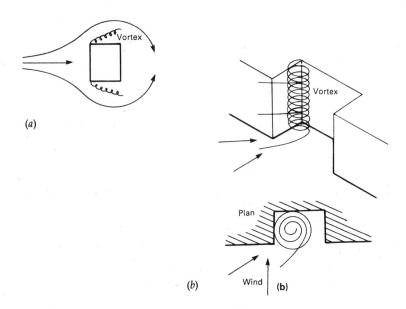

(a)

(b)   Wind   **(b)**

(d) Special effects – roofs: Changes in direction/speed and vortices

When the wind flow is normal to a *flat roof*, vortices are created near the edges of the roof. These are normal to roof edge (Diagram D3.2/12(a)). However, when the wind flow is at an angle, 'delta-wing' vortices are formed because of the corner that 'creates' them (Diagram D3.2/12(b)). The effect of the vortex is strongest at the corner creating it – the strength of the vortex decreases as it moves away from the corner. In general, the vortices are not strong and do not interfere with the flow of water over the surface towards the edges of the roof, as such water can get round the vortices. It should, however, be noted that in more complex cases, particularly with pitched and flat roof combinations, a vortex may not only prevent water on the surface of the roof flowing but may actually cause the water to be driven against its 'natural' flow. When this happens, details of joints and junctions (such as overlaps of flashings or welts in jointed metal roofs) capable of excluding the normal flow of water will be rendered completely ineffective. In these cases, an 'anti-vortex' detail is required.

Diagram D3.2/12 *Wind vortices at end of a flat roof with wind in two different directions. (a) Normal; (b) at an angle – see text. (From 'Typical wind patterns', by Tom Lawson, Fig. 10, p. 1426)*

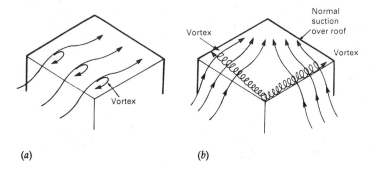

(a)   *(b)*

In the case of *pitched roofs*, wind flowing normal to the roof simply follows the shape of the slope (Diagram D3.2/13(a)). If the wind is at an angle to the roof the apex of the roof (a corner) creates a vortex (Diagram D3.2/13(b)). In practice, the vortex will have the effect of shedding more water in its zone of influence. Particular problems arise when pitched roofs are arranged in certain ways. For example:

1. Two parallel low buildings with pitched roofs (Diagram D3.2/13(c)). In this case the apex of one of the roofs initiates a vortex that flows down the open space of the buildings, giving rise to high air velocities and turbulence. If this open space is a gutter, then the vortex would impede the flow of water and put at risk the watertightness of any joints along its length. (The effect of the vortex diminishes if the width of the space between the buildings is greater than a quarter of the length of the building.)

2. A typical el-shaped pitched roof (Diagram D3.2/13(d)). The basic effect of the vortex initiated at the apex is to cause a local suction. There may be more water shedding or other effects on flow. The vortex may damage the roof covering if it is light in weight.

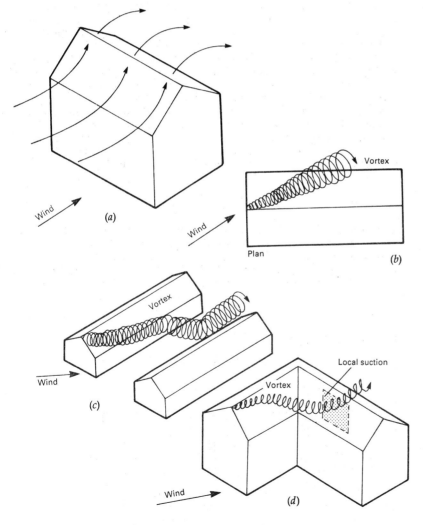

Diagram D3.2/13 *Wind vortices relative to pitched roofs. (a) Normal to roof edge; (b) at an angle; (c) effect of gap between two buildings; (d) 'knock-on' effect with pitched roofs at right-angles – see text. (From 'Typical wind patterns', by Tom Lawson, AJ Handbook,* Building environment, Information sheet, *Air movement 3 and 4, AJ, 11 December 1968, Figs. 11, 7 and 13, pp. 1426, 1429 and 1430)*

(e) Special effects – walls: run-off and splashing

The run-off from vertical surfaces will add to the load of water on any abutting horizontal or sloping surfaces, as indicated in Diagram D3.2/9(2). However, wind may assist in causing the water flowing over the sloping or horizontal surfaces to be 'whipped up', and in this way cause splashing of the vertical surface at or near the junction with the horizontal or sloping surfaces. Pavings abutting vertical surfaces at ground level must also be included in this consideration. Splashing does, therefore, tend to increase the exposure of parts of vertical surfaces.

(f) Special effects – roofs: uphill flow

The pressure exerted by wind is capable of supporting a column of water, the height of which is dependent on the wind speed. The relationship between wind speeds of from 5 to 30 m/s and the height of a water column which can be supported at various speeds in the range is given in Chart C3.2/4. From the values in this chart it is possible to calculate the size of lap or upstand necessary during rain accompanied by a particular wind speed. If, for example, the design condition is taken to be 20 m/s, then the laps in the pitched roof or the upstand joints in a flat roof would have to be of sufficient size to avoid a head of water of 26 mm to cause leakage. In other words, a head of water of more than 26 mm would have to be capable of moving upwards before leakage occurred. It is perhaps unnecessary to emphasize that the worst condition to be expected (including sudden variations which may take place over a short period of time) must be used as the design condition. For convenience, it may be noted here that the dimensions of laps and upstands are governed by tradition. However, there is some evidence to suggest that these may be excessive and therefore changed, in which case the future use of the values given in the chart, together with other available information may be fully exploited.

3 The basic exposure of a site

The current methods for assessing exposure to wind-driven rain from BSI are described in the Draft for development, DD93: 1984 (see *3.1 General considerations* under 'Assessment 4. Examples (a) Rain penetration', p. 159, for the background). In due course that Draft will be amended. In this section the principles on which the Draft is based are

Chart C3.2/4 *Relation between wind speed and height of water column which can be supported. (Data from* Principles of Modern Building, *Vol. 2, HMSO, London, 1961, p. 103, Crown Copyright)*

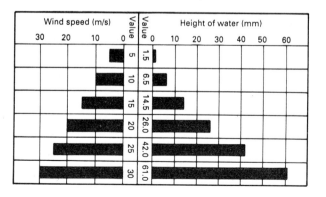

explained. Reference is made to the detailed worksheets described in the Draft and the results of worked examples are given to illustrate the principles. (It is likely that the final version of the draft will be published at about the same time as this book. Readers should therefore check on the progress of the Draft and use the latest version of it.)

(a) The methods and their uses

Two methods are described for assessing exposure of a building on a given site to exposure to wind-driven rain:

1. One is for assessing the resistance of a *wall* to water penetration or when designing windows. This method is based on the *local spell index*. This is a measure of the maximum intensity of wind-driven rain on a vertical surface of given orientation in a given period. A spell is of variable length and may include several periods of wind-driven rain interspersed with periods of up to 96 hours without appreciable wind-driven rain.
2. The other is for considering the average moisture content of exposed building materials or when assessing the likely growth of mosses and lichens (i.e. when considering or assessing probable changes in appearance of external surfaces of the envelope of a building). This method is based on the local annual index that has similarities with the BRE annual omnidirectional driving rain index (DRI).

Both indices are expressed in litres per square metre. The BRE driving rain index was expressed in square metres per second. (Prior compares the schemes given in BSI publications such as CP 121 and BS 5618 that used the BRE omnidirectional DRI with DD93.)

(b) The maps and rose values

To give comprehensive and detailed data for sites throughout the UK, the country has been divided into regions and sub-regions for driving rain using meteorological data from 50 Meteorological Stations (Prior's map in Diagram D3.2/14). In the Draft, maps of all the sub-regions are given and alongside each map the annual and spell rose values for each sub-region (simplified in Prior's map (Diagram D3.2/15), in which the rose values are referred to as 'average annual driving rain indices' and '1 in 3 year driving rain spell indices'). The scale for each rose value is logarithmic but note that the scales for each are different – see the relevant scales given with each index in the map.

A logarithmic scale has been used because the driving rain amounts expressed as square metres per second (product of rainfall and windspeed) have values as low as $1\,\mathrm{m^2/s}$ for the average DRI, and such low values could be significant. The scale chosen puts six units equal to $1\,\mathrm{m^2/s}$ and each additional six units corresponds to a multiplying factor of two. To obtain the spell index expressed as $\mathrm{l/m^2}$, the map values have to be converted by a factor. In the Draft, there are two tables to facilitate the conversions, one for the spell value (Table 1 in the Draft) and one for the annual value (Table 2 in the Draft).

Each sub-regional map shown gives the correction to be applied to the rose values. Thus in Diagram D3.2/15, for example, the values for Peterborough, PB1 would be as shown; for Cambridge, the PB1 values would be reduced by one unit; and for Boston they are all increased by one unit. In the same way, the values for PB2 would be used for Bedford and for Norwich the PB5 values, both increased by one unit.

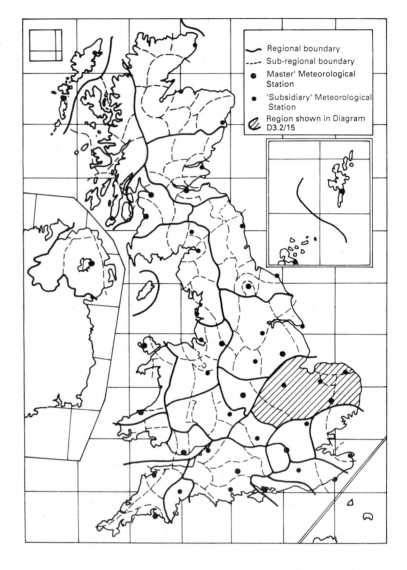

Diagram D3.2/14 *Regions and sub-regions for driving rain index in DD93. (From Prior, M. J., Directional driving rain indices for the United Kingdom – computation and mapping (Background to BSI Draft for Development DD93), BRE Report, 1985, Fig. 1, Building Research Establishment, Crown Copyright)*

Regional boundary
Sub-regional boundary
'Master' Meteorological Station
'Subsidiary' Meteorological Station
Region shown in Diagram D3.2/15

As each rose has wind directions for twelve different orientations of wall it is not possible to have a map showing all. Prior has illustrated in two comparative driving maps the annual driving rain indices for sub-regions (Diagram D3.2/17) and the spell index (Diagram D3.2/16), both for south- and north-facing walls. Using the technique of the BRE rain roses, Diagram D3.2/18 compares the annual and spell indices for all twelve orientations for selective locations.

(c) Corrections for local and other features

As with all wind data, it is necessary to correct the basic values given for the indices. The base, so to speak, for wind data is 10 m above unobstructed ground. In DD93 the corrections are given in tables.

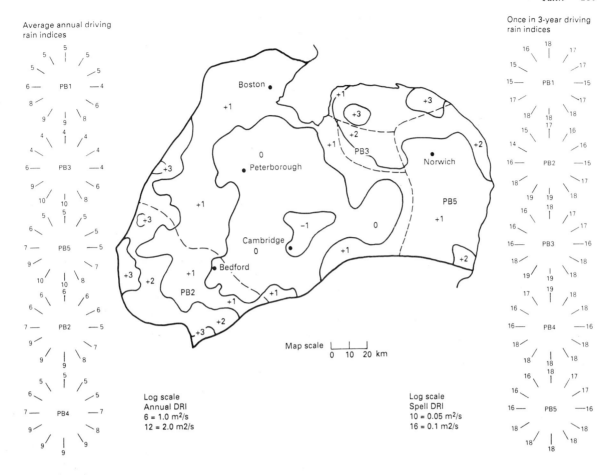

Average annual driving rain indices

Once in 3-year driving rain indices

Log scale
Annual DRI
6 = 1.0 m²/s
12 = 2.0 m²/s

Log scale
Spell DRI
10 = 0.05 m²/s
16 = 0.1 m²/s

Map scale
0 10 20 km

Diagram D3.2/15 *Driving rain indices for the Peterborough Region in DD93. (From Prior, Fig. 3, Building Research Establishment, Crown Copyright)*

Summarized, the corrections and the tables in which they occur in DD93 are:

1. The nature of the surroundings and the protection they give to the site, described in terms of *protection category* and given in terms of *protection height* (Table 3). The most exposed locations are those having long stretches of open level or nearly level country with no effective shelter which are within 8 km of a coast or large estuary. Consequently, they require no correction. Locations with similar terrain but more than 8 km from the coast or large estuaries are less exposed. The exposure of sites reduces as obstructions such as other buildings, walls, hedges and trees increases.
2. The roughness of the ground, *the roughness factor* (Table 4). This is related to the exposed height of the wall and the protection category (from Table 3).
3. The nature of the topography, *the topographical factor* (Table 5). Valleys or groupings of buildings can cause the wind to funnel, thereby increasing its exposure – by a factor of 1.2. Where valleys provide shelter for sites on steep sides of a valley, for example, a factor of 0.8 applies.
4. Values of the spell factor. The correction is related to the product of the roughness factor ((2) above) and the topography factor ((3) above) (Table 6).

Diagram D3.2/16 *Once in 3-year spells for P = 1.0 for sub-regions: lowest values for (a) south-facing walls and (b) north-facing walls (logarithmic scale). (From Prior, Figs 8 and 9, Building Research Establishment, Crown Copyright)*

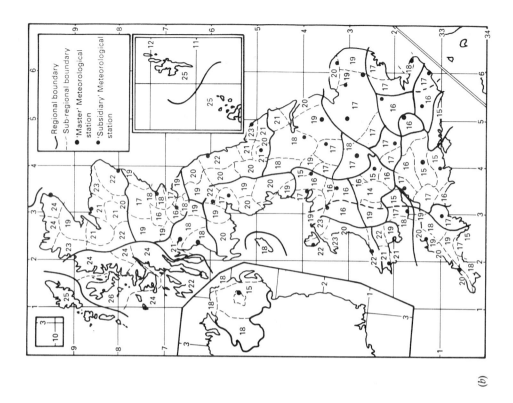

(b)

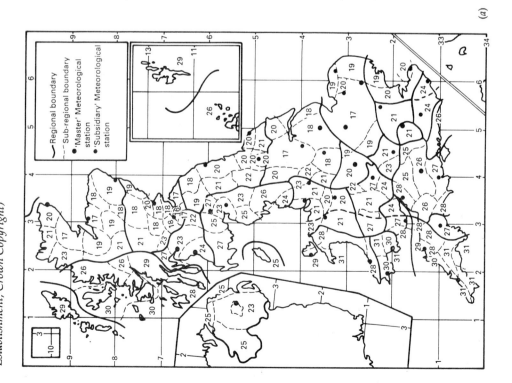

(a)

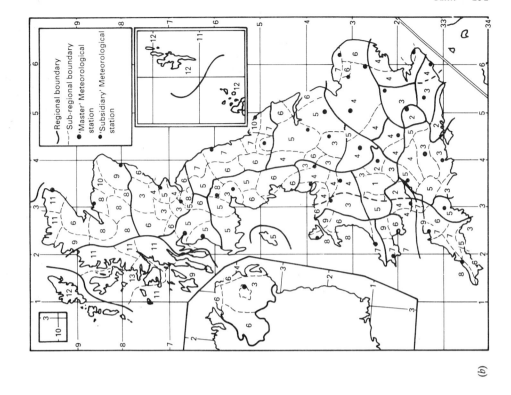

(b)

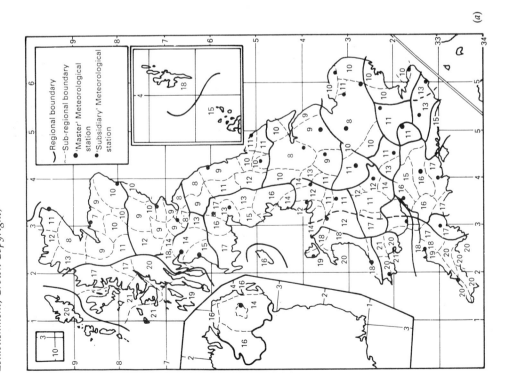

(a)

Diagram D3.2/17 *Average annual driving rain indices for sub-regions: lowest values for (a) south-facing walls and (b) north-facing walls (logarithmic units). (From Prior, Figs 4 and 5, Building Research Establishment, Crown Copyright)*

Diagram D3.2/18 *Comparison of annual and spell driving indices (for DD93) for selective locations in the UK using the BRE 'rain rose' technique. (Values from DD93)*

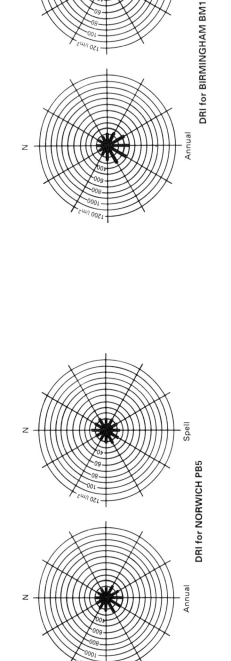

DRI for NORWICH PB5

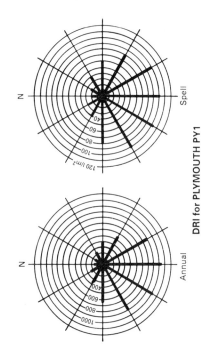

DRI for BIRMINGHAM BM1

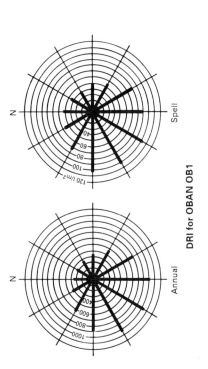

DRI for PLYMOUTH PY1

DRI for OBAN OB1

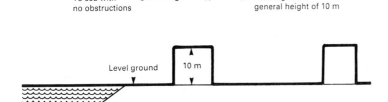

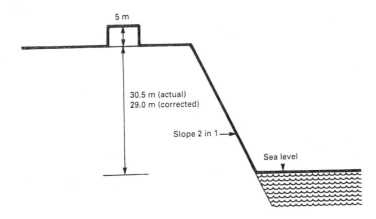

Diagram D3.2/19 *Application of DD93: section for effect of orientation – see text*

5. The height of the local ground. Three different types of slopes are given in Figure 5; one for an escarpment, one for a cliff and one for a hill. In each case, the equation for calculating the ground-correction height along parts of the slope are given. The type of slope therefore affects the value of the correction in particular cases.

(d) Effects of changes in orientation, protection and topography

The four examples that follow are abstracted from DD93 and illustrate the effects of changes:

1. *Effect of orientation (Diagram D3.2/19)*
   Location: Hove, Sussex, on seafront, map sub-region PH1.
   North- and south-facing walls, both 10 m high. Level site to the north and south – topographical factor is 1.0. To the north the general height of buildings is 10 m – protection category is 3 and protection height is 10 m. To the south and the sea there are no obstructions – protection category is 0 and protection height is 0 m. The assessed indices (l/m^3) are:

	Annual	Spell
North-facing wall	86.8	9.4
South-facing wall	717.5	79.9

2. *Effect of abrupt change in local ground height* (Diagram D3.2/20):
   Location: near Swanage, Dorset, on the coast, map sub-region SY3. Wall facing south-east, 5 m high. Level site near the top of an escarpment, 30.5 m above sea level, with slope down to the sea approximately 2 in 1. No obstructions to the sea – topographical factor is 1.0.

Diagram D3.2/20 *Application of DD93: section for abrupt change in local ground height – see text*

Diagram D3.2/21 *Application of DD93: section for steeply sloping escarpment – see text. (From DD93, Appendix A)*

A correction is needed for the local ground height. This is assessed to be 29.0 (i.e. 1.5 less than the actual height of 30.5 m). As there are no obstructions looking towards the sea, the protection category for that direction is 0. The assessed indices (l/m^2), for the site described above and (for comparison) if the site was level as in example (1) earlier:

	Annual	Spell
With escarpment	616.5	77.6
On level ground	481.5	56.5

The effect of the abrupt change in ground height is therefore significant.

3. *Effect of a steeply sloping escarpment* (Diagram D3.2/21)

Location: Truleigh Hill, Sussex, map sub-region GK1. North-facing wall, 7.5 m high, at different locations on the escarpment.

Corrections are required for the ground height, in accordance with the equations for an escarpment, and the exposed height of the wall for each location. The ground height corrections are: 0 m for locations 1 to 3, 80.5 m for location 4 and 95.0 m for locations 5 and 6. The assessed indices (l/m^2) for each location are:

	Annual	Spell
Location 1	169.2	22.4
2	169.2	22.4
3	169.2	22.4
4	266.4	38.8
5	284.4	43.6
6	284.4	43.6

Relatively small increases in the steepness of the ground for the corrected ground height/exposed wall height have no significant effect. The first significant change occurs at location 4 – the slope between location 3 and 4 is steep, thus increasing its relative height significantly. Similarly, the relative height of location 5 is increased sufficiently to increase the indices, whereas the relative height of location 6 is not increased sufficiently to change the indices.

4. *Effect of a complex cliff profile* (Diagram D3.2/22)

Location: St Lawrence, Isle of Wight, Map reference SY2. North-facing wall, 10 m high, at different locations of a cliff. The cliff faces the open sea.

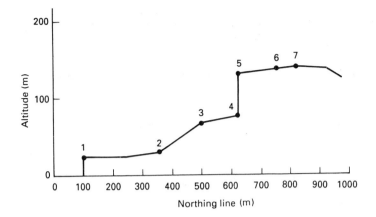

Diagram D3.2/22 *Application of DD93: section for effect of a complex cliff profile – see text. (From DD93, Appendix B)*

As with example (3), corrections are required for the ground height for each location but in this case for a cliff. The corrected heights are: 23.0 m for location 1; 4.0 m for location 2; 29.0 m for location 3; 23.5 m for location 4; 78.5 m for locations 5 and 6; and 65.0 m for location 7. The assessed indices (l/m²) are:

	Annual	Spell
Location 1	1233	138.8
2	1080	112.3
3	1233	138.3
4	1233	138.3
5	1404	155.4
6	1404	155.4
7	1350	137.3

The reduction in exposure between locations 1 and 2 reflects the reduction in the corrected ground height and similarly between locations 3 and 4 and locations 6 and 7. Increases in the corrected ground height are reflected in increases in exposure between locations 2 and 3 and locations 4 and 5. There is no change between locations 5 and 6. The reductions are indicative of the sheltering effect to be expected on a cliff; and conversely for the increases. For example, the location receives the full brunt of the wind, the slope between locations 1 and 2 having little effect, whereas the strength of the wind is reduced by the slope behind it (i.e. that between locations 2 and 3). The same applies to locations 3 and 4.

Sources

BRE
Principles of Modern Building, Vol. I, 3rd edn, 1959 and Vol. II, 1961, HMSO, London
Lacy, R. E., *Climate and Building in Britain*, BRE Report, HMSO, London, 1977
Prior, M. J. *Directional driving rain indices for the United Kingdom – computation and mapping*, (background to BSI Draft for development DD93), BRE Report, 1985
Digest 127, *An index of exposure to driving rain*, revised 1971
Digest 236, *Cavity insulation*, April 1980, minor revisions 1984

BSI
DD93: 1984, *Draft for Development, Methods for assessing exposure to wind-driven rain*, 30 March 1984

BS 8208: Part 1: 1985, *British Standard Guide to Assessment of suitability of external cavity walls for filling with thermal insulants. Part 1. Existing traditional cavity construction*, 31 July 1985

PSA
PSA Feedback Digest 50, *Brickwork rain and trouble*, Construction, pp. 2–3, Spring 1985

Others
ACBA, BDA, C&CA, Eurosil-UK, NCIA and STA, *Cavity insulated walls*, Practice Note 1, September 1984
ACBA, BPF, BDA, C&CA, Eurosil-UK and NCIA, *Cavity insulated walls, Specifiers guide*, January 1987
BBA and others, *Cavity insulation of masonry walls – dampness risks and how to minimise them*, November 1983

Ground

The elements or parts of a building which are exposed to groundwater* are essentially those which are either (1) *placed in*, that is, buried, in the ground and consisting of foundations (all types), lower parts of walls and columns (generally from foundation level to dpc level), retaining walls (in basements, on sloping sites, in manholes, etc.), or service components such as pipes, conduits, cables, etc., or (2) those in *close contact* with the ground, such as floor slabs. In all cases any form of direct connection between any of the elements with other parts of the building must be taken into account. Although suspended timber floors occurring immediately above ground level should not be connected directly or indirectly with the ground itself, they may nevertheless be exposed to groundwater in the vapour state. Such exposure occurs if there is inadequate ventilation to the outside air of the plenum space between the ground and the underside of the floor and may, in the long term, be sufficient to cause serious rotting of the timber.

Although the elements which are exposed to groundwater are limited, a number of conflicting reasons make quantitative assessments of exposure extremely difficult. Accordingly, it is only possible to indicate, in general terms, the relationship between the factors involved and upon which assessments have to be made.

1 Significance

Unlike most other conditions of exposure, exposure to groundwater is not only perennial but also lasts throughout the whole life of the building. Exposure to groundwater is, therefore, the most severe condition encountered in building when compared to other exposure conditions, particularly when the importance of most of the elements involved in exposure to groundwater are taken into account.

The severity of conditions is dependent not only on the fact that water is, in general, always present, but also on the fact that the water is usually polluted to some extent. In many cases the knowledge that the water, or the ground for that matter (which in this context can usually be regarded as damp), is polluted will dictate the precautions which must be taken in the selection and use of materials.

*The term 'groundwater' must, in the present context, include not only 'free-flowing' water but also water contained in or held between the particles of any substance making up the ground as a whole. The latter is perhaps more general, but at the same time the former cannot be excluded. In this section both are implied unless the term is specifically qualified.

2 Variable factors

Although the precise quantities of water together with the degree of pollution which may constitute the worst condition in any given situation would assist in the selection of materials to be associated with groundwater, there are a number of variable factors involved which make this extremely difficult. Of these, the following are the most important:

1. Variations in the composition of the substances making up the ground, in which the differing water-transfer or water-holding capacities of different types of soil and rocks are significant;
2. Variations in the actual source of the groundwater, that is, whether entirely from local rain or whether derived from other source or sources which could be near or far from a given site (see *3.1 General considerations*, 'Water cycle', p. 137);
3. Climatic conditions affecting rainfall and evaporation.

Experience of local conditions and the results of site surveys (the latter always being advisable) may help in determining the effect of most of these factors on the actual exposure condition which may be anticipated. In general, it may be noted in connection with soil types that the cohesive soils, such as clays, are always likely to transfer and/or hold considerably more water than the non-cohesive soils, such as the gravels, sands, etc. The latter generally permit fairly rapid drainage to lower depths of, for examples rainwater, although they are capable of holding water in the interstices between the particles. (A description of the main classes of soil types is included in *2.1 Cracking*, p. 71.)

3 Pressure

Special mention should be made of 'free-flowing' groundwater which may be encountered on a site during a survey or the dig for foundations. Groundwater such as this usually implies that the source is at some distance from the site, in either a horizontal or downwards direction. In this the water table is significant. If the 'flow' is 'dammed up' by the construction of the building there is the possibility of pressure being exerted on those elements causing the obstruction. In most cases pressure build-up, which may be overcome by diverting the watercourse or by adequate pumping, is significant because of its structural implications. However, it does also accelerate the rate at which groundwater may be transferred through porous materials associated with the ground.

4 Capillarity and vapour

The manner in which the groundwater in close contact with the relevant elements of a building is depleted and subsequently replenished (a continual process) is, as already noted, complex. However, transfer of water into and through materials by means of capillarity plays an extremely important part. In this, cognisance has to be taken of the fact that most of the structural elements referred to earlier are composed of porous materials, all of which, in varying degrees, are capable of 'encouraging' water transfer, either horizontally or vertically, by capillarity. Some of the materials used, and concrete is a notable example, may be highly resistant to water transfer in the liquid state even under pressure, but they may not be sufficiently resistant to the transfer of water in its vapour state. In all cases the control of the transfer of water, in any of its states, is effected at a predetermined level (or levels) by the incorporation of an impervious layer of some kind or another (dpcs and membranes). Subject to conditions, the 'route' of such a layer may be quite complex. One of its main requirements is that it should be continuous in order to avoid local leakages – **continuity**.

It may be useful here to quote an example which should illustrate the importance of transfer of groundwater in the liquid state, particularly in floor slabs. Where concrete ground-floor slabs have been laid on hardcore, without the incorporation of a damp-proof membrane, on what were considered to be dry soils, dampness has often occurred underneath essentially vapour-resistant floor finishes such as linoleum. In these cases vapour has permeated through the concrete only to condense when it has come into contact with the vapour-resistant material, which has acted as a damp-proof membrane. This unfortunately has been 'incorporated' in the wrong position.

5 Pollutants

(a) Generally

In any consideration of the effects of groundwater on materials buried in or in close proximity to the ground, account should be taken of the wide variations that may exist from site to site (or even within a site) as to the nature, distribution and degree of pollution. Consequently, each case should be treated separately. An analysis of soil conditions for both source and pollution of groundwater is always advisable.

Water plays two important roles in the pollution of groundwater:

1. It transfers pollutants in solution comparatively easily. Hence the source of the water and the pollutants are interrelated.
2. As rainwater, it is one of the primary agents which result in the pollutants found in groundwater. Free carbon dioxide or sulphur dioxide are the more important corrosive agents that are present in rainwater. These, together with others, react with constituents of rocks, soils, etc. resulting in the groundwater containing a variety of chemical compounds. Those that are important include: carbonates, sulphates, nitrates and chlorides of calcium, magnesium, potassium or sodium. Groundwater may also be corrosive due to its acidity (see below), and this should also be taken into account.

(b) Specific problems

The pollution of groundwater is more commonly related to the three specific problems explained below.

(1) Sulphate attack
Sulphate attack is confined mainly to cement-based products. In solution sulphate salts cause considerable expansion and loss of strength. The sulphates of magnesium (Epsom salt) and sodium (Glauber's salt) are the chief destructive agents; calcium sulphate (gypsum or selenite), also harmful, is less soluble. Although clay and other soils may contain any or all of the sulphates, it must be noted that these may also be derived from other materials used, directly or indirectly, in association with the ground, such as bricks or hardcore. The major areas of the UK known chemically and geologically to contain sulphates are indicated in Diagram D3.2/23.

Diagram D3.2/23 *Sulphate conditions in the UK.* Note: *The shaded areas of the map indicate major areas known chemically and geologically to contain sulphates in the soil or water. Unshaded portions, however, are not necessarily free from sulphates: local conditions must be taken into account and each site examined accordingly. (From map produced by the Cement Marketing Company Ltd)*

In conditions other than those encountered relative to the ground, sulphate attack may be controlled by ensuring that the elements or materials concerned do not remain damp for long periods of time. Other factors are also relevant, but these are discussed more fully in *3.6 Chemical attack*. Apart from damp-proofing, the solution to the problem may, subject to the concentration of sulphates, lie in the suitable selection and use of cements, such as sulphate-resisting, blastfurnace slag, or high alumina, which are specifically made to resist or otherwise have special characteristics which resist sulphate attack. The degree of resistance provided by the various special cements available covers a fairly wide range. In addition, the quality of the product is significant.

In the case of concrete work, greater resistance is *generally* provided by precast rather than cast-*in-situ* members.

The severity of conditions can be obtained from the concentration of sulphur trioxide either in *groundwater* (measured in grams SO_3 per litre of water) or in soil (measured in percentage SO_3 or as grams SO_3 per litre of water extract). BRE's classification of sulphate soil conditions affecting concrete is given in Table T3.2/1. The related precautionary measures have been omitted here but are included as appropriate in *3.6 Chemical attack* under 'Sulphates', p. 412 ff. The classification is intended to act as a *guide*.

(2) Efflorescence

On the whole, the crystallization of soluble salts on the surfaces of materials such as brick are not destructive. As in the case of sulphate attack, salts may be derived from the materials themselves. However, the presence of nitrates and chlorides is usually indicative of some contamination from groundwater. Although the dpc is the 'line' above which there should be no transfer of water, the use of unsuitable materials or the physical bridging of the dpc very often allows the soluble salts to be eventually deposited in uncontrollable positions, thus in some ways negating one of the principal reasons for using a dpc (see *3.5 Efflorescence*, p. 364, for examples).

(3) Acidity

The metals are the main materials to be affected by the acidity of the ground in which they may be buried or with which they may be in close

Table T3.2/1   *BRE classification of soils and groundwater according to concentrations of sulphates (SO_3)*

Class	In soil		In groundwater (g/l)
	Total SO_3 (%)	SO_3 in 2:1 water:soil extract (g/l)	
1	Less than 0.2	Less than 1.0	Less than 0.3
2	0.2–0.5	1.0–1.9	0.3–1.2
3	0.5–1.0	1.9–3.1	1.2–2.5
4	1.0–2.0	3.1–5.6	2.5–5.0
5	Over 2	Over 5.6	Over 5.0

See Table T3.6/3 for related requirements for concrete.
From BRE Digest 250, *Concrete in sulphate-bearing soils and groundwaters*, minor revisions 1986, Building Research Establishment, Crown Copyright.

contact. The acidity or alkalinity of water or soils is measured by the pH value; a value below 7 indicates that the water or soil is acid while that above 7 that either is alkaline. The method of measurement and a broad classification of both acidity and alkalinity relative to metals in particular is included in full in 'Water supply', p. 203 ff.

In addition to metals, lime and particularly cement may be adversely affected by acids. As in the case of sulphate attack, the most successful means of reducing acid attack is by choice of suitable cements relative to particular conditions. Although emphasis has been given to the possibility of the deterioration of the concrete itself, due regard must be taken of the fact that reinforcing steel is vulnerable to the corrosive effects of acids and other substances which occur in the ground (see *3.7 Corrosion*, p. 482). Adequate protection of the steel is important and in this, quality control of the concrete itself is extremely so.

Sources

BRE
Principles of Modern Building, 3rd edn, Vol. 1, HMSO, London, 1959
Digest 89, *Sulphate attack on brickwork*, revised 1971.
Digest 245, *Rising damp in walls: diagnosis and treatment*, revised 1981
Digest 250, *Concrete in sulphate-bearing soils and groundwaters*, Minor revisions, 1986

During construction

1 Background

It is perhaps all too obvious that water must be introduced during construction involving all the wet-trade processes in brick and block laying, concreting, plastering, screeding, some painting, etc. However, the quantities of water necessary to make any of these processes chemically and physically possible in the first instance are, in a sense, a nuisance after any of the operations concerned have been completed. If the water remains behind too long it can also cause inconvenience by way of delay. Under winter conditions, too much water in the elements may, under certain conditions, lead to frost action and related deterioration. For a number of different reasons, therefore, drying out commensurate with other requirements, such as avoidance of cracking (see *2.1 Cracking*, p. 56), should be accomplished as soon as possible. Where the drying-out time cannot be speeded up sufficiently it is sometimes possible to use forms of construction which will allow drying out to proceed after a particular operation has been completed, and thus reduce delay. In winter, special precautions are necessary.

2 Drying and wetting

Part of the problem of controlled drying out is better resolved under controlled factory conditions, usually off the site, as happens with large panel systems and many components such as lintels, sills, floor planks, etc. However, wherever the drying out is to take place it should be properly controlled. In many cases units or elements which may have been properly dried out are then left unprotected, either during storage prior to use or while actually in position in a building under construction, only to be rewetted during rain or by careless spilling during cleaning and other operations. These all aggravate an already difficult and time-consuming task. Materials such as timber, timber products or other fibrous products, not normally associated with the water-bound problems of the wet-trade processes are, however, also highly susceptible to moisture content changes. They, too, should not be left exposed to rain or other sources of water both on or off the site.

3 Quantities used

It is not always appreciated that fairly large quantities of water are, in fact, using during the construction of most buildings. Despite the increased use of 'dry' forms of construction, it has been considered advisable in this book to give some indication of the probable quantities of water which may be used with the wet-trade processes. Table T3.2/2 sets out the quantities of water introduced during construction. From this it will be seen that in a traditional three-bedroom house, for example, it is estimated that between 6364 and 8183 litres are actually used in the construction. This is equivalent to about 6364 and 8183 kg of

Table T3.2/2   *Estimated quantities of water introduced during construction*

Material/construction		Quantity of water (l)
Description	Unit	
Brickwork, average[a]	m^3	Up to 178.38
Brickwork, average, 229 mm thick[a1]	m^2	Up to 40.8
Brickwork, average, 114 mm thick[a1]	m^2	Up to 20.4
Brickwork, average[a1]	Per brick	Up to 0.38
Concrete, average[a]	m^3	Up to 178.38
Concrete, 1:2:4 mix, water/cement ratio 0.45[b]	m^3	140.63
Concrete, 1:2:4 mix, water/cement ratio 0.65[b]	m^3	200.00
Plaster undercoat, 6.4 mm thick[a]	m^2	About 2.72–4.08
Plaster scim coat[a]	m^2	About 1.36
Screed, lightweight aggregate or aerated cement, 76.2 mm thick[c]	m^2	About 5.44–10.87
Traditional three-bedroom house[d]	Complete	About 6364–8183
2 tower blocks of flats, each 13 storeys high (combined floor area = 969 m^2 combined cube = 716 761 m^3)	Complete	About 3.705×10^6

References:
[a] National Building Studies, Bulletin No. 2, *Painting new plaster and cement*, HMSO, London, 1948. *Important note*: Quantities are only rough estimates.
[a1] Calculated from (a).
[b] Calculated from values given in *Report on concrete practice*, Part one, table 12, p. 56 (Cement & Concrete Association), 1963.
[c] *Developments in roofing*, BRS Digest (2nd Series), No. 51, p. 1, October 1964.
[d] From the BRS. *Note*: Total quantity used, i.e. including cleaning, spilling etc. estimated to be between 8092 and 22 730 l.
[e] From Widnell and Trollope, Surveyors, London. *Note*: The total quantity of water used on the job, i.e. including waste, was metered as 7.410×10^6 l. The figure of 3.705×10^6 l quoted in the table assumes 100% waste (see (d) above).

water or a weight of between 62.4 and 80.2 kN. These figures, it should be noted, do not include the quantities of water spilt, used for cleaning (cement mixers, etc.) or a variety of other site uses. The total quantity including these contingencies is estimated in the example quoted to be between 8092 and 22 730 litres. There is, therefore, a considerable amount of 'waste' water, the destination of which cannot be realistically accounted for.

4 Drying-out time

An indication of the significance of some of the quantities of water introduced during construction relative to drying-out times might be given by taking an extreme example of a screed or concrete slab laid on an impervious membrane as would occur in a ground-floor slab. It is estimated that, under normal conditions, the rate of drying out would be at least 1 month per 25.4 mm.* Thus, a 50.8 mm screed would take two months to dry out; a 152.4 mm combined screed and concrete slab six months.

On the other hand, the importance of protection from rainfall in order not to delay drying out can be obtained by considering the fact that 13 mm of rain is equivalent to 13 l/m². In the case of a lightweight concrete screed (aggregate or aerated cement) to a flat roof it is estimated that only a small proportion of this quantity would be shed, except in heavy showers. In an experiment carried out at the BRS† it was shown that an unprotected dry aerated concrete screed, 76.2 mm thick, laid to a fall of 1:60 picked up 9.24 l/m² of rainwater in four wet days. Of the 9.24 l/m², only 1.08 l/m² could be drained off. In eight days of excellent summer weather half of the remaining water, that is, 4.08 l/m², dried out while the screed was protected from rain and the surface freely ventilated. (The eight summer days are equivalent to about 48 normal winter days.) For complete drying out, 36 good summer or 180 winter days were needed.‡

5 Value of drying-out guidance

The rate of drying out is dependent on a number of factors, but the guides now available are useful in that they are indicative of the period that should reasonably be left before it is worthwhile actually measuring the degree of dampness with measuring devices (see *3.1 Dampness*, p. 152). On the other hand, they may indicate when an alternative form of construction or special precautions are necessary. It may also be noted here that greater attention is now being paid to moisture contents than hitherto, and consequently more quantitative information may be expected.

6 Pollutants

Pollutants derived from the materials used in many of the wet-trade processes must always be taken into account, especially when considering following trades. Cement is notable, for instance, for producing alkaline solutions which may have deleterious effects on paints and adhesives. In brickwork efflorescence may occur during drying out which may cause plaster or paint to deteriorate.

Sources

Principles of Modern Building, 3rd edn, Vol. 1, HMSO, London, 1959

Damp-proofing solid floors, BRS Digest (2nd series) No. 54, p. 4, January, 1965, 'Drying out time'. *Note:* The exact metric equivalents of the Imperial units used by BRS have been retained. Nowadays the values would be rounded to 25, 50 and 150 mm.

†*Developments in roofing*, BRS Digest (2nd series) No. 51, p. 1, October, 1964, 'Drying wet screeds'. See note above for Digest 54.

‡Methods of providing ventilation so as to reduce the time before the impervious roofing is laid and to avoid 'entrapped' water problems are now being used. These are described in the digest referred to above and are also included in *3.4 Exclusion*.

Water supply

1 Relevance and scope

Water as supplied by a water authority or obtained from other 'private' sources such as wells, rain-collecting tanks, etc. is required in buildings for a number of different reasons, among which are included human consumption (cooking, drinking, etc.); cleanliness (ablutions, etc.); comfort (heating, cooling, etc.); and industrial and other processes. All the water required is contained, distributed and eventually carried away by means of some form of piped system. In addition to pipes, many other component parts are also used.

Despite the number and diversity of pipes and components which may be involved in the piped system as a whole, the problems of exposure are at least confined. On the other hand, the surfaces of the materials used (metals or plastics) are *always* exposed and, in a sense, are subjected to the severest conditions of exposure found in a building. More importantly, the permanency of the exposure means that in the final analysis the severity of exposure in individual cases is determined by the degree to which the water is polluted (i.e. as far as materials are concerned) as supplied. Further pollution may take place during the distribution of the water. It is also necessary to take into account the nature of pollutants to be found in 'wastes' drained away from a building.

Consequently, emphasis is given here on pollution relative to the piped system in a building. However, the explanations of the nature and derivation of the pollutants, together with the descriptions of terms commonly used, can be applied to aspects of the building process in which water is used. Those processes in which lime and cement are utilized are of particular interest. Water for use with either should be fit for drinking or taken from an approved source. In short, the water must be free from such impurities as suspended solids, organic matter and dissolved salts. (A method for checking the effect of any given water on concrete is described in BS 3148, *Tests of water for making concrete*.)

Despite the widespread use of plastics for piping and other components in a piped system, classifications of the suitability (better, reliability) of metals relative to particular water supplies persist. Generally, plastics have been found so far to be more resistant to the deleterious effects of pollutants in water supplies than metals. The latter have proved on the whole to be weak in resisting the corrosive agencies in water supplies. As yet, there is no authoritative source that has included plastics in a meaningful or helpful way. Consequently, in this book the existing classifications related to metals are used.

2 Substances in the water

(a) Purpose of treatment

There is a fairly wide range of substances to be found in all water supplies. Treatment of 'raw' water by a water authority is primarily directed at meeting human needs (health and taste), rather than the specific requirements of the durability of materials in general and metals in particular. Some consideration is, however, given to the latter. During the treatment processes (and there are many of them) substances other than those occurring in the natural water prior to treatment may be added; on the other hand, some substances may be reduced in concentration or removed completely. Thus, there is likely to be a wide variety in the 'quality' of water which can be expected in any particular area of the country. The range becomes even wider if those supplies which make direct use of 'raw' water from wells, catchment tanks and underground sources are included. It is, therefore, essential that the precise chemical nature of any water supply, whether from a water authority or other source, be ascertained prior to the selection of materials to be used in association with the water supply. The quality of any particular water supply serves as a useful, albeit not conclusive, guide on the probable behaviour of metals most commonly used.

(b) Contamination during distribution

When dealing with the use of materials relative to a particular water supply it is also necessary to give consideration to the possibility of the water becoming contaminated during its passage through the piped system. This may be particularly significant in composite systems, that is, those in which more than one material is employed, even if each of the materials is resistant to the corrosive agencies contained in the basic water supply when used in isolation. In a composite system traces of one of the materials (usually metals) may provide the required corrosive agency to deteriorate one of the other materials. (This particular point, which is primarily associated with the *juxtaposition* of metals of different kinds, is dealt with in greater detail in *3.7 Corrosion*.) Contamination of the water supply by the component parts of a piped system do, therefore, increase the range of the quality of water which can be anticipated. Thus it is necessary to distinguish between the quality of water as supplied (ascertained from an analysis) and the quality after it has passed through the piped supply system (this has to be anticipated from known facts). Where necessary, the quality of the water as supplied is referred to, in this section, as the 'direct water supply', that is, the water which reaches the commencement of the piped system of a building; the quality of water after it has passed through the piped system is referred to as the 'indirect water supply'. (Pollution of water which discharges into wastes and drains is better included under 'Processes and cleansing agents'.)

(c) Groups of water

Although there is a wide variation in the substances to be found in water supplies, together with the variable factors involved which cause this variation, water supplies may be classified into three broad groups according to their degree of (1) hardness, (2) acidity and (3) alkalinity. (Acidity and alkalinity are more conveniently considered under the heading of pH value.) The main aspects of these three groupings are outlined here, with particular emphasis on the substances involved and methods of measurement. Some indication is also given in Table T3.2/3 of the significance of some of the substances found in water. Only those substances which normally affect the metals used in water supply systems are included. Where possible, some reference is made to those broad geological areas in the UK in which waters of a particular class are likely to be encountered.

(d) Using local experience

Finally, the limitations of any broad classifications of water supplies, particularly in relation to the effects of such supplies on metals, must be

Table T3.2/3 *Significance of some substances found in water*

Substance	Source/Cause	Occurrence	Significance
1. Aluminium	1. Use of aluminium products, e.g. cooking utensils, tanks and pipes 2. Excess use of aluminium sulphate in treatment works	1. Seldom found in natural water 2. Should not occur in direct water supply (limit of aluminium 0.15 ppm) 3. May occur in indirect water supply due to source/cause (1)	1. Corrosion of aluminium products from copper-bearing waters 2. Occasionally alkaline waters or water containing sodium carbonate cause corrosion of aluminium – pitting takes place
2. Calcium	Action of free carbon dioxide, CO_2, a weak acid in rainwater, on calcium strata	1. Present as (a) bicarbonates (causing temporary hardness) and (b) chlorides and sulphates (causing permanent hardness) 2. Waters from chalk or limestone areas	1. Cause of hardness 2. Calcium bicarbonate one of the most usual causes of alkalinity Notes: 1. Calcium carbonate is insoluble in water, i.e. settles out or held in suspension 2. Calcium bicarbonate (often present in large quantities) is soluble and predominant cause of hardness
3. Carbon dioxide	Normal constituent of atmosphere: usually mixed with rainwater	Often present in large quantities in waters from iron-bearing formations and in moderate quantities from moorland and shallow well waters	1. Significance when present as free CO_2, i.e. uncombined with other substances such as calcium or magnesium 2. One of the main causes of acidity of water 3. Waters containing more than 20 ppm of free CO_2 are likely to be corrosive Note: The free CO_2 content of water is significantly reduced by (1) the presence of calcium which helps to absorb the CO_2 or (2) aeration of water – a simple means of reducing the CO_2 content
4. Copper	1. Copper products, e.g. piping, tanks, etc. 2. Use of copper sulphate to reduce algae growth in reservoirs	1. Seldom found in natural waters 2. May occur in direct water supply; more likely to occur in indirect water supply due to source/cause (1)	Causes corrosion, pitting of aluminium or galvanized products, e.g. pipes, tanks, cisterns, etc.
5. Iron	1. Underground formations 2. Use of iron piping	Frequently found in raw waters	1. Often corrosive 2. Brown staining of sinks, baths, basins, laundry.

Table T3.2/3   *(Continued)*

Substance	Source/cause	Occurrence	Significance
			(On exposure to air, water takes up oxygen; iron may precipitate and the deposits cause the brown stains)
6. Lead	Lead piping in distribution systems by the action of plumbo-solvent waters	Acid waters; frequently waters from upland and moorland gathering grounds; some well waters	1. Lead is a cumulative poison – not more than 0.1 ppm should be present 2. Each water authority should check its water supply for plumbo-solvency and forbid the use of lead piping if necessary
7. Magnesium	Action of free carbon dioxide, CO_2, a weak acid in rainwater on magnesium strata	Present as bicarbonates (causing temporary hardness) or chlorides and sulphates (causing permanent hardness)	Most waters which are hard contain a proportion of magnesium compounds, but this is usually less than the proportion of calcium compounds
8. Potassium	Sulphate: indirectly from sulphur content of polluted atmospheres	In 'sulphated water' together with sodium and magnesium sulphate	Potassium sulphate, a salt, usually only significant compound in water supply. (Potassium nitrate usually associated with sewage pollution)
9. Sodium	Chloride: pollution by sea water or sewage	Some salt found in most 'pure' waters (quantities of 10–20 ppm)	–
	Sulphate: indirectly from sulphur content of polluted atmospheres	In 'sulphated water'; sodium sulphate chief constituent	–
10. Zinc	Use of galvanized iron piping and tanks	1. Rarely found in natural waters 2. Direct water supply from hard chalk waters carried in galvanized piping – up to 3.0 ppm of zinc in solution 3. In indirect water supply 4. Collection of drinking water from galvanized iron roofs for isolated supplies	1. Some waters readily take up zinc, especially in cooking processes 2. Formation of loose deposits of zinc carbonate as a result of action of hard chalk waters on galvanized products

From Twort, A. C., *A Textbook of Water Supply*, Edward Arnold, London, 1963, Part I – 'The significance of various substances found in water', pp. 153–165.

Definitions: 'Direct water supply' used under 'Occurrence' means water as supplied by a water authority to a building.
   'Indirect water supply' used under 'Occurrence' means water in the distribution system of a building after it has been taken from the water authority's main. Thus the effect of the materials of tanks and piping is taken into account.

emphasized. None of these classifications can take into account variations due to local conditions. Accordingly, advantage should always be taken of local experience; where necessary, as in the case of new supplies, suitable tests for corrosive effects should be carried out.

3 Hardness

The term 'hardness' when associated with water is, in some senses, indicative of the 'feel' of the water. However, the term is more commonly associated with the soap-consuming power of water. Soap reacts with certain minerals in water. This reaction causes a precipitation which appears as a scum or curd on the surface of the water. No lather can be formed until all the minerals have reacted with the soap. Waters which behave in this way are said to be 'hard'. (As synthetic detergents do not behave like soap, that is, they do not cause a precipitation, the inconvenience of hard water is not felt so keenly.) In addition to the problem of soap lather, precipitation of the hardness compounds causes a hard scale on the surfaces of boilers and the interior of pipes in heating systems. This has been put forward as a further reason against the use of hard waters.

(a) Measurement

Water may then be classified according to its degree of hardness, covering a range from soft to hard. The measure of the hardness can be obtained by the 'soap-consuming power' of a water, but this may not, however, be very precise. A more accurate method would be to measure the total of the hardness producing constituents. Use of the latter method has resolved itself into the expression of these constituents in the chemically equivalent amount of calcium carbonate ($CaCO_3$). The amount of calcium carbonate may be measured in different ways, two of which are:

1. Grains of calcium carbonate per gallon of water. These have been termed degrees of hardness or, more precisely, 'degrees Clarke'.
2. Parts of calcium carbonate per million parts (ppm) of water. This method of measurement appears to be more generally favoured than the previous one. (As grains per gallon = parts per 70 000, a hardness of 100 ppm = 7 degrees Clarke.*)

The degree of hardness as experienced by the consumer in relation to the ppm equivalent amount of calcium carbonate is given in Table T3.2/4. The table also includes, for convenience, degrees Clarke, calculated from the formula given in (2) above. Table T3.2/5 indicates the usual range of natural hardness of raw waters.

(b) Types of hardness

There are two types of hardness: *temporary* and *permanent*. Temporary hardness is destroyed by boiling while permanent hardness is unaffected by boiling. The compounds causing either type of hardness are given in Table T3.2/6. From this it will be seen that calcium and magnesium are the main cause of hardness, with the bicarbonates causing temporary hardness, and the chlorides and sulphates causing permanent hardness.

In the case of temporary hardness the bicarbonates may decompose slowly at ordinary temperatures. The rate is much more rapid when the water is heated. Decomposition results in the formation of carbonates

*As 'degrees Clarke' is not measured in SI its use is likely to be completely discontinued. However, it has been considered useful to retain these values in the tables.

which are insoluble and deposited on surrounding surfaces – boilers, pipes, etc. Although the coating or scale so formed may result in some protection of metals, waters of moderate temporary hardness may contain, in appreciable amounts, other additional substances such as sodium bicarbonate, sulphate and chloride which may pit galvanized products, especially hot water storage tanks. In general, the loss in performance due to the reduction in the bore of pipes, for instance, is probably far more significant than the liability of the system to corrosion. The latter, of course, cannot be overlooked.

Table T3.2/4   *Classification of water hardness*

Description	Quantitative limits (equivalent amount of $CaCO_3$)	
	Measured in parts per million (ppm)	*Measured in degrees Clark*[a] *(grains/gallon)*
Soft	0–50	0.0–3.5
Moderately soft	50–100	3.5–7.0
Slightly hard	100–150	7.0–10.5
Moderately hard	150–250	10.5–17.5
Hard	over 250	over 17.5

From Twort, *Hardness*, pp. 158–159, for description and related ppm measurements.

[a] Values of *degrees Clarke*, a common method of measurement in the past, have been added for convenience from 7 degrees Clarke = 100 ppm. (Degrees Clarke are grains/gallon or parts per 70 000.)

Table T3.2/5   *Usual range of natural hardness of raw waters*

Source of water	Range		Classification[a]
	ppm	*degrees Clarke*[a]	
Moorland catchments	10–50	0.7–3.5	Soft
Shallow clays and gravels	Varying	Varying	Varying
Sandstones	100–300	7.0–21.0	Moderately soft–hard
Chalk	200–300	14.0–21.0	Moderately hard–hard
Oolitic and carboniferous limestones	200–400	14.0–28.0	Moderately hard–hard
Magnesium limestones	400–500	14.0–35.0	Hard

From Twort, Table 14, p. 212.

[a] 'Degrees Clarke' and 'Classification' have been added for convenience.

Table T3.2/6 *Substances causing hardness*

Temporary hardness	Permanent hardness
Calcium bicarbonate. Ca(HCO₃)	Calcium carbonate. CaCO₃ (to a limited extent)
Magnesium bicarbonate. Mg(HCO₃)	Calcium sulphate. CaSO₄ Calcium chloride. CaCl₂ Magnesium sulphate. MgSO₄ Magnesium chloride. MgCl₂

From Twort, Table 13, p. 211.

The largest portion of the hardness in the majority of waters is temporary. However, a high percentage of permanent hardness is usually the cause of water being excessively hard.

In general, and as can be seen from Table T3.2/5, most of the hard waters in the UK are to be found in chalk and limestone areas. Both calcium and magnesium are constituents of chalk and limestone. The hardness compounds are primarily formed when rainwater containing free carbon dioxide (forming a weak acid) percolates underground through the chalk or limestone. (See also *3.1 General considerations*, 'Weathering – limestone', p. 146.) The deeper the water is drawn from underground, the harder it is likely to be. On the other hand, waters are always soft when they are obtained from surface run-off from impermeable catchments, where there is little or no contact with chalk or limestone. (Soft waters should always be checked for plumbo-solvency.) It may be noted that although calcium and magnesium bicarbonate both cause temporary hardness, it is calcium bicarbonate which is often present in large quantities in water and accordingly is regarded as the predominant cause of hardness.

(c) Removal of hardness

Removal of hardness by a water authority may be done by (1) chemical means, in which case the hardness compounds are removed, or (2) 'base-exchange' method – the hardness compounds are changed. Hardness may also be removed by the consumer as part of the distribution system of a building. Various methods are available.

4 pH value (acidity and alkalinity)*

The pH value is a measure of the acidity or alkalinity of water (also applicable to soils).

(a) Measurement

Pure water, H_2O, also consists of a relatively small number of positive hydrogen ions, H^+, and a relatively small number of negative hydroxyl ions, OH^-. The concentration of hydrogen ions in 1 litre of pure water is 1/10 million ($1/10^7$ or 10^{-7}) of a gram, which is found by electrical measurement. For convenience, the pH value is expressed as:

$$\text{pH value} = \log\frac{1}{1/10^7} \text{ or } \log\frac{1}{10^{-7}}$$

$$= 7 \text{ for pure water}$$

*Twort, A. C., *A Textbook of Water Supply*, Edward Arnold, London, 1963, pp. 153–154 and 161–162.

Generally, when the concentration of hydrogen ions *exceeds* the concentration of hydroxyl ions, the solution has *acid* characteristics, and the pH value of such a solution would be *below* 7. On the other hand, waters are described as *alkaline* when the pH value is *above* 7, that is, there is an excess of hydroxyl ions.

Neutral waters, implying non-corrosive qualities, are generally those with a pH value within the range 6.80–7.20.

Natural waters seldom have an acidity below a pH value of 5.50; a range of 6.00–7.00 is generally applicable to softwater from moorland areas, although the value may be lower when peat is present. For comparative purposes it may be useful to note that hydrochloric acid has a pH value of 3. Alkalinity is often associated with waters which have much carbonate hardness but with little free carbon dioxide. These waters may have a pH value of 8 or over. A strong alkali such as caustic soda, when dissolved in water to form an alkaline solution, has a pH value of 14.

(b) Causes of acidity

There are three main causes of acidity:

1. *Free carbon dioxide (CO_2)*. This excludes any carbon dioxide which may be combined with other substances, such as the carbonates. The carbon dioxide is picked up from the atmosphere during rain. Most natural waters contain some carbon dioxide.
2. *Decomposing organic matter*. The origin of the organic matter may be vegetable, but peat is of particular significance. A number of different acids are produced. Most *moorland* waters are acid.
3. *Sulphates*. These are formed either as a result of the action of rainfall in polluted atmospheres (sulphur dioxide content important) or during the transfer of groundwater through soils of high sulphate content, such as some clays and some iron-bearing formations. Sulphuric acid is normally produced in the water.

Most acid waters are corrosive (see 'Corrosive quality' below).

(c) Causes of alkalinity

Alkalinity is caused mainly by the presence of bicarbonates, carbonates and hydroxides of calcium, magnesium, potassium and sodium.

Calcium and magnesium bicarbonate are the most usual constituents of waters which are alkaline. The same constituents are also one of the primary causes of temporary hardness. Thus, when alkalinity and hardness figures are equal, temporary hardness is known as *carbonate hardness*. On the other hand, where the total hardness figure is less than the alkalinity, then the presence of potassium and sodium salts are indicated.

(d) Corrosive quality

Although there are many variable factors to be considered in assessing the corrosive quality of water relative to metals commonly used in connection with water supplies, the following three characteristics of water are normally indicative of the possibility of consequential corrosion:

1. A low pH value (that is, acidity)
2. A high free carbon dioxide content – in itself a cause of acidity;
3. An absence of temporary hardness or alkalinity.

Waters, frequently corrosive, are normally derived from the following sources: moorland waters; shallow well waters of low pH value, little temporary hardness but much permanent hardness; water from iron-bearing formations; chalk and limestone waters having a high CO_2 content; waters from greensands; waters from coal measures; waters having a high oxygen content or a high chloride content.

From the point of view of health, it is always imperative that water is checked for its plumbo-solvency. Lead is a cumulative poison and, above all, is, when in solution, essentially colourless and thus difficult to detect visually. Where a water is plumbo-solvent, then it is normal for the use of all lead piping to be prohibited in any distribution system.

Solution of building materials

1 Context

In this context solution of building materials is intended to imply the 'leaching' of pollutants from a material as water flows over it. This form of pollution is mainly related to the problems associated with the *juxtaposition* of materials of different kinds (see Photographs P3.2/1–P3.2/3). Although the results of the incorrect juxtaposition of materials are usually manifest externally, in which rainwater as a source is important, they may also occur within the thickness of a structure or internally from rainwater, condensation, maintenance or usage. Accordingly, the range of pollutants and their corrosive quality is wide. Each situation must be considered on its own merits. Some indication may, however, be given of the materials which are likely to provide sources of pollution.

2 Sources and effects

Traditionally, stonework (particularly Portland stone) is notable for the staining and sometimes deteriorating effects of lime washing on brickwork and other types of stone such as sandstone. Lime washing may also originate from cement-based products (see *3.5 Efflorescence*, p. 360 and p. 361). The alkaline nature of these products is also capable of causing leaching of oils from bituminous products. When wet, cement and lime have a corrosive effect on some of the non-ferrous metals, such as lead and aluminium (see *3.7 Corrosion*, p. 484).

On the whole, the surfaces of most metals are dissolved to some extent. Copper and copper alloys are notorious for green staining of

Photograph P3.2/1 *Lime washing from Portland stone onto brickwork below*

Photograph P3.2/2 *Lime washing from concrete onto brickwork below during construction because of lack of adequate protection of the brickwork*

Photograph P3.2/3 *Rust staining from corrosive products of ferrous railings onto mosaic facing*

many materials. This only occurs when the water first flows over the metal. The oxidized surfaces of metals are also capable of being dissolved and later deposited over other materials. Rust staining is a notable example (see *3.7 Corrosion*, p. 480).

Finally, certain types of timber, such as oak and western red cedar, are capable under damp conditions of exuding acids which may be harmful to metals (see *3.7 Corrosion*, Table T3.7/1, p. 485).

Faulty services

1 Generally

Water, which may leak from any of the component parts of the piped system of the water supply *or* from pipes, gutters and other component parts used to drain rainwater from horizontal or sloping surfaces, constitutes conditions of exposure which may seem, at first, to be completely fortuitous.

However, this does not detract from the fact that a condition of exposure exists when a leak has occurred. The severity of the exposure will be dependent on the position in which the leak takes place, the rate of leakage and the speed with which the requisite repair is carried out. Subject to the relationship between these factors, faulty services may be a major cause of dampness. This means, then, that some serious consideration must be given at the design stage of the manner in which such a contingency is to be dealt with during the life of a building – **creative pessimism**. This involves basically the relationship between materials selected and maintenance. It may be noted here that pipes, for instance, which may be buried within the fabric or under the structure of a building, present special cases, particularly as far as effects of leaks and replacement of parts are concerned.

2 Design faults

Faulty services are also associated with faults in design. In this, the detailed positioning, size and projection of overflow pipes, which are in any case specifically incorporated into a piped system to deal with faults in some of the water-containing components, are important. A number of examples are to be found in buildings where this aspect has not received the attention it deserves. The result is invariably a change in appearance of materials underneath the overflows caused by local concentrations of water (see Photograph P3.2/4).

The severity of exposure conditions resulting from faulty services may be significantly increased by pollutants. Services as such are, of course, liable to contain a variety of these.

Maintenance and user requirements

1 Generally

Maintenance, especially regular cleaning of surfaces of materials in buildings, may involve the use of water. Similarly, there are a number of user requirements, such as occur in kitchens, bathrooms, swimming baths, and many industrial and other processes, which necessitate the use of water. In all cases the quantities of water and the relationship between these and the materials concerned are subject to wide

Photograph P3.2/4 *An example of the intense disfigurement of brickwork because of wrongly sloping overflow*

variation. However, what is significant is the fact that the continual or occasional presence of water is, or should be, known in advance. This is important, and of some considerable advantage in the final selection and use of materials to meet particular conditions. However, it should be noted that a careful analysis is required of the conditions which are likely to prevail, with particular attention paid to the possibility of pollutants being involved in any of the processes concerned.

2 Pollution

(a) Generally

As in the case of solution of building materials, there is an extremely wide range of pollutants which may be derived from various processes and the use of cleansing agents. Each particular condition must, therefore, be considered on its own merits. However difficult it may be to establish the precise degree of severity of conditions which are likely to prevail, cognisance must be made of any circumstances which are likely to lead to 'special' conditions so that suitable precautionary measures may be taken.

(b) Industrial processes

The most vulnerable processes are usually those associated with various types of industry in which highly corrosive acids or alkalis may be produced. Many materials are affected by either of these. However, some mention might be made of the care required in the selection and use of metals and cement-based products. Both are important as they are common materials for either carrying the corrosive agents or for the enclosing structure and general fabric. Mention has already been made of the effects of acids on cement. It should also be noted that some industrial processes may provide a source for sulphate attack from spillage or splashing of materials containing sulphates.

(c) Wastes

Waste water may be highly contaminated in buildings other than industrial, such as hospitals, laboratories, etc. A certain amount of contamination may also occur in waste water from domestic and other similarly apparently harmless sources. The latter do not, on the whole, constitute severe conditions of exposure. Nevertheless, some consideration must be given to the suitability of materials which may be used.*

(d) Building processes

Cleaning of stonework, particularly sedimentary stones, is most effectively, albeit more laboriously, carried out by using clean water. The use of alkaline detergents, although quicker, often leads to subsequent efflorescence or decay of the stone.

(e) Cleansing materials

General cleaning of surfaces of building materials also requires special consideration if cleansing agents are to be used. Many of these may contain substances which may be corrosive relative to a particular material; on the other hand, the inclusion of abrasives may have erosive effects.

*The note on the suitability of materials for wastes and drains, even in those circumstances when the water to be carried away is unlikely to be severely contaminated, has been included in order to emphasize the importance of checking the likely performance of the newer materials. This performance may then be usefully compared with the materials traditionally used.

Entrapped moisture

1 Scope

Entrapped moisture (sometimes also known as 'entrapped water') is a relatively new problem and describes water, usually as vapour, held in constructions, often for long periods. Accordingly, entrapped moisture is a source of water *within* constructions – a hidden source. With the other sources considered earlier, water attempts to get into a construction; with entrapped moisture it attempts to get out. In trying to make this transfer, the resistance that materials within a construction offer to water in both vapour and liquid are significant. Importantly, most vapour control layers are also impervious to liquid water. All this means that the problem of reducing or eliminating entrapped moisture has to be considered and resolved differently from moisture from an external source.

The problem was first noticed and investigated in flat-roof constructions, particularly those having lightweight cement-based insulating screeds. However, moisture may also become entrapped in constructions for pitched roofs and walls, especially in those having substantial amounts of thermal insulation. In many cases a vapour control layer, used intentionally or not, may be present. The problems associated with entrapped moisture can be related directly to the substantial increase of thermal insulation in constructions.

In some (isolated) cases where there has been no specific provision for moisture to escape from within the construction, the entrapped moisture may eventually escape but only after a protracted period of drying out that may involve years rather than months after the completion of a building. In flat roofs, for example, it is more common for water to leak out of the construction as liquid at weak points such as cracks, rainwater outlets or electrical conduits. While the moisture remains entrapped there is the risk that materials within the construction may deteriorate. At the very least, the entrapped moisture reduces the thermal insulation properties of many thermal insulation materials; it is likely to cause organic materials to decay.

The principles involved are clear; it is the practical details (i.e. the rules and precautions) that need to be developed further. Authoritative quantitative data (or any method) for a realistic assessment of the severity of conditions and thus the particular solutions that should be adopted are still lacking.

As explained in *Materials for Building*, Vol. 2, under 'Entrapped water', p. 156, much of the theoretical and practical groundwork was carried out in Scandinavia, other parts of Europe and in America, where conditions are somewhat less severe and less hostile from those in the UK.* Nevertheless, the experience gained in other countries provides a useful basis for this study.

Additionally, minds have been concentrated on the problems in flat roofs. Consequently, it is the latter that are emphasized in this section.

2 Sources of the moisture

There are five possible sources of the moisture; two during construction and three afterwards:

(a) During construction

1. Water used in mixing and cleaning and spillage generally;
2. Rainwater – including, importantly, materials *en route* to and on-site until fixed.

*In most other countries in Continental Europe or North America temperatures during the winter may be much lower than those in the UK but they are consistently lower. In contrast, in the UK temperatures and moisture may fluctuate considerably during the winter, varying from moderately warm/moist to bitterly cold/dry, sometimes over a short period of time. Also, the wind might be intense at times. Accordingly, this makes the UK climate more severe and more hostile than in Continental countries. The difference has important consequences.

(b) After construction

1. Interstitial condensation;
2. Thermal pumping by which water is 'sucked' into a construction; and
3. Water penetration.

All sources of water are important because the problems associated with entrapped moisture are not only dynamic but are also applicable to both wet (inherently the most severe) and dry forms of construction. Failure to recognize the significance of all sources of water can lead to a misunderstanding of the necessity for adopting certain rules or taking precautions in particular circumstances.

3 Reasons for the entrapment

(a) Generally

The reasons why water becomes entrapped are complex. In the main, the entrapped moisture has difficulty in finding its way to the free surface of a construction so that evaporation and eventual drying out can take place. The location of impervious layers of material (e.g. waterproof membranes) or of materials providing high resistance to moisture vapour transfer (and then acting as impervious layers) influences significantly the extent to which moisture can escape from the construction (Diagram D3.2/24(a)).

The transmission of moisture from the interior of a construction to its free surface is influenced primarily by the structure of the materials within the construction (see *3.3 Moisture content*, '1 Capillarity (a) Drying out of a porous body', p. 230). However, thermal and vapour

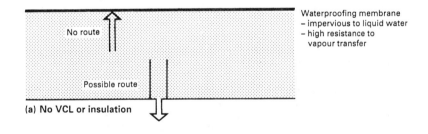

Waterproofing membrane
– impervious to liquid water
– high resistance to vapour transfer

No route

Possible route

(a) No VCL or insulation

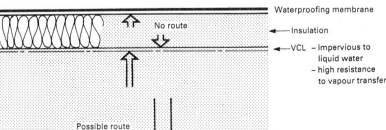

Waterproofing membrane

Insulation

VCL – Impervious to liquid water
– high resistance to vapour transfer

No route

Possible route

(b) With VCL and insulation

Diagram D3.2/24 *Possible routes for the transfer of moisture in liquid and vapour forms in a flat roof (a) without insulation and (b) with insulation and a vapour control layer*

pressure effects, the related gradients in particular, are also significant. These tend to reduce the rate of drying out by causing moisture to migrate from one place to another within a construction and to encourage condensation of the moisture already entrapped (this type of condensation is sometimes referred to as 'secondary condensation'). The secondary condensation therefore converts the vapour to a liquid. It is the latter that collects in pools in flat roofs and subsequently leaks out under gravity.

(b) Flat roofs

The unprotected horizontal surface of flat roofs are inherently more suceptible to 'collect' water during construction. This water may not have dried out before the roof covering is laid. Much water can be contributed by most *in-situ cement-based* lightweight screeds.

However it has become entrapped, the water has much difficulty in getting out because in most cases the waterproofing significantly resists the transfer of water vapour through it – some of the newer plastics membranes do breathe, so not all membranes have the same effect. The only route for the moisture is towards the internal surface. But the route to this surface may be 'blocked' if there is a vapour control layer within the roof construction (see Diagram D3.2/24(b)).

(c) Pitched roofs

Entrapment of moisture in pitched roofs relates primarily to those constructions in which the insulation is immediately under the roof covering. In principle, these can be considered as flat roofs of warm deck design but laid on a slope. It should be noted that a form of entrapped moisture may occur in pitched roofs of cold deck design (i.e. with the insulation at ceiling level) where the roof void has been inadequately ventilated.

In pitched roofs with profiled metal claddings there is usually a vapour control layer underneath the insulation. This also acts as a waterproofing layer. Consequently, moisture becomes entrapped between the metal cladding and the vapour control layer. In pitched roofs finished with tiles or slates the insulation may be placed below the sarking felt. If this does not breathe, the roof has a layer impervious to liquid water and moisture becomes entrapped between the sarking and the vapour control layer. Thus in both cases, as in a flat roof, the inclusion of a vapour control layer below the insulation results in another waterproofing layer. Entrapped moisture would therefore have almost as much difficulty in getting out as it would in a flat roof.

Unlike flat roofs, there is the possibility of incorporating a 'second' line of defence if the water that initially reaches the underside of the roof covering can flow out to a gutter at the eaves (i.e. the principle used with sarking felt). Where inorganic and non-absorbent insulating materials are used, it may be advisable, if not necessary, to provide a 'third' line of defence for water to drain to a gutter (Diagram D3.2/25). If a vapour control layer is used for this, then its ability to drain to a gutter needs special consideration.

(d) Walls

In general, moisture is less likely to become entrapped within walls or, if it does, is less likely to cause problems except in those walls having impervious claddings or incorporating impervious layers within them. In masonry wall constructions water absorbed during construction or

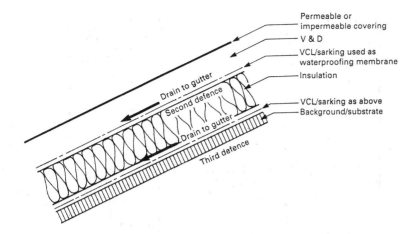

Labels in diagram:
- Permeable or impermeable covering
- V & D
- VCL/sarking used as waterproofing membrane
- Insulation
- VCL/sarking as above
- Background/substrate
- Drain to gutter
- Second defence
- Drain to gutter
- Third defence

Diagram D3.2/25 *Likely routes for water in an insulated pitched roof with a vapour control layer*

water from interstitial condensation during the use of the building is more likely to dry out in time, in the external leaf especially (Diagram D3.2/26(a)). Efflorescence may occur during the early days as a result of entrapped moisture; the risk of frost damage of the outer leaf may be increased.

Where walls have impervious claddings, water may become entrapped as the result of water trying to escape from the supporting background or from interstitial condensation. Generally, it can be expected that the consequences of moisture entrapped during construction trying to escape or of interstitial condensation is for water to run off the back surface of the cladding or insulation attached to it (Diagram D3.2/26(b)). This water should be drained away, a relatively simple matter in a wall. If the water is allowed to collect at the bottom of the cladding (i.e. at window or door heads or at the base of the wall) there is a risk of local damage, deterioration or decay of the components of the cladding or its supporting structure.

4 Flat roof predecent

(a) Scope

The background to the problem of entrapped moisture in flat roofs is covered in detail in *Materials for Building*, Vol. 2, p. 156. That background and the details of all the precautions that were considered appropriate at the time are not repeated here. Instead, the influencing factors and the principles underlying the rules that should be adopted and precautions that should be taken are described.

(b) Quantity of water

The quantity of water that may become entrapped influences the nature and extent of the precautions that should be taken in individual cases. Generally, the greater the quantity of water involved, the more elaborate the precautions required to ensure that the moisture can escape. The area of the roof and its shape are other influential factors.

It is, however, difficult to evaluate with any precision the quantity of water that may be derived either during or after construction.

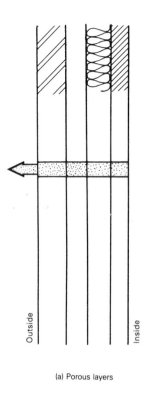

(a) Porous layers

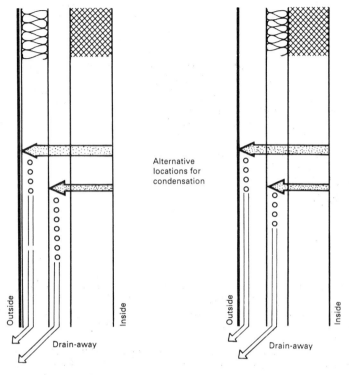

Alternative
locations for
condensation

(b) Impermeable outer layer

Diagram D3.2/26 *Comparison of vapour flow and possible locations for condensation and its drainage in walls with (a) porous layers and (b) an impermeable outer layer*

*Interstitial condensation, the factors that influence it and methods of predicting the risk of it occurring are covered in *Materials for Building*, Vol. 4, 4.06 *Condensation*, p. 124. More up-to-date guidance on predictive techniques and other related matters are given in *Building Failures*, 2nd edn, by Lyall Addleson, Butterworths, London, 1989.

Predictions of the risk of interstitial condensation may provide a basis for assessment.* Moisture from condensation is likely to behave in much the same way as that entrapped during construction. Solutions for both will therefore be similar. Where screeds are to receive the waterproofing system, the amount of moisture that may become entrapped will depend on:

- The amount of water used in mixing the screed (lightweight cement-based screeds may use up to $11\,l/m^2$, a considerable quantity).
- The amount of rainwater allowed to be absorbed by the screed and the deck before the waterproof covering is laid. (A dry screed with a fall of 1 in 60 may absorb about $8\,l/m^2$ during four days of rain. This is almost as much as the mixing water needed in an aerated cement screed.)
- The amount of water that is allowed to evaporate before the waterproof covering is laid. (Such water would include either mixing water or rainwater.)

It should be noted that all forms of substrate for waterproof coverings (i.e. both *in-situ* screeds and dry forms of background, the latter often used specifically to overcome entrapped moisture problems) are susceptible to absorbing rainwater unless suitably protected.

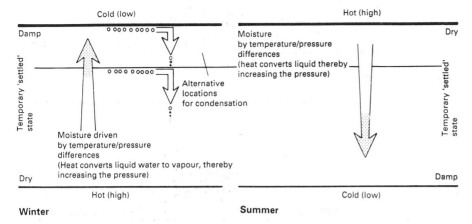

Diagram D3.2/27 *Moisture migration in a roof slab in winter and summer*

(c) Moisture transmission

Moisture trapped in a screed or concrete deck will at first be more or less evenly distributed (i.e. a condition of equilibrium). The even distribution or equilibrium is broken by the action of temperature and/or vapour pressure gradients. The effect is to cause the moisture to form in pools or ponds on the surface of the deck (or vapour control layer) and eventually leak out at day joints (in reinforced concrete roof slabs), at rainwater outlets or from electrical conduit boxes (or light fittings).

The moisture tends to migrate from **high to low**, usually from the warmer to the cooler zones of the construction (Diagram D3.2/27). In the warmer 'zone' water present as liquid vaporizes, thereby causing the pressure in this zone to increase. This, in turn, produces another **high to low**, thereby encouraging moisture to transfer to the cold/lower pressure zone where the moisture vapour condenses.

Thus in summer the moisture will tend to be pushed to the inner face of the roof or to any other area that may be shaded, particularly when the sun is shining. In winter the moisture goes the other way, generally condensing on the underside of the waterproofing covering. The condensed water might leach out bitumen in the waterproof covering. The bitumens provide the staining that is often seen when water leaks out of a concrete slab. Moisture from interstitial condensation would have the same effect.

The seasonal variations result in the lower parts of the construction being wetter in summer and the upper parts wetter in winter (Diagram D3.2/27). This means that organic materials within the roof construction should be carefully positioned to reduce the risk of their decay, more so if no provision has been made for the entrapped moisture to escape. Blistering of asphalt and built-up felt roofing membranes may be caused by moisture entrapped immediately below them (or within their layers) or in the substrate.* The sun's heat vaporizes the moisture. The intense pressure so created is sufficient to expand the waterproofing so that it forms a blister (Diagram D3.2/28). In extreme cases the blister may burst. It is more usual for the blister to be torn by foot traffic on the roof.

*The mechanism of blistering is complex. Entrapped moisture is the root cause. But moisture may also become entrapped below or within the covering by being drawn through the material (i.e. a pumping action).

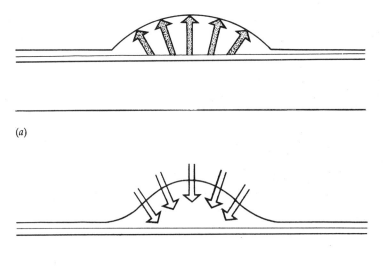

(a)

(b)

Diagram D3.2/28 *Processes that give rise to blistering in roof membrane (built-up felt illustrated). (a) Rapid expansion of trapped moisture (or air) forms blister; (b) partial vacuum at night when the blister is stiff draws additional water vapour (or air) into blister. The process in (a) occurs again, thereby enlarging the size of the blister*

(d) Drying out

Ventilation is the means, in principle, that is required to allow entrapped moisture to evaporate (Diagram D3.2/29(a)). This, in turn, means that there must be an inlet and an outlet to each and every ventilation channel incorporated within the roof. Ensuring that there is **continuity** between inlet and outlet is a difficulty that has to be overcome if the system used is to have a chance of success. It is, of course, also important that both the channels and the inlet/outlets are sufficiently large. A number of purpose-made 'roof ventilators' are available.

Various alternative methods have been proposed to effect ventilation (Diagram D3.2/29(b)–(f)). In the simplest case, built-up felt systems use a partially bonded first layer. The channels are, of necessity, small, so these systems are usually only sufficient to deal with small quantities of moisture. A development of this system is a series of larger channels formed into a two-way channel system incorporated into a special mat or on the underside of an insulation board. In the more sophisticated systems, large ducts are incorporated within the screed. These may be formed in no-fines or lightweight bitumen-bonded screeds. Alternatively, channels, much like a land drain system, may be cast into a screed.

In those cases where natural ventilation is unlikely to be effective the roof ventilators would have to incorporate fans to achieve mechanical ventilation. This means specially designed ventilators.

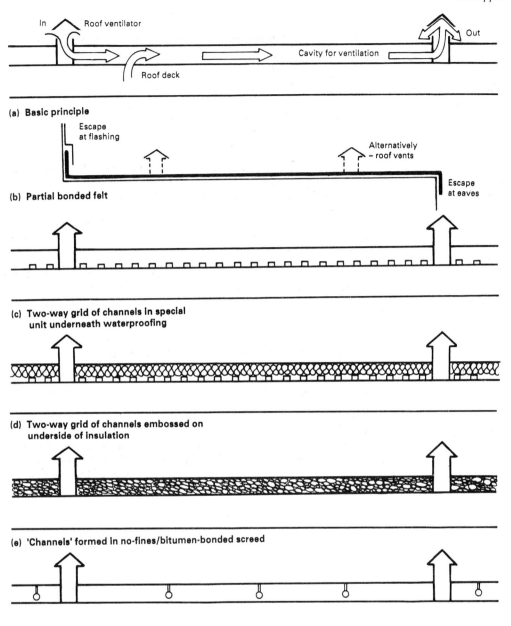

(a) **Basic principle**

(b) **Partial bonded felt**

(c) **Two-way grid of channels in special unit underneath waterproofing**

(d) **Two-way grid of channels embossed on underside of insulation**

(e) **'Channels' formed in no-fines/bitumen-bonded screed**

(f) **Channels within screed**

Diagram D3.2/29 *The basic principle for ventilation below the waterproofing membrane (a) and alternative solutions (b) – (f)*

(e) Precautions

The precautions that may be taken to reduce the risk of entrapped moisture occurring or from causing damage include:

1. The structural deck should have holes at all low points to enable rainwater falling on the deck or screed to drain away (Diagram D3.2/30).

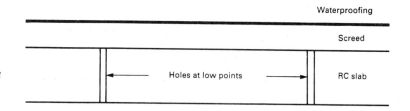

Diagram D3.2/30 *Providing holes in a reinforced conrete slab at low points to allow water to drain out – see text*

2. Where cement-based screeds are used, the amount of mixing water should be reduced to a minimum.
3. Instead of cement-based screeds, treated aggregates of dry insulants should be used.
4. Whatever base is used for the waterproof membrane the base should be adequately protected from the weather before and during the laying of the membrane.
5. If entrapped moisture and/or interstitial condensation is unavoidable, there must be provision to enable the moisture to dry out by incorporating an appropriate system of ventilation.

Sources

Principles of Modern Building, Vol. II, HMSO, London, 1961
BRE Digest 51, *Developments in roofing*, October 1964
Tolhurst, A. J., 'Trapped moisture in roofs', articles in the *Journal* of the Mastic Asphalt Advisory Council, January 1963, pp. 31–33 and July 1963, pp. 33–37
Billington, N. S., *Thermal Properties of Building*, Clever-Hume, London, 1952
McInnes, H. W. (translator), 'Waterproofing problems', *The Builder*, 22 November 1963, pp. 1081–1082 (translation of an article in German)
March, Francis, *Flat roofing, A guide to good practice*, Tarmac, 1982

Moisture content

Scope

Changes in the moisture content of porous materials may have physical and/or chemical effects. There may be two physical effects: one relates to dimensional fluctuations of a material, generally referred to as moisture movements; the other concerns the mechanical properties of a material, it being generally the case that the higher the moisture content, the lower the strength of a material. Chemical effects result from the interaction of water and the solids of which the materials are composed, the latter including solids that may be absorbed by, transferred into or in close contact with porous materials.

One of the commonest, if not the most difficult, problems to be faced in building practice is how to resolve moisture movements that result from changes in moisture content. This is because restrained moisture movements can result in cracking, as explained in *2.1 Cracking*.

In this chapter detailed consideration is given to the mechanisms involved relative to the properties of materials, the extent of the movements likely to be encountered, and the practical significance of the movements. These aspects are relevant in other problems such as cracking (*2.1 Cracking*) and the exclusion of water (*3.4 Exclusion*) or are needed for an understanding of phenomena such as efflorescence (*3.5 Efflorescence*), chemical reactions including corrosion of metals (*3.6 Chemical attack* and *3.7 Corrosion*) and frost action (*3.8 Frost action*).

Moisture-content changes and the related movements may take place in some non-porous materials such as plastics – they never occur in metals. However, the extent and, more importantly, the effects of changes are more pronounced in the porous materials. Thus these materials are primarily considered here. The special problems that occur under 'normal' conditions of exposure with timber, timber-based and cement-based products are given special emphasis.* This emphasis should not, however, detract from the consideration that should be given in practice to the movements which are likely to occur in other porous materials (and, where relevant, non-porous plastics) in the relevant extreme cases of exposure.

In dealing with moisture content on a broad basis and by concentrating on dimensional fluctuations there is some overlapping. It will be seen that many of the factors involved are closely interrelated.

Sources

The following and those included later under 'Timber', 'Timber-based products' and 'Cement-based products' apply to this and the next three sections.

BRE
Stradling, R. E., 'Effects of moisture changes on building materials', *Building Research Bulletin*, No. 3, HMSO, 1928 (among the earliest results of building research)

*The detailed consideration included with these classes of materials should help to clarify many of the interrelated factors that are outlined in general terms only for other materials included in this chapter. Furthermore, and importantly, timber and cement-based products represent not only extremes of movement relative to other materials but also completely different behaviour during movement. Timber is, in terms of moisture movement, anisotropic, and this results in the movement not being the same in all directions. In contrast, cement-based products, like most other materials, are isotropic, and this results in the movements being roughly the same in all directions.

Schaffer, R., *Weathering of natural building stones*, Building Research Special Report No. 18, HMSO, London, 1932. Reprinted 1972 as facsimile with minimum changes and up-to-date information included in Appendix II – cleaning of stonework
Principles of Modern Building, Vols 1 (3rd edn) and 2, HMSO, London, 1959 and 1961

Definitions and their application

The definitions commonly used in connection with moisture content are derived from timber technology. It was in this field that the greater part of the early work on moisture content and the effects of changes in it was undertaken. The definitions have not changed as the field of study has broadened to include most, if not all, porous materials (and recently the non-porous plastics). Consequently, the term 'timber' has been omitted from the definitions.

1 Moisture content

(a) Definition

Moisture content is the amount of moisture which a material contains at a given time and is expressed as a percentage of its *dry* weight (strictly speaking, its 'oven-dry' mass, that is, after all moisture has been expelled from the material) and may be written as a formula:

$$\text{Moisture content (\%)} = \frac{\text{Weight of water present}}{\text{Dry weight of material}} \times 100$$

As the moisture content is expressed as a percentage of the *dry* weight of the material rather than its total weight, it is possible to have moisture contents of well over 100%, with timber especially.

(b) Measurement

There are a number of ways in which moisture content may be determined (see *3.1 General considerations*, 'Dampness', p. 152). The standard and most accurate of these is the oven-drying method.* Briefly, in this method a sample is weighed (giving its initial weight), then it is dried in an oven and then reweighed (giving its dry weight). The loss in weight during the drying indicates how much water was present in the sample. Its moisture content can therefore be calculated as follows:

$$\text{Initial moisture content (\%)} = \frac{\text{Initial (wet) weight} - \text{Dry weight}}{\text{Dry weight}} \times 100$$

*Details of the method as applied to timber are described in *Moisture content determination by the oven-drying method*, FPRL Leaflet No. 7, HMSO, London, July 1949. As this is now out of print see Pratt, G. H., *Timber Drying Manual*, 2nd edn, revised by C. H. C. Turner, BRE Report, 1986, p. 2; or *Moisture content of timber, its importance, measurement and specification*, TRADA Wood Information, Section 4, Sheet 14, October 1985 (revised) under 'Oven dry method'.

For example, if the sample weighed 40.31 g initially and 30.77 g when dried, the difference of 9.54 g is the weight of the moisture initially in the sample, and its moisture content would be:

$$\frac{40.31 - 30.77}{30.77} \times 100 = \frac{14.53}{20.77} \times 100 = 31.0\%$$

To carry out the division only, the formula may be written as:

$$\text{Moisture content (\%)} = \left(\frac{\text{Initial weight}}{\text{Dry weight}} - 1 \right) \times 100$$

$$= \left(\frac{40.31}{30.77} - 1 \right) \times 100 = 31.0\%$$

It should be noted that the moisture content obtained as described above is the average for the sample. The actual moisture contents at different locations within a sample may vary considerably from the average value, depending on the size of the sample. Generally, the smaller the sample, the smaller the variation of moisture content from place to place within it. A profile of the moisture of a sample can be obtained by oven drying smaller pieces of it and working out the moisture contents of each piece.

2 Equilibrium moisture content (emc)

(a) Definition

Equilibrium moisture content is that which a material will achieve when it is in equilibrium with the moisture content of the surrounding air (i.e. in **balance**). It is a variable value which is governed mainly by the temperature and relative humidity of the air.

(b) Adjustments and time

All materials capable of taking in or giving off water will adjust their moisture contents according to the moisture content, that is, the humidity, of the air surrounding them, until moisture equilibrium is attained. Thus, depending on conditions, materials may absorb moisture from the atmosphere or vice versa. The process is, however, not instantaneous, the length of time required for equilibrium to be reached being primarily dependent on the nature of the material concerned. This particular aspect is important, because *momentary changes* in the humidity of the atmosphere are unlikely to have very much, if any, moisture movement effect on a material.

The adjustments which materials make in their moisture contents so that equilibrium with the surrounding air may be attained has an important bearing on the moisture contents of materials saturated by liquid water. Saturation, whether partial or complete, can only persist while environmental conditions of humidity so allow. In other words, materials exposed to air will not 'tolerate' an imbalance between their moisture contents and that of the surrounding air. The rate at which the adjustment of moisture content takes place when materials are saturated will depend on those factors which influence evaporation. Thus, under unfavourable conditions, equilibrium moisture content may only be achieved after a considerable time.

3 Moisture movement

(a) Definition

Moisture movement is very often mistakenly, although understandably, interpreted as the movement of moisture in and about a material. It is, in fact, concerned with the movement of the material itself, that is, the alteration in its physical dimensions due to changes in its moisture content. Moisture movement may then be defined as the change in unit dimensions of a material due to changes in its moisture content and expressed as a percentage of its original dimension. It is usual to refer to the length of a material. Moisture movement may be written in formula form as:

$$\text{Moisture movement (\%)} = \frac{\text{Change in length}}{\text{Original length}} \times 100$$

The original length, even in drying shrinkage tests, is usually indicative of the dry length, so that the formula may be rewritten as:

$$\text{Moisture movement (\%)} = \frac{\text{Wet length} - \text{dry length}}{\text{Dry length}} \times 100$$

(b) Types of movement

Values of moisture movement may indicate either an expansion or a contraction of a material. However, a distinction has to be drawn between two types of movement, one of which is irreversible and usually, though not always, implies a contraction, while the other is reversible and implies movement in either direction. Cement-based products are good examples of materials which undergo both types of movement. These are shown in Diagram D3.3/1.*

(1) Irreversible movement
This movement is concerned entirely with those materials requiring water for their use or manufacture and most, though not all, of it takes place during the the drying-out period.† Thus the movement only takes place once during the life of the material, and is of particular significance in all the wet-trade processes undertaken on a building site. Precautionary measures are always advisable to avoid undue shrinkage cracking. In those cases where materials, such as clay- or cement-based products, are made off the building site, the drying shrinkage should be complete before the products are built into a building.

(2) Reversible movement
This movement may occur in all materials. Reversible movement is closely related to the moisture content of a material, and thus expansion takes place when water is taken in and contraction when water is given off. In general, and taking into account practical application, the extent of movement in either direction is roughly the same. Any departures from this should be stated in published data.

(c) Interpretation of values

(1) Type of movement
Practical application, in terms of allowing for movement, is restricted mainly because of the complexities of the interrelationship of the many factors involved, plus the fact that the laboratory tests necessary to

*It should be noted that in the case of bricks there is an irreversible expansion that takes place over a long period of time, being greatest during the first day after removal from the kiln and subsequently reducing until a limiting value is reached after several years. Importantly, this expansion, having a curve similar to that of the expansion of concrete maintained in continuously moist condition in Diagram D3.3/1, is superimposed on the reversible expansion and contraction due to wetting and drying. See (d) 'Clay brickwork' later, p. 271.

†Timber may be said to undergo irreversible moisture movement during its own initial drying-out process. However, in practical building this is of little significance.

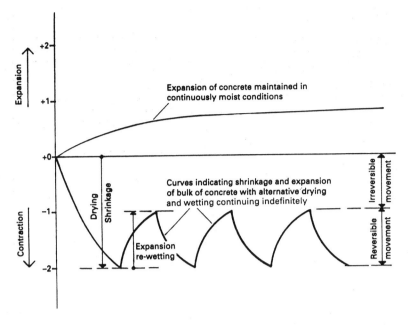

Diagram D3.3/1 *Illustration of irreversible and reversible movements in concrete. Principles are equally applicable to other porous materials — see text*

evaluate the extent of movement in any material usually employ, as a basis, moisture content extremes, while comparatively small samples of a material are used. In many cases these tend to give exaggerated movements, which may seldom be encountered in practice. This is probably more significant with reversible rather than irreversible movements. As far as the latter are concerned, it is normal to measure the total drying shrinkage of a material (wet to oven dry) which, although this may include some reversible movement, is a valuable guide when assessing precautions which should be taken.

In the case of reversible movements, the oven-dry condition is often used as the condition upon which the original length would be based. Thereafter, the material may be totally immersed in water for a sufficient period of time to cause complete saturation, or exposed to controlled temperature and humidity conditions for a specified period of time. Thus, unless the range of moisture content or the method of calculation is stated in published data, values may be misleading, particularly for detailed design application.

(2) Range of movements
Chart C3.3/1 illustrates the general range of moisture movements in various materials based on a variety of calculation methods and moisture content extremes. It is intended to serve as a comparison of the behaviour of the range of materials in terms of moisture movement and not for detailed design application. There are available values of movements for timber (given later) which can be applied in design for particular circumstances. Generally, rule-of-thumb methods must still be used in the majority of cases. (See also '6 Precautions' given later, p. 253.)

Despite the lack of comprehensive design guidance, some reliance may often be placed on reasonable interpretation of available values. In order to do this, the importance of the relative humidity of the air surrounding a material, as opposed to the amount of liquid water absorbed by a material, together with the relationship between the two must be understood. In practice both are important.

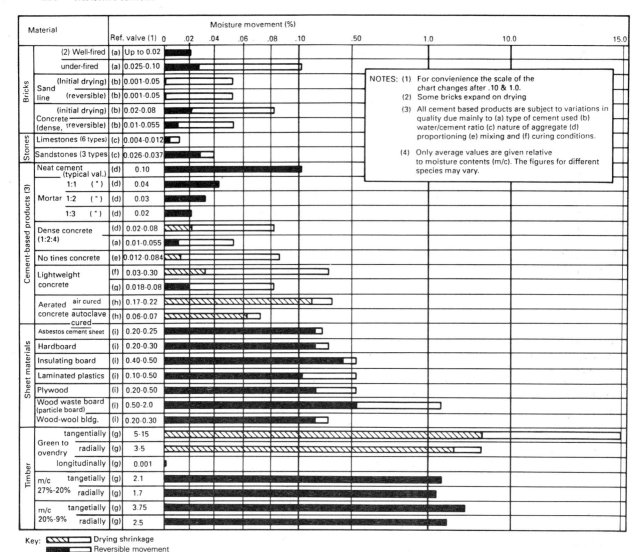

Chart C3.3/1 *Range of moisture movements of various materials. Note: The chart is intended to give an comparative overview of moisture movements. The values have been taken from the following references, some of which have been overtaken by time. For practical working values refer to BRE Digest 228,* Estimation of thermal and moisture movements and stresses: Part 2, *August 1979.* References: *(a) Butterworth, R.,* Clay building bricks, *NBS Bulletin No. 1, HMSO, London, 1948; (b) Sand-lime and concrete bricks, *NBS Bulletin No. 4, HMSO, London, 1948 and Bessey, G. E., Sandlime-bricks, NBS Special Report, No. 3, HMSO, London, 1948; (c) Schaffer, R. J.,* The weathering of natural building stones, *Building Research Special Report No. 18, HMSO, London, 1932; (d) Lea, F. M.,* The Chemistry of Cement and Concrete *(revised edition of Lea and Desch), Edward Arnold, London, 1956; (e) BRE Digest (2nd series) No. 35,* Shrinkage of Natural aggregates in concrete, *new edition, 1968; (f) BRE Digest (2nd series) No. 25,* Lightweight aggregate concrete – 1: Materials and properties, *August 1963; (g)* Principles of Modern Building, *Vol. 1, 3rd edn, HMSO, London, 1959; (h) BRE Digest (2nd series) No. 16,* Aerated concrete – 1: Manufacture and properties, *November 1961; (i)* House construction, *Post-war Building Studies No. 1, HMSO, London, 1944, Crown Copyright*

(3) Exposure

For the maximum moisture movement over a given range of moisture content to take place, the moisture involved must be associated with the maximum amount of the solids of which the material is composed. (The nature of this association and the conditions under which it will persist are discussed in 'Presence of water; 3 Sorbed water', later, p. 234.) In general, it is extremely difficult to ensure that the desired association between water and solids takes place if a material under test is, for example, only allowed to absorb sufficient liquid water after immersion to correspond to the range of moisture content required. The reasons for this are complex but, in the main, they are connected with the porosity characteristics of the material which may not readily allow water to permeate throughout the material. On the other hand, if a material is immersed for long enough, saturation can occur, in which case it is then possible for the maximum amount of solids to be associated with water.

This, of course, implies, as already stated, the extreme moisture content range for the material and is indicative of the situation that is likely to occur to materials which are used in positions (exposure to the weather, for example) where saturation may take place. However, even in these cases it is unlikely that, in practice, the material would have started off, so to speak, in an oven-dry condition before saturation, so some allowance would have to be made for this. Thus, the movement is unlikely to be as great as the values normally given. It may be noted here that, where partial saturation of a material occurs, there will be a certain amount of differential movement between the wet and dry zones. This may be important in some situations (see '4 Moisture gradient').

As all materials are not necessarily exposed to liquid water, a range of movements based on exposure to specified relative humidities can provide useful guidance. Again, it should be noted that most values computed on this basis do, however, use original length based on oven-dry conditions. Some allowance must therefore be made for this. On the other hand, the magnitude of movement may not necessarily be the same for all increments in moisture content. Such differences may not always be relevant, but nevertheless detailed values for each material should be considered.

4 Moisture gradient

(a) Definition

Moisture gradient is defined as the variation of moisture content between the outer and inner parts of a piece of material.

(b) Significance

The gradient is of particular significance in timber, as its existence may cause stresses to be set up within the material which, if of sufficient magnitude, may eventually lead to rupturing or distortion. (Compare, by way of a useful analogy, the effects of either heating or cooling a material in *2.1 Cracking*, Diagrams D2.1/5 and D2.1/6.) The gradient is 'set up' during the seasoning process if the moisture from the outer layers is allowed to evaporate at a rate which does not allow moisture from the interior to reach the surface. When the outer layers set in a stretched condition due to the moisture gradient, which in this instance would be steep, the piece of timber is said to be *case hardened*.

Similar gradients may also be set up when wet-trade products dry out. Stresses may also occur, but the defects are normally not as pronounced as in the case of timber, although cracking of the surface may take place.

5 Moisture transmission

(a) Definition

Moisture transmission, or transfer as it is also sometimes known, may be defined as the movement of water in and through a material or combination of materials. The water may be in either the liquid or vapour state.

(b) Significance

The mechanisms involved in moisture transmission, together with the factors which influence it, are of particular importance when considering the drying out of materials. It is also useful in gaining an understanding of the significance of the transfer of water between two adjoining materials relative to such phenomena as efflorescence.

General influences

The properties of materials and factors which generally influence moisture content, or are often referred to in connection with some of the phenomena involved, are discussed below.

1 Capillarity

Capillary action or the capacity of water to rise in narrow channels is extremely important in porous materials which, by their nature, may contain a variety of channels, including many which are narrow. The functioning and the moisture content of virtually all the porous materials is partly dependent upon capillary action. Size and arrangement of the pores within a material are associated with the formation of capillaries, and thus influence both the amount and rate of moisture transmission in and out of a material.

The importance of capillarity may be underlined by considering the mechanisms and factors involved in (a) the drying out of a porous body, and (b) the transmission of water between two adjoining materials.

(a) Drying out of a porous body

Two interrelated factors govern the drying out of a porous body. These are:

1. *The environmental conditions* surrounding the body and made up of the temperature and humidity of the atmosphere and the velocity of air flow past the surface. All these are concerned with evaporation.
2. *The structure of a material* which controls the transmission of water to the surface of the material.

A porous solid of appreciable thickness will only dry out if water is evaporated from the surface. Thus, water must travel, in one form or another, from the interior to the surface before it can escape as *vapour* into the surrounding air. The water lost by evaporation from the surface may be replaced by water flowing under the action of capillary forces or the evaporating surface may fall *below* the surface of the solid, in which case the water must 'jump across' a certain length of pore space before it can escape at the surface.

The rate at which evaporation of surface water of a *saturated* material initially takes place is solely dependent on the environmental conditions and is thus the same for any free surface water exposed to the same conditions. Provided that compensating water is brought to the surface so as to maintain it in a *wet* condition, evaporation will continue at a constant rate. If, however, the amount of water brought to the surface is insufficient to keep the surface wetted, then there will be a fall in the rate of evaporation. Thus, in materials in which water can move easily, most of the drying out takes place readily at the surface. However, in a dense material the initial constant rate only persists for a short period of the total drying-out time. The surface may, therefore, be dry long before all the pores have been emptied, with the result that at some point *below* the surface the material may still be saturated. For the significance of surface dryness in practice see *3.1 General considerations*, 'Dampness', p. 231.

The effect of the density of a material or surface relative to two building applications may be noted. In brickwork having dense cement mortar pointing, such as a 1:3 cement:sand mix, little transmission of water can occur through the pointing; thus practically no drying out takes place from the mortar joints. Similarly, trowelled finishes to renders, etc. may produce dense surfaces which will impede evaporation.

(b) Transmission between two adjoining materials

The capillary properties of materials influence the suction force which is exerted to draw water from within a material to the surface. They also control the transmission of water, which may include soluble salts, for instance (important for efflorescence), from one material to another.

Capillary suction is greater in *uniformly* fine-pored materials than in *uniformly* large-pored ones. Thus, if different materials are placed in contact with one another, a uniformly fine-pored material will draw water from a uniformly large-pored one. However, materials normally contain pores of various sizes, and so the suction force which is developed will depend on *moisture content*. Thus, subject to the relative pore structure, the *direction* of moisture transmission between two particular materials alters with a change in the degree of saturation. This will, in part, be influenced by capillary attraction, particularly if the material concerned faces onto a source of water. In general, it is extremely difficult and complex to predict the direction of transmission of water between any two materials because the capillary suction of a material varies with its moisture, while the rate of change of moisture content under wetting and drying conditions varies with different materials.

2 Absorption

Absorption is the property of permitting water (or any other liquid) to enter the *pores* of materials. In this sense, absorption does not imply any particular association between the absorbed water and the solids of which a material is composed, although a particular association may

Chart C3.3/2 *Comparison of porosity, absorption and saturation coefficient of some materials. References: (a) Butterworth, R., Clay building bricks, NBS Bulletin No. 1, HMSO, London, 1948; (b) Schaffer, R. J., The weathering of natural building stones, Building Research Special Report No. 18, HMSO, London, 1932; (c) Warnes, A. R., Building Stones, Benn, London, 1926; (d) Sand-lime and concrete bricks, NBS Bulletin No. 4, HMSO, London 1948 and Bessey, G. E., Sandlime – bricks, NBS Special Report, No. 3, HMSO, London, 1948; (e) House construction, Post-war Building Studies No. 1, HMSO, London, 1944 Crown Copyright)*

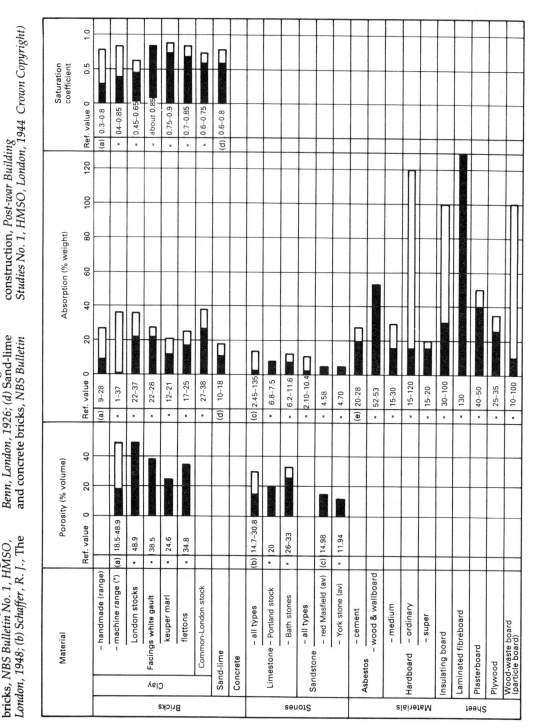

occur. The amount of water absorbed is usually expressed as a percentage of the weight of the material.* It may be written in formula form as:

$$\text{Absorption (\%)} = \frac{\text{Weight of water absorbed}}{\text{Weight of material}} \times 100$$

Absorption of some materials, together with porosity, is given in Chart C3.3/2.

In practice, the rate of absorption is usually far more significant than the amount actually absorbed. Usually, the rate of drying out is the same as the rate of absorption. It may also be noted that dry porous materials initially absorb water rapidly by capillary action – known as suction – but the rate soon decreases until saturation is reached. The rate of absorption may be increased significantly if the source of water is under pressure (see '4 Permeability' below).

3 Saturation coefficient

The saturation coefficient is the ratio of the volume of water absorbed and the total volume of voids in a material. It may be written in formula form as:

$$\text{Saturation coefficient} = \frac{\text{Volume of water absorbed}}{\text{Total volume of voids}}$$

Thus, when all voids are filled with water the coefficient equals 1.0. Few materials have coefficients of 1.0 (coefficients for some materials are included in Chart C3.3/2) which partly supports the theory that not all voids in a material are necessarily filled when a material is saturated.

Saturated coefficients are sometimes considered to give some guidance as to frost resistance of materials (see *3.8 Frost action*, p. 548).

4 Permeability

Permeability may be defined as that property which permits the passage of a liquid *through* a material and is thus distinct from the penetration of moisture into a substance by means of absorption due to capillary action. In some senses permeability only commences after a material has been saturated.

The passage of water may be caused by (a) capillary action and (b) pressure due to force of gravity, driving rain, head of water, or thermal expansion of water already absorbed at a lower temperature. Until water has penetrated through a material the rate of flow results from combined pressure and capillary action, but when penetration is complete this rate is entirely dependent on pressure.

The rate of flow is a measure of the permeability of a material. Thus, permeability must be due to the existence of continuous passages *right through* the material. Permeability may, of course, also be applicable to composite materials, in which case water may penetrate at the interface of joints between units making up the composite. This implies breakdown of adhesion. Joints in brickwork, for example, often provide through-routes for the complete penetration of water, and such penetration usually occurs at the interface of the jointing medium, i.e. the mortar and the bricks.

*Absorption may also be computed on a percentage volume basis.

The presence of water in materials

Water can be present in building materials in three ways: (1) in chemical combination, (2) as free water, and (3) as 'sorbed' water, each of which may influence moisture content but not moisture movement. To understand the significance of these relative to moisture movement they are discussed separately.

1 Water in chemical combination

The chemical combination of various materials to form compounds, with properties peculiar to the compounds, requires specific molecular proportions of the individual materials. When water is essential to the existence of the compound, it is said to be in chemical combination, and can only be removed by breaking down the compound under certain definite conditions. Thus, for example, one molecule of calcium hydroxide ($Ca(OH)_2$), slaked lime, can only be formed when one molecule of calcium oxide (CaO), quicklime, unites with one molecule of water (H_2O). This may be written as:

$$CaO + H_2O = Ca(OH)_2$$

Unless the required proportions of quicklime and water are present, slaked lime cannot be formed. However, slaked lime can still be formed, and this is important, even if the quantities of water present *exceed* the fixed amount required. Such excess water is not then in chemical combination.

It may be noted here that the reaction of water with cement, although following the same principle, is extremely complex as, among other things, water reacts with the individual constituents of which the cement is composed.

2 Free water

Water in a material is said to be free when it is associated with a solid in such a way that its properties still remain, for all practical purposes, unchanged. In general, such water wets the material and remains in association with the material only during external conditions which maintain complete moisture saturation. As an example, consider a good-quality brick saturated with water. Almost all the water is 'free' and the association is maintained by keeping the brick either immersed in water or surrounded by air saturated with water vapour.

Timber provides another and more practical example. In newly felled (i.e. green) timber, the water in the cavities of the cells is free. It is this water that is the first to leave the timber when it is being dried, usually without causing the timber to shrink (see '3(c) Drying out' later, p. 236).

3 Sorbed water

In building materials, 'sorbed' water (also known as *bound moisture*) in association with a solid is a condition intermediate between chemically combined water and free water. Such water is held more firmly to the solid than free water, but is not held so as to form a definite chemical compound. The most important factor regarding 'sorbed' water is that the *association* with the solid material depends entirely upon external humidity conditions. Provided sufficient time is allowed (this is extremely important), the amount of water 'held' changes as the humidity changes. However, the moisture equilibrium of the solid, that is, the amount of water in equilibrium with the solid, is dependent on both temperature and relative humidity of the surrounding air. As previously stated, in practice, relative humidity is usually of greatest value. In addition, different materials have different moisture equilibria.

(a) Gels

Sorbed water is associated with *gels*. Those encountered in building materials are non-rigid gels, which, among other things, means that they do not go into solution but have definite saturation levels. In timber, fibre saturation point is reached when the *cell walls* contain between 25% and 30% of their dry weight.

When water is sorbed by the gels there is an increase in the volume of the material, while there is a corresponding decrease when the moisture is liberated. Thus, the presence of gels makes materials hygroscopic.*

The evidence available suggests that the expansion (and corresponding contraction) which takes place may often be ascribed to the swelling (or contraction) of colloidal constituents in the material. However, it is also thought that the action of surface tension forces in the pores may also be of some significance. Thus, it is possible for moisture movements to take place when water is held in the pores of materials, whether the presence of such water implies complete or partial saturation. On the other hand, under certain conditions, some of the water held in the pores may be sorbed into the hygroscopic solids of the material. Water sorbed in this way can only remain in association with the solids for as long as the environmental conditions allow.

(b) Colloidal constituents in some materials

The presence of colloidal constituents is more commonly, but not exclusively, associated with timber and cement.

1. In the case of *timber* and other natural fibrous materials, the gels are 'incorporated', so to speak, during growth and are in a sense an integral part of the material.
2. With *cement* the products of the chemical combination of water and the cement are present as gels.
3. *Clay bricks* (and other clay products for that matter) are formed of typically colloidal material, clay. However, the colloidal character is destroyed by firing and so bricks do not exhibit appreciable reversible moisture movements. It is for this reason that thorough firing must be ensured during manufacture – underfired bricks are liable to large expansions on becoming wet.†
4. *Sandlime* bricks, which are manufactured by heating mixtures of sand and lime with steam under pressure, have fairly high moisture movements. This may be due to the presence of colloidal silicates of lime.
5. *Sandstones* show the most pronounced moisture movement of the natural building stones. This may be due to the presence of colloidal minerals acting as the cementing medium between the silica grains. Such movement may be significant when sandstone is used as an aggregate in concrete, particularly during initial drying out.

*Hygroscopic materials are those that expand and contract as moisture is gained and lost.

†Well-fired bricks undergo a long-term expansion, as already noted (see footnote, p. 226). The large expansion of underfired bricks on being wetted is therefore a separate and different problem.

(c) Drying out

In general, the water held freely in the capillaries or pores of materials is less important than sorbed water in considerations of moisture movement. Such free water will, of course, contribute to the moisture content of the material.

During the drying-out process it is usually the free water which is first liberated. Once this water has been removed by evaporation, then it is the sorbed water which is given off into the atmosphere until moisture equilibrium is reached. Thus initial, and perhaps substantial, drops in moisture content may not always indicate corresponding contractions in a material if the drop in moisture content is due to the loss of free water. Contractions, as explained earlier, are associated with the loss of sorbed water. This is of particular significance in timber in which the loss of sorbed water is usually not uniform.

(d) Irreversible movement

The considerations dealt with above have all been concerned with changes in moisture content which are reversible. However, the behaviour of rigid gels, formed under supersaturated conditions, is significant when considering initial drying shrinkage, particularly in cement-based products.

When, for example, water is added to cement, it is at first free, then some water enters into chemical combination with the cement; some of the products of this reaction are present as gels. When gels are formed in the presence of excess water, it is thought that more water may initially be associated with the gel structure than is necessary for equilibrium even under conditions of water storage.

The excess water is gradually forced out from the gel and may hydrate further particles of cement which have previously not been reached by water, or may leave the whole mass and pass out into the surrounding atmosphere. As this water cannot be replaced, a shrinkage of the mass occurs when the excess water passes out. Thus, the greater the amount of excess water, the greater the shrinkage.*

After equilibrium conditions have been reached, the cement is in the condition previously described, that is, capable of undergoing reversible moisture movement.

Exposure

1 Generally

The importance of the humidity of the air surrounding materials as regards both moisture content and moisture movement has been emphasized earlier. The main factor in this is the ability, if not the 'desire', of all materials liable to moisture movement to adjust their moisture contents to that of the surrounding air until moisture equilibrium has been attained. However, at the same time, the rate at which this adjustment takes place when water is being given off to the surrounding air will be dependent not only on the structure and size of the material but also (and this is important) on the prevailing environmental conditions which encourage evaporation. These include temperature and air movement in addition to relative humidity.

Thus, it is necessary to take into account the importance of moisture content changes and moisture movements which may result from saturation (complete or partial) by water in the liquid state. Under

*It is for this reason that the amount of mixing water for all cement-based products should be carefully measured and controlled. Such control includes taking into account the amount of moisture that may be present in the aggregates. The latter are seldom completely dry when used, on building sites in particular.

unfavourable conditions of evaporation (that is, high relative humidity, low air temperature, or low rate of air movement, separately or in combination), materials previously saturated by water in the liquid state may undergo fairly extensive movements. The maintenance of high moisture contents may be undesirable anyway for reasons other than movement. For example, chemical action is particularly relevant.

The purpose of this section is (1) to outline the practical significance of saturation by water in the liquid state; (2) to develop certain aspects of the practical significance of relative humidity previously excluded; and (3) to draw attention to the interaction of thermal and moisture movements.

2 Liquid water

(a) Sources

All the sources of water explained in *3.2 Exposure* are capable of causing materials to become saturated. In general, rainwater presents severe conditions for most external elements; water introduced during construction is particularly relevant for initial drying shrinkage in the wet-trade processes, while condensation may present rather severe conditions internally. The other sources, such as water supply, faulty services and maintenance and usage, require individual assessment. Finally, groundwater usually implies perennial saturation and the materials used in association with the ground should be selected specially for this condition.

(b) Extent and rate of movements

The extent of the linear expansion which can be expected in a material as a result of saturation will be dependent on two factors. These are: (1) the nature of the material and (2) the length of immersion or exposure to water. Chart C3.3/3 illustrates the linear expansion of some materials commonly used externally when small samples of each (except for the clay brick) were immersed in water over a period of days. In all cases the longer the period of immersion, the greater the expansion.

However, what is significant is the fact that the rate of expansion varies quite considerably with the various materials. Expansions *approaching* the maxima shown on this chart and also in Chart C3.3/1 may, within limits, be attained in external elements exposed to rainwater. In this, account has to be taken of the prevailing wet and dry conditions (see *3.1 General considerations*, Chart C3.1/1). In winter, for example, external elements can be expected to be saturated over a long period of time. Drying out will commence during the spring and generally dry conditions will normally prevail during the summer.

(c) Special design considerations

From available information, it would appear that materials with movements of 0.04% and above require special design consideration in terms of allowing for expansion, providing sufficient restraint or complete impervious protection which will prevent the ingress (and egress too!) of water.

If reference is made to Chart C3.3/1 it will be seen that many materials, including most sheet materials, have expansions well in excess of 0.04%. In general, the latter materials (including 'composites') require special consideration if they are to be exposed externally.

Material	Immersion (days)	Value	Linear expansion: percentage of original length.
Neat cement	1	.009	
	7	.039	
	14	.055	
1:1 Cement screed	1	.0025	
	7	.014	
	14	.0175	
1:3 Cement screed	1	.002	
	7	.010	
	14	.014	
1:2:4 Concrete	1	.031	
	7	.031	
	14	.0335	
Breeze	1	.071	
	7	.096	
	14	1085	
Art.Port. stone	1	.028	
	5	.0375	
	–	–	
Portland stone	1	.008	
	4	.009	
	–	–	
Sandstone	1	.037	
	7	.039	
	14	.004	
Clay brick (fletton)	1	.004	
	5	.007	
	–	–	

Scale marks: 0, .010, .020, .030, .040, .050, .060, .070, .080, .090, .100, .110

Chart C3.3/3 *Comparison of linear expansion of some materials immersed in water.* Note: *All samples immersed on water after drying at 50°C. Samples, except clay bricks, were 6 × 6 × 3 in prisms (equal to 152.4 × 152.4 × 76.2 mm exactly). (Data from Stradling, R. F. Building Research Bulletin No. 3, HMSO, London, 1928, Graphs in Fig. 9, p. 11)*

Although joint design and restraint may offer some solution to the problems involved, it is usually imperative to ensure that as little movement as possible takes place by the provision of an impervious coating. The success of such a coating in reducing changes in moisture content will largely be dependent, in addition to its own characteristics, on complete **continuity** of the coating. Thus edges must also be protected. At the same time, it should also be remembered that the 'backs' of these materials, although not exposed to rainwater, may 'draw' water from backing materials, while cavities behind the material may provide sources of high humidity, if not condensation. All these reasons underline the necessity for ensuring that 'all-round' protection is provided. Care must also be taken to ensure that excessive quantities of water are not trapped within the material during the application of the coating.

The principles outlined above in connection with rainwater are also applicable to other sources of water, although many of these will usually be of a localized nature. Thus, special consideration may only have to be given to the selection and use of certain sheet materials, for example, where there is a risk of excessive condensation.

3 Water vapour

Emphasis is given here to the significance and effects of water vapour on the moisture movements of materials (see *3.2 Exposure*, 'Atmosphere', p. 165, for an explanation of the sources, measurement and variability of water vapour in the atmosphere).

Once the moisture equilibria of materials are known, and provided these can be related to moisture movement, then it is possible to assess approximately the range of movements which are likely to occur due to either diurnal or seasonal variations in relative humidity (see 'Definitions of terms, 3 Moisture movement (c) Interpretation of values', p. 226). The graph in Chart C3.3/4 gives moisture equilibria for some materials. The values, it should be noted, are averages and do not take into account variations which do occur within one type of material.

It is important to note that the magnitude of changes in relative humidity need to be related to the time over and during which the changes occur. For example, the average diurnal and annual variations of relative humidity (i.e. between maxima and minima) recorded at Kew Observatory over a number of years give a low of 9% in January, a high of 30% in July with an annual value of 22%.

After making due allowances, the maximum ranges of relative humidity, whether diurnal or seasonal, are unlikely to cause movements of the magnitude of those given in Chart C3.3/1. In fact, the actual movement will be significantly less. This, however, would be the situation in general. The special cases mentioned earlier could make very important differences (continuous exposure to a heat source may cause excessive drying; continuous exposure to water vapour producing sources may lead to excessive wetting), in which case the magnitude of movements given in Chart C3.3/1 could be attained.

4 Interaction between thermal and moisture movements

A somewhat complex situation arises when materials with significant moisture contents are exposed to solar radiation (or other intense sources of heat). In cases such as these, movements caused by temperature and moisture changes may often – perhaps usually – act in opposition, and thus hold a fortuitous **balance**. This tends to be an additional safety measure against trouble due to extreme dimensional fluctuations in some circumstances.

Timber

1 Generally

(a) Scope

The moisture movements of timber and the factors that influence the movements are discussed or explained comprehensively here. There are, however, detailed differences in the properties and behaviour of all the available species of timber, as explained below. It is beyond the scope of this book to consider every species. However, there is a general pattern to the behaviour of timber, and this can be useful in gaining an understanding of the special problems encountered in its use. The pattern also provides a basis upon which variations can be assessed. Thus *average* values or *typical* patterns of behaviour are used.

Timber-based products such as plywood, blockboard, laminboard, laminated construction, fibreboards, particleboards and woodwool slabs have been developed, in part at any rate, in attempts to overcome some of the inherent weaknesses of timber, not least those associated with moisture movement. In general, these products have moisture movements of a smaller magnitude than normally converted timber. For this and other reasons, timber-based products are better considered separately.

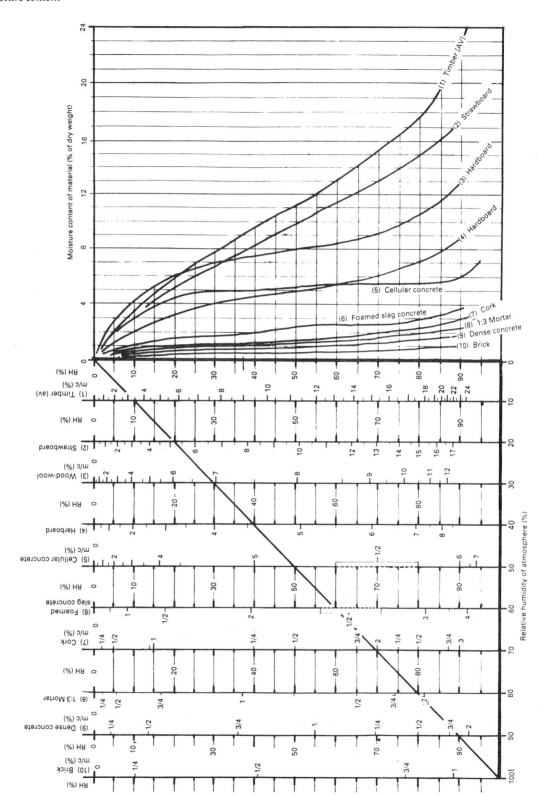

Chart C3.3/4 *Comparison of equilibrium moisture contents of some materials at various humidities.* Note: All values interpolated from graphs in the following references. References: (1) FPRL Leaflet No. 9, Moisture content of timber, HMSO, London, revised November 1963; (2)–(4), (6), (8)–(10), Ragsdale, L. A. and Raynham, E. A., Building Materials Practice, Edward Arnold, London, 1964, Fig. 26, p. 96; (5), (7) Billington, N. S., Thermal Properties of Buildings, Cleaver-Hume, London, 1952, Fig. 5.3, p. 117

(b) Significance of variations

When compared to other materials for building, moisture movements in timber presents special problems. Most of these arise because timber is a natural product with a complex cellular structure, being cut down for use while still *alive*. In addition, there is a variety of species available, each with its own characteristics. There are variations not only in the properties and behaviour between species but also between samples of the same species. Apart from moisture movements in timber being far greater than any other material (see Chart 3.3/1, referred to earlier), there are usually significant differences between tangentially and radially cut timber, whatever the species. Variations in moisture movement arise because different species have different fibre saturation points, while hygroscopicity also plays an important part.

Examples of variations in moisture content in living trees may also help to underline the fact that timber cannot really be considered as one material but rather as a group of materials, the individual members of which have characteristics which differ from one another. The moisture content in living trees may vary from 40% to 200%, while further variations in the moisture content of sapwood and heartwood, particularly in softwoods, do occur. For example, the sapwood in long-leaf pitch pine has a moisture content in excess of 100% while the heartwood ranges from 30% to 40%.

(c) Classification of moisture movements

BRE Technical Note 38 (PRL), *The movement of timber*, May 1969 (revised November 1982), quotes values of moisture movement for over 180 different species of timber. Inevitably, few of the values are the same, although in some cases the differences may have little practical significance. For purposes of comparison, BRE have classified timbers according to their degree of movement. There are three classes (small, medium and large), each based on the sum of the percentage tangential and radial movements as follows:

Class	Amount of movement	Sum of movements (%)
1	Small	Less than 3.0
2	Medium	Between 3.0 and 4.5
3	Large	Exceeds 4.5

Chart C3.3/5 illustrates the movements of a selected number of species from the BRE Note, grouped in the three classes described above. The wide variation between species and type of conversion is immediately apparent, not only between the classes but also between the species in the same class.

(d) Movements and distortion

It should be noted that there is not necessarily a direct relationship between the movement classification and the way a timber will distort in service. Thus a timber classified as having a small movement value does not mean that it will distort less than one in the medium or large class. Table T3.3/1 compares moisture movements in common timbers as to their tendency to distort and to swell and shrink. Factors that influence distortion include grain direction, the presence of knots and other structural features. The influence of these is greater for distortion than for movements in the tangential and radial directions. The factors that influence shrinkage and movement of timber are explained below under '2 Effects of conversion'. Suffice it to note here that, in general,

Class	Timber	Tangential Value	Radial Value
1 – Small movement values	African walnut	1.3	0.9
	Afrormosia	1.3	0.7
	Agba	1.3	0.6
	Balsa	2.0	0.6
	Cedar, South American	1.3	0.9
	Douglas fir	1.5	1.2
	Hemlock, Western	1.9	0.9
	Iraka	1.0	0.5
	Mahogany, African *	1.5	0.9
	Mahogany, C American	1.3	1.0
	Obeche	1.25	0.8
	Opepe	2.0	1.1
	Pine, yellow	1.7	0.9
	Rosewood, Indian	1.0	0.7
	Rhodesian, teak	1.6	1.0
	Spruce, British	1.3	0.9
	Teak	1.2	0.7
	Utile	1.8	1.6
	Western red cedar	0.9-1.9	0.45-0.8
2 – Medium movement values	Ash	2.5	1.5
	Elm, English	2.4	1.5
	Jarrah	2.6	1.8
	Keruing	2.5	1.5
	Mahogany, African **	1.8	1.3
	Oak, English	2.5	1.5
	Parana pine	2.5	1.7
	Pine, Carribean pitch	2.6	1.4
	Pine, Scots	2.1	0.9
	Poplar, Black Italian	2.8	1.2
	Red wood	2.1	0.9
	Sapele	1.8	1.3
	Spruce, European	2.1	1.0
	Sycamore	2.8	1.4
	Walnut, European	2.0	1.6
3 – Large movement values	Ash, Japanese	3.5	1.5
	Beech	3.2	1.7
	Gurjun	3.3	2.0
	Oak, Turkey	3.3	1.3
	Olive, East African	2.9	1.7
	Ramin	3.1	1.5
	Wattle, black	3.5	1.2

Moisture movement (% original length range 90%–60% R.H. at 25°C)

*   (Khaya invorensis)
** (Khaya grandifolioial)

Chart C3.3/5 *Comparison of tangential and radial moisture movements in a selection of timber species grouped in three classes of movement (small, medium and large) – see text. (Values from BRE, Princes Risborough Laboratory, Technical Note 38, Building Research Establishment, Crown Copyright)*

Table 3.3/1 *Moisture movements of common timbers*

Tendency to distort:	Tendency to swell or shrink:		
	Small	*Medium*	*Large*
Small	Douglas fir WR cedar Guarea Mahogany (America) Makore Obeche Walnut (African)	Ash Greenheart Jarrah Parana pine Sapele Utile	Birch
Medium	Afrormosia Idigbo Mahogany (African) Teak	Elm Keruing Oak Opepe	Beech
Large	Agba Hemlock Iroko	Lime Redwood (Europe) Sycamore Whitewood (Europe)	Ramin

the smaller the difference between the tangential and radial movement values for a species, the less the cross-sectional distortion consequent upon changes in moisture content.

(e) Correct designation

It should be recognized that, in practice, it is important for care to be taken when selecting a timber for specific characteristics of movement (or any other property, for that matter). The full designation of the timber required should be used rather than generalized descriptions such as 'mahogany', 'oak', 'pine', 'spruce' and 'teak'.

2 Effects of conversion

(a) Definition and recognition

There are a number of different ways in which a log may be sawn so as to provide sections of timber convenient and suitable for building and other uses. The process is known as conversion.* However, there are basically two different sections or types of 'cut' which may result.

One is *tangential* (derived from cutting basically at right angles to the rays and producing what is commonly referred to as flat-sawn or through and through sawn timber) and *defined* as timber converted so that the annual rings meet the face at an angle of *less* than 45°. The other is *radial* (derived from cutting basically parallel to the rays and producing what is commonly referred to as quarter- or rift-sawn timber) and *defined* as timber converted so that the annual rings meet the face at an angle *greater* than 45°. Tangential and radial sections of timber are illustrated in Diagram D3.3/2.

The diagram also shows the manner in which both types will shrink after drying. Not only is the shrinkage greater in the tangential

*Conversion takes place prior to seasoning. Subsequent sawing or shaping of converted and seasoned timber is more commonly known as 'manufacture'.

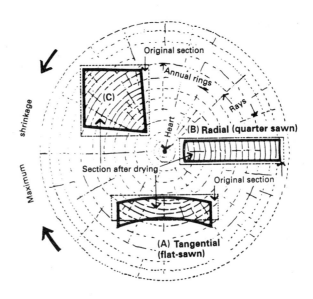

Diagram D3.3/2 *Illustration of the conversion of a log into 'true' tangential and radial sections (A and B) and the associated distortions of the sections compared with a square section (C) – compare with Photograph P3.3/1. (Based on* Principles of Modern Building, *Vol. 1, HMSO, 1959, Fig. 11.3, p. 123, Crown Copyright)*

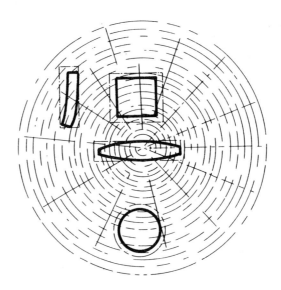

Diagram D3.3/3 *Distortion of timber cut from different parts of a log – compare with Diagram D3.3/2*

direction, that is, in the direction of the annual rings, but there is also notable distortion because the annual rings tend to bend away from the heart.

(b) Basic distortions in section

Thus, the wide faces of the rectangular tangential section (A) deform with a noticeable curvature, while the narrow end faces of the rectangular radial section (B) are, by comparison, only moderately curved and, although the magnitude of shrinkage is still greatest in the tangential direction, the section does not distort. In the case of the square section (C), the net result of the tendency of the annual rings to bend away from the heart is to cause significant uneven shrinkage on the two 'radial' faces. These are the basic and most important effects of conversion in terms of movement across the section of a piece of timber. In practice, the annual rings will not always follow the orientation shown in Diagram D3.3/2. In some cases they are quite different, as illustrated for comparison in Diagram D3.3/3. Distortions that may occur along, rather than across, a timber section are discussed later.

(c) Practical implications

The differences in the magnitude of movement in the tangential and radial direction have already been included in the preceding charts. However, as radially cut timber is far more stable, in terms of distortion particularly, than tangentially cut timber, it is important to note the practical implication of the distortion of tangentially cut timber. In this the relationship between the curvature of the annual rings and the backgrounds to which timber may be fixed are relevant. Possible results which may occur when tangentially cut timber is used in such situations as skirtings, cover fillets and floors are illustrated in Diagram D3.3/4. It should be noted that, in order to illustrate the principle involved, the distortions have been *greatly exaggerated*, while the effects of restraint by any fixings which, within limits, could mitigate the bending effect, have been generally disregarded.

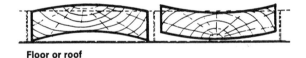

Floor or roof

Cover mould

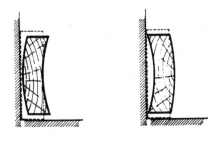

Skirting

Diagram D3.3/4 *Distortion due to drying shrinkage of timber used in three typical locations – compare with Diagrams D3.3/2 and D3.3/3. (Dashed lines represent original section before drying)*

3 Equilibrium moisture content (emc)

(a) Effect of relative humidity and temperature

(1) Generally
The emc of wood is dependent on the moisture content of the surrounding air. The latter is, in turn, the result of the combination of relative humidity and temperature. Chart C3.3/6 illustrates how relative humidity and temperature affect the moisture content of wood. It will be seen that there is a significant difference in the emc between the temperature ranges 15–40°C and 40–70°C.

As is common with almost all test results on wood in which average values are given, there can be and usually are variations, sometimes significant ones, between different pieces of timber of the same species or between different species. As regards the latter, for example, species such as teak and afromosia have values of emc that are from 2% to 3% below the average values for all species.

(2) Effect of drying
The values in Chart C3.3/6 were obtained from test specimens during a drying cycle. These would have been about 2% lower if the wood had first been dried to a low moisture content and then allowed to absorb moisture and come into equilibrium with air at higher humidities. This phenomenon is referred to as hysteresis: when desorption and absorption curves are drawn on the same axis of a graph they form what is known as a hysteresis loop.

(3) Effect of temperature and steaming during drying
The emc of wood dried from a green condition by high-temperature and steaming conditions may be 2% or 3% less than that of air-dried

Chart C3.3/6 *Comparison of the effect of both relative humidity and temperature on the equilibrium moisture content of timber for a selected range of relative humidity and temperature. (Values interpolated from graph in BRE Report,* Timber drying manual, *revised by C. H. C. Turner, 1986, Fig. 2, p. 5, Building Research Establishment, Crown Copyright)*

RH %	Temp °C	Value	Moisture content-%
20	15	6	40
	25	5.5	70 ‖ 15
	40	5	
	70	3	25
40	15	9.8	
	25	9.2	70 40 ‖ 15
	40	8	
	70	5	25
60	15	14.2	
	25	1.3	70 40 15
	40	11.8	
	70	7	25
80	15	20	25 15
	25	18.4	70 40
	40	16.4	
	70	10.5	

Note: Values are approximate only.
Observations made during a drying cycle.
Average data from six species.

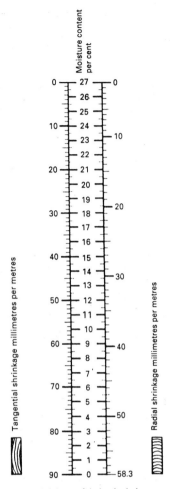

Average shrinkage of timber in drying
(shrinkage varies considerable according to species).
Millimetres per metre of original length

Chart C3.3/7 *Conversion chart given the shrinkage movement of timbers relative to moisture content.* Note: *Values for different species vary. The chart is based on* average values. *(Values interpolated from graph in* Principles of Modern Building, *Vol. 1, HMSO, London, 1959, Fig. 11.2, p. 112, Crown Copyright)*

timber. This accounts in part for the comparatively low emc of plywood. High temperatures are used in the manufacture of plywood.

(4) Movements
As a guide, Chart C3.3/7 gives the shrinkage of tangential and radial sections of timber relative to moisture content – importantly, the chart gives average values only. For most practical purposes, the amount of swelling which will take place during increases in moisture content can normally be taken as the same as the values given in the chart. Some jointing methods that allow for movements are illustrated in Diagram D3.3/5.

(b) In service – 1: Generally

In practice, account should be taken of the fact that the moisture content of timber is affected by the temperature and relative humidity of the air surrounding it, as explained earlier. Ideally, the moisture

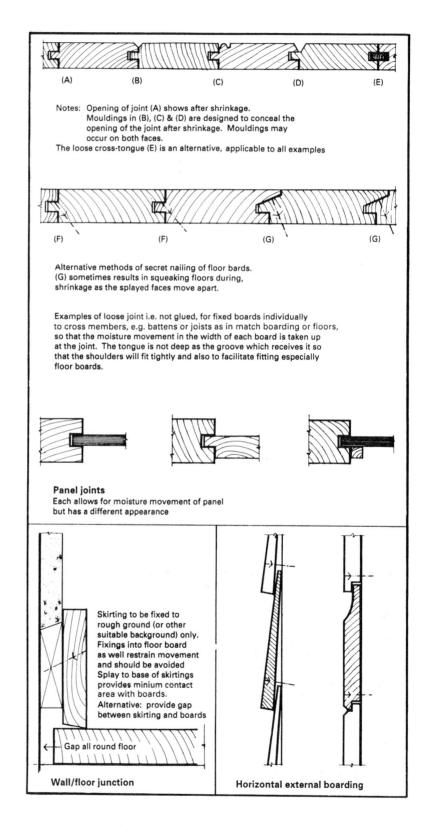

Notes: Opening of joint (A) shows after shrinkage.
Mouldings in (B), (C) & (D) are designed to conceal the
opening of the joint after shrinkage. Mouldings may
occur on both faces.
The loose cross-tongue (E) is an alternative, applicable to all examples

Alternative methods of secret nailing of floor bards.
(G) sometimes results in squeaking floors during,
shrinkage as the splayed faces move apart.

Examples of loose joint i.e. not glued, for fixed boards individually
to cross members, e.g. battens or joists as in match boarding or floors,
so that the moisture movement in the width of each board is taken up
at the joint. The tongue is not deep as the groove which receives it so
that the shoulders will fit tightly and also to facilitate fitting especially
floor boards.

Panel joints
Each allows for moisture movement of panel
but has a different appearance

Skirting to be fixed to
rough ground (or other
suitable background) only.
Fixings into floor board
as well restrain movement
and should be avoided
Splay to base of skirtings
provides minium contact
area with boards.
Alternative: provide gap
between skirting and boards

← Gap all round floor

Wall/floor junction

Horizontal external boarding

Diagram D3.3/5 *Allowing for moisture movements in timber in typical examples*

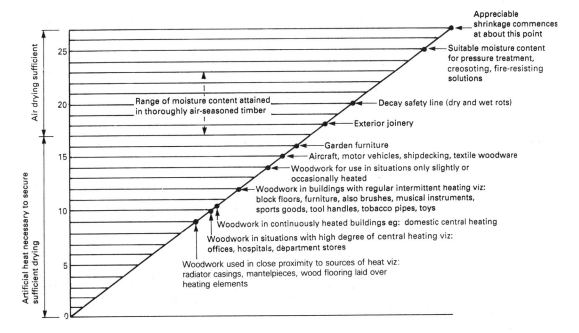

Diagram labels:

Appreciable shrinkage commences at about this point

Suitable moisture content for pressure treatment, creosoting, fire-resisting solutions

Range of moisture content attained in thoroughly air-seasoned timber

Decay safety line (dry and wet rots)

Exterior joinery

Garden furniture

Aircraft, motor vehicles, shipdecking, textile woodware

Woodwork for use in situations only slightly or occasionally heated

Woodwork in buildings with regular intermittent heating viz: block floors, furniture, also brushes, musical instruments, sports goods, tool handles, tobacco pipes, toys

Woodwork in continuously heated buildings eg: domestic central heating

Woodwork in situations with high degree of central heating viz: offices, hospitals, department stores

Woodwork used in close proximity to sources of heat viz: radiator casings, mantelpieces, wood flooring laid over heating elements

Air drying sufficient

Artificial heat necessary to secure sufficient drying

content to which timber should be dried before it is machined and used is the average or somewhat below the average that the item will attain in use. However, it is not always essential to match the moisture content accurately to conditions that are likely to prevail in service. This is particularly true of the specification of moisture contents for certain structural components. On the other hand, for many uses (for example, as flooring, doors and furniture) it is necessary to specify moisture contents close to those that are expected to be achieved in service if a satisfactory performance of the timber is to be obtained.

For basic guidance, Diagram D3.3/6 illustrates typical moisture contents of timber in various environments. The moisture contents should be treated as 'a first approximation' only. Importantly, the points shown are averages; actual values may vary by 1% or 2% moisture content from the point values indicated.

Diagram D3.3/6 *Moisture contents of timber in various environments.* Note: *The points show the averages; actual values may vary by 1% or 2% moisture content from the point values shown. (From BRE Report,* Timber drying manual, *Fig. 3, p. 6, Building Research Establishment, Crown Copyright)*

(c) In service – 2: Specification

(1) The service environment

Although problems associated with changes in the moisture content of timber may arise because the timber was placed in an environment that was either too damp or too dry for it, the main aim when specifying moisture contents is to ensure that problems will not arise because the timber is placed in an environment that is too dry. The environment that determines the emc of the timber may not be the one in which the timber is finally used. Consider, for example, garden tools with wooden handles. The latter will attain their emc with conditions in a garden shed but may be stored in a heated supermarket for many months. Their moisture condition during manufacture should have been adjusted to the conditions in service (i.e. in the garden shed).

(2) Objectives
A realistic specification would state:
• The average moisture content of the batch of timber;

- The tolerance limits on the average moisture content within individual pieces either at different depths (i.e. moisture content gradients) or at different positions along the grain;
- The method of measurement used to determine the above.

(3) Categories

Currently, there are at least 50 British Standards that specify the moisture content of timber and timber products. However, as yet the Standards differ in both requirements and in the way these are defined. According to TRADA, three categories of BS specification of moisture content can be identified, depending on the reason for need to control the moisture content of timber. For the present purposes, the first category that relates to products for uses such as furniture, wood flooring blocks and joinery is the most relevant. The other two categories relate to specifications for processing (e.g. preservative treatments, gluing, machining and fabrication) and the control of moisture content in timber or panel products.

(4) Constructional timber

Timber used for carcassing, roof timbers and joists will ultimately reach moisture contents that range between 12% and 20%, depending on their location and the type of heating in the building (see Table T3.3/2).

Table 3.3/2   *Moisture content for different uses*

Position of timber in building	Average moisture content attained in service conditions (%)	Moisture content which should not be exceeded at time of erection (%)
External uses, fully exposed	18 or more	–
Covered and generally unheated	18	24
Covered and generally heated	16	21
Internal in continuously heated buildings	14	19

From BS 5268.

In these items a limited amount of shrinkage and distortion due to drying *in situ* can be tolerated. BS 5268: 1989, *Structural use of timber: Part 2 Code of practice for permissible stress design, materials and workmanship*, limits the moisture content for *two* conditions of service:

- Not exceeding 18% for *dry exposure* (i.e. including most covered buildings and internal use); and
- Exceeding 18% after construction for *green exposure* (i.e. when timber is used outside).

(5) Joinery

The range of moisture contents for joinery in service is between 6% and 20%. The moisture content likely to be achieved depends on the location of the item and the heating in the building. Chart C3.3/8 illustrates the range of moisture contents to be expected in certain locations/conditions of heating.

In BS 1186: Part 1: 1986, *Timber for and workmanship in joinery*, a tolerance of ±3% is allowed for external joinery but this is reduced to

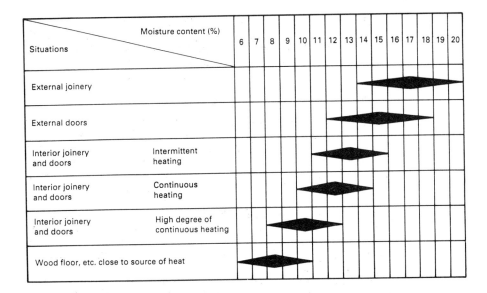

Situations \ Moisture content (%)	6	7	8	9	10	11	12	13	14	15	16	17	18	19	20
External joinery									◆						
External doors								◆							
Interior joinery and doors — Intermittent heating						◆									
Interior joinery and doors — Continuous heating				◆											
Interior joinery and doors — High degree of continuous heating			◆												
Wood floor, etc. close to source of heat	◆														

Chart C3.3/8 *Range of moisture content likely in flooring and joinery. (From BRE Princes Risborough Laboratory, Technical Note No. 12,* Flooring and joinery in new buildings, How to minimise dimensional changes, *February 1966, Fig. 1, p. 3, Building Research Establishment, Crown Copyright)*

±2% for internal joinery. The average moisture contents for joinery specified in the Standard are as follows:

External joinery

Floor-level hardwood sills and thresholds	19 ± 3%
All other external joinery	16 ± 3%

Internal joinery

For buildings with intermittent heating	15 ± 2%
For buildings with continuous heating providing room temperatures of 12–19°C	12 ± 2%
For buildings with continuous heating providing room temperatures of 20–24°C	10 ± 2%

4 Control of hygroscopicity

(a) Treatments

As the hygroscopic nature of wood is responsible for causing the variations in the moisture content of timber and the moisture movements that result from them, attempts have been made to eliminate this inherent 'disadvantage'. The treatments developed do not eliminate hygroscopicity completely but do reduce it permanently to some extent. In these, the accessibility to water of the matrix constituents of wood is reduced. Three methods are available:

1. The substitution for the active bonding links of the matrix which are less attractive to water. The process is known as acetylation. It is very effective but little practised.
2. The impregnation of the cell walls with chemicals which hold them in a swollen state after the removal of water. This minimizes further dimensional movement. Both phenol formaldehyde and polyethylene glycol (PEG) have been used successfully.
3. The impregnation of the wood with low-viscosity liquid polymers. These are then polymerized *in situ* to convert them to solid polymers. Accessibility of water is virtually eliminated because the

polymer impregnates the cell walls as well as coating the cell cavities. Wood treated in this way is referred to as a wood–plastic composite. It is four times as expensive as wood but is being used commercially for one or two specialized purposes where dimensional stability and increased abrasion resistance are required.

(b) Coatings

Oil- and alkyd-based paints and varnishes are claimed to slow down or even remove the movement of wood underneath them. Such claims are true in the short term when daily changes in the moisture content of the surrounding air will have little effect on the wood, as explained earlier. As none of the coatings are completely impermeable to water vapour, the wood will absorb such vapour if it is exposed to high moisture contents in the surrounding air for a long period of time.

In contrast, special paints known as 'microporous paints' are available to enable the wood to breathe, thereby it is claimed, reducing the tendency of the more vapour-resistant coatings to premature failure. Microporous paints may be said to encourage changes in moisture content.

There are conflicting views about the need or otherwise to coat timber or the effectiveness or otherwise of such coatings in reducing the amount of moisture movement or the risk of decay.* Nevertheless, it appears to be advantageous to provide sealers and/or coatings in the following circumstances:

- The end grain of timber in joints in joinery prior to the joints being formed, and
- Those parts of timber units or elements that are in close contact with damp porous materials, during the early life of a building in particular.

Where on-going dampness can be expected, such as occurs around openings with timber window and door frames, a material impervious to water in both liquid and gas states (i.e. a dpc material) is essential.

5 Other distortions

In addition to distortions along the section of tangential or radial converted timber (see '2 Effects of conversion' earlier, p. 243), a piece of timber may also distort along its length due to moisture movements of the wood. Three characteristic distortions are illustrated in Diagram D3.3/7. Their causes and occurrences are briefly:

1. *Twist*: This is caused by the spiral interlocking grain which is common in some tropical hardwoods and may occur in isolated cases in imported softwoods. It can be minimized by stacking the converted timber flat under weights before and during drying. To be serviceable, the dimension X should not exceed 1 mm per 25 mm width in any length of 3 m.
2. *Bow*: Bow is generally caused by bad stacking (that is, the supports for the timber, known as stickers, were spaced too widely, thereby allowing the timber to sag under its own weight) before or during drying. It can sometimes be removed by wetting and redrying. The dimension X should not exceed one half of the thickness in any length of 3 m.
3. *Spring*: This is caused by the release of internal stresses when the log is sawn. It is not uncommon in boards from near the 'core' or 'heart' of a log. Certain species are more susceptible than others to spring. It can only be corrected by reducing the width of the section. The dimension X should not exceed 15 mm in any length of 3 m.

*It is important to note that preservative treatments are used essentially to reduce the risk of decay of wood. Most have little effect on the inherent moisture movements of the timber.

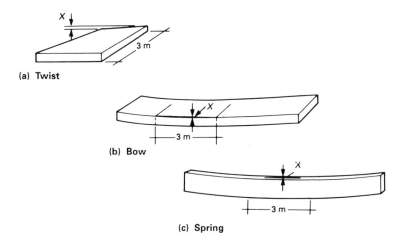

(a) Twist

(b) Bow

(c) Spring

Diagram D3.3/7 *Distortions along the length of a piece of timber. (Based on* Specification 87*, Architectural Press, London, published annually)*

6 Precautions

In practice, a number of requirements influence the precautions to be taken to minimize the effects of moisture movements in timber. The principal aim should be to ensure that the timber is specified with the appropriate moisture content and that due allowance has been made for such movements as might occur before the emc of the wood has been attained.

Summarized, the precautions that should be taken to minimize the magnitude of changes in moisture content, *after* timber has been converted, seasoned and/or preserved are one or more of the following:

1. Careful storing of the timber after manufacture;
2. Protecting the timber during delivery to the site;
3. Careful storing on-site – before and after fixing – avoiding conditions of exposure to moisture, as liquid or vapour, which are materially different from those expected during service;
4. Allowing for moisture movements that may occur until the timber attains its emc – notably, but not exclusively, during the drying out of wet constructions;
5. Sealing the end grain of timber in joints before the joints are formed;
6. Ensuring that joinery to be painted subsequently on-site is primed immediately after its manufacture or, failing that, before it leaves the joinery works;
7. Avoiding reliance on paint coatings to timber that may be exposed to damp conditions for a long period of time;
8. Ensuring that there is a dpc wherever timber will be in contact with damp porous materials for long periods of time (e.g. at window- and door-frame abutments with masonry).

Sources

Desch, H. E. (revised by Dinwoodie, J. M.), *Timber, its Structure, Properties and Utilisation*, 6th edn, Macmillan, London, 1981

BRE
Defect action sheet
11, *Wood windows and door frames: care on site during storage and installation*, December 1982

Digests
261, *Painting woodwork*, May 1982
286, *Natural finishes for exterior timber*, June 1984
287, *Specifying structural timber*, July 1984
304, *Preventing decay in external joinery*
321, *Timber for joinery*, May 1987
Princes Risborough Laboratory, Technical Notes
12, *Flooring and joinery in new buildings*, February 1960 (reprinted April 1980)
38, *The movement of timbers*, May 1969 (revised November 1972)
46, *The moisture content of timber in use*, October 1970 (revised September 1982)

Reports
Pratt, G. H. *Timber drying manual*, (2nd edn revised by C. H. C. Turner), 1986
McIntyre, J. S., *Moisture conditions in the walls of timber frame houses – the effects of holes in vapour barriers*, 1986

TRADA
TRADA Wood Information, Section 4, Sheet 14, *Moisture content of timber – its importance, measurement and specification*, October 1985 (revised)

Timber-based products

1 Generally

(a) Scope

This section includes a brief summary of the characteristics and factors related to the moisture movements of board materials made of timber that have been cut up and then reformed often, but not always, with the aid of a synthetic resin. These materials are also known as timber panel products. Much of the coverage given here is also relevant for the decay of these products (see *3.6 Chemical attack*, p. 431 ff).

(b) Background

(1) Overcoming disadvantages of solid timber
Timber-based boards have been developed to overcome three disadvantages of solid timber:

- Its variability in performance;
- Its availability in limited widths; and
- The marked difference in its properties along and across the grain.

(2) Characteristics of boards
These disadvantages are removed if the wood is reduced in size and then reassembled in a particular way. The main board materials are therefore characterized by:

- Much-reduced moisture movement in the plane of the board compared to solid timber;
- Such movements as do occur are the same or very similar both along and across the length of the board;
- Greater movement across the thickness of the board; and
- Availability of large sizes, 2400 × 1200 mm being common.

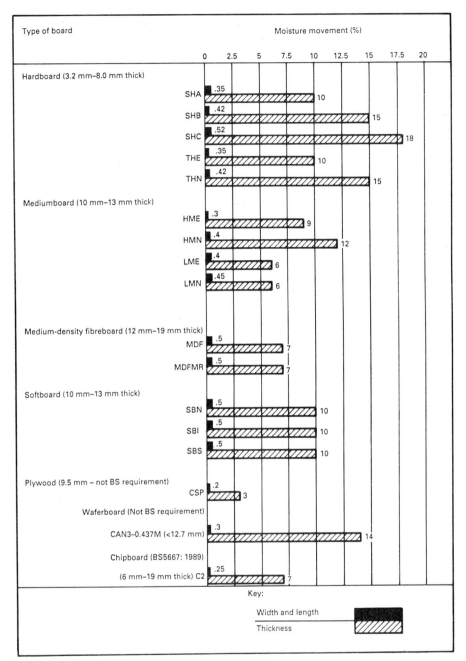

Chart C3.3/9 *Comparison of moisture movements in timber-based boards. (Values from BRE IP12/91, Uses of fibre building board, by J. M. Dinwoodie and B. H. Paxton, June 1986, Table 1, Building Research Establishment, Crown Copyright)*

(c) Moisture movements compared

Chart C3.3/9 illustrates the moisture movement of typical board materials along their length, width and thickness. It will be seen that for most of the boards the movement along their length and width is the same, the range varying between 0.10% and 0.40%. In contrast, all have greater movements across their thickness, varying between 6% and 23%. However, put into context, both directions are or could be significant in terms of actual dimensional change. For example, after absorbing moisture for the range stated in the chart the change in dimension of a board 2400 mm long would be 6 mm for a value of 0.25% and 9.6 mm for a value of 0.40%. Similarly, a board 20 mm thick would swell 2 mm for a value of 10%.

2 Plywood

(a) Definition

According to Desch, plywood is defined internationally as 'panels consisting of an assembly of plies bonded together with the direction of the grain in alternate plies usually at right angles. In general, the outer and inner plies are placed symmetrical on both sides of a central ply or core' (**balance**).

(b) Subdivision

Again according to Desch, plywood is subdivided internationally into two distinct groups: *veneer plywood* and *core plywood.*

(1) Veneer plywood
Veneer plywood is made up of plies of veneer up to 7 mm thick orientated with their plane parallel to the surface of the panel.

(2) Core plywood
Core plywood has a core that may consist of solid wood, veneers, a cellular construction or materials other than solid wood or veneers. The group is subdivided into:

1. *Boards*
- *Battenboard*, in which the core may consist of strips of solid wood more than 30 mm wide which may or may not be glued together;
- *Blockboard*, in which the core is made of strips of solid wood more than 7 mm wide but not wider than 30 mm which may or may not be glued together;
- *Laminboard*, in which the core is made of strips of solid wood or veneer not wider than 7 mm placed on edge and glued together;
2. *Cellular plywood*: The core consists of a cellular construction with at least two cross-banded plies on both sides of the core.
3. *Composite plywood*: The core (or certain layers) is (are) made of materials other than solid wood or veneers with at least two cross-banded plies each side of the core.

(c) Production

Only certain species of timber are suitable for the production of plywood. Those that are appropriate need to have certain peeling characteristics, veneer smoothness and bending properties. The veneers are peeled from a log pretreated by steaming or boiling – a 'softening' process – and dried and graded into one of three qualities. They are then glued, being assembled at right-angles to produce a **balanced** construction. The assembled packs of veneers are then pressed in heated multi-daylight presses. The normal sheet size for plywood is 2400 × 1200 mm, but larger sizes can be produced using special processing techniques.

Plywood may be produced in one of three forms: unsanded, sanded or scraped. Other layers may be added in further processing to produce plywood with faces of foil, metal, plastic, phenolic film, paper and hardboard or the face (or faces) may be prime painted, texture coated or printed. Plywood may be impregnated (ideally, at the veneer stage) with preservatives and flame retardants.

(d) Adhesives and their effects

The characteristics and end use of a plywood is determined by the adhesive used. For some uses (e.g. furniture) the adhesive need not necessarily be resistant to moisture; for others (e.g. as external cladding) it needs to be capable of withstanding all types of weather without breaking down. In contrast, plywood used within a construction may have to withstand short periods of wetting and subsequent drying.

Four types of adhesive may be used:

1. *Type WBP: Weather and boil proof*: the bond is highly resistant to weather, microorganisms, cold and boiling water, steam and dry heat. It is important that the plywood selected be also capable of resisting decay.
2. *Type BR: Boil resistant*: the bond is resistant to weather and to a boiling-water test but fails under the very prolonged exposure to weather that type WBP will survive. The bond will resist cold water for many years and is highly resistant to attack by microorganisms.
3. *Type MR: Moisture resistant and moderately weather resistant*: the bond is sufficient for short-term exposure to the weather (i.e. for a few years only) and will withstand cold water for a long period but hot water for a limited time, failing under a boiling-water test. The bond is resistant to attack by microorganisms.
4. *Type INT: Interior*: the bond is resistant to cold water and is not required to withstand attack by microorganisms.

3 Fibre building board

The term 'fibre building board' includes a range of boards with quite different properties and appearance.

(a) Production

With one exception (medium-density fibreboard – MDF), fibre building boards are made of reconstituted timber bonded together, *without* the aid of adhesives. The basic strength and adhesion is obtained from the felting together of the (wood) fibres themselves and from their own inherent adhesive properties.

Small chips of wood, usually from small-diameter logs but sometimes from waste material from timber processing, are defibrated to bundles of wood fibres by either mechanical grinding (the 'Defibrator' method) or by exploding (the 'Masonite' process). *One of two* processes may be used to bind together the fibres: the traditional 'wet' or the newer 'dry' process:

1. *In the 'wet' process* the defibrated wood is mixed with water to produce a pulp to which various chemicals may be added, depending on the board type being produced. The wet pulp is then laid on a moving wire mesh that encourages the interlocking of the fibres (felting) to form a wet sheet, the matt. This is allowed to dry partially and is then cut into sheets. The final drying is done in drying ovens. The sheets may be pressed. After leaving the presses, the boards are subjected to a conditioning treatment in humidifying chambers to increase the moisture content to 4–8%. Heat treatments

and oil-tempering are carried out to produce boards of improved strength, moisture resistance and dimensional stability.

2. *In the 'dry' process* the fibre bundles are dried to a low moisture content and, in the case of medium-density fibreboard (MDF), are impregnated with a resin adhesive, usually urea formaldehyde, at about 10% by weight. The impregnated fibres are laid in a matt and then cut up and hot-pressed to produce a board with two smooth faces. The type and amount of adhesive used govern the performance of the board.

(b) Characteristics

By being the only timber-based board which is reconstituted rather than consisting of veneers or particles bonded together, fibre building boards have the following favourable characteristics:

1. The total absence of surface defects. This makes the material ideal for painting and surface finishing.
2. A greater resistance to timber pest infestation and fungal decay because sugars and starches (i.e. much of the food) are removed during the production and pressing of the fibres.
3. The attainment of a lower equilibrium moisture content.

(c) Types

Fibre building boards are grouped for convenience according to their density. There are, however, several types within each grouping. Each type can vary in strength and resistance to moisture. From Chart C3.3/9 it will be seen that differences in moisture movment in length and width between the types is relatively small (i.e. from 0.25% to 0.40%). In contrast, the differences are more marked across the thickness (i.e. from 5% to 15%). However, when applied in context, the potential movements can be significant (see 1(c) earlier).

The types, with the designation given to them in BS 1142, are:

1. *Hardboards* (density exceeding $800 \, kg/m^3$)
   S: Standard grade
   TN: Tempered hardboard; Oil-impregnated hardboard – having increased density, strength and moisture resistance
   TE: Tempered hardboard: Oil-impregnated hardboard – having even higher levels of density, strength and moisture resistance
2. *Medium boards* (density $350–800 \, kg/m^3$)
   High-density medium boards ('panelboard') – density $560–800 \, kg/m^3$
   HMN: Standard high-density medium board
   HME: High-density medium board – with improved moisture resistance
   Low-density medium boards – density $350–560 \, kg/m^3$
   LMN: Standard low-density medium board
   LME: Low-density medium board with improved moisture resistance. (Note: The industry has produced another grade: *Sheathing medium board*. This is a special 9.0 mm thick medium board, not specified in British Standards, with defined levels of vapour permeance and racking resistance.)
   MDF: A medium-density fibreboard: A medium-density board produced by the 'dry' process and having a density range of $640–860 \, kg/m^3$
3. *Insulating boards* (density below $400 \, kg/m^3$)
   Softboard – as panels or tiles with a density less than $350 \, kg/m^3$
   Insulating board – panels with a density less than $350 \, kg/m^3$ and thermal conductivity not exceeding 0.059 W/mK

Bitumen-impregnated insulating board – softboard impregnated with bitumen at the time of manufacture in amounts ranging from 5% to 25% by weight. (*Note*: The industry has introduced another grade of bitumen-impregnated board with a defined level of vapour permeance as well as satisfying a water-absorption test, known as *sarking and sheathing grade*. This is not yet specified in British Standards.)

(d) Precautions in use

1. In common with all timber-based materials, fibreboards require moisture conditioning before being fixed in a building. This is to reduce distortion of the board. (Hardboard, for example, leaves the mill with a moisture content of between 5% and 8% – an emc with a relative humidity of about 65%.)
2. FIDOR (Fibre Building Board Development Organization) advises two alternative methods, depending on the conditions within a building:

- *In occupied heated buildings* exposing the surfaces of the boards for a few days is generally sufficient; and
- *In buildings that have been unoccupied for some time or where 'wet' trades are active* the boards should be conditioned with water. Wavy boards can be flattened by water *before* being fixed.

3. Careful handling, stacking and storage is necessary, protecting sheets from abnormal exposure to moist or damp conditions especially.
4. Allowance for movement at joints between boards is generally necessary, but particularly in applications such as overlays in flooring.

4 Chipboard

(a) Production

Chipboards, or particleboards as they are also known, are produced generally by pressing together, under heat and pressure, small softwood chips together with some type of synthetic resin. The latter comprises about 8% of the dry weight of the board. Coarser chips are used in the centre of the board with smaller chips for the surface layers. Boards with particular requirements are produced by adjusting the geometry of the chips, the relative proportion of resin in the central and surface layers and the relative thickness of these layers. In addition, several different types of resin are used, the most common in the UK being urea formaldehyde.

There are two other types of wood-based particleboard:

1. *Waferboard*, produced from very large thin flakes of poplar wood bonded together with about 3% by weight of phenol formaldehyde; and
2. *Wood-cement particleboard* that has a similar weight of softwood chips as used in chipboard but bonded together with cement.*

(b) Moisture sensitivity

In general, chipboards are characterized by their tendency to swelling in the thickness of the board on absorption of either liquid water or water vapour.† The swelling often results in a loss of strength and is

*This board should not be confused with woodwool slabs, that are made from long wood shavings coated with cement and then compressed.

†They do, of course, also change dimensions in length and width. In the context of their use and the defects commonly associated with chipboards, these are less important, but have nevertheless should be taken into account in practice.

usually due to exposure of chipboard to relative humidities above 65% for long periods, or excessive wetting that might occur from interstitial condensation or spillage by users. Importantly, on drying there may be some residual swelling and the full original strength may not be regained – but see the note later on urea formaldehyde chipboards.

On delivery, the moisture content of chipboard is generally between 7% and 13%. The emc of chipboard in normally heated buildings is about 10%. The inherent sensitivity of chipboards to moisture depends on the way they are made (platten-pressed boards being more sensitive than extruded ones), the type of adhesive used to bond the chips together and certain additives or surface coatings.

It is important to note that, as yet, no chipboard is suitable for use externally. The apparent success of chipboards covered with certain finishes or treated with chemicals and considered as 'exterior chipboard' in other countries has not been repeated in the UK. Difficulties of ensuring sufficient edge protection or treatment of the boards appear to be the main reasons for this.

(c) Types

BS 5669 specifies four types of chipboard:

1. *Type I Standard.* Boards of this type are bonded with urea formaldehyde and are regarded as general-purpose boards more suited for the furniture industry.
2. *Type II Flooring.* Boards of this type are also bonded with urea formaldehyde but have a higher performance in regard to all properties, an additional requirement being impact resistance. They are used mostly in domestic flooring.
3. *Type III Improved moisture resistance.* In addition to having even higher values for strength and stiffness than type III, boards of this type have *a measure* of resistance to moisture. The latter is achieved by using a mixture of urea and melamine formaldehyde. Phenol formaldehyde may be used as a binder in addition. The boards are used in flat-roof construction.
4. *Type II/III Combining the strength properties of type II with the improved moisture resistance of type III.* Boards of this type are intended for use as flooring where there is a risk of high moisture contents, as may occur from spillage in kitchens and bathrooms.

It should be noted that boards bonded with *urea formaldehyde* (i.e. types I, II and III, about 95% of the chipboard used in the UK) give satisfactory performance when used and maintained under dry conditions. Moisture causes deterioration of the resin and thereby a marked reduction of performance. When wetted by water vapour or liquid water as mentioned earlier, the urea formaldehyde chipboard swells appreciably. Importantly, the changes are irreversible, so on drying, the board hardly recovers its original dimensions (i.e. it remains virtually in its swollen state) and does not regain its original strength (there is a considerable loss in its strength properties).

(d) Precautions

In common with all timber-based products, but perhaps more so with chipboard, it is *essential* that the boards are handled carefully, stacked correctly and protected properly from exposure to excessive moisture before and after installation in any part of the fabric of a building.

1. *Handling*: Boards should be handled with care, bearing in mind the vulnerability of the boards and their edges to impact damage. The

surfaces of prefinished boards should be protected from damage or contamination. Surfaced boards should be lifted from a stack and never slid.

2. *Stacking*: Boards should be piled on a level surface with all four edges flush and not stacked on edge. Slatted pallets should be so arranged to prevent boards from sagging under their own weight.

3. *Storage*: Boards should be stored in an enclosed dry building. Where this is not possible, the stacks should be covered with polythene or a tarpaulin with all boards kept well clear of the ground.

4. Only use *flooring grade* chipboard (type II) in those situations where the absence of moisture can be guaranteed throughout the life of the building.

5. Where *floors* are likely to be exposed to moisture adopt one or other of the following solutions:

- Tank the floor with a continuous plastics sheet, taken about 75 mm up the wall and seal around all pipes passing through the sheet;* or
- Use type II/III grade that incorporates a fungicide to avoid fungal attack. This results in loss of strength.

6. *In pitched roofs* it would appear better to use type III boards, possibly with a protective coating of wax. Further experience is needed before a firmer recommendation can be made.

7. *For sarking* use type II boards with a protective coating such as wax or bitumen (i.e. prefelted boards), making sure that newly cut edges are effectively sealed and the joints in prefelted boards taped.

8. *In flat roofs* ensure that type II/III chipboard, preferably with a fungicide, is used, but nevertheless check that the boards are not unduly wetted before the waterproof membrane is laid and that there is effective ventilation in cold deck designs **(continuity)**.

9. Ensure that *veneered chipboards* are **balanced**.

Sources

Desch, H. E. (revised by Dinwoodie, J. M.), *Timber, its Structure, Properties and Utilisation*, 6th edn, Macmillan, London, 1981

BRE
Defect action sheets
31, *Suspended timber floors: chipboard flooring – specification*, July 1983
32, *Suspended timber floors, chipboard flooring – storage and installation*, July 1983

Digests
323, *Selecting wood-based panel products*, September 1987
329, *Use of chipboard*, July 1980

Information papers
4/83, *Wood–cement particle board – a technical assessment*, by J. M. Dinwoodie and B. H. Paxton, April 1983
3/85, *Wood chipboard – recommendations for use*, by J. M. Dinwoodie, March 1985
4/85, *Specifying plywood*, by J. F. S. Carruthers, March 1985
5/85, *Waferboard and OSB: their composition, properties and use*, by J. M. Dinwoodie, March 1985
23/86, *Uses for fibre building board*, by J. M. Dinwoodie and B. H. Paxton, December 1986
20/87, *External joinery: end grain sealers and moisture control*, by E. R. Miller, J. Boxall and J. K. Carey, December 1987

*Do not rely on the apparent waterproof properties of some floor finishes such as vinyl tiles to provide a seal. Water penetrates at the joints between the tiles and is then absorbed by the chipboard. The absorbed water is prevented from drying out by the tiles.

BSI
BS 1105: 1981, *Specification for wood wool cement slabs up to 125 mm thick*
BS 1142: 1989, *Specification for fibre building boards* (in three parts)
BS 5669: 1979, *Specification for wood chipboard and methods of test for particle board*. In due course this standard will also include specifications for waferboard, OSB and cement-bonded particleboard
BS 6566: 1985, *Plywood* (in eight parts)

Chipboard Promotion Association
Wood chipboard data sheets, revised edition 1976. The CPA appears to be no longer in existence.

Fibre Building Board Development Organization (FIDOR)
Application data sheets AD/4, AD/5 and AD/6
Design data sheets DD/1, DD/2, DD/3 and DD/4
Sitework recommendations SR/3 and SR/5
Technical bulletins T001, T002 and T002A

Cement-based products

1 Generally

(a) Scope

Cement is used in a large range of building products. These include concretes for panels, planks, columns, beams, slabs, bricks and blocks (in dense or lightweight; plain or reinforced; cast-*in-situ* or precast forms); screeds, mortars, plasters and renders. It may also be used in timber-based products such woodwool slabs or in other composites such as glass-reinforced concrete (grc) and asbestos-free cement.* Emphasis is given here to those factors that influence the products as a group. Examples of the influencing factors and their effects on moisture movement are taken mainly from concrete technology. For completeness, the factors specifically related to individual types or groups of products are included later under 'Notes, rules and precautions', p. 268.

(b) Moisture movement characteristics

All the products are characterized by their comparatively large *initial* drying shrinkage. Most of the latter is *irreversible*, being about 50% *greater* than any subsequent *reversible* movements due to wetting and drying of the product (see Diagram D3.3/1). Controlling the drying shrinkage is therefore important if the risk of cracking is to be reduced. However, there is still a need for the effects of reversible movements to be considered. Allowances for such movements may be important in reducing the risk of cracking.

Like most hygroscopic materials, cement-based products rarely attain a completely stable condition. The moisture environment around the product will be changing constantly, although daily variations are unlikely to be significant. There is always a need to take into account the effects of exposure of any product to the weather (see 'Exposure' earlier, p. 236).

(c) The importance of quality and quality control

Unlike timber, cement-based products are manufactured. Consequently the magnitude of their moisture movements is dependent mainly on the quality of the 'ingredients' and of the method of manufacture.

*Asbestos-free cement products have organic fibres instead of asbestos fibres. The properties of the sheets now made have virtually the same properties as the asbestos cement they replace.

Cement can be said to be the root cause of the movements that occur, the initial shrinkage in particular. Aggregates help to restrain the shrinkage of the cement, the amount of restraint they provide being dependent on their source, size, shape and distribution.* Some aggregates may increase shrinkage; most make a contribution to reversible moisture movements. Water is needed to enable the cement to react chemically and to make the wet mix workable. The amount of water used influences the amount of shrinkage that will take place initially, and the quality of the water can sometimes also be influential. Quality of the materials and the degree of quality exercised throughout the whole 'production' process is a fundamental requirement for all the products. In general, the standard of quality control achieved under factory conditions is likely to be higher than can be reached normally on-site.

(d) Variations

Like timber, cement-based products have a wide range of movements, even when strict quality control has been exercised at all stages of their production. The variations arise because there are significant differences in the properties of the types of cement and aggregate available which, among other things, can be combined in a variety of proportions. As stated above, the amount of water used (i.e. the water/cement ratio) can be influential. In recent years the properties of cements have altered, mostly because of changes in their manufacture.

(e) Interrelationship of factors

To summarize, the four factors that influence moisture movements (and, of course, other important properties) of cement-based products are:

1. Type of cement used;
2. Type(s) of aggregate used;
3. Mix proportioning; and
4. Curing.

All are closely interrelated. In respect of moisture movements, drying shrinkage in particular, all are important. It should also be noted that the values for shrinkage and moisture movements quoted later are mostly based on the testing of small samples under controlled conditions. The latter rarely occur in practice. When related to practical conditions, the laboratory results may be an *underestimation* (i.e. the movements may be greater in practice).† Nevertheless, the data available provide useful guidance. They also illustrate the pattern of movements that can be expected.

2 Influencing factors

(a) Type of cement

The cement is or may be said to be the *primary* cause of shrinkage of cement-based products. As a rough guide, the shrinkage of ordinary Portland cement is about 0.03%. Shrinkage of different types of Portland cement varies by as much as 50%. The magnitude of moisture movements quoted later are for ordinary Portland cement unless otherwise stated.

*Moisture movements may, of course, be restrained by reinforcement. This aspect is excluded here but see 2.3 *Cracking*, p. 53.

†Where precise information is not available on a particular product it is wise, as a rule, to use the maximum value from a quoted range.

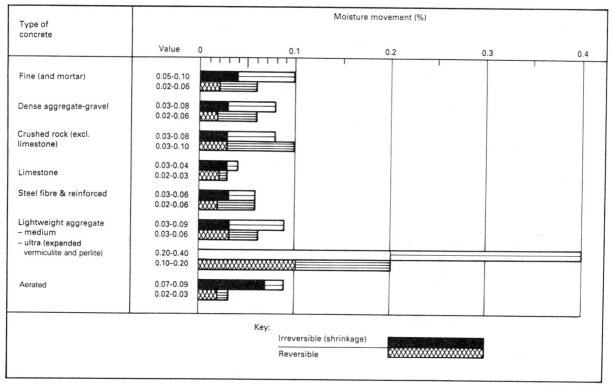

Type of concrete	Value	Moisture movement (%)
Fine (and mortar)	0.05-0.10 0.02-0.06	
Dense aggregate-gravel	0.03-0.08 0.02-0.06	
Crushed rock (excl. limestone)	0.03-0.08 0.03-0.10	
Limestone	0.03-0.04 0.02-0.03	
Steel fibre & reinforced	0.03-0.06 0.02-0.06	
Lightweight aggregate – medium – ultra (expanded vermiculite and perlite)	0.03-0.09 0.03-0.06 0.20-0.40 0.10-0.20	
Aerated	0.07-0.09 0.02-0.03	

Key:
Irreversible (shrinkage)
Reversible

Chart C3.3/10 *Comparison of irreversible and reversible movements in concrete with different aggregates and densities. (Values from BRE Digest 228,* Estimation of thermal and moisture movements and stresses – Part 2, *August 1979, Table 1, Building Research Establishment, Crown Copyright)*

(b) Type of aggregate

(1) Generally

In general, the aggregate provides a restraint to the shrinkage of the cement. Grading and shape may have an indirect effect on the magnitude of shrinkage due to the quantity of cement and the water/cement ratio. However, some aggregates may themselves have significant drying shrinkages after being wetted by the mixing water. This will almost inevitably increase the shrinkage due to the cement. Aggregates that undergo volume changes on wetting and drying will contribute to reversible moisture movements. In addition, not all aggregates provide the same degree of restraint.

Concretes are classified as either *lightweight* or *dense,* according to the bulk density of the aggregate used. The irreversible and reversible movements of different types of concrete are compared in Chart C3.3/10 to which reference is made below.

(2) Irreversible movements

The contribution that an aggregate may make to drying shrinkage, most of which is irreversible, depends on whether it undergoes volume changes on wetting and drying. The primary source for the wetting is, of course, the mixing water.

Most natural aggregates (such as gravel and crushed rock) for concrete used in the UK do not expand or shrink. These are used mainly, though not exclusively, for a range of *dense* concretes (and most screeds, mortars and renders), whose shrinkage varies between 0.02% and 0.08% (see Chart C3.3/10). Notably, the small range for concrete with limestone aggregate (i.e. between 0.02% and 0.04%) is a reflection of the stone's low reversible moisture movement (about 0.01%).

Lightweight concretes as a class have a much larger variation of shrinkage (i.e. between 0.03% and 0.40%) than dense concretes. However, it will be seen from the chart that ultra-lightweight concretes using expanded vermiculite or perlite aggregates are exceptional. Aerated concrete, itself a special form of lightweight concrete,* has a large minimum value (i.e. 0.07%). If both these types are excluded, then, as a general rule, the shrinkage of most lightweight concretes is roughly the same as for dense ones.

(3) Reversible movements

The contribution that an aggregate may make to reversible moisture movements is dependent on the volume changes it may undergo due to wetting or drying. Such changes, in turn, depend on the inherent potential for reversible moisture movement of the aggregate and the extent to which it is accessed by water. The latter will be influenced by the exposure of the product to water (see 'Exposure' earlier, p. 236 ff) and the extent to which moisture may be transmitted through the product (see 'Definitions and their application, 5 Moisture transmission', p. 230 and 'General influences', p. 230 ff).

It will be seen from Chart C3.3/10 that, generally, the minimum moisture movements are less than 0.03% with the maximum as high as 0.10%. Again, the ultra-lightweight concretes are exceptional, with a range of 0.10% and 0.20%. Aerated concrete also has a small range (i.e. between 0.20% and 0.30%).

(c) Mix proportioning

(1) Scope

The type of aggregate apart (see (2) above), the quality of a cement-based product is largely dependent on the quantities of cement and water used in the mix. Both of these need to be controlled to produce a product with required properties of density, strength and durability rather than to reduce the amount of drying shrinkage or subsequent reversible movements. For the present purposes, emphasis is given to the influence that the quantities of cement and water have on drying shrinkage. It should be noted that the examples in the accompanying charts are used solely for comparative purposes and to illustrate principles.

(2) Quantity of cement

The quantity of cement used in a mix is related to the aggregate, whether fine or coarse or a combination of both. Hitherto, mix proportions have been more commonly specified as 'nominal' volumetric portions such as 1:2:4 of cement/fine aggregate/coarse aggregate (for concrete) or 1:3 cement/fine aggregate (for screeds, mortars and renders). The modern method is to specify the cement content in kilograms per cubic metre of finished concrete. Statistical principles are used in the development of an approach for compliance of specifications.

From Chart C3.3/11 it will be seen that drying shrinkage increases significantly as the quantity of cement in the mix increases. Excessive drying shrinkage results, almost always, in unacceptable cracking of relatively thin products such as mortars and renders. The same may be true in the case of screeds.

Mix-proportion (by weight)	Drying shrinkage (%)				
	Value	0	0.04	0.08	0.12
1:7	0.03				
1:6	0.04				
1:5	0.06				
1:4	0.085				
1:3	0.12				

Chart C3.3/11 *Comparison of the variation in drying shrinkage with cement content of concrete or mortar. Values are for specimens with a section of 5 × 5 in (equal to 127 × 127 mm exactly), water/cement ration 0.5 after 6 months' exposure in air at 70°F (21.1°C) and 50% rh. (Abstracted from Chart C3.3.12)*

*Autoclaved aerated concrete, a product made from siliceous material (e.g. sand and/or pulverized fuel ash) and cement or cement and lime, is special in that it has a uniform cellular structure. The latter is achieved by a special process. Usually, fine aluminium powder is incorporated into the mix. This reacts with lime or other alkaline substances that are formed by the setting of the cement, or added deliberately, to produce bubbles of hydrogen. The cells walls are composed of siliceous particles bound together by a crystalline form of calcium hydrate. The material has good thermal insulation properties. In the present context it should be used with care.

(3) Quantity of water

The relationship between the quantity of water to the quantity of cement used in a mix is known as the *water/cement ratio*. Theoretically, water sufficient to enable the chemical reactions to take place in the cement is all that is required. However, for practical reasons (e.g. to enable the wet mix to be 'worked'), it is necessary to use more than the minimum 'chemical reaction' quantity.

The excess water needed to make the product workable in its wet state contributes to the drying shrinkage (see 'The presence of water, 3 Sorbed water, (d) Irreversible movement', p. 236). In principle, the greater the excess water, the greater the shrinkage (see Chart C3.3/12 for the variation of drying shrinkage with water/cement ratio).

The methods normally used to determine and achieve the correct water/cement ratio are outside the scope of this book. Suffice it to say that the modern trend is to use slightly more water than has been the

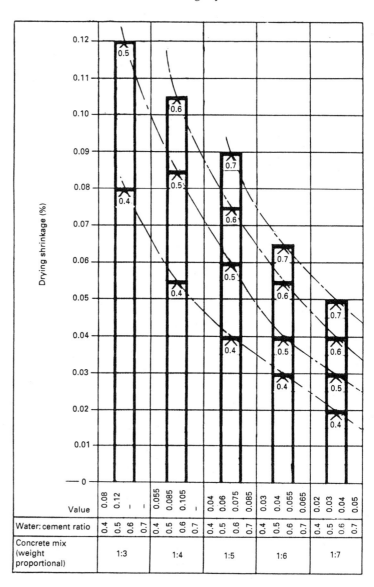

Chart C3.3/12 *Comparison of the variation in drying shrinkage with water/cement ratio. Values are for specimens with a section of 5 × 5 in (equal to 127 × 127 mm exactly), water/cement ratio 0.5 after 6 months' exposure in air at 70°F (21.1°C) and 50% rh. (Values from Lea, F. M.,* The Chemistry of Cement and Concrete, *revised edition of Lea and Desch, Edward Arnold, London, 1956, Table LIV, p. 253)*

Value	0.08	0.12	–	–	0.055	0.085	0.105	–	0.04	0.06	0.075	0.085	0.03	0.04	0.055	0.065	0.02	0.03	0.04	0.05
Water: cement ratio	0.4	0.5	0.6	0.7	0.4	0.5	0.6	0.7	0.4	0.5	0.6	0.7	0.4	0.5	0.6	0.7	0.4	0.5	0.6	0.7
Concrete mix (weight proportional)	1:3				1:4				1:5				1:6				1:7			

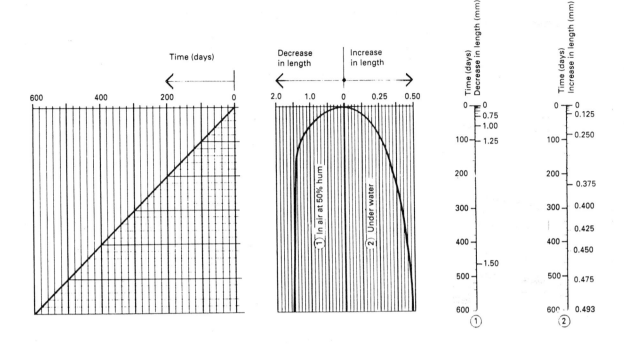

Chart C3.3/13 *Comparison of changes in length of 1:5 crushed granite concrete, 3 × 3 in sections (76.2 × 76.2 mm exactly), under different conditions of curing: (1) dried in air at 50% rh and (2) under water. (Data from Lea, Fig. 109, p. 352)*

case, for concretes in particular. Brick and block masonry, it should be noted, present a special case in that water may be added to the fresh mortar by wetted units during their laying, or to fresh render applied to the units, thereby increasing the water/cement ratio. The opposite is, of course, true if the units are too dry. In practice, an appropriate balance has to be achieved (see '3 Notes, rules and precautions' later, p. 268).

(4) Curing
Curing is the stage during which the cement in a mix completes its chemical reactions. Some of these can take a long time to complete, but generally a concrete mix develops most its ultimate strength after about 28 days. The water not used in the chemical reactions (i.e. the 'free water' – see p. 234) has eventually to leave the concrete. The *hygrothermal conditions* under which concrete is cured will affect the magnitude of shrinkage.*

When drying takes place in air there is a progressive *increase* in the shrinkage as the relative humidity of the air is *decreased*.† Greater shrinkage is produced if the concrete is dried under high temperature and low humidity of the air. In contrast, if concrete is cured *under water* then an expansion takes place. Changes in length of small sections cured in air and under water are compared in Chart C3.3/13. For the air drying, the relative humidity was 50%.

In addition to the hygrothermal conditions, the time during which curing takes place influences the magnitude of shrinkage, both initially and in the long term. Inevitably, the conditions and time are related. For example, and referring to Chart C3.3/12: the sections were dried for six months at 21.1°C and 50% relative humidity. If the relative humidity had been between 65% and 70%, the values would have been *decreased* by about two-thirds, but would have been *increased* by about one-third

*In order to reduce cracking in relatively thin sections such as renders and screeds in particular, the slower the curing during the first week of being applied or laid, the less the magnitude of shrinkage is likely to be. This is because the product is able to develop sufficient strength in its early life, thereby enabling it to resist better the stresses induced in the material during subsequent drying out.

†With some products, such as aerated concrete, the moisture content of the surrounding air is increased by steam curing ('autoclaving') for precast units in order to reduce the excessive shrinkage that occurs with air drying.

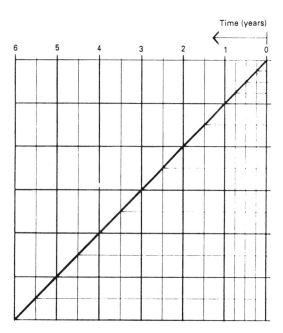
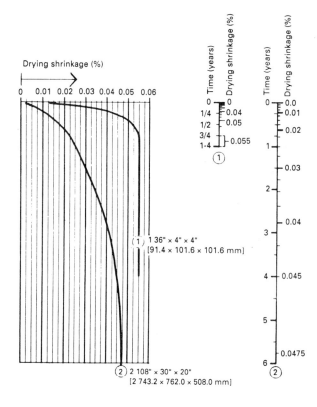

Chart C3.3/14 *Comparison of the effect of size on drying shrinkage. Specimens (1:2:4 concrete) exposed to air at 64°F (17.8°C) and 65% rh. (Data from Lea, Fig. 110, p. 353)*

if the exposure had been prolonged with no change in the humidity. For a cross section smaller than 127×127 mm (i.e. the section used in the test) the *maximum* value would have been obtained in six months.

Size is another factor that influences the magnitude of drying shrinkage (see Chart C3.3/14). From the chart it will be seen that the smaller section has greater drying shrinkage and that the shrinkage takes place more rapidly.

3 Notes, rules and precautions

(a) Scope

Consideration is directed here to those factors that relate to individual types or groups of cement-based products. Generally, in each case there is a short general note followed by the rules that should be adopted in design and the precautions that should be adopted on-site. All this is preceded by notes that apply in general to rules and precautions.

(b) Generally

(1) Effects of variations
The precautions that should be taken with cement-based products to minimize drying shrinkage and subsequent reversible moisture movements obviously vary according to the use, location and exposure

of the product. The main objective of any precautions should be to reduce the risk of cracking as far as is practically and economically possible. In this, account should be taken of the possibility of cracking leading to water penetration or the inconvenience and expense of filling cracks or other remedial work as and when the cracks occur. What follows therefore is a list of suggested measures that should be considered.

(2) Quality control
Quality controls should be exercised in the specification of the ingredients (cement, aggregates and water), their mixes, curing and method of test for compliance with the specification.

(3) Mixed in-situ *and precast units*
All precast concrete products:

- Should meet the relevant requirements of British Standards as to limits for drying shrinkage; and
- Should be well matured before use.

(4) Mixes for mortars and renders
Avoid the use of strong cement mortars for mortars and renders. Such mortars and renders have large drying shrinkages that lead to cracking and often the penetration of water is facilitated at the cracks. Such water may then have difficulty in evaporating through a render.

(c) Masonry – generally

NOTES

(1) The mortar and the units
In all types of masonry, drying shrinkage is normally the result of the shrinkage of the mortar unless the units themselves also shrink as part of their inherent reversible movements as they dry out (i.e. they were laid wet and therefore in an expanded volume). After the masonry has dried out, subsequent reversible moisture movements will result from the combined movements in mortar and the units. Chart C3.3/15 compares the range of irreversible and reversible moisture movements for masonry consisting of selected units.

(2) Allowing for moisture movements
From the values in Chart C3.3/15 it will be seen that the range of moisture movements lies between 0.02% and 0.09%. For relatively short lengths of masonry without openings or changes in thickness, the maximum movements that could occur are insignificant, for most practical purposes. The same is not necessarily true of relatively long lengths, even for small values of moisture movements, as the table below illustrates.

Length (m)	2.5	5.0	10.0	15.0	20.0	30.0	40.0
Moisture movement (%)		Change in length (mm)					
0.02	0.5	1.0	2.0	3.0	4.0	6.0	8.0
0.04	1.0	2.0	4.0	6.0	8.0	12.0	16.0
0.06	1.5	3.0	6.0	9.0	12.0	18.0	24.0
0.08	2.0	4.0	8.0	12.0	16.0	24.0	32.0
0.10	2.5	5.0	10.0	15.0	20.0	30.0	40.0

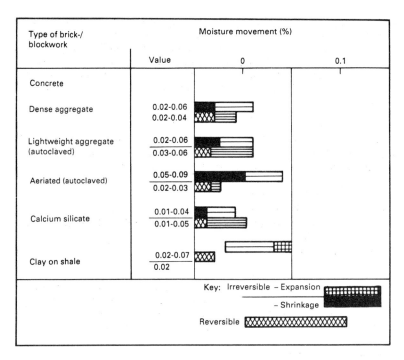

Chart C3.3/15 *Comparison of irreversible and reversible moisture movements in brickwork and blockwork. (Values from BRE Digest 228,* Estimation of thermal and moisture movements and stresses – Part 2, *Table 1, August 1979, Building Research Establishment, Crown Copyright)*

For long lengths of wall, therefore, provision for moisture movements should be made or reinforcement added to the mortar to accommodate the stresses induced.

(3) Allowing for moisture and thermal movements
In practice, except in special cases, allowance is not normally made for moisture movements alone but for the combined effects of moisture and thermal movements. The latter are normally dominant but their magnitude may often be reduced by moisture movements (see '3 Moisture movement', (c) Interpretation of values', p. 226).

PRECAUTIONS

(4) Protection
All bricks and blocks should not be allowed to become excessively damp before being used in masonry. They should be protected from the weather with polythene sheets or tarpaulins. The water absorbed by the units may add to the water/cement ratio of the fresh mortar, thereby increasing the likelihood of cracking in the dried mortar or render.

(5) Wetting dry units
Wetting of units that are too dry may be necessary to reduce water from the mortar or render being absorbed by the units, thereby weakening the mortar or render.

(6) Drying wet units before use
Dry out units delivered to the site with high moisture contents. The shrinkage in drying from delivery to equilibrium moisture content may be 0.02%. That shrinkage would, of course, be increased if the units are

allowed to become excessively wet on-site from rainwater (or other sources, sites being what they are).

(7) Controlling drying out
Mortars and renders should not be allowed to dry out too rapidly. This means that:

- The masonry should be protected from the weather after being laid;
- In some cases, it may be advisable during hot weather to 'shade' the masonry so as to delay the curing of either mortar or rendering;
- In extreme cases, renders newly applied in hot weather may need to be sprayed with water or otherwise wetted.

(8) Protection after laying units
After it has been laid all masonry should be protected from being allowed to become excessively wet by rainwater penetrating at the top of incompleted walls.

(d) Clay brickwork

NOTES

(1) Moisture expansion and its effects
As noted earlier (see footnote under 'Definitions and their implications, 3 Moisture movment, (b) Types of movement', p. 226), clay brickwork undergoes an irreversible expansion that takes place over a long period of time. This needs to be taken into account.

The greatest expansion occurs during the first day after the manufacture of the bricks. Subsequently the rate of expansion decreases, reaching a limiting value after many years. According to BRE, a typical brick would be expected to expand about 0.8 mm per metre in the *first eight years*, of which *about half* occurs in the first week. Bricks of different origins vary in their expansion behaviour considerably. Firing could be responsible for some of the variations. For example, some engineering bricks, if only moderately fired, can have large expansions (up to 1.6 mm/m) whereas, if well fired to a low absorption, the same bricks will usually give low expansions.

As with reversible movements, the expansion of the brickwork as a whole needs to be considered. BRE suggest that the ratio of *brickwork* expansion to *brick* expansion be about 0.6, provided no other source of expansion (such as sulphate attack) is also present. Superimposed on this moisture expansion is the comparatively smaller reversible movement from wetting and drying. Such movement is unlikely to exceed 0.2%, as seen in Chart C3.3/15. There is no evidence that the reversible moisture movements lessen with time.

RULES

(2) Movement joints
Provide sealed vertical movement joints that will allow for the combined effects of thermal and moisture movements.* In unreinforced clay brick walls the spacing of the movement joints should not exceed 15 m. Such joints may also be required horizontally in framed buildings.

PRECAUTIONS

(3) Other precautions
The precautions given earlier for 'Masonry generally' in respect of protection and mortar strength also apply.

*BS 5628: Part 3: 1985 (clause 20.3.2.2, p. 32) advises that an unrestrained or lightly restrained unreinforced wall (e.g. parapets and loaded spandrels built on membrane-type dpcs) will expand 1 mm/m during the life of the building due to changes from thermal and moisture movements.

(e) Calcium silicate (sandlime, flintlime) brickwork

NOTES

(1) Care in design storage and during building
Care is needed in design, storage and during building, unless the bricks are to be used in permanently damp conditions. Drying shrinkage of the bricks used above a ground-level dpc can present problems that lead to cracking.

RULES

(2) Movement joints
- Provide sealed vertical movement joints to accommodate both thermal and moisture movements above the dpc at intervals of about 7.5 m, depending on the shape of the panel and the disposition of openings. Movement joints should never be more than 9.0 m apart. (Movement joints are normally not required below dpc level.) Horizontal movement joints may be required in addition.
- Also provide movement joints along lines of changes of height or where there are variations in the thickness of the wall or where there are low horizontal panels of brickwork, as usually occur above and below windows (see *2.1 Cracking*, D2.1/8).
- If temporary movement joints are used* the joints associated with them should have a mortar of low strength (e.g. 1:3 to 1:4.5 mix – by volume – of *non-hydraulic lime* and sand.

PRECAUTIONS

(3) Other precautions
The precautions given earlier for 'Masonry generally' in respect of protection and mortar strength also apply.

(f) Concrete masonry

NOTES

(1) Special care
Special care in design, storage and on-site generally is needed with most concrete units. Autoclaved lightweight concrete blocks should be given special care as reversible movements in some types can be relatively large (see Chart C3.3/15).

(2) Movement joints
- Design the concrete masonry as a series of panels separated by sealed vertical joints.
- Space the vertical movement joints at intervals not exceeding 6.0 m. The risk of cracking increases if the length of a panel exceeds *twice* the height.
- With external walls having openings, either provide vertical movement joints more frequently or restrain the movements above and below the opening(s).
- Also provide movement joints along lines of changes of height or where there are variations in the thickness of the wall or where there are low horizontal panels of brickwork as usually occur above and below windows (see *2.1 Cracking*, D2.1/8).

*Such joints may be left until the initial 'settling down and drying' period is over and the jointing removed. They would then be sealed with a mortar of the same strength as the surrounding brickwork.

PRECAUTIONS

(3) Other precautions
The precautions given earlier for 'Masonry generally' in respect of protection and mortar strength also apply.

(g) Renders

RULES

(1) Restricting use of rich mixes
Rich render mixes (i.e. strong in cement) should be restricted to strong backgrounds and to those walls having a severe exposure to driving rain or for application during winter conditions, bearing in mind that such mixes are strong and relatively impervious with a high susceptibility to cracking.

(2) Grading mixes in coats
The mix for each successive coat should be weaker (i.e. less rich in cement) than the coat to which it is applied.

PRECAUTIONS

(3) Condition of backing coat(s)
- Ensure that the background for the render is neither too wet nor too dry.
- Allow all backings and undercoats to dry and shrink before finishes are applied.

(4) Protection
- Protect newly applied renders from the weather (wet, heat or cold) until they have dried.
- In hot weather it may be necessary to damp a newly applied render to control its rate of drying and reduce the risk of shrinkage cracking.

(h) Floor screeds

NOTES

(1) Need for special care
The risk of cracking and curling of floor screeds can be reduced significantly provided well-established principles, rules and precautions are followed. Investigations into floor screeds* demonstrate that special care is needed in their design, workmanship and supervision. The starting point is the recognition that the floor screed is part of the whole floor as an element. Thereafter, the rules and precautions applicable to concrete in general need to be applied completely and skilfully.

(2) Forms of dense concrete floor screeds†
There are five different forms of floor screed as illustrated in Diagram D3.3/8. The main difference between them is the thickness of the screed. This is related to the state and nature of the base on which the screed is laid, i.e. its age when the screed is to be placed and the strength of the bond likely to be achieved. The main considerations for each form is as follows:

- *Monolithic construction*: A complete bond between a relatively thin screed and an *in-situ* concrete slab can be obtained provided the

*It is not the intention of this book to focus attention on failures. However, in the case of screeds it should be noted that the remedial work for floor screed failures has cost many tens of thousands of pounds. The BRE has concluded from their investigations that published guidance on design, workmanship and supervision has too often gone unheeded.

†Screeds may be made with other materials such as synthetic anhydrite, modified cement and sand or lightweight aggregates. None of these are included here.

Monolithic-screed placed within three hours of base

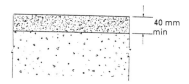

Separate-more than three hours' interval

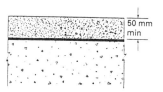

Unbonded-on damp-proof membrane

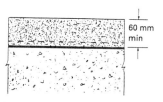

Unbonded and with heating cables

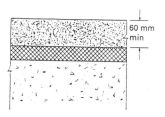

Floating-on compressible material

Diagram D3.3/8 *Five different forms of screed derived from the nature of the substrate. (From BRE Digest 104, Floor screeds, Fig. 1, Building Research Establishment, Crown Copyright)*

*All these substances interfere with the process of good adhesion.

†Cellular concrete may not be able to provide sufficient restraint.

screed is laid before the base has set (i.e. within 3 hours of its placing). The achievement of the bond ensures that the screed is restrained as much as is needed and that the base and the screed can shrink together, thereby reducing the potential shrinkage between them. A screed thickness of only 10 mm is all that is needed. If thicknesses greater than 25 mm are used, it will not be possible for the shrinkage forces in it to be resisted.

- *Separate and bonded construction*: The screed is laid after the base has set and hardened. The bond between the screed and the base will therefore depend on the thoroughness with which the base has prepared. For maximum bond the base should be thoroughly hacked by mechanical means, cleaned and damped to reduce suction, then grouted immediately before the screed is laid (a bonding agent may be used as an alternative to grouting). With a good average standard of preparation, the screed should not be less than 40 mm.
- *Unbonded construction*: Where it is not possible to achieve a bond between the screed and its base (e.g. damp-proof membranes, concrete containing water-repellant treatments, concrete that has been impregnated with oil and grease* or cellular concrete†), the screed should be designed and laid as if it were unbonded. The absence of a useful bond means that the screed has to be thicker – a minimum thickness of 50 mm generally and at least 60 mm if it contains heating cables.
- *Floating*: With a screed laid on a compressible layer of sound-insulating material there is no meaningful adhesion between the two while the screed has 'to span over' the support given to it by the insulation. Consequently an even thicker screed is required – at least 60 mm thick generally and at least 75 mm if it contains heating cables.

(3) Division into bays

Hitherto it was the practice to lay screeds in bays to avoid random cracking and reduce curling. It is now *not recommended* to lay screed in bays *except* where they contain underfloor heating cables. In those cases the bays should not exceed 15 m²; the ratio between the lengths of the sides should be no greater than 1:1.5. Long narrow bays should be avoided.

(4) Mix design

As explained earlier for concrete generally, mix design is one of the factors that can minimize the effects of shrinkage. Curing is another. With screeds, additional factors include the restraint achieved (i.e. bond strength), the limitation of differences in shrinkage that can occur between layers of concrete of different mix proportions or states of dryness. For screeds:

- *Up to 40 mm thick* cement and sand is suitable with a mix of 1:3–4.5 *(by weight)*, the weaker mixes (i.e. less rich in cement) being likely to have lower shrinkage.
- *Over 40 mm thick* cement, fine aggregate and coarse aggregate is suitable with a mix of 1:1.5:3.

(5) Water/cement ratio

When being laid, screeds should be neither too wet nor too dry. Screeds that are too wet shrink more; those that are too dry tend to break up, especially under impact. Thus the water cement/ratio should be limited to that required to achieve good compaction. Workability aids should be used with care to avoid air-entrainment.

(6) Compaction
All screeds should be thoroughly compacted. If, as in the case of screeds thicker than 40 mm, thorough compaction is unlikely to be achieved, then the screed should be laid in *two layers*, each *at least 20 mm thick* and the first layer should be thicker than the second. The mix must be the same for both layers with the second laid as soon as the first has been compacted. This is to ensure that both layers bond together to form a homogeneous mass.

(7) Curing and drying
- It is important to allow the screed to cure properly during the first week after it has been laid. If this is achieved, then the screed is better able subsequently to resist shrinkage stresses because the shrinkage process has been delayed while it has been developing strength. Consequently, the screed should be covered with a *waterproof* sheet soon after it has hardened.
- Drying out of the screed should be slow to minimize cracking and curling and thorough, so that there is little or no residual water in it that may affect flooring finishes to be laid on it. The rule of thumb of an allowance for drying of one month for every 25 mm thickness of screed cannot be relied upon. It is better to measure the dampness/dryness of the screed (see *3.1 General considerations* under 'Dampness', p. 152).

RULES

(8) Overall view
Consider the screed as one of the layers in the overall design of the floor and decide at any early stage which form the screed is to take (i.e. whether monolithic, separate, unbonded or floating). The thickness of the screed can affect other aspects.

(9) Bays and expansion joints
- Follow the current recommendation *not to divide* screeds into bays *except* where they contain underfloor heating elements.
- Where bays are required, limit their size to 15 m^2 with the ratio between the length of their sides not greater than 1:1.5. Avoid narrow strips of screed.
- Provide expansion joints only where similar joints occur in the main structure.

(10) Thickness and mix
- Relate the thickness to the form of screed (i.e. for monolithic not thicker than 10 mm; for separate at least 40 mm; for unbonded at least 50 mm without embedded heating pipes and at least 60 mm with such pipes; and for floating at least 60 mm without embedded heating pipes and at least 75 mm with such pipes).
- Specify appropriate preparation of the base before laid (e.g. for monolithic, brushing the surface of the concrete base and for separate, hacking of the concrete base).
- Relate the mix to the thickness of the screed (i.e. for thickness up to 40 mm a mix of 1:3–4.5 (cement and fine sand) and for thickness greater than 40 mm a mix of 1:1.5:3 (cement:fine aggregate:coarse aggregate).
- Specify the mix *by weight*.
- Specify the minimum water/cement ratio compatible with workability/compaction.

(11) Curing and drying
- Encourage methods for slow curing and drying out and discourage those that cause rapid drying out naturally or with the aid of air blowers, etc.
- Specify seven days' curing (e.g. under polythene sheet) and then adequate drying time before finishes are laid.
- Ensure that the screed is dry enough before floor finishes are laid.

PRECAUTIONS

(12) Mixes
- Weigh the cement and aggregates.
- Ensure that the fresh mix is neither too wet nor too dry.

(13) Compaction
- Compact the screed thoroughly. If the screed is too thick (i.e. normally thicker than 40 mm), layer it in two courses at least 20 mm thick but with the first layer thicker than the second.

(14) Laying
- *For monolithic construction* ensure that the screed is laid within 3 hours after placing the concrete base and that the screed is no more than 10 mm thick.
- *For separate and bonded construction* ensure that the base is thoroughly prepared so that the screed will bond to the concrete base.
- *For unbonded screeds* ensure that the surface is clean and that the dpc is undamaged.
- *For floating screeds* ensure that there is no bridging anywhere.

(15) Curing and drying
- Ensure that the screed is cured (e.g. under polythene) *for at least 7 days*.
- Avoid measures that will cause rapid drying out of the screed.
- Plan to allow one month per 25 mm thickness of screed for drying but do not lay moisture-sensitive floor finishes until the screed is dry enough by testing.

Sources

Lenczner, D., *Movements in Buildings*, 2nd edn, Pergamon Press, Oxford, 1981

BRE
Defect action sheets
51, *Floors, cement-based screeds – specification*, May 1964
52, *Floors: cement-based screeds – mixing and laying*, May 1964

Digests
104, *Floor screeds*, new edition 1973, minor revision 1979
157, *Calcium silicate (sandlime, flintlime) brickwork*, new edition 1981
160, *Mortars for bricklaying*, December 1973
164, *Clay brickwork: 1*, 1980 (with minor revisions)
165, *Clay brickwork: 2*, May 1974
178, *Autoclaved aerated concrete*, July 1975
196, *External rendered finishes*, December 1976
325, *Concrete Part 1: Materials*, October 1987
326, *Concrete Part 2: Specification, design and quality control*, October 1987

Report
Design of normal concrete mixes, revised edition 1988

BSI
BS 12: 1978, *Specification for ordinary and rapid-hardening cements*
BS 882: 1983, *Specification for aggregates from natural sources for concrete*
BS 1047: Part 2: 1974, *Specification for air-cooled blastfurnace slag coarse aggregate for concrete*
BS 2028, 1364: 1968, *Precast concrete blocks*
BS 3892: Part 1: 1982, *Specification for pulverised-fuel ash for use as a cementitious component in structural concrete*
BS 5328: 1981, *Methods for specifying concrete, including ready-mixed concrete*
BS 5628: 1985, *British Standard Code of practice for use of masonry* Part 3. *materials and components, design and workmanship*
BS 6699: 1986, *Ground granulated blastfurnace slag for use with Portland cement*
BS 8110: 1985, *The structural use of concrete*

Exclusion

This chapter deals with the exclusion of water from buildings. It is divided into two broad parts, the first covering the theoretical principles* and the second describing the practical methods of achieving exclusion in the light of these principles. The first section identifies three basic methods by which exclusion is achieved. In the second section each of the main elements forming the external envelope of a building – roof, walls and structure at and below ground – are discussed in turn with reference to the relevant methods of exclusion. In addition, both the theoretical and practical importance of excluding water at joints and junctions is described. Rainwater collection and disposal is outside the scope of the chapter.

Principles of exclusion

1 Introduction

(a) Definition

The function of excluding water from buildings is secondary only to providing the building's overall stability.† This follows normal expectations: shelter from the elements is a more fundamental requirement than, for example, good sound insulation. However, there is a problem of precise definition.‡ Porous materials which are capable of holding water are not expected to remain completely dry either during construction or throughout the life of the building. Brickwork and timber, for example, have significant values of moisture content under normal conditions.§ Materials used externally in the UK may be so exposed to long periods of wet weather that at certain times of the year those parts of the external fabric which are made of porous materials may be close to saturation.

Water used during construction accounts for the high initial moisture contents of porous materials, but this water generally evaporates in a comparatively short time until a point of equilibrium is reached. Water trapped in the construction so that drying out is prevented is a separate problem and is discussed in *3.2 Exposure*, p. 214 ff.

Thus exclusion may be defined as the control of damaging penetration of moisture in the building fabric or interior. This definition applies to moisture in all its forms and refers equally to the fabric of the building and the dampness of the internal atmosphere. In the same way that porous materials have equilibrium moisture contents in normal use (an example of the principle of **balance** – see *3.3 Moisture content*, p. 225 ff, Chart C3.3/4 ff), the requisite dryness of the internal atmosphere of a building is relative, and depends on the degree of comfort and amenity that the users expect, and the tolerance of furnishings, fittings and equipment to humidity. The source of such humidity is often the users themselves, their activities and processes.

Hitherto, the task of exclusion has been considered primarily as one of keeping rainwater out of buildings. However, because of the risk of

*The term 'principle' is used here in the colloquial sense and does not refer specifically to the principles for building.

†This is the priority given in the classic work on the scientific approach to building construction, *Principles of Modern Building*, by R. Fitzmaurice, Vol. 1, *Walls, Partitions and Chimneys*, HMSO, London, 1938, from which the term 'exclusion' is taken.

‡'It is not even possible at this stage to define what is meant by exclusion of rain' – *Principles of Modern Building*.

§These are described fully in *3.3 Moisture content*.

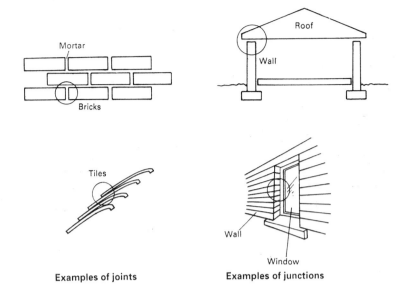

Examples of joints	**Examples of junctions**

Diagram D3.4/1 *Joints and junctions*

interstitial condensation, exclusion must be treated as a twofold task, from the outside dealing with rainwater or ground moisture and from the inside* controlling the effects of water vapour. The two aspects must be considered together because, as is discussed in more detail below, the practical ways of achieving one (the exclusion of rainwater in particular) often have a direct effect on the other.

(b) Continuity

Unlike some aspects of building performance, effective exclusion of moisture requires total continuity of performance. However, in practice this must always be achieved by an assembly of discontinuous elements. This is the nub of the problem. Even a minor leak – a discontinuity in performance – may cause damage to the fabric and inconvenience to the users. Yet a building, by its nature, is constructed of discrete units which have to be put together in sequence. Even if the units themselves are watertight, at all those points where there are junctions with other units or components, continuity of performance must be maintained.

It is essential to remember that since continuity of the performance of exclusion is achieved generally by discontinuous units there will always be two aspects to consider. First, there are the properties of the individual units themselves, as pieces of homogeneous material exposed to water. Second, there are joints and junctions. In this chapter the term 'joint' is used to refer to the joints between units of the same material which together make up building elements such as walls and roofs. The word 'junction' is used to describe the interface between different components or elements, such as windows in walls or the meeting of wall and roof (see Diagram D3.4/1). Exclusion is only achieved by the effectiveness of both aspects, although joints and junctions are usually more critical because the function of exclusion is often in conflict with other requirements, such as allowance for movement, renewal and the practicalities of the construction process – all aspects of the principle of **separate lives**.

*This is true for the majority of buildings in the winter when both temperature and vapour pressure gradients (**high>low**) fall from inside to outside. At other times of the year, these gradients may fall the other way. Condensation is rarely a problem in the summer because the temperature of the fabric is usually above dewpoint. There are, of course, exceptions, such as cold stores and other special-purpose buildings.

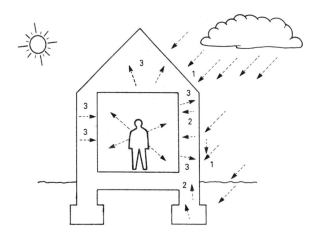

1 Flowing water:
    Sources:
        • Precipitation
        • Service
2 Capillary action
    Sources: • Groundwater
        • Water absorbed by
          porous materials
        • Water between
          sheet materials
          in close contact
3 Water vapour
    Sources: • Evaporation of water
        • Human activities
         and processes

Diagram D3.4/2 *Sources and transfer of water*

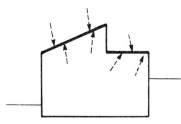

(1) Structures above ground: roofs

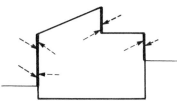

(2) Structures above ground: walls

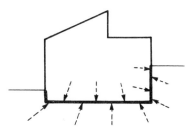

(3) Structures at or below ground

Diagram D3.4/3 *Sub-division of building elements as used in the discussion of methods of exclusion*

(c) Sources and transfer of water

In the various parts of the building the problem of controlling damaging water penetration differs in character because of the different ways in which water is encountered. As far as a building is concerned, water comes from several sources and travels in different ways, each of which requires alternative strategies to achieve exclusion. These sources are: precipitation (rain and snow), the ground, and the atmosphere (see *3.2 Exposure* and Diagram D3.4/2). These sources relate closely to the three basic ways in which water travels:

1. Flowing downwards under the effect of gravity, over the surface of the building or material;
2. Travelling against the pull of gravity vertically and laterally by capillary action through porous materials such as soil, brickwork, plaster and timber, and between sheets of non-porous material in close contact;
3. Diffusing in its gaseous state as water vapour through air or porous materials under the influence of a vapour pressure gradient.

Clearly, rainwater is the main source of downward-flowing water which, when absorbed by the ground or porous materials, can travel by capillary action, whereas atmospheric moisture comes into the third category. This range of sources and ways that water travels underlies all aspects of exclusion met in practice and determines the solutions required. In most situations exclusion of water in more than one of its forms is needed.

In the second section of this chapter, which deals with the practical methods of achieving exclusion, parts of the external building envelope are considered in turn (see Diagram D3.4/3). The sub-division used is somewhat arbitrary, since the principles governing the performance of the materials are the same. However, differences in the balance of the sources of water and their methods of transfer between each part make it convenient to divide the building as follows:

1. The building above ground, i.e. the external envelope of walls and roofs. Walls and roofs are considered separately because, despite areas of overlap, there are important distinctions. In both cases the task is primarily to exclude precipitative moisture from the outside and vapour from the inside. Thermal pumping is also considered.
2. The building at or below ground, where the problem is one largely of excluding water rising by capillary action. This area includes the use of dpcs in walls above ground.

2 Background

(a) Use of permeable materials

Historically* building fabrics used porous, permeable materials for the structure and reserved the few, expensive impermeable materials for special situations such as windows (glass) and, to a limited extent, roofs (lead and copper). Even in timber-framed building, which in some ways foreshadows modern framed construction, the infill panels were made of porous materials.

Permeable materials were used with an empirical but deep knowledge of both their performance and of the micro-climate of the locality. A limited range of materials in a particular area produced local (vernacular) solutions to the problem of exclusion. The combination of variations in available materials and climate accounts for the very wide range of vernacular building techniques, even in a country the size of England. In the Yorkshire Dales, for example, exclusion was achieved by thick walls of solid masonry, with small deep-set windows, whereas in Essex the use of render and timber weatherboarding on timber frames made thinner walls possible. In the Surrey and Sussex Weald, where timber was readily available, framed structures were used which kept rainwater out by the application of tile hanging or brick infill. In general, successful performance relied on frequent maintenance.†

(b) Use of impervious materials

From what has been said so far about continuity of performance, it might be thought that the ideal method of achieving exclusion, economically feasible with modern impermeable materials, would be to clad the building completely in an impervious skin that would keep water out in whatever form it occurred. One of the reasons the impervious 'raincoat' is appealing is that, theoretically, it can go anywhere, regardless of local exposure. As will be seen, this is one approach, but it is not necessarily ideal, and indeed care is needed to resolve a number of inherent problems.

There are several reasons why, even in contemporary construction, materials which are completely impermeable to water have not been used as the chief barrier to downward-flowing water, in spite of the apparent attractiveness of the idea. First, in building practice it is extremely difficult to make the building envelope a continuous impervious 'raincoat'. Second, an impervious outer skin is often in conflict with the need to allow internal vapour to permeate (or 'breathe') through the fabric. The implications of these points are developed in subsequent sections.

*More general treatment of historical background is given in *Introduction*, 'Changes and their effects', p. 4 ff.

†For a full survey of the range of vernacular materials and building techniques see Clifton Taylor's *The Pattern of English Building*, 4th edn, Faber and Faber, London, 1987. This outstanding book describes the source and use of materials rather than how they work.

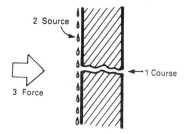

Diagram D3.4/4 *The three preconditions for moisture penetration*

3 The three methods of exclusion

(a) Water and the building fabric

(1) Preconditions for moisture penetration

The mechanisms of moisture penetration must be described before considering how different constructions resist it. In his pioneering research in Canada,* Garden identified three basic conditions that must be met before water penetration can begin to occur. These are:

1. A *course* for water to take into and through the construction;
2. A *source* of water to be present at the beginning of this course; and
3. A *force* which will drive the water along the course. (See Diagram D3.4/4.)

Remove any one of these conditions and the task of exclusion is achieved. Even if moisture is present at an opening, penetration will not occur unless a force or combination of forces is available to drive the moisture through the opening. Although the Canadian Building Digest was written to describe the mechanisms of rain penetration, the same three preconditions apply to penetration of moisture vapour into the building fabric. The forces that lead to penetration of liquid moisture are the kinetic energy of the raindrop, capillary suction, gravity, air pressure differences, and hydraulic pressure (the latter affecting standing water on flat roofs and water penetration of basements). A vapour pressure gradient is the force that leads to penetration of water vapour. In addition, if there is an air pressure gradient across the fabric and a route through, vapour in the air will be carried along the route by the movement of the air of which it forms a part (see Diagram D3.4/5).

Diagram D3.4/5 *Forces which lead to moisture penetration and the critical aperture size. t = opening size; ΔP = air pressure difference in mm of water. (From CBD 40,* Rain penetration and its control)*

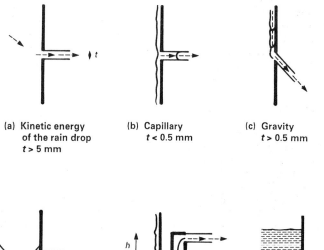

(a) **Kinetic energy of the rain drop**
$t > 5$ mm

(b) **Capillary**
$t < 0.5$ mm

(c) **Gravity**
$t > 0.5$ mm

(d) **Air pressure gradient**
0.5 mm $< t <$ 6 mm

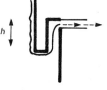

(e) **Air pressure and capillary**
$t < 0.5$ mm
$h > \Delta P$

(f) **Hydraulic pressure**
$t > 0$

*NRC DBR Canadian Building Digest 40, *Rain penetration and its control,* by G. K. Garden, April 1963.

(2) Mechanisms of rain penetration

When rain is driven by wind against a wall or roof of a building the raindrops have kinetic energy which will carry them through openings over a certain size: about 5 mm seems to be a lower limit. If they strike the surface the drops are shattered by the force of impact, and they may then be forced through gaps which do not directly face the direction of exposure. The extent of penetration depends on the speed of the wind carrying the drops, and the droplet size. Rain penetration by this means can be prevented by laps, baffles and splines – in other words, the seal against water is achieved by the geometry of the opening or joint. The depth of cover needed is based on empirical observation. Thus in slated roofs the laps provide the geometrical seal, and experience has shown, for example, that a 65 mm lap is generally adequate where the pitch is 45°, whereas at 30° pitch a lap of 75 mm is necessary.*

Capillary forces will suck water at the surface into and through porous construction where the size of the openings is less than about 0.5 mm. Penetration of liquid water will not occur until the moisture storage capacity of the wall is exceeded, although if moisture is concentrated at a particular point it is the local absorptivity which is critical rather than that of the whole wall. Capillary suction can work together with other forces, including gravity and air pressure differences, and capillary forces can be combined with a wind-induced pumping action of two sheets which are close together and able to move. Capillarity can be prevented by discontinuities in the capillary paths. Water will flow freely downward under the effect of gravity over surfaces or within materials whose pore size is greater than 0.5 mm, below which size capillary forces come into play.

An air pressure difference across an opening will exert a force on a stationary raindrop which may be sufficient to cause it to move through the opening (Diagram D3.4/6). Where the pressure difference is working against the force of gravity on the drop, it can be seen that smaller drops are more easily carried than larger ones (Diagram D3.4/7). The detailed effects of air pressure differences also depend on the size of the openings. With openings between 5 mm and 0.5 mm in size, air pressure gradients across the construction will force water through. With openings below 0.5 mm, air pressure gradients must work in conjunction with capillarity.

(3) Seals against water and air

Since an air pressure gradient is one of the forces which will drive water across an opening, in achieving exclusion a seal against air infiltration is just as important as one against water. Each seal corresponds to a sharp

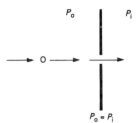

$P_o = P_i$

A raindrop carried by the wind has kinetic energy which can carry it through an opening even if the air pressures are equal on both sides

P_o = outside air pressure
P_i = inside air pressure

$P_o > P_i$

A stationary raindrop at an opening may be sucked across the opening if the air pressure difference is large enough

Diagram D3.4/6 *Forces acting on a raindrop at a lap*

Diagram D3.4/7 *Kinetic energy and air pressure differences acting on a raindrop*

P_o P_i

ΔP

O

g

Force down is proportional to the volume (i.e. r^3)
Force up is proportional to the area (i.e. r^2)
Therefore the force needed to carry the drop over the lap increases linearly with the radius

P_i = internal air pressure
P_o = outside air pressure
ΔP = Pressure difference x area
g = gravitational force

*See BS 5534: Part 1: 1978, *Code of practice for slating and tiling.*

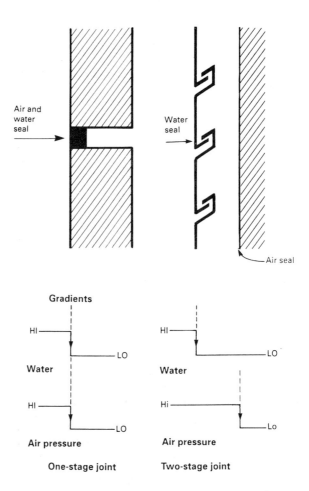

Diagram D3.4/8 *One-stage and two-stage joints. The dampness and air pressure gradients for each are shown below*

gradient, the water seal being the wet/dry boundary and the air seal the point at which the air pressure gradient occurs (Diagram D3.4/8). The two seals can occur at the same point or they can be separated by moving the air seal behind the water seal. Separating the two removes the air pressure gradient away from the beginning of the water path. This makes the physical continuity of the water seal less critical, whereas when the seals occur at the same point, there must be physical continuity to achieve exclusion.

The concept of air and water seals forms the basis of joint design from the point of view of exclusion. It should be remembered that air pressures are not static and therefore the gradient may not necessarily be high to low from outside to inside. However, this is the worst-case condition as far as exclusion of rainwater is concerned, and is therefore normally taken as an assumption.

(4) The mechanisms of moisture vapour penetration
The three preconditions for moisture to penetrate apply equally to vapour, although here the source (vapour in the air) and the force (a vapour pressure gradient) are effectively the same. The course is any path through permeable materials or through joints between units of either permeable or impermeable materials. A vapour pressure gradient may operate in conjunction with an air pressure gradient.

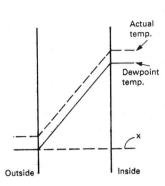

(a) Homogeneous construction

The gradients show typical temperature and moisture vapour* conditions

* The dewpoint temperature at any point is governed by the amount of moisture vapour (i.e. vapour pressure) at that point

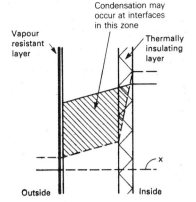

(b) Multi-layer construction with risk of condensation, within shaded area

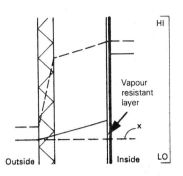

(c) Multi-layer construction without risk of condensation

Diagram D3.4/9 *The effect of different types of construction on the risk of interstitial condensation. The broken line (x) shows the dewpoint temperature gradient where the vapour pressure inside is the same as that outside, achieved by ventilation or air conditioning. It can be seen that this removes the risk of condensation occurring*

The overall performance of the building fabric should not be damaged by water vapour which penetrates to the interior, nor by condensation resulting from the presence of vapour. Interstitial condensation may occur if at any point the temperature of the fabric falls below the dewpoint of the vapour (Diagram D3.4/9). Given normal conditions in which there are gradients of both temperature and vapour pressure across the fabric, falling from inside to outside, the basic rules to observe to prevent condensation are:

1. The thermal resistance of the fabric should be highest at the cooler side, thereby keeping the fabric warmer than the dewpoint;
2. The vapour pressure gradient across the fabric should be kept to a minimum, by removing the source of vapour and thereby the force driving it (see × in the diagram);
3. The vapour resistance of the fabric should be at its highest at the damper side, thereby limiting the flow of vapour into the fabric (Diagram D3.4/10).

If water vapour or condensate is allowed into the interior of the fabric and prevented from dispersing harmlessly it may accumulate and cause the degradation of the materials and their performance. The rate at which this happens depends on the vulnerability of the materials to moisture: timber or timber-based products and insulations are more susceptible to damage than masonry. Diagram D3.4/11 illustrates four ways in which vapour movement and accumulation within the fabric may be controlled.

The relationship between air pressure and vapour pressure as driving forces for vapour needs elaboration. Water in its gaseous state is one of several component gases of air. Air exerts a physical pressure: the water vapour in the air contributes to this overall pressure, but the pressure exerted by the vapour can be described separately. This is sometimes called a partial vapour pressure, since it is only part of the

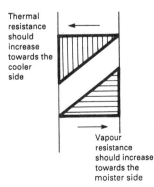

Thermal resistance should increase towards the cooler side

Vapour resistance should increase towards the moister side

Diagram D3.4/10 *Basic rules for preventing interstitial condensation*

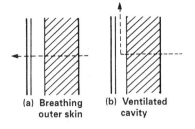

(a) Breathing
    outer skin

(b) Ventilated
    cavity

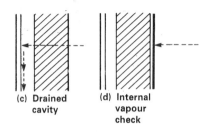

(c) Drained
    cavity

(d) Internal
    vapour
    check

Diagram D3.4/11 *Four ways of
controlling moisture accumulation
within the building fabric*

air pressure as a whole. The amount of water that air can hold at any
given temperature is limited, and the vapour pressure at that
temperature is therefore the saturation vapour pressure. Vapour
pressures correspond exactly to the mass of water in the air, expressed
as grams of water per unit mass of dry air.

A vapour pressure gradient occurs between two points where there
is a difference in the quantity of vapour in the air. By the process of
diffusion, which is the mixing of different gases as a result of their
molecular motion, the distribution of moisture will tend to become
even and thus the pressures will be equal. This happens purely as a
result of the difference in partial vapour pressures and occurs where
the air pressure is constant. Diffusion can take place through permeable
materials, though at a rate governed by their resistance to moisture
diffusion.

A similar mechanism takes place where there is a difference in air
pressure, giving rise to an air pressure gradient. As a result of
molecular motion, air moves from an area of higher pressure to one of
lower pressure. Clearly, since vapour forms part of the air, the flow of
air across an air pressure gradient is a mechanism for driving water
vapour in addition to diffusion. There are some cases where this
mechanism can be used to reduce the flow of moisture into the building
fabric, such as the pressurization of a roof cavity above a swimming
pool. In other cases it may aggravate problems. Pratt* gives as examples
factory buildings in which the method of air conditioning involves a
slight increase in the internal air pressure, thereby causing an
additional amount of moist air to enter any cavities in the fabric by
streaming through gaps in the lining, and tall buildings in which the
stack effect adds to the air pressure internally. A corollary of this is that
the risk of condensation might be lower on walls facing the prevailing
wind where there is an air pressure gradient from outside to inside.

The movement of moisture in air under the influence of an air
pressure gradient is often greater than by diffusion. Therefore the task
of exclusion in relation to vapour is as much concerned with controlling
the infiltration of moist air into the building fabric as with ensuring the
effectiveness of vapour control layers, whose primary purpose is to
limit diffusion. In either case the impact of vapour movement is
lessened by making the vapour pressure gradient shallower, in other
words, by reducing the source of moisture. Both air infiltration and the
accumulation of moisture internally are best controlled by means of
ventilation, i.e. by ensuring that the route that vapour follows down
the vapour pressure gradient is through intended paths such as
windows and ducted extract systems rather than through the fabric of
the building. It can be stated immediately that moisture driven by a
vapour pressure gradient can only be resisted by continuous layers
impervious to vapour.

(5) Exposure and protection

The geometry of the building form and its details play an important
part in determining the amount of water on the surface of the building
which must be excluded. The task of exclusion becomes less onerous as
the amount of water at the beginning of potential routes is reduced.
Thus, following the principle of **high>low**, an overriding objective is to
shed water from the envelope of the building as directly as possible.
Diagram D3.4/12 illustrates practical applications of this rule.

(6) The three methods of exclusion

Given these basic mechanisms by which moisture may penetrate the
building envelope, there are three basic methods by which exclusion is

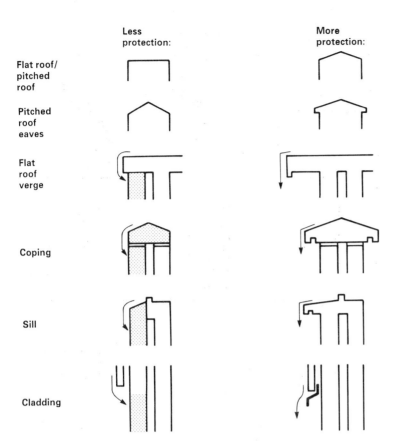

Diagram D3.4/12 *The effect of protection on a building's exposure to rainwater*

achieved. The principles of each of these are discussed before turning to consider in detail how they apply to specific forms of construction used in the various parts of the building.

The first method of exclusion is to use completely impermeable materials and joints. This is sometimes called the 'raincoat' principle when applied to walls (Diagram D3.4/13). The second method applies only to porous or permeable materials and is the controlled penetration of water into the fabric which is thick enough to prevent complete penetration. Again, for walls this is sometimes known as the 'overcoat' principle. The third method works by arranging layers in a construction so that water is drained away (**high>low**: shedding) and capillary paths are broken, either by continuous cavities or by discontinuities between components. This method relies on the fact that liquid water flows downwards unless blown by the effect of an air pressure gradient. In walls it is known as the 'rainscreen'.

(b) Exclusion method 1: complete impermeability

Penetration of both liquid and gaseous water may be prevented by using a layer of material that is completely impermeable to water, either at the face of the building or within the thickness of construction. The latter applies to those situations where a dpc, membrane or vapour control layer is necessary.

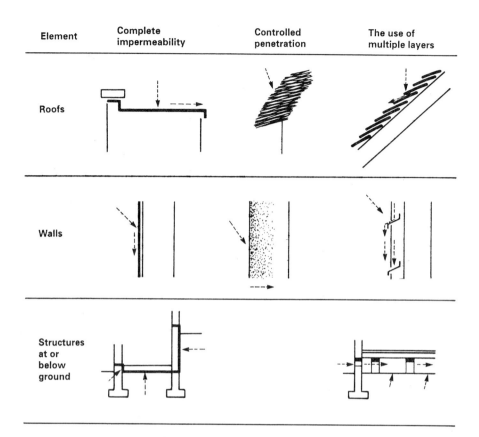

Element	Complete impermeability	Controlled penetration	The use of multiple layers
Roofs			
Walls			
Structures at or below ground			

Diagram D3.4/13 *The three methods of exclusion applied to the major building elements*

There are inherent difficulties in using this method of exclusion, which are discussed in this chapter and referred to again when specific rules are described under 'Methods of exclusion'. These difficulties do not preclude its use, but instead underline the need for appropriate compensatory measures to achieve a successful result. These will vary from rigorous control of dpc laps to careful design by specialists of seals in curtain walling and sheeted roofs.

Examples of impermeable materials include sheet metals, glass, plastics (including thin membranes), bituminous products such as asphalt, and high-quality cementitious materials. All these are refined materials and as a result are expensive in bulk form, and have therefore to be formed into thin sheets to be used economically in building. The tendency of particular refined materials to revert to their natural state* (for example, the corrosion of metals, the degradation of plastics by ultraviolet light, etc.) coupled with their thinness means that an understanding of their performance over time is essential. These are characteristically (if not exclusively) modern materials, requiring sophisticated technology in their manufacture. Because of their thinness, they generally perform only the one function of exclusion, so that other materials and components are required to meet other performance requirements. As a result, they invariably form part of multi-layer construction.†

Some materials are impermeable to water in its liquid state, but not to water vapour. This is because the molecules of water are larger in its liquid form than in its gas state, and the size of the openings through

*See *Introduction*, p. 14.

†See *Introduction*, p. 7.

the material at a molecular scale is critical. Thus certain paints (known as microporous paints) and PVC sheet are effective as waterproofing membranes but not as vapour control layers.

It has already been pointed out (under the mechanisms of moisture vapour penetration) that to prevent damaging penetration of moisture vapour into the building fabric the vapour resistance of the construction should be greatest at the higher end of the vapour pressure gradient, which is normally at the inside. This is clearly in conflict with the idea of having an impervious skin to exclude rainwater on the outside, and therefore some kind of compensation is required to reduce the risk of interstitial condensation. Such compensatory measures include an internal vapour control layer, a drained and ventilated cavity to clear condensate,* or an outer skin which is permeable to vapour. When the external envelope uses this method of exclusion it can, as it were, defy gravity. As a result, in buildings where this method of exclusion is adopted, the typological differentiation of roofs and walls is not important (Diagram D3.4/14).

Successful exclusion relies on the material and its joints and junctions being functionally continuous. Even for a notionally continuous material, joints and junctions are inevitable, and for this kind of construction joints are particularly critical, for several reasons. Thin, impervious materials such as metals and plastics have far greater thermal movements than traditional heavyweight porous ones, which must be allowed for in the detailing of the joint. The tendency towards using larger units amplifies the effect of thermal movement further, and puts greater stress on the effectiveness of joints which are less frequent. 'Styles' of external detailing which favour clean, flush junctions are much harder to make watertight than the traditional loose-fitting methods of laps and cover mouldings. Again, prefabrication and other new techniques rely on a much tighter first fit than before, with little tolerance of dimensional variations. Finally, standards of workmanship in the construction industry have not kept pace with the demands for finer tolerances. This applies not only to dimensional fit but also to the tolerance of materials, particularly sealants and adhesives, to dirt and moisture.

Whether by design or necessity, there is often only one line of defence. Once this has been breached it is often in the nature of the construction that the effects of water ingress are likely to be extensive. The reasons for this are as follows. Where the impermeable layer is used within a multi-layer construction, water may be able to track a long way from the point of entry, diverted along its route by other layers. For several impervious materials (particularly for roofing) it is often hard to actually find the point of entry. Once water has entered the construction the impermeable skin prevents or delays its drying out.

Modern buildings have a higher ratio of window to wall than traditional ones (culminating, of course, in the curtain wall). Water falling against windows is shed immediately rather than absorbed so that the amount of water flowing over joints and junctions is increased. Because further stress is put on the joints, both by thermal and moisture movements and the demands of dimensional tolerance, it can be seen why the ideal properties of this method are hard to achieve and maintain in practice.

(c) Exclusion method 2: controlled penetration

This method of exclusion applies only to water in its liquid form and to porous materials used in massive construction. The permeability of

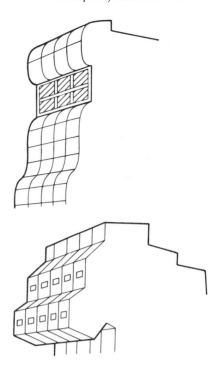

Diagram D3.4/14 *Characteristic form of buildings which exclude water by complete impermeability*

*Note that this will usually mean that the air seal becomes separated from the water seal, so that the exclusion method becomes the third type.

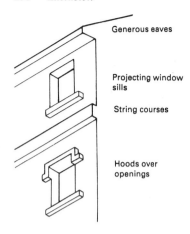

Generous eaves

Projecting window sills

String courses

Hoods over openings

Diagram D3.4/15 *Characteristics of buildings which exclude water by controlled penetration.* Note: *similar water-shedding features should be used on rendered walls*

porous materials is variable and depends on the structure of the pores and matrix at the microscopic (though not atomic) scale.

For a given porous material to be effective in excluding water in a particular circumstance its permeability, thickness and the degree of exposure must all be taken into account. Successful exclusion relies on the fact that there are cycles of wetting and drying, so that there is always an opportunity for water that has penetrated some distance into the material's porous structure to dry out. The thickness is chosen so that, during the wetting part of the cycle, water penetrates a certain amount but not the whole thickness of the construction. It is essential that the absorbed water starts evaporating once the wetting stops. There has, therefore, to be a **balance** between the rate of flow of liquid water into the material and its rate of evaporation.

This method of exclusion relies almsot entirely on empiricism and, while research can examine the factors which affect performance, such as the absorbency of bricks, there are so many local factors that experience and observation of local practice remain the best guide. Thus, for example, traditional brickwork when used 230 mm (9 in) thick is found to provide a weathertight construction under the conditions of exposure that London and parts of south-east England experience. On thinner walls, or on walls in exposed locations, additional weatherproofing (such as tile hanging or rendering) may be observed.

Severity of exposure is critical for the success of this method exclusion (see *3.2 Exposure*, p. 180 ff). As with the exclusion method of complete impermeability, there is only a single line of defence, albeit a thick layer. Shelter provided by adjacent buildings and trees can be significant. On walls, features such as projecting eaves, cornices, or string courses are beneficial because these throw water away from the face of the building rather than allowing it to concentrate at exposed points. As a result, buildings with walls which exclude moisture by this method are characterized by means of providing shelter (Diagram D3.4/15).

The heating pattern within the building influences the rate of evaporation of the absorbed moisture from the outside surface of the construction. A well-heated interior will drive liquid moisture within the building envelope towards the colder parts. On the other hand, the same mechanism in reverse can cause a condensation problem at the inside of the solid walls which have been thermally upgraded by battening out the interior and lining with thermal plasterboard* (see Diagram D3.4/16). The heat of the sun drives moisture from outside to inside, where it condenses on the back of the vapour check or insulation. The phenomenon occurs on south-facing facades in late spring when walls are damp and temperatures high.

Exclusion by controlled penetration applies primarily to walls resisting rainfall. In roofs, only thatch works on this principle,† and even in this case it is debatable whether it is really an inefficient type of multiple-layer construction (see below). Although thatch is widely used throughout the world, it represents a very small proportion of roofs in the UK. As regards resisting water from the ground, controlled penetration is too unpredictable to be used as a general method, and the principle of complete prevention in the form of a dpc is therefore normally adopted.

*BRE IP 12/88, *Summer condensation on vapour checks,* by J. R. Southern.

†This is true for the UK. In drier climates flat roofs are constructed of porous materials without a waterproof membrane.

(d) Exclusion method 3: the use of multiple layers

The third method of exclusion applies only to exclusion of downward-flowing water. Whereas the two methods of exclusion described above rely on continuity of material to achieve **continuity** of performance, the

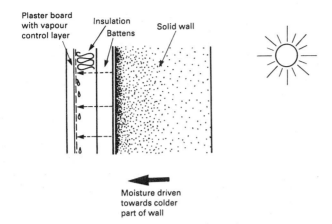

Plaster board with vapour control layer
Insulation
Battens
Solid wall

Moisture driven towards colder part of wall

Diagram D3.4/16 *The effect of sunshine on vapour movement within a solid wall*

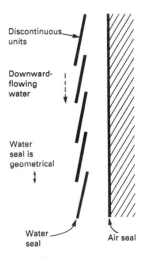

Discontinuous units

Downward-flowing water

Water seal is geometrical

Water seal

Air seal

Gradients:
1. Dampness:

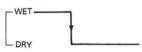

WET

DRY

2. Air pressure:

HI

LO

Diagram D3.4/17 *Exclusion by the use of multiple layers. The dampness and air pressure gradients are shown below*

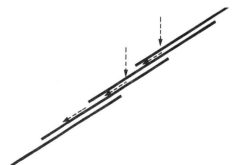

Diagram D3.4/18 *With overlapping units the assembly of units as a whole works as the water seal, not just the outermost surface*

third method is characterized by discontinuity of material. To achieve watertightness, discrete units are used in conjunction with the basic principle that liquid water flows downwards under the effect of gravity. The key characteristics of the method are that:

1. The seal against water is geometrical, such as a lap;
2. The use of discontinuous units enables capillary paths to be broken; and
3. The two seals which are required – against water and air – are separated (Diagram D3.4/17).

It is important not to mistake this method for the principle of two lines of defence. Using two lines of defence is a general method of compensation in the light of **creative pessimism**, and can apply to all three methods of exclusion.

Both traditional techniques such as tiling, slating and weatherboarding and modern rainscreen cladding fall into this category. In the case of tiling, for example, water is allowed into the thickness of the construction but is then drained out. It is important not to confuse this with the exclusion method of controlled penetration described above. In that method, water is allowed to penetrate by capillarity, whereas this technique breaks any capillary paths. The surfaces of the tiles under the laps onto which the water drains form, in effect, part of the outer layer (Diagram D3.4/18).

The degree of exposure is again critical, but for reasons which differ from absorbent construction. Whereas in the latter the amount of water falling on the wall or roof between drying cycles is critical, with discontinuous units the angle and velocity of the driving rain or snow are important. The angle is partly governed by the droplet size, small droplets being more readily blown upwards than larger ones. A tiled roof may be found to leak in wind-driven drizzle rather than in heavy rain in still conditions.

By placing the air seal behind the water seal, a cavity between the two is created which is ventilated to external air and, as a result, the pressure there tends to approach that of external air. The cavity may be designed actually to be pressure equalized or merely ventilated and drained. The advantage of pressure equalization is that there is no air pressure gradient across the water seal so that water is not dragged across it. The air pressure gradient occurs over the air seal at the back of

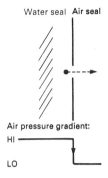

Water seal Air seal

Air pressure gradient:

HI

LO

**(a) Pressure-
equalized cavity:**
the air seal must
be continuous

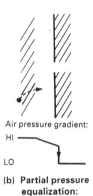

Air pressure gradient:

HI

LO

**(b) Partial pressure
equalization:**
there is a risk
of water being
dragged across
the water seal

Diagram D3.4/19 *The effects of full
and partial pressure equivalization of
the cavity between the water and air
seals*

*Cavity walls are discussed in more detail under
'Methods of exclusion', pp. 317 and 320.

the cavity. If water manages to get into the airspace, there remains a risk of it being sucked across any discontinuities in the air seal (Diagram D3.4/19).

It can be seen that there is actually a basic difference in principle between tiled roofs which have a layer of sarking felt below and those which do not (Diagram D3.4/20). Where sarking felt is not used, exclusion relies solely on the size of the laps of the tiles being sufficient to prevent water being blown across them. The laps thus form both water and air seals. Where sarking felt is used the interface between the felt and the roof space becomes the air seal, so that the air seal is moved away from the water seal. In addition, sarking felt acts as a second line of defence against rain that is forced over the laps, draining it out at the eaves. When this does occur, the laps in the felt become both the secondary water seal and the air seal. It is important, therefore, that the felt itself is adequately and tightly lapped.

Since the joints between the units are fundamental to this method, they have to be thought of as the key to successful functioning of the system. The units themselves may be totally water impermeable (such as slates) or absorbent (such as plain clay tiles). As the unit size increases, the construction changes from one in which the general areas use either of the first two exclusion methods (a continuous impermeable layer or controlled penetration) but the joints are geometrical water seals. Whereas with small units the characteristics of the materials and the joints go together to form a complete system of water exclusion, at the other end of the scale the material itself and the joints within it can be considered separately. This sliding scale explains some of the difficulty in classifying certain constructions. Inevitably, there are grey areas where a particular construction shares characteristics of at least two different types. This is important, because understanding the principles of each type may be critical in deciding whether and where to put seals.

The cavity wall* is a construction type that does not fit consistently into any of these categories. Its classification depends on whether or not the cavity is ventilated to the outside. If it is, then it may be classified as the rainscreen type: the outer skin is the water seal, the cavity is ventilated and drained to the exterior and the inner leaf acts as the air seal. Of course, cavities are rarely (if ever) ventilated with this principle in mind. What little ventilation there is is normally fortuitously provided by weepholes, which, with cavity trays, form part of the elaborate second line of defence. Cavities should be positively ventilated to outside air. If the cavity is not ventilated, then the wall falls into the controlled penetration category, in which total penetration is prevented by two lines of defence, comprising both the cavity itself, which breaks the capillary paths, and the dpcs, which drain water back to the exterior.

The filled cavity wall *does* work on the principle of controlled penetration, with certain thermal insulation materials designed to make the risk of rain penetration negligible. Again, in this arrangement, the dpcs act as a second line of defence.

4 Joints and junctions

(a) Introduction

As stated above, effective exclusion requires total continuity of performance which must be achieved by an assembly of discontinuous elements. Therefore how discontinuities are designed lies at the heart

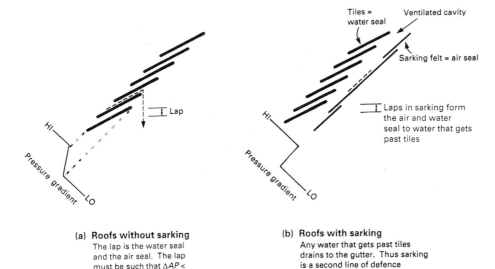

(a) **Roofs without sarking**
The lap is the water seal
and the air seal. The lap
must be such that $\Delta AP <$
force of gravity on the drop

(b) **Roofs with sarking**
Any water that gets past tiles
drains to the gutter. Thus sarking
is a second line of defence

Diagram D3.4/20 *Principles of how roofs with overlapping units exclude water*

of designing for effective exclusion, and this demands an understanding of jointing theory. The resolution of many of the problems with modern building revolves around joints. The theory of jointing is discussed in *2.2 Strength and the use of materials*, p. 95 ff, and only its application to the task of exclusion is considered here.

(b) Joints and exclusion

Experiments have shown that the flow of water in narrow vertical depressions in a wall face can be many times greater than the average over the wall.* In other words, vertical joints are particularly vulnerable to rain penetration. The reason is that a film of water running over a facade is blown laterally by the wind† so that vertical depressions act as gutters (Diagram D3.4/21.) To some extent, this can be combatted by having an upstand beside the joint. An upstand of only 5 mm will reduce the water load on the joint by two thirds.‡ Research has also shown that the majority of water actually entering vertical slots drains within a zone 50 mm back from the face of the wall.¶ The same research also found that the roughness of the surface affected the side flow but not consistently.

Various ways have been devised to categorize joint types, depending on the purpose of the classification. Classification according to geometry and ways of allowing for deviations is described in *2.2 Strength and the use of materials*, p. 99.

From the point of view of exclusion, it is useful to relate joint types back to the three basic methods of exclusion. In constructions which achieve exclusion by controlled penetration or a completely impermeable layer, the joints are sealed and known as the *one-stage* or *closed* type, so called because rain and air are excluded at the same physical point. With construction relying on controlled penetration such as solid brickwork, the joint filling material should perform in the same way as the joined units. Similarly, on a roof clad in built-up felt, the joints between the sheets must provide the same seal as the sheets

*Canadian Building Digest 40, p. 1.

†This is similar to the angle of creep applicable to roofs with small units described below.

‡BRE CP 86/74, *Window to wall joints*, 1974.

¶BRE CP 81/74, *Some observations on the features of external walls*, 1974.

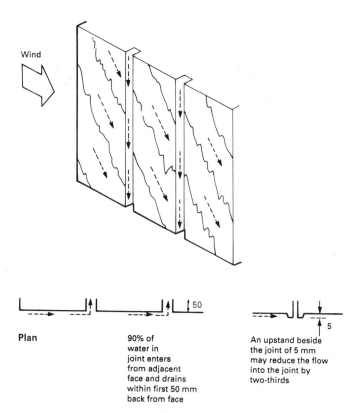

Wind

Plan

50

90% of
water in
joint enters
from adjacent
face and drains
within first 50 mm
back from face

5

An upstand beside
the joint of 5 mm
may reduce the flow
into the joint by
two-thirds

Diagram D3.4/21 *Flow of water in vertical joints*

themselves. The third method of exclusion, by means of multiple layers, uses joints of the *two-stage* or open type, in which the water seal is achieved geometrically (such as by a lap) and the seal against air infiltration is separate. (See Diagrams D3.4/8 and D3.4/17.) The water seal, or wet/dry boundary, is kept as far forward as possible and the air seal (where the pressure gradient occurs) is kept as far back as possible. Continuity of performance of the water seal is achieved by overlaps, baffles and carefully planned drainage.

(1) One-stage joints
One-stage joints depend entirely on the effectiveness of the jointing products to seal against water and air penetration simultaneously. Therefore every factor which affects their performance throughout the construction and life of the building must be considered carefully, and a second line of defence provided if possible.

Correct choice of the jointing product is essential. There are two aspects to consider, the interface between the seal and the joint profile and the nature of the sealing material itself. Sealing materials come in two types, unformed materials (sealants) and formed materials (gaskets). Guidance on the choice and application of sealants is given in *2.2 Strength and the use of materials*, p. 117 ff.

(2) Two-stage joints
Two-stage joints are so called because the position of the seals against water and air are separated. This approach requires the component to have sufficient thickness for both seals and the gap separating them. In

large-panel wall construction (in contrast to claddings composed of small overlapping elements) the joints have to be carefully designed to achieved correct performance in accordance with this principle, and it is these purpose-made joints which are described here.

One benefit of the two-stage joint compared with the one-stage joint is that the jointing product of the air seal is protected from the elements, because the water seal prevents water from reaching it. The water seal is a loose-fitting joint, to minimize the air pressure gradient across it. Downward-flowing water is excluded by means of overlaps and drips and wind-driven water is prevented from entering the joint by means of baffles, splines, sufficient depth and suitable profile of the joint mouth. As BS 6093 states:* 'The effectiveness of the air seal is the most important factor in preventing rain penetration through the joint.' This is because, where the air pressure on both sides of the water seal is equalized, rain will not be sucked across it.

Because the two-stage system accepts rather than fights against downward-flowing water, the horizontal and vertical joints are differentiated. Horizontal joints rely on lapping the panel elements and the baffles, usually by means of a dpc. At all junctions the continuity of the air seal must be maintained. To achieve this in practice, careful consideration must be given to the erection sequence.

(c) Junctions

Although much of this chapter concentrates on describing the principles of how different constructions perform, the reality of building is an assembly of units, components and elements which have different performance characteristics. As a result, there are innumerable junctions, and in many respects it is at these points that the task of achieving exclusion becomes most onerous. The junctions between different materials, elements and components are therefore fundamental to how the building as a whole performs. Many of the problems facing designers lie in this field, particularly on sites where the proximity of existing buildings imposes additional constraints. A common problem at junctions is how to achieve exclusion when constructions which work on different principles meet.

(1) Definitions
In this study of exclusion, the words 'joints' and 'junctions' have been used with distinct meanings. Whereas *joint* refers to the interface between the basic units which make up a component or element, such as the mortar-filled joints between bricks in a brick wall, the dry-lapped joints between slates which go together to form a roof, or the sealed joints between glass and aluminium mullions which make up a curtain wall, *junction* refers to the interface between two elements, such as a roof and a wall, or between a component and an element, such as window or door in a wall or a chimney in a roof.

Within the broad category of junctions there are several important sub-groups. These are *openings, projections, abutments* and changes of material or type of construction within an element (Diagram D3.4/22).

(2) Principles
The principles for building which are most relevant here are **continuity** and **separate lives**. Junctions are physical discontinuities – a break in material or construction, or a major change in geometry. In spite of the discontinuity, continuity of the performance requirement of exclusion *at least* is essential. In most situations it is an ideal, which is rarely met, to achieve continuity of all aspects of performance across the junction.

*BS 6093: 1981, Code of practice for design of joints and jointing in building construction.

Diagram D3.4/22 *Types of junction*

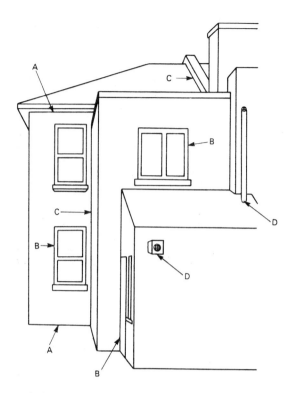

A: Junctions between elements
B: Openings
C: Abutments
D: Projections

Inside

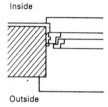

Outside

(a) Recessed into wall

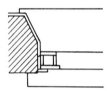

(b) Flush with face of wall

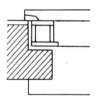

(c) Set into rebate

Diagram D3.4/23 *Three different ways of fitting a window into a solid wall*

*This true for the box itself and the sliding sashes: the joints between the sliding sashes and the box exclude moisture by having a labyrinthine water seal which is ventilated and drained.

Thus, for example, whereas it is essential that windows exclude rain to the same degree as the surrounding wall, it is also recognized that the same level of performance of either thermal or acoustic insulation cannot be achieved except by major compensatory measures. A test of a good detail for a junction is how well it succeeds in achieving continuity of all the performance functions of the element.

A major difficulty in detailing junctions occurs where components or elements which use different exclusion methods meet. The detail for forming the junction is usually based on experience in such cases. The problem is found particularly at openings in elements which exclude moisture by controlled penetration. For example, a sash window box is effectively a piece of impermeable membrane* which is thin in relation to the wall as a whole. There is, in abstract, a question as to where within the wall thickness the window should be placed (Diagram D3.4/23). In practice, the window has been placed both at the front and the back of the wall, in flush reveals and rebated ones, with reasonable success in all arrangements. A similar example, this time of an abutment, is the case of a slated roof abutting a higher wall (Diagram D3.4/24). In solid construction (working by controlled penetration), a simple step flashing is adequate where the exposure is not too great. If the wall is of cavity construction (working by the use of multiple layers)

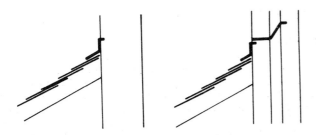

Diagram D3.4/24 *Abutment of roofs to walls*

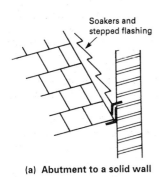

Soakers and stepped flashing

(a) Abutment to a solid wall

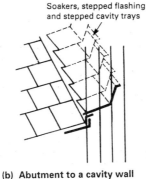

Soakers, stepped flashing and stepped cavity trays

(b) Abutment to a cavity wall

Diagram D3.4/25 *Isolated asphalt kerb allows for differential movement between wall and roof*

to omit proper stepped damp-proof cavity trays is to invite rain penetration via the cavity to some point below.

The principle of **separate lives** must also be taken into account at junctions. Examples of how it can apply include:

- Will exclusion continue to be achieved when the effects of differential movements are taken into account? (See Diagram D3.4/25.)
- If exclusion relies on a sealed joint of shorter life than the adjoining elements, what happens when the sealant begins to break down?
- The buildability of the detail (in other words, how it can be achieved for a process of assembly) affects the performance of exclusion.

Consideration of these questions shows that the lessons from the three methods of exclusion can be applied to junctions. Where movements and dimensional deviations must be allowed for (as is often the case at junctions), a joint of the open type will be more suitable than the closed type. This gives rise to an 'ideal' building section in which the roof overhangs the wall, and the upper storeys overhang the lower ones (Diagram D3.4/26). Applied to openings, it means that window and door components are best fitted into recesses which allow for a lapped joint (Diagram D3.4/27).

Many of the solutions to the problem of achieving exclusion at junctions are, by nature, heuristic – that is, based on experience and rules of thumb – whereas others can be found by applying the principles for building. Examples are given in 'Methods of exclusion, Junctions', p. 334 below.

Diagram D3.4/26 *Ideal section of a building which reconciles water shedding and* **separate lives**

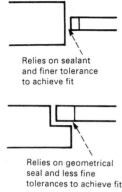

Relies on sealant
and finer tolerance
to achieve fit

Relies on geometrical
seal and less fine
tolerances to achieve fit

Diagram D3.4/27

Methods of exclusion

1 Introduction

Exclusion illustrates, perhaps more cogently than any other function of the building fabric, the system of principles, rules and compensation described in the introductory chapter of this book. It will be remembered that either these principles are followed or, where this is neither possible nor desired, steps must be taken to compensate for non-compliance. The practical measures which derive both from the principles and experience take the form of positive rules and precautions, and these change with experience and as techniques and standards develop.

This section looks at the key parts of the building in turn (i.e. roofs, walls, construction at ground level and junctions) and considers for each the various methods by which moisture is excluded. For convenience, the methods of exclusion applicable to liquid moisture are considered first, and then for each method the implications for the exclusion of moisture vapour are discussed. The intention generally is to analyse existing solutions and to relate these to the principles of exclusion discussed above, rather than to extrapolate from the principles to postulate new ways of achieving exclusion.

The rules and precautions which make up these methods of exclusion are a mixture of empirical knowledge (based on experience and observation) and the result of scientific research. With traditional construction the rules are almost always *heuristic*, that is, they are rules of thumb which usually take account of factors beyond what is immediately apparent. Adopting realistic rather than optimistic expectations of the shortcomings of workmanship is an example: rules based on laboratory studies often overlook such aspects.

To date, authoritative information in the form of British Standard Codes of Practice, BRE publications and Agrément Certificates are the best source of rules and precautions, not only for exclusion but for all the performance requirements of building. It is emphasized that within the scope of this book there is only space to consider key rules and precautions which illustrate the general principles, and reference should be made to the former for full guidance on each particular subject.*

One point must be emphasized: if failure occurs, it will not be because the principle was at fault but because the rules followed and precautions taken were inadequate for the circumstances or were wrongly applied.

2 Roofs

Roofs are exposed to moisture in the form of downward-flowing water and internally produced vapour. All three methods of exclusion are used to control the damaging penetration of external moisture, but only a completely impermeable layer is effective in controlling the movement of vapour into the fabric to a point where interstitial condensation can occur.†

The pitch of the roof is critical in determining which of the three exclusion methods is feasible (Diagram D3.4/28). At pitches above 50° all methods may be used. As the pitch decreases, roofs with overlapping elements become impractical, until at very low pitches only systems which work by means of a continuous impermeable layer are practicable.‡

*See the Further reading at the end of this chapter.

‡This is true for the control of vapour moving through the fabric under the influence of a vapour pressure gradient. The best way of controlling condensation is to make the vapour pressure gradient more shallow by effective ventilation.

*This is true for the UK. In drier climates, flat roofs often rely on controlled penetration alone.

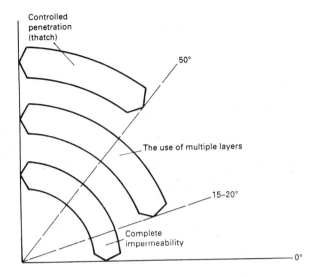

Diagram D3.4/28 *Relationship between roof pitch and exclusion method*

(a) Impermeable sheets and membranes

The range of materials of this type is wide and covers asphalt, built-up roofing, single-layer membranes such as PVC and EPDM, traditional fully supported sheet metals, low-pitched self-supporting metal standing seam systems, and profiled metal and fibre-cement sheets whose joints are sealed. They all exclude rainwater by the method of complete impermeability.

It is debatable whether the last three systems listed work by this method or by the use of multiple layers. The test is whether the air seal and water seal coincide. Hence it is important to distinguish between systems with sealed joints which have either no cavity behind or an unventilated void (these work by complete impermeability) and those with unsealed joints and a ventilated cavity beneath the sheets (Diagram D3.4/29).

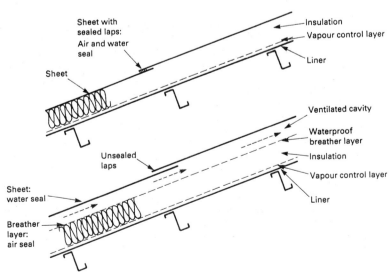

Diagram D3.4/29 *The method by which sheeted roofs exclude moisture depends on whether the sheets are sealed*

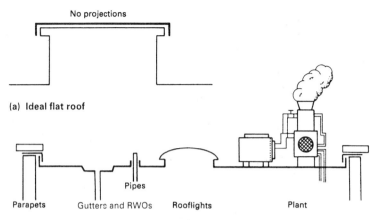

(a) Ideal flat roof

Pipes

Parapets   Gutters and RWOs   Rooflights   Plant

Diagram D3.4/30

(b) Typical flat roof as found, in conflict with 'separate lives'

The optimum roof construction using impermeable sheets and membranes has a waterproof membrane and vapour control layer (if there is one) which are physically continuous. All physical discontinuities in the membrane, such as abutments, parapets, rooflights, outlets, pipes or mechanical plant compromise this objective and compensation must be made (Diagram D3.4/30). These features are also constrained by the principle of **separate lives**, from a number of points of view. First, the effects of differential movement between the membrane and the projection itself must be anticipated and allowed for. Second, the maintenance requirements of the different parts vary. For example, ventilation plant is often placed on a flat roof. Invariably, this puts at risk the **continuity** of the membrane either when the plant is replaced (the membrane being damaged in the process) or when the membrane is renewed (when it is very difficult to get proper access to the roof beneath the plant).

(b) The effects of adding insulation

Condensation risk is a particular problem with roofs of this type. As noted previously, the rules for reducing the risk of interstitial condensation are to increase the thermal resistance towards the outside in order to keep the fabric warm, and to increase the vapour resistance towards the inside, thereby keeping the fabric dry. If the layer most impermeable to vapour is on the outside of the roof, the second rule is broken. To compensate for this, there are a number of possible solutions. One is to provide a vapour control layer on the warm side of the construction whose vapour resistance is at least as high as that of the main waterproofing membrane. A second is to provide and ventilate to outside air a void beneath the membrane. A third is to put waterproof insulation on top of the membrane (Diagram D3.4/31).

These three forms of compensating for the basic weakness inherent in the exclusion method form the basis for the three methods of categorizing flat roofs,* and it can be seen from the diagram that they correspond to the different possible positions for placing the thermal insulation layer. The first is called the *cold deck* roof. The others are both *warm deck* roofs, the second being the *sandwich* roof and the third the *inverted* roof.

*See BRE Digest 312, 1986, *Flat roof design: the technical options.*

Roof type	Advantages	Disadvantages
Cold deck roof: Solar protection Waterproof membrane Deck Ventilated cavity Thermal insulation Vapour control layer Ceiling	Membrane not subject to **separate lives** stresses from insulation Easy to upgrade existing roofs Membrane can be replaced without damage to insulation	Difficult to ensure adequate ventilation Ventilation may be in conflict with fire cavity barriers Difficulty to ensure continuity of VCL particularly at service points
Warm deck 'sandwich' roof: Solar protection Waterproofing membrane Insulation Vapour control layer Deck Ceiling	Continuity of VCL easily achieved Roof as a whole can be made thin and light	Movement of insulation puts stress on membrane Both membrane and VCL must be perfect Insulation must carry imposed loads Insulation damaged when membrane renewed Both membrane and insulations must resist wind forces High risk of entrapped moisture
Warm deck 'inverted' roof: Ballast/solar protection Insulation Waterproof membrane Deck Ceiling	Membrane protected from extremes of temp. and traffic Only one continuous membrane needed	Membrane cannot be inspected easily High weight of roof due to ballast Drains poorly Insulant must be weather resistant Liable to wind scouring and plant growth

As shown in the diagram, there are advantages and disadvantages for using each type, and thus there is no standard solution. The constraints on and requirements of each roof must be considered individually. For example, cold roofs require adequate cross ventilation to clear vapour entering the roof cavity. Openings to give a total of at least 0.4% of the plan area should be put on opposing sides of the roof, or up to twice this for sheltered conditions. A clear 50 mm cavity must be maintained. However, on wide roofs* providing such ventilation may be in conflict with the need for fire cavity barriers.

Diagram D3.4/31 *Cold deck, warm deck and inverted roofs and their advantages and disadvantages*

(c) Thermal pumping

Sheeted roofs may be vulnerable to water penetration by the mechanism known as thermal pumping. This can occur where the joints between the sheets are tightly lapped but not completely airtight,

*The 1985 Building Regulations require fire cavity barriers at a maximum of 8 m intervals where the materials facing the cavity are combustible.

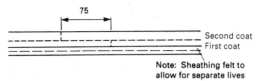

Joints in asphalt

Joints are used during application of the asphalt
Once cooled, the asphalt is effectively homeogenous

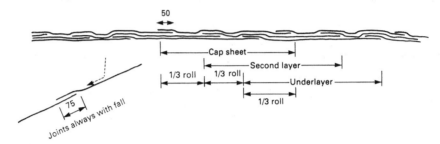

Joints in built-up three-layer felt

Diagram D3.4/32 *Compensation by
means of multi-layer systems*

and where there is a cavity below the sheets which is sealed to air
except at the laps. It has been shown that, when the roof is subject to
sudden changes in temperature, water which is held in the laps by
capillarity can be sucked over them. Showery days in the summer,
when the roof is exposed to alternate sunshine and rain, provide
suitable conditions for thermal pumping to occur. As rain falls on the
warm sheets, the temperature of the roof surface drops, causing a
partial vacuum in the cavity which may be a sufficient force to suck
water over the laps.

(d) Notes on the systems

(1) Fully supported bituminous and polymer membranes
These membranes are laid either without joints or with fully sealed
joints. They require 100% **continuity** of material to achieve 100%
continuity of performance: all laps must be physically fully sealed.
However, **creative pessimism** suggests that 100% is difficult to achieve
in practice – particularly for the life of the building. The precautions for
compensating for this risk are fairly limited in scope (for example,
insistence on good standards of quality control such as installation only
by licensed contractors, regular inspections and preventive maintenance). Multi-layer systems compensate for the risk of discontinuity by
reducing the reliance on just one lapped joint (Diagram D3.4/32).

(2) Fully supported metal roofs
In traditional fully supported metal* roofs exclusion of rainwater is
achieved by joints which are lapped and welted tightly to form a
geometrical water seal between the sheets (Diagram D3.4/33). These
types of joint are effectively watertight at the pitches used (minimum 1
in 80). The clips which hold down the metal (if used) are hidden within
the joints. Since the metal is supported on a boarded timber deck with
any insulation placed below, these roofs are of the cold deck type, so
that the rules governing the avoidance of interstitial condensation (i.e.
adequate ventilation of the cavity and the provision of a vapour control
layer) must be followed.

*The principles are the same but the details for fixing
and jointing lead, aluminium, copper and zinc differ
and are covered by BS 6915 – for lead – and the
various parts of BS CP 143, Code of practice for sheet roof
and wall coverings.*

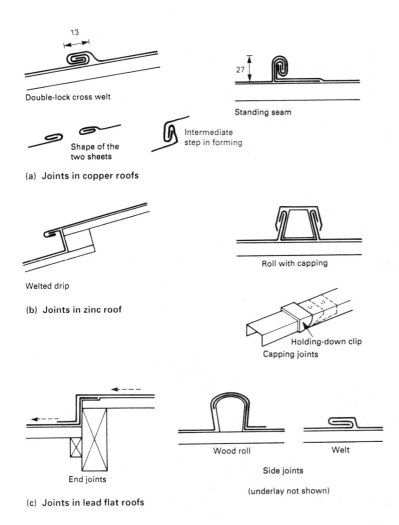

13
Double-lock cross welt

27
Standing seam

Shape of the
two sheets

Intermediate
step in forming

(a) Joints in copper roofs

Welted drip

Roll with capping

(b) Joints in zinc roof

Holding-down clip
Capping joints

End joints

Wood roll   Welt

Side joints

(underlay not shown)

(c) Joints in lead flat roofs

Diagram D3.4/33 *Joints in
continuously supported metal roofs*

Attempts to place the metal directly above a layer of insulation in the manner of a warm deck roof have not always been successful. There have been problems with lead roofs built in this way where thermal pumping has occurred, leading to premature failure of the lead by corrosion (see *3.7 Corrosion*, p. 503). The current advice of the Lead Development Association is to provide a ventilated air space above the insulation. This requires one deck to support the vapour control layer and insulation and a further layer of substructure to carry the boards which support the lead sheet. In addition, a further effect of a better insulated roof is to increase the temperature range of the lead in service. This will increase the stresses on the metal so that the size of sheets should be reduced.

(3) Self-supporting profiled sheets
Forming corrugated profiles in thin sheets of metal or fibrous cement enables them to span between intermediate supports without a separate layer of substructure. Thus with self-supporting profiled sheets the deck is itself the waterproof layer (Diagram D3.4/34).

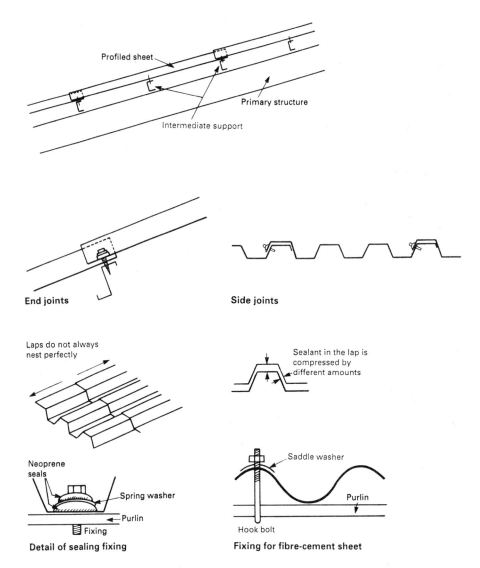

Profiled sheet

Primary structure

Intermediate support

End joints   **Side joints**

Laps do not always
nest perfectly

Sealant in the lap is
compressed by
different amounts

Neoprene
seals

Spring washer

Purlin

Fixing

Detail of sealing fixing

Saddle washer

Purlin

Hook bolt

Fixing for fibre-cement sheet

Diagram D3.4/34 *Direct fixing of profiled sheets*

Historically, these types of profiled sheet were first used extensively for the roofs of industrial and agricultural buildings where the standards of exclusion could be less than 100%. Sheets of corrugated iron* and, later, asbestos cement were fixed through the crowns using direct fixings to purlins below. Generally, no thermal insulation was provided.

This approach of using direct fixings flies in the face of the principle of **high>low** by creating potential paths for water penetration all over the roof. Nevertheless, direct fixings have been used for sheets used on other building types where exclusion is more critical and higher standards of thermal insulation are required. Methods have been developed to compensate for the risks of rain penetration and to incorporate thermal insulation into the construction. The effects of these two aspects on exclusion are now considered in turn.

*An early use of which was on the boiler house of the Crystal Palace in 1851.

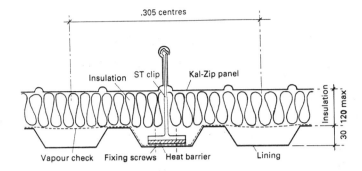

Diagram D3.4/35 *Example of a secret-fix system. ('Kalzip', courtesy Hoogovens Aluminium UK Ltd)*

(e) Fixings

Direct fixings are used for both corrugated and trapezoidal profiled sheets. On corrugated sheets, the fixings are through the crown and large saddle washers are used to seal the fixing hole. Steel and aluminium sheets with a trapezoidal section are attached directly to purlins with self-tapping fixings which incorporate a neoprene seal held under a compression washer, together with special tools which help to place the fixings at the required torque and alignment. Nevertheless, the basic risk of direct penetration remains.

At the sides and ends of trapezoidal and corrugated sheets the joints are made by simple lapping. Whether or not additional sealant is used depends on the pitch. Sealants tend to be employed in a very uncontrolled way in these situations: end laps can never nest perfectly because of the effect of the sheet thickness and any minor misalignment. Due to the geometry of the profile, the sealant is compressed unevenly. It is prudent, therefore, not to rely on sealants, which means, in effect, that profiled sheets with simple lapped joints should not be used below a pitch of about 15°, above which the lapping on its own will be sufficient to achieve exclusion.*

Thermal pumping (see p. 301) is thought to contribute to water penetration at the laps of low-pitched roofs.†

Indirect fixings are always to be preferred from the point of view of exclusion, and there is a growing range of systems which use them, generally derived from the fixing and jointing methods of traditional fully supported metal roofs (Diagram D3.4/35). Many of these have the added benefit of recognizing the principle of **separate lives** and allow for thermal movement to take place over the fixings. This also enables sheets to be used in a single length from eaves to ridge (or even from eaves to eaves at very low pitches) so that end laps are not required. The joints between the sheets in secret-fix systems are more tightly engineered and enable the sheets to be used at lower pitches, which may be as small as 1.5°.

(f) The effects of adding insulation

The story of adding thermal insulation to profiled sheet roofs is something of a sorry one, fresh problems becoming apparent with each development. Diagram D3.4/36 shows the outline of this development pictorially. The common theme is the risk of condensation.

With self-supporting sheets it is normal for the insulation to be placed below the sheet, in effect making a cold roof. Following the rules

*Note that this means that the sheet now works by the exclusion method of multiple layers, and therefore the size of the lap will be determined by how the various forces driving water over it will be resisted. In particular, consideration must be given to pressure gradients across the construction and the effectiveness of the air seal.

†See AJ article by Peter Falconer, 'Metal industrial roofs – moisture problems', 12 June 1985 and 7 May 1986.

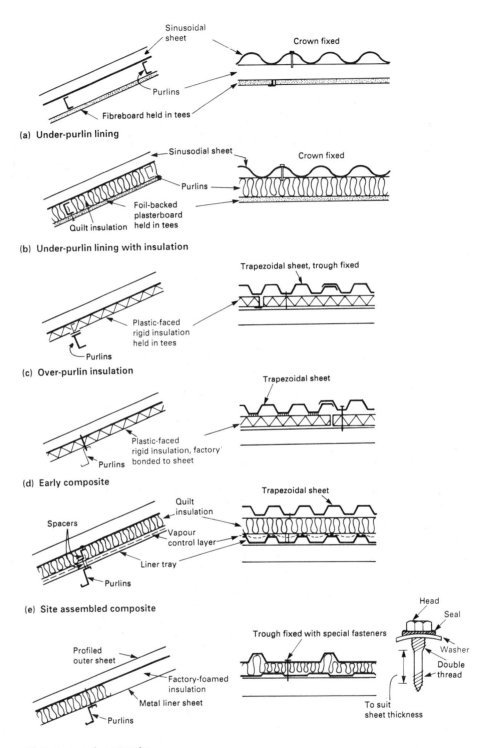

(a) Under-purlin lining

(b) Under-purlin lining with insulation

(c) Over-purlin insulation

(d) Early composite

(e) Site assembled composite

(f) Factory-made composite

Diagram D3.4/36 *Development of direct-fixed profiled sheet roofs*

set out above for cold roofs, it would seem that the insulation should have beneath it a vapour control layer and a ventilated cavity above it. However, with this kind of construction exclusion of condensation moisture presents special problems. On the one hand, there are difficulties in providing continuity of both the insulation and the vapour control layer, because of the penetrations caused by fixings. Further, the thin metal sheets are subject to the effect of supercooling by radiating their heat to clear night skies. This causes the surface temperature of the sheet to fall below the dewpoint of the external air so that ventilating cavities may actually encourage an accumulation of condensation on the underside of the sheet, rather than removing excess moisture as intended. Because the cavity volume is small a large air-change rate in the cavity is required to remove the condensed moisture, which is difficult to achieve by natural means with roofs laid at very low pitches. In the absence of published research in this matter, it is difficult to give quantified guidance on the minimum free areas at each side of the cavity.

One way round the problem is to use composite panels in which the outer sheet and an inner liner tray are bonded together by polyurethane insulation foam formed under factory conditions. It is important that the method of jointing the panels avoids penetrations and achieves **continuity** of all functions, particularly thermal insulation. These panels are still technically cold decks, but the liner sheet is intended to act as a continuous vapour control layer. As there is no air path for vapour to reach the underside of the main sheet, ventilation is neither possible nor required, and the risk of damaging condensation caused by night-sky radiation is reduced. There are potential problems of differential stressing across the panel due to the temperature gradient (an aspect of *imbalance* in conditions and the **separate lives** of the various parts not being allowed for) and recent experience shows that these have yet to be resolved.

Warm roofs in this form of construction would place the insulation on the outside of the deck. This approach has been used as a remedial technique for rehabilitating uninsulated asbestos cement roofs, where polyurethane foam is sprayed over the top of the existing sheet. It is possible to envisage a waterproof insulation mat laid on top of profiled metal roofs which would overcome the condensation problems – a sort of hi-tech thatch!

(g) Tiles, slates and shingles

Traditional methods of roofing comprising overlapping small units, whether tiles, slates or shingles, exclude rainwater by the use of multiple layers. The rules for using the various materials to achieve exclusion apply to the number and size of laps required to ensure that water finding its way to the edge of the unit is drained to the outside again, for given pitches and exposure. *Single-lap* tiles include traditional pantiles and Spanish tiles, together with modern interlocking concrete tiles that can be used down to pitches as low as 17.5° (Diagram D3.4/37).

Slates and plain tiles require a *double lap*. In section the effective lap is the portion where there are three layers, and this lap must be maintained at the ridge and eaves by means of an undercloak course and the lap of the ridge tiles. The weather resistance is also governed by the angle of creep, which is the angle between the top of the exposed joint and the nearest point of entry to the roof below (Diagram D3.4/38). It will be seen that the narrower the unit, the smaller the angle of creep, and therefore the less resistant the roof will be to rain penetration. The lap controls penetration of water driven by the direct effect of gravity

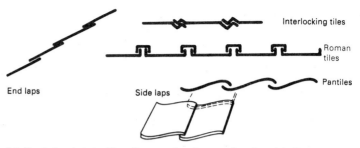

(a) Single lap: interlocking tiles, pantiles, Roman tiles, Spanish tiles

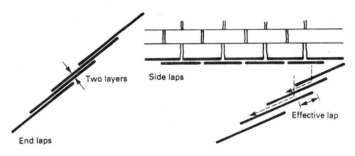

Diagram D3.4/37 *Overlapping roof units*

(b) Double lap: slates, plain tiles

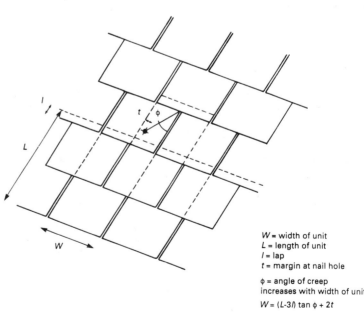

W = width of unit
L = length of unit
l = lap
t = margin at nail hole

ϕ = angle of creep
increases with width of unit

$W = (L - 3l)\tan\phi + 2t$

Diagram D3.4/38 *Angle of creep*

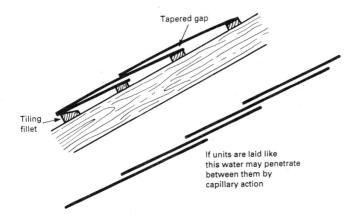

Tapered gap

Tiling
fillet

If units are laid like
this water may penetrate
between them by
capillary action

Diagram D3.4/39 *The importance of
tilting fillets in slated roofs*

and by wind up the roof, and the angle of creep controls penetration of
water driven by wind across the roof. Water travelling between the
units by capillarity is prevented by tiles having a camber in two
directions and by slates being set off from a tilting fillet so that they only
touch at the bottom edge (Diagram D3.4/39).

Shingles made from oak or western red cedar are *triple lapped*,
basically because the irregularity of the units makes the size of the bond
and therefore the adequacy of the angle of creep harder to guarantee.

The lap size in relation to pitch which is required to prevent rain
being driven over the units is established empirically for the different
materials, depending on their exposure. The relationship between
pitch and lap is illustrated in Diagram D3.4/40. It has always been
recognized that the system of overlaps can only be economically
provided to cater for the majority of weather conditions, and that the
kinetic energy of droplets thrown against the roof may still carry them
past the laps in windy conditions. Further measures are needed to
achieve a 'drop-tight' roof. Traditionally, these included the practices of
bedding the heads of the units on mortar, and torching, which is the

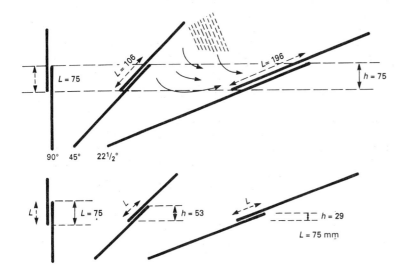

$L = 75$

$L = 106$

$L = 196$

$h = 75$

90°   45°   22$\frac{1}{2}$°

$L = 75$ $h = 53$ $h = 29$

$L = 75$ mm

Diagram D3.4/40 *The relationship
between pitch, lap and head*

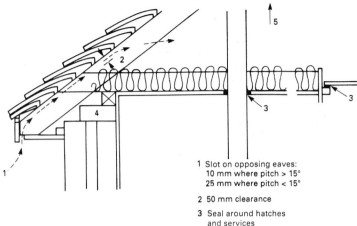

1 Slot on opposing eaves:
   10 mm where pitch > 15°
   25 mm where pitch < 15°

2 50 mm clearance

3 Seal around hatches
   and services

4 Block tops of cavities

5 Ridge vents only used in
   conjunction with eaves vents, to
   avoid negative pressure in
   roof space

Diagram D3.4/41 *Rules for controlling moisture accumulation in roof spaces*

term for pointing the back of the tiles or slates.* Today a second line of defence in the form of sarking felt is used, which drains any moisture which travels past the laps back to the gutter. Sarking felt has another function, which is to tend to allow the air pressure behind the units to equalize with external air so that water is not so readily sucked over the laps (see Diagram D3.4/20). Because the felt is not sealed nor the size of the cavity controlled, this is rather crude.

As exclusion is achieved by discontinuous units, the system tends, as a matter of course, to work with the vapour pressure gradient, with the construction able to breathe towards the outside. However, the use of sarking felt does hinder this: to avoid condensation in the roof space the sarking felt should be vapour permeable and, where the roof is insulated at ceiling level, the roof space must be ventilated to the exterior from the eaves. This is particularly important if the roof pitch is low. The rules to adopt are to provide a continuous ventilation slot along opposing eaves with a total free area of 0.3% of the plan area of the roof, to hold the insulation back at the eaves to provide a through path for ventilation and to draught-seal openings into the roof void at ceiling level (Diagram D3.4/41). Where the insulation is on the line of the rafters the cavity between the units and the insulation requires more positive ventilation (Diagram D3.4/42).

Because the roof is constructed of small units, the principle of **separate lives** is followed as a matter of course without need for compensation. In terms of maintenance this has the advantage that it is simple to identify and replace individual units.

The small units used are vulnerable to wind uplift, particularly at roof margins. Although this generally does not affect exclusion as such, it is interesting to note that in many parts of the UK where exposure to winds is severe it is characteristic for verges to be finished with a parapet wall which protects the ends of the tiles or slates from wind uplift, at the cost of a greater risk of rain penetration at these points.

*Neither of these practices is recommended today. With movement of the units the mortar works loose and drops out, and it may retain water.

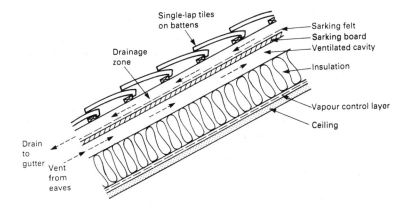

Diagram D3.4/42 *Insulating and ventilating tiled or slated roofs at rafter level*

(h) Thatch

Of all materials used for roofing in the UK, only thatch achieves exclusion by controlling the penetration of water into it.* Although it might appear at first sight to be a material only of the past, in the light of the principles for building there are a number of reasons why it has advantages over apparently more sophisticated techniques.

The first point is that it is the perfect partner, as a roofing material, to walls of monolithic construction, i.e. solid brickwork or stone, in that (apart from structure) all the functions of the envelope are performed by the single thickness of material. This means that at the junction between the wall and roof there is a **continuity** of similar systems.

Pitch and thickness are chosen to achieve exclusion to suit the exposure of the site. Because of the absorption of rainwater by the thickness of the material, gutters are not usually necessary: although sometimes seen, they are the exception rather than the rule.†

As regards the temperature gradient across the roof, the thickness of thatch required to achieve exclusion provides a high level of thermal insulation compared with other traditional roof materials. From the point of view of the vapour pressure gradient, the whole construction can breathe towards the outside. These two critical gradients run in such a way for there to be a low risk of condensation in the construction. In terms of the material's durability (i.e. an aspect of *reversion*) Norfolk reed is said to last 60 years. Obviously, there are problems with flammability. Because of its loose-bound nature there are no difficulties with differential movements.

This is not to advocate a revival of traditional thatching, only to point out that the concept of a method of roofing using controlled penetration as the exclusion method has certain attractions. It may be that the idea could be adopted using a modern material.

3 Walls

Walls are exposed to moisture in all its forms and from all sources, as downward-flowing water, wind-driven rain, rising damp and internally produced vapour. All three methods of exclusion are used to control the damaging penetration of moisture into the fabric, usually in combination, since openings are an integral part of wall construction.

*Clifton-Taylor describes at some length the range of materials and techniques in thatching used in the UK.

†There is an interesting corollary to this in that, whereas on walls of absorbent materials gutters are not provided, on walls of non-absorbent materials such as curtain walling it is important to consider the run-off of water hitting the facade.

(a) Applied coatings

An impermeable coating applied to the outside of a wall made of porous material will exclude rainwater by virtue of its physical **continuity**. The risk of poor performance lies in the possibility of there being discontinuities, so that the relevant rules and precautions spring from ways of achieving continuity of the coating, preventing discontinuities and compensating for the risks that arise from reliance on 100% continuity of material. As with roofs which have impermeable membranes, there is an inherent conflict between the need for impermeability of the outer layer to achieve exclusion of rainwater and the need for the vapour resistance of the wall to decrease towards the outside in order to prevent the possible accumulation of damaging moisture within the wall.

(1) Paints and water repellents

Absorption of water by a porous material takes place when the open ends of the capillaries formed by the arrangement of pores in the material occur at the exposed surface. Water penetration, though, can only occur if there are *complete* through-routes from the outside to the inside surface. If a porous material is to be made impermeable one of two methods may be used. Both are illustrated in Diagram D3.4/43. Either the open ends of the capillaries may be sealed by the application of a thin liquid coating, such as paint, or the walls of the capillaries may be lined with a water-repellent substance which will prevent water from being transmitted into the wall by capillary action. In the latter case the water repellent acts in a manner analogous to grease on a glass surface which prevents the wetting of that surface. It is important to note that water under pressure may be driven along through-routes larger than capillary size, or even where the water-repellent force is weaker than the driving force of a difference in air pressure across the wall.

Despite differences in the behaviour of paints and water-repellent treatments, there are two important factors common to both. First, there is an increased run-off of rainwater from the treated surfaces which immediately increases the risk of water penetration through cracks, defective joints and other weak points, and, second, periodic renewal of the applied coating is necessary. Renewal periods may be as

Diagram D3.4/43 *Two basic methods of making porous materials impermeable by the application of thin coatings*

External face

'Open' ends of capillaries sealed. Moisture as liquid or vapour cannot enter capillaries, nor can there be any evaporation from the surface. Moisture entering material through cracks or from elsewhere becomes trapped

Impervious coating

External face

'Open' ends of capillaries lined. Moisture as liquid can enter capillaries but cannot travel along them. Water vapour can be transferred along capillaries, thus evaporation from surface is possible

Water-repellent coating

short as two years. The restriction imposed by the necessity for periodic renewal of applied coatings has discouraged their use as the *primary* means of defence against rainwater. In many cases they have been used as a precaution additional to the controlled penetration of water that solid masonry achieves, or to improve appearance.

Thin applied coatings may reduce the rate of evaporation from the surface so that the quantity of water in the wall may actually increase. In extreme cases this may be sufficient for frost damage to occur.

The surfaces of permeable materials may be sealed by a wide variety of paints, including oil paints, plastic-based emulsions, bituminous emulsions, and paints based on chlorinated rubber. Most paints have a high resistance to the passage of liquid water and can be regarded as impervious. Resistance to water vapour varies with the formulation. Alkyd paints have a sufficiently high resistance to the diffusion of water vapour to prevent entrapped water from evaporating. Microporous paints are designed to be considerably more vapour permeable, and are therefore more tolerant of being applied to surfaces which have not fully dried out. On the other hand, by allowing moisture vapour to breathe the other way the timber will be able to respond more freely to variations in external conditions and may therefore have greater fluctuations in moisture movement. (See *3.3 Moisture content*, p. 239 ff.)

Modern colourless water-repellent treatments are based on silicone resins, polyoxo aluminium stearate or monomeric alkyd alkoxy silanes. Protection is achieved by lining the pores of the substrate with a thin film of hydrophobic material rather than by fully sealing them. To be effective, substrates must be free from cracks larger than 0.15 mm: these and other defects must be made good before the repellent is applied.*

(2) Rendering

Historically, rendered finishes to walls are important because they provide a *relatively* impermeable coating to very absorbent substrates such as rammed earth and wattle and daub. Renderings themselves may be strong and dense or weak and porous, depending on the proportions of Portland cement, lime and sand used. Although dense renderings with a high cement content will provide greater resistance to water penetration if they remain homogeneous, they are more vulnerable to cracking from drying shrinkage and moisture movement than weaker, more porous compositions. Clearly, it is at the cracks that water finds a route through to the substrate.

To prolong the life of render, good shedding of water is important. In addition, the leading edges of render at parapets, verges, eaves and around openings should be protected by generous overhangs† (see Diagram D3.4/15).

It was noted previously that the thermal resistance of the envelope should increase towards the outside to reduce the risk of interstitial condensation. Special renders incorporating insulation granules and renders applied to metal lath over a layer of insulation are available which enable this objective to be achieved.

The principle of **separate lives** should be observed when detailing render. To prevent the render from cracking or delaminating from the substrate, the mix and thickness of each layer must be carefully controlled, and at changes in substrate the render should be reinforced or separated by a movement joint.‡

(b) Tiling and mosaics

Glazed ceramic tiles have been used to provide a high-quality finish that may be decorative as well as more durable than the substrate for

*BS 6477, *Specification for water repellents for masonry surfaces.*

†It will be seen that these are the same rules for protecting walls which exclude rain by controlled penetration.

‡See BS 5262, *Code of practice for external rendered finishes.*

many centuries.* Essentially, they exclude water by the method of complete impermeability. Modern examples seem to have been fraught with problems. In a typical construction tiles are bonded by adhesive to a render substrate, itself applied to a masonry wall.

While the tiles themselves are impervious to water, the joints between them are usually less so. At these points, and at any cracks that occur as a result of movement, water may find its way behind the tiles or render. Consequential damage may be rain penetration, sulphate attack or frost damage, causing the tiles to be forced off the substrate. As with render, it is often more difficult for water that finds its way in to find its way out.

Because of the various layers bonded together, the principle of **separate lives** must be recognized here, and a number of rules follow from it. To achieve lasting adhesion between the layers, a combination of allowing for differential movement and providing compensation in the form of a mechanical key is required. The risk of water getting behind the tiles is reduced if measures to shed water frequently from the facade are adopted, such as those shown in Diagram D3.4/15.

(c) Fully sealed curtain walling and cladding

Cladding panels, which themselves are non-structural, are generally fixed to a structural building frame. Framed structures are not new. Many primitive building types comprise a simple timber framework covered in thatch or sods, and timber frames were developed to a high degree of sophistication in the medieval buildings of northern Europe. However, the development of concrete and steel structural frames in France and the USA towards the end of the nineteenth century was one of the principal technological developments of modern architecture. Framed structures, by definition, separate the parts of the construction which have different functions: this, as discussed in the introductory chapter, is a characteristic of modern construction, and one that must persist as long as monolithic construction is unable on its own to meet rising performance standards. The function of exclusion is achieved by impervious cladding panels which are fixed to the structural frame either directly or via a secondary framework. Cladding based on lightweight materials such as glass and aluminium is known as curtain walling, although the principles of achieving exclusion may be the same as that used by heavyweight panels.

The cladding panels are usually prefabricated, and made as large as possible to reduce work on-site, their size being limited by the constraints of manufacturing, transport, storage, handling and installation. The detailing of the joints determines whether the method of exclusion is by complete impermeability or by the use of multiple layers. Cladding of the latter type is known as rainscreen cladding, and is discussed separately at (i) below. In either case, the design and performance of the joints are critical to the success of the construction. Much of the discussion of jointing theory in *2.2 Strength and the use of materials*, p. 95 ff, and under 'Principles of exclusion' in this chapter is especially relevant here.

Cladding panels of this type are made of materials such as glass, metals (aluminium in flat or profiled panels, or vitreous enamel coated steel), GRP, precast concrete or glass-reinforced concrete, all of which are effectively impervious to flowing water. Where they are used in a wall that excludes moisture on the principle of complete impermeability the joints between the panels must be fully sealed using either sealants or gaskets (Diagram D3.4/44).

One distinctive characteristic of buildings clad in this way is that there is no need to respect the sense of gravity: horizontal and vertical

*An early example is the remarkably well-preserved blue glazed tiles decorated with lions which lined the entrance route to Babylon at the time of Nebuchadnezzar in the sixth century BC, now in the Pergamon Museum in Berlin.

joints can be detailed identically, and the form is freed so that cladding can flow from the top of the parapet to the ground using sloping areas, vertical facades and undercuts without a change in the cladding or its jointing system – indeed, just like a raincoat.

Continuity of exclusion relies on total functional **continuity** of material, which in the light of the principle of **creative pessimism** is an objective with attendant risks of poor performance. One method of compensation is to provide a second line of defence, and this is more practical to achieve with walls than with flat roofs.

With impervious materials water accumulates on the facade, increasing towards the bottom as the film flows downwards, and putting a very large load of water on joints. Consideration should be given to draining this water away at intermediate points on the facade.

Despite the fact that exclusion can be achieved in defiance, as it were, of gravity, the effect of disregarding the way that water will run over the facade is often to cause unsightly weathering. Examples are given in *3.9 Flow and changes in appearance*, p. 562 ff.

Condensation can be a high risk in this type of construction because the impermeable layer on the outside of the building restricts breathing, and usually prevents thermal insulation being placed in its optimum position towards the outside of the construction. There is more scope for ventilating cavities behind the cladding than there is within flat roofs, because the stack effect* of convection currents is relatively simple to harness to clear condensation moisture on the back of the impermeable layer. However, if the source of ventilation is external air, the effect will be to separate the air seal from the water seal, thereby changing the way that exclusion is achieved from complete impermeability to the use of multiple layers.

Since continuity of material is so important, the implications of the principle of **separate lives** have to be considered carefully. Thermal and moisture movements, differential structural movement between the panel and the frame will all put stress on the sealed joints. These are what BS 6093 calls the *induced deviations*. The *inherent deviations*, such as tolerances in panel and frame dimensions, and the whole process of assembly must similarly be taken into account. Finally, the sealant may well have a different durability to the panels themselves. Successful exclusion throughout the life of the building relies on the correct choice of sealant or gasket for the job and design of the joint and the assembly sequence so that the seal can be installed and replaced during the life of the building.

(d) Solid masonry

This group includes the traditional monolithic wall constructions using brick, stone and more primitive materials such as rammed earth (pisé), which exclude rainwater by controlled penetration. The porous materials used to construct walls are generally used in small units, which means that the joints between them have to perform in a way similar to that of the units themselves. Mortars used to form the joints perform one half of the task of making the wall, as a whole, watertight. Traditionally, solid brickwork was laid using lime mortar, which is weak, absorbent and relatively flexible. The absorption could be controlled by the choice of mortar mix to match that of the bricks themselves. However, if strong mortars are used in which Portland cement largely replaces the lime, the mortar is susceptible to shrinkage, so that there are a large number of capillary paths through the wall at the interface between the brick and the mortar (Diagram D3.4/45).

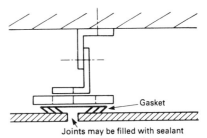

Flat panel system (e.g. Redland's 'Petrarch')

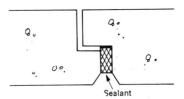

Pre-cast concrete

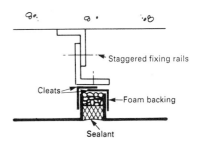

Tray panels

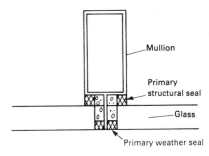

Structural silicone glazing curtain walling

Diagram D3.4/44 *Some examples of fully sealed cladding (schematic)*

*Note that the stack effect is another example of the **high>low** principle, in this case the mechanism being air flowing from a zone of high pressure to one of low pressure.

Diagram D3.4/45 *Water may find its way through masonry at hairline shrinkage cracks between the units and the mortar*

Water may find its way along these cracks, and this process will be greatly aggravated if the wall is exposed to strong winds. Put in terms of the principles for building, to achieve **continuity** of the function of exclusion, the physical **separate lives** of the two parts, i.e. bricks and mortar, should be designed so that they are compatible.

The absorption of the unit material can be known. Within limits, the greater the capacity of the materials for absorbing water, the longer will be the period before the rain reaches the inner face. This is why it is harder to exclude water with dense masonry units (such as engineering or concrete bricks, or granite) laid in cement mortar than it is with absorptive bricks and soft mortar. Other factors affect the resistance of the wall to rain penetration. Walls without straight-through joints – a brick-and-a-half wall, for example – present a lower risk. Raked-out joints increase the possibility of rain penetration.

Direct penetration of water through plain brickwork is not as common as is often supposed. More often, dampness is associated with local defects such as cracked sills, badly designed parapets and other features which are exposed to greater quantities of flowing water. Associated with this method of exclusion is a whole range of architectural features, such as string courses, copings, projecting window hoods and sills, designed to shed water away from the surface of the building.* The benefits of these are relative rather than absolute. Such projections certainly play a very definite role in controlling (though not preventing) surface staining, as discussed in *3.9 Flow and changes of appearance*, p. 571. To assist in predicting the effect of these architectural features studies are needed to show what proportion of rain is driven against a wall at particular angles. Rendering is often used over walls that would otherwise fall into this category.

Particular care is needed when fitting openings into walls of this form of construction because of the inevitable discontinuity at the junction between the elements which work by different exclusion methods where the thickness of the wall is effectively reduced. This is an area of

*This is generally true, although there are exceptions. An example is the use of parapet party walls where issues of fire spread override exclusion and create an entry point for water into the top of the wall.

detailing where, to compensate for the risk taken of poor performance, designers have to apply rules of thumb as to what will actually work in the specific locality of the building (see Diagram D3.23).

Exclusion by the use of multiple layers

The group of constructions which exclude rainwater by this method is interesting, because it includes both traditional techniques such as tile hanging and weatherboarding and the most up-to-date rainscreen claddings, as well as domestic cavity wall construction.

(e) Cavity walling

Whereas in a solid masonry wall the risk of rain penetration is minimized, for given materials and conditions, only by increasing the thickness of the construction, the introduction of a continuous cavity provides 'an efficient and certain safeguard against the penetration of walls by rain'.* In masonry construction the basic principle of the cavity is that the capillary paths are broken. In practice, the successful application of this principle means that there should be no bridge of solid material capable of transmitting water across the cavity.

Contrary to appearances, there is a significant difference in the behaviour of a traditional nineteenth-century 230 mm (9 in) brick wall and the modern 103 mm outer leaf of a brick cavity wall. In the former, exclusion is achieved by controlled penetration, and the detailing of the wall and its openings is such as to prevent concentrations of water at particular points. However, for a number of reasons, the modern single leaf of brickwork does not simply act as a thinner version of the solid wall. The absorption of the brickwork is important in determining what proportion of the water is held sponge-like within the wall and how much finds its way through. Modern bricks tend to have lower absorptivity because of both the use of perforations and the different pattern of firing. Modern cement mortars have greater drying shrinkage than lime mortars, resulting in more extensive cracking at the brick/mortar interface. Most importantly, the perpends in the half-brick wall are rarely fully filled and provide many routes for water through the wall.

In practice, this means that it must be assumed that under severe conditions there will be a film of water flowing down the inner face of the outer leaf. Such water must not be allowed to splash across the cavity, while adequate provision must be made for its escape to the outside at the base of the wall and where openings bridge the cavity.

Clearly, wall ties, which are essential structurally, compromise the objective of having no bridges across the cavity. Not only do the ties provide potential bridges which will allow water to run from the outer to the inner leaf, they also provide obstructions on which mortar droppings may lodge and so form a better bridge than the tie itself. To compensate for these risks, the ties, which are designed with kinks to force water on them to drip off, should not be laid on a slope towards the inner leaf, and the cavity should be kept scrupulously clean during the construction of the wall.

Because of the possibility of flowing water within the cavity, proper design of damp-proof trays, which act as a second line of defence, is essential. In fact, dpc detailing is a fundamental part of the design of the cavity wall. Complexities on the drawing board mirror those on-site, and faulty design and installation of dpcs is a common source of failure in modern building, underlining the need for **creative pessimism**.

Principles of Modern Building, Vol. 1, p. 203.

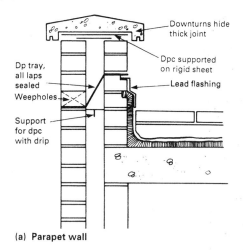

(a) Parapet wall

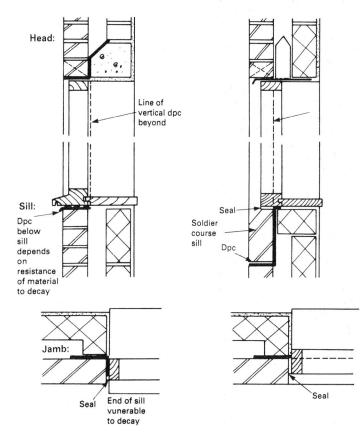

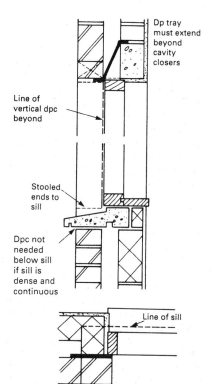

(1) Frame flush with wall face

(2) Frame straddles dpc

(3) Frame recessed

(b) Windows

Diagram D3.4/46 *DPCs and cavity trays used to exclude moisture from cavity walls. (c as BRE DAS 15, Building Research Establishment, Crown Copyright)*

The following are the key rules for the correct design and installation of dpcs (see Diagram D3.4/46):

1. A cavity dpc should be provided over every opening unless the brickwork above is entirely protected, and the outer edge of the dpc should be as near the opening as possible.
2. The cavity should never be bridged with a horizontal dpc.
3. Cavity dpcs over openings should be long enough to prevent the water discharged from them flowing back to the jambs. Tall openings thus require longer trays.
4. Cavity dpcs should be of a material that can be formed to the shape required. The ends should be stopped in order to discharge the water against the outside leaf of the wall.

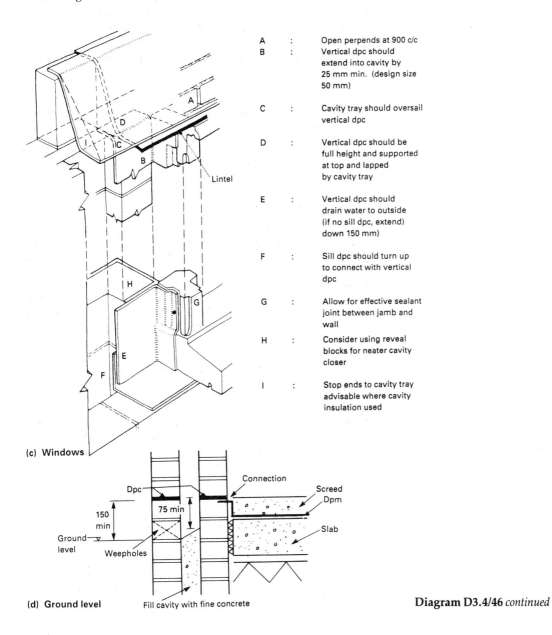

A : Open perpends at 900 c/c
B : Vertical dpc should extend into cavity by 25 mm min. (design size 50 mm)

C : Cavity tray should oversail vertical dpc

D : Vertical dpc should be full height and supported at top and lapped by cavity tray

E : Vertical dpc should drain water to outside (if no sill dpc, extend) down 150 mm)

F : Sill dpc should turn up to connect with vertical dpc

G : Allow for effective sealant joint between jamb and wall

H : Consider using reveal blocks for neater cavity closer

I : Stop ends to cavity tray advisable where cavity insulation used

(c) Windows

(d) Ground level

Lintel

Dpc

Connection

Screed
Dpm

150 min

75 min

Ground level

Weepholes

Slab

Fill cavity with fine concrete

Diagram D3.4/46 *continued*

5. Outlets should be provided to discharge any water collected in the cavity. Open vertical joints or weepholes are usually employed in brickwork or masonry.
6. Vertical dpcs should be provided at openings to break all contact between the outer and inner leaves.
7. Where a cavity wall is used partly as an external wall and also becomes an internal wall at a lower level, below a roof projecting from the side of the building, the cavity must be stopped with a cavity dpc at the lowest level of the exposed wall. Outlets should be provided at suitable places to discharge any collected water.
8. The cavity should start at least 75 mm below the dpc of the inner leaf, and outlets for any collected water should be provided in the outer leaf below the dpc. The space thus formed is a precaution against water rising above the inner dpc. It should be clear of rubbish and mortar droppings. The dpc of the inner leaf may well be at a higher level than that of the outer leaf.

If the cavity is designed to be unventilated, the outer leaf will act as the air seal, so that the air pressure gradient occurs across it. In conditions of severe exposure, with rain being driven against the building, at any small opening in the outer leaf, such as a poorly filled perpend, water can spray across the cavity. If the cavity is ventilated the air seal occurs at the inner leaf. This tends to make the cavity pressure equalized with external conditions so that this effect should not occur. This aspect is rather empirical and has not been quantified (for example, as to what the optimum compartment size of the cavity should be, or the size and spacing of ventilation openings).

(f) Insulated cavity walling

It has become common in recent years to improve the thermal insulation of cavity walls by putting a layer of insulation material in the cavity, either fully or partially filling it. The cavity wall in which the cavity is totally filled ceases to exclude water *by means of* its multiple layers: the fully filled cavity wall excludes water *in spite* its multiple layers. Full-fill cavity insulation can either be installed as the wall is built, in which case slabs of mineral wool or foamed plastic board are used, or added to an existing cavity wall by injecting polystyrene beads or mineral wool.*

The following factors are critical to achieving exclusion:

- The exposure of the building. This should be quantified in the form of the driving rain index as described in DD93† and explained in *3.2 Exposure*, p. 187 ff.
- Choice of insulation material and method of installation to suit the exposure. Full-fill insulation is not suitable at all in buildings which are severely exposed to driving rain. Agrément Certificates of materials being considered should state their limitations.
- The extent to which the architectural details throw water off the facade or concentrate it at points vulnerable to rain penetration.
- The general quality of the workmanship in the construction of the wall. Walls fully filled with insulation are less tolerant of poor workmanship than plain cavity walls. Detailed requirements such as cavity cleanliness, wall ties being level or sloping outwards, laps and stop ends to dpcs and filling of perpends, though usually spelled out in specifications for unfilled cavity walls, become critical for filled walls and must be followed (Diagram D3.4/47).

If insulation batts are used it is important that they are not compressed: the cavity width should not be less than the thickness of the insulation.

*The practice of injecting urea-formaldehyde foam has declined.

†British Standard Draft for Development DD93: 1984, *Methods of assessing exposure to wind-driven rain.*

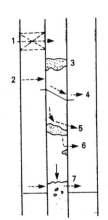

Routes for moisture penetration:

1   Open perpends

2   Hairline cracks between bricks and mortar

3   Capillarity through blockages

4   Ties sloping downwards

5   Run-off across blockages

6   Drips collecting on snots

7   Rubbish at base of cavity bridging dpc

(a) Unfilled cavity

Water 'jumps' laminations at joints, via mortar snots

Water 'jumps' slabs by mortar snots and misaligned slabs

(b) Fully filled cavity

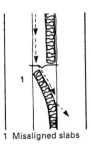

1 Misaligned slabs

2 Ties sloping downwards
3 Protruding mortar snots

(c) Partially filled cavity

Diagram D3.4/47 *Water-penetration mechanisms through cavity walls, with and without thermal insulation*

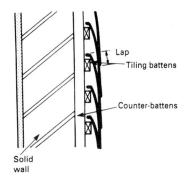

Lap
Tiling battens

Counter-battens

Solid
wall

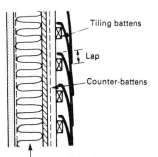

Tiling battens

Lap

Counter-battens

Timber frame wall lined
internally with plasterboard,
externally with sheathing
and breathing paper

Diagram D3.4/48 *Methods of fixing
tile cladding to solid and framed
backgrounds. The illustrations show
only the basic principles. The lap of the
tiles will depend on the exposure*

Diagram D3.4/49 *The two types of
rainscreen cladding*

*See *Rainscreen Cladding: a guide to design principles and
practice*, by J. M. Anderson and J. A. Gill, CIRIA/
Butterworths, London, 1988.

(g) Cladding with small units

(1) Tile and slate hanging

This method of achieving exclusion is, in essence, the same as for roofs. The rules differ slightly: because the pitch is so much steeper, the lap between the units may be reduced. Where the battens are fixed to a solid background such as brickwork or blockwork there is no need for a sarking felt, since the masonry acts as the air seal. On timber-framed walls a sheathing felt is used in the same way as sarking felt (Diagram D3.4/48).

(2) Weatherboarding

Weatherboarding, a vernacular cladding common in East Anglia, works on the same principle as tiling and slating. Because of the few vertical joints the angle of creep does not have to be considered.

(h) Rainscreen cladding

Rainscreen cladding is a term applied to two distinct types of cladding which have in common the characteristic that the water seal is separated from the air seal. The effect of this is to minimize the air pressure differential across the water seal. Pressure equalization is an example of the principle of **balance**.

The first type has been described as *back ventilated and drained* cladding* (Diagram D3.4/49). The outer layer or *rainscreen* is designed to allow for being wet on both sides, although it is intended that most of the incident water runs off the cladding. The cavity wall is of this type. In a lightweight system the cladding may comprise panels fixed some distance away from the building structure with open vertical joints and horizontal joints with flashings. It has been found that relatively small open joints (less than 2.5 mm vertically and less than 5 mm

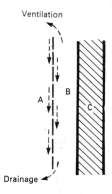

Ventilation

A B C

Drainage

A :   Relatively impermeable screen
      with baffled joints. It is intended
      that most, though not all, of the
      incident water will run off

B :   Air space, not designed to be
      pressure equalized

C :   Structural wall/air seal

**Back ventilated and
drained cladding**

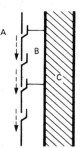

A B C

A :   Impervious screen with
      labyrinthine joints

B :   Air space, designed to be
      pressure equalized by
      means of cavity closers

C :   Structural wall/air seal

**Rainscreen with a
pressure-equalized cavity**

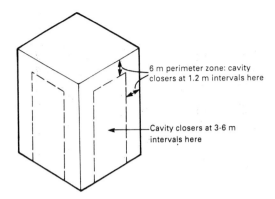

6 m perimeter zone: cavity
closers at 1.2 m intervals here

Cavity closers at 3-6 m
intervals here

Diagram D3.4/50 *Spacing of cavity
closers in rainscreen cladding with
pressure-equalized cavity*

horizontally) reduce driving rain across the cavity to negligible
quantities if the cavity is at least 25 mm deep. The cavity is ventilated at
top and bottom in addition to the joints but it is not designed to be
pressure equalized. Thus some water may reach the inner face of the
cavity and the fabric there should be tolerant of small amounts of
moisture. To enable full drainage, the cavity must be kept clear of
horizontal obstructions such as fixing rails.

The second type is described as *rainscreen cladding with a pressure-
equalized cavity*. The key difference is that the air space is designed to be
pressure equalized with external air, to improve the effectiveness of the
water seal at the joints in the cladding. In other words, the cladding is
intended to be wet on the outside only. To achieve this, the joints
themselves must be geometrically more complex to 'baffle' raindrops
flung and shattered against the joint. They are thus referred to as
labyrinthine. Because air pressures on the exterior of the building vary,
the cavity behind the rainscreen should be interrupted at suitable
intervals to control pressure gradients *within* the cavity, and thereby
minimize lateral and vertical air movement. The spacing of the cavity
closers should be such that the variation of air pressure outside any
compartment is kept to an acceptable minimum. Thus many cavity
closers are required both vertically and horizontally, particularly at the
top and sides of a facade where the air pressure gradients are largest.
Garden* suggests cavity stops at 1.2 m centres parallel to the corners
and the top of the building in a 6 m perimeter zone, and at 3–6 m
centres elsewhere (see Diagram D3.4/50). A benefit of the pressure-
equalized system over the back-drained and ventilated type is that
these cavity barriers can be exploited to act as fire cavity barriers.

It is important to remember that the wall and not the cladding takes
the wind load, although the cladding is subject to wind effects.

Rainscreen claddings are generally non-structural, and may be
formed from the same materials used for fully sealed cladding. In other
words, the basic panel material itself is impermeable to water. The
discontinuity occurs at the joints where, as noted previously in the
descriptions of two-stage joints, to achieve continuity of performance
the horizontal and vertical joints must be different in character. The
vertical joints may be butted or lapped or formed into a vertical channel
with baffles (Diagram D3.4/51), but the horizontal joints must
incorporate some form of flashing to drain to the exterior the water that
gets past the rainscreen.

*In Canadian Building Digest 40.

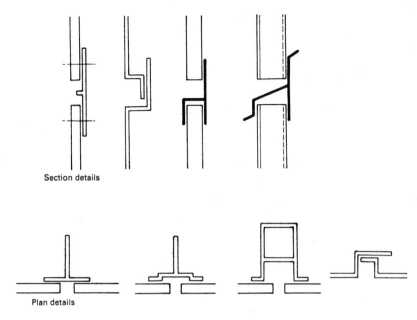

Section details

Plan details

(a) Drained and back-ventilated rainscreen

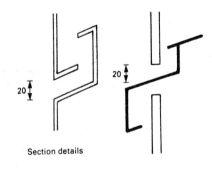

Section details

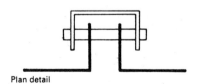

Plan detail

(b) Pressure-equalized rainscreen

Diagram D3.4/51 *Examples of rainscreen cladding joints. (From Anderson and Gill)*

(1) Exclusion of condensation

Because of the extensive ventilation behind the rainscreen panels, and of the need for a thorough air seal at the back of the cavity, it is possible to achieve the optimum locations for thermal insulation and vapour-resisting material. This can be done by fixing the insulant to the face of the inner lining (it needs to be water resistant but not waterproof) at the back of the cavity. This is effectively the outside of the building in thermal terms, as the vented cavity itself has negligble thermal resistance. Similarly, the air seal tends to function as an effective vapour seal, and beyond it the construction can breathe to outside easily. On both these points the rainscreen cladding works *with* rather than *against* the normal thermal and vapour pressure gradients, thereby minimizing the risk of interstitial condensation.

(2) Separate lives

By virtue of its loose fit and unsealed character, rainscreen cladding presents less problems arising from differential thermal and moisture movements than similar non-structural panels which have fully sealed joints. A corollary of this is that there should be fewer problems achieving fit on-site since joint dimensions tend to be less critical. Not only is assembly easier but, since the jointing product is effectively air, there is no differential durability between the cladding panels and the jointing material. For all these reasons, it can be seen why Garden said 'with the many advantages of the open rainscreen, its full development should be pursued by all building designers'.

(i) Pressure-equalized glazing and curtain walling

The principle of separating air and water seals may be applied to the design of curtain walling and glazing. The glass- or panel-to-frame joint is detailed in such a way that a pressure-equalized cavity is formed within the frame which is designed to remain dry. Diagram D3.4/52 illustrates the difference between this approach and a fully sealed system which incorporates secondary drainage.

4 Structures at and below ground level

The parts of the building which sit on or below the ground are subjected to moisture in the form of groundwater under hydraulic pressure, groundwater travelling by capillary action through porous materials, or vapour driven by the influence of a vapour pressure gradient. Water under hydraulic pressure can only be excluded by means of a *continuous* impermeable layer. Dampness rising by capillary action can only be safely excluded by the same method, although, historically, the method of controlled penetration has been normal, particularly for floors, in the absence of suitable continuous impermeable membranes. The same is true for the movement of water vapour through floors. This section considers, in turn, basements, floors and walls.

(a) Basements

(1) Methods of exclusion

The same basic preconditions for water penetration discussed earlier in this chapter (i.e. a source of water, a course for the water, and a driving force) apply. The first rule for controlling water penetration from the ground is to reduce the source of water. This is done by one or both of

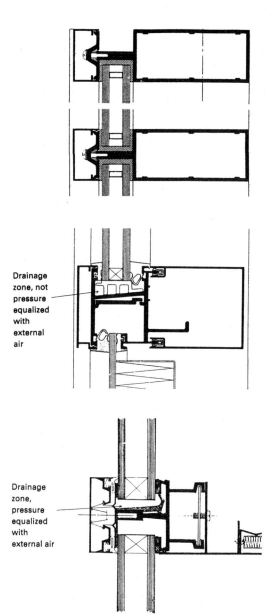

Drainage zone, not pressure equalized with external air

Drainage zone, pressure equalized with external air

Diagram D3.4/52 *Three types of curtain walling system.* Top: *fully sealed system (Mellowes PPG's Archital Isobar);* centre: *gasket-glazed and drained system (Mellowes PPG's Archital Isotherm);* bottom: *gasket-glazed, drained and pressure-equalized system. (Mellowes PPG's barrier curtain wall). (Courtesy Mellowes PPG Ltd)*

two methods. These are, first, to reduce the surface water entering the ground adjacent to the building, and, second, to provide sub-surface drainage around the basement where the groundwater table is likely to rise above the top of the footings. All basement structures should be designed on the basis that water pressure will need to be resisted at some stage in their life.

Structures below ground are constructed mainly of reinforced concrete. However, in some circumstances plain concrete, brickwork or blockwork may be used. The impermeability of good-quality dense concrete is such that it can, in suitable conditions and configurations, be used without the application of a separate impervious membrane.

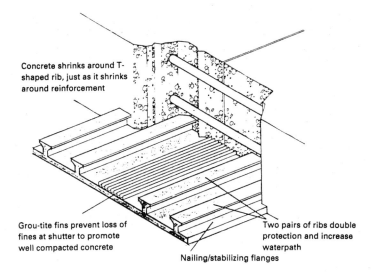

Concrete shrinks around T-shaped rib, just as it shrinks around reinforcement

Grou-tite fins prevent loss of fines at shutter to promote well compacted concrete

Nailing/stabilizing flanges

Two pairs of ribs double protection and increase waterpath

Diagram D3.4/53 *An example of a waterbar. (Courtesy Servised Ltd)*

Additives may be utilized to improve the workability and cohesiveness or the mix, and thereby its water resistance. The details which require special attention are construction joints, movement joints and junctions between floors and walls. Specially designed waterbars and waterstops are used at these points (Diagram D3.4/53). To achieve continuity of exclusion such components must be continuous themselves. In this type of concrete construction visible penetration of liquid water can be prevented, but it should be emphasized that *the transmission of water in the form of vapour may not be wholly excluded.* Where masonry is used below ground, and where complete protection is required against the penetration of water in both liquid and vapour form through reinforced concrete, it is necessary for a completely impermeable membrane to be applied, in the form known as 'tanking'.

(2) Range of materials
Materials suitable to form an impermeable membrane are restricted to mastic asphalt and bituminous sheeting bonded and sealed at its joints.

(3) Rules and precautions
Membranes of either material are basically sandwiched within the construction. To resist the effects of hydraulic pressure, the membrane, weak in itself, has to be loaded by a mass of material. To ensure continuity it must not be inadvertently damaged. Hence for floors, the membrane is first protected by a screed, and then covered by a reinforced concrete loading coat (Diagrams D3.4/54 and D3.4/55).

There are basically two methods by which the membrane may be applied, namely (1) as externally applied tanking which is defined (in BS 8102) as an 'external application of an impervious membrane to the structural walls' and (2) as internally applied tanking which is defined as 'an application of an impervious membrane to the inner surface of walls and the upper surface of floors before placing of structural walls or floors or protecting or loading coats'. The different constructions are illustrated in Diagrams D3.4/54 and D3.4/55. In externally applied tanking it is necessary to protect the vertical membrane from being

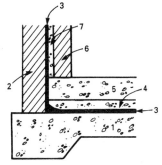

Diagram D3.4/54 *Internally applied tanking. (As BS 8102). Here and in Diagrams D3.4/55 and D3.4/56 the numbers indicate sequence of operations*

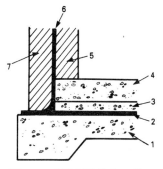

Diagram D3.4/55 *Externally applied tanking. (As BS 8102)*

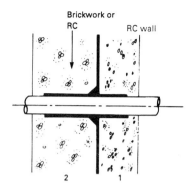

Diagram D3.4/56 *Treatment of pipes passing through tanking. (From CP 102)*

punctured during backfilling or subsequent excavation. Such protection may be provided by a wall of brickwork built against the membrane. To ensure continuity, openings and protrusions should be kept to a minimum. Pipes must be treated as shown in Diagram D3.4/56.

(b) Floors

The methods for achieving exclusion of moisture from floors depend on whether the floor is solid or suspended. Modern solid floors rely primarily on a continuous impermeable membrane, although in the past controlled penetration worked so long as the ground was well drained and water-tolerant materials such as bricks, tiles or stone flags were used. Suspended floors work on the principle of breaking capillary paths by means of a cavity, although at the points of support a dpc is required. The rules applicable to solid and suspended floors are considered separately.

(1) Solid floors
Solid floors consist essentially of oversite concrete and a screed which acts as background for the floor finish. Apart from site and environmental conditions, the method of damp-proofing required is dependent on the floor finish to be laid.

Site conditions: Wherever possible, solid floors should not be subjected to hydrostatic pressure: in many cases adequate drainage of the subsoil will ensure this. Where it cannot be avoided, the floor should be tanked. The following discussion is applicable to solid floors which must resist capillary moisture only.

Principles: Good-quality concrete can be almost impervious to the transmission of water. Oversite concrete and floor screeds as commonly laid, however, cannot be expected to prevent the transmission of water in either liquid or vapour form. The type of dpm which should be used depends on the floor finish, particularly its tolerance to moisture. BS CP 102 grades floor finishes into four categories, A, B, C and D, on this basis. These grades of finish and their tolerance to moisture are given in Table T3.4/1.

Damp-proofing materials may be applied in the form of membranes either on the surface of the screed or sandwiched below the screed. In either case it is essential that **continuity** of the membrane is achieved (Diagram D3.4/57). Two areas where in both design and construction particular care is needed are (1) the junction between the dpm in the floor and the dpc in the walls (see Diagram D3.4/58), and (2) around pipes, service entry points and other unavoidable perforations. Achieving continuity in practice is not always easy: the work has to be done in often the worst conditions on a waterlogged, wintry site with materials which require a good degree of care and skill. Good quality control on-site is therefore essential.

Temperature *gradients* through the floor tend to increase the moisture content in the colder zones, as shown in Diagram D3.4/59. Where the surface of the floor is colder than the ground below it a higher degree of protection than that recommended in Table T3.4/1 will be required, particularly for moisture-sensitive finishes.

(2) Suspended floors
Suspended floors are commonly of timber, and this part relates to this type of floor (Diagram D3.4/60). In general, timber should be maintained at a moisture content of 20% or less to avoid conditions

Table T3.4/1   *Resistance of floor finishes to ground moisture*

Group	Material	Properties
(A) Finish and dpm combined	Pitch mastic flooring Mastic asphalt flooring	Capable of resisting rising dampness without dimensional or material failure
(B) Finishes that may be used without extra damp protection	Concrete Terrazzo Concrete or clay tiles	Capable of transmitting rising dampness without dimensional, material or adhesion failure
	Cement/latex Cement/bitumen	Capable of partially transmitting rising dampness without dimensional or material failure and generally without adhesion failure
	Wood composition blocks (laid in cement mortar) Wood blocks (dipped and laid in hot pitch or bitumen)	Capable of partially transmitting rising dampness without material failure and generally without dimensional or adhesion failure. *Only in exceptional conditions of site dampness is there risk of dimensional instability*
(C) Finishes that are not necessarily trouble-free without damp protection	Thermoplastic flooring tiles (BS 2592) PVC (vinyl) asbestos floor tiles (BS 3260)	Capable of partially transmitting rising dampness through the joints without dimensional failure and generally without adhesion or material failure. Water penetration at the joints may result in decay at the edges in some conditions when ground water contains dissolved salts or alkalis
(D) Finishes which require reliable protection against damp	Magnesite	Capable of transmitting rising dampness but adversely affected by water (softens and disintegrates under wet conditions)
	Flexible PVC flooring in sheet or tile form (BS 3261)	Impervious, but the flooring adhesive is sensitive to moisture (may expand under damp conditions)
	PVA Emulsion cement	Impervious, but dimensionally sensitive to moisture (adhesive for tiles also sensitive to moisture)
	Rubber	Impervious, but prone to adhesion failure mainly through sensitivity of its adhesive (may expand under damp conditions)
	Linoleum	Sensitive to alkaline moisture attack through breakdown of bond and adhesive film (may expand under damp conditions)
	Cork (carpet and tile) Wood (block in cold adhesives and strip and board flooring) Chipboard	Acutely sensitive to moisture with dimensional or material failure

Important: With floor warming a damp-proof sandwich membrane is necessary, whatever the floor finish.

The degree of protection to be provided by the dpm will depend on:
1. Wetness of the site.
2. Temperature gradient through the floor.
3. Contamination of aggregates, hardcore and fill with sea salts.

Order of protective value of sandwich membranes (see text):
1. Mastic asphalt.
2. Bitumen sheeting.
3. Hot applied pitch or bitumen.
4. Cold applied bitumen solution and coal tar pitch/rubber emulsion or bitumen/rubber emulsion.
5. Polythene or polyethylene film sheeting.

Sources: CP 102, Tables 1 and 2, pp. 49–50: BRE Digest (2nd Series) No. 54, Tables 1 and 2, pp. 2–3.

Note: CP 102 is being revised and reissued as BS 8102. A revised version of Tables 1 and 2 should be included in a forthcoming part of BS 8102 which suit the establishment of dry rot. Rules aimed at reducing the likelihood of moisture contents in excess of 20% are:

- *Preventing direct contact* between timber and porous materials below dpc level. Wall plates on sleeper walls must have a dpc below them.
- *Providing ventilation* below the floor. This is achieved by means of vents in the external walls, provided on opposite sides of the building to enable a through-draught. Sleeper walls should be of honeycomb construction. There should be a minimum gap of 150 mm between the joists and the ground below.
- *Covering the ground* with a layer of concrete, to minimize the evaporation of water into the air space below the floor. The surface of this oversite concrete should not be below the level of the adjacent ground. The void should be free of debris, particularly timber waste, which could begin to rot and start a chain of decay.

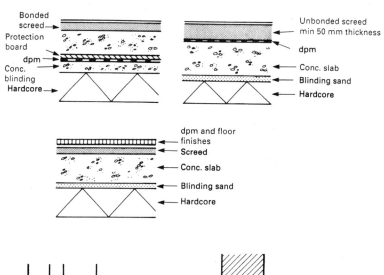

Diagram D3.4/57 *Alternative positions for the dpm in a solid floor. (As BS 8102)*

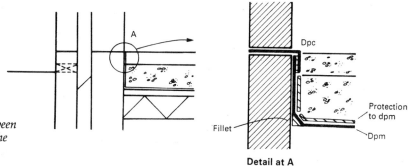

Diagram D3.4/58 *Continuity between the dpm in the floor and the dpc in the wall*

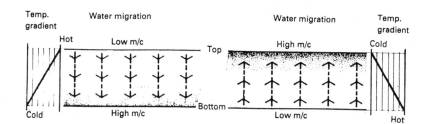

Diagram D3.4/59 *The effect of temperature gradients on moisture content in a ground-floor slab*

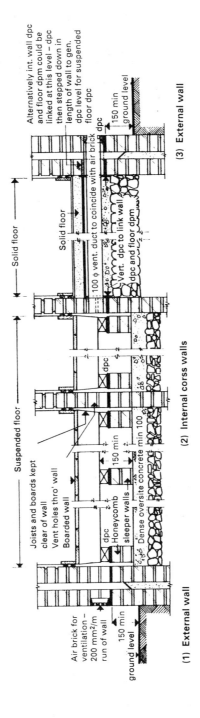

Air brick for ventilation – 200 mm²/m run of wall

ground level

150 min

(1) External wall

Joists and boards kept clear of wall
Vent holes thro' wall
Boarded wall

dpc
Honeycomb sleeper walls
150 min
Dense oversite concrete min 100

Suspended floor

dpc

(2) Internal corss walls

Solid floor

Solid floor

100 φ vent. duct to coincide with air brick
Vert. dpc to link wall
dpc and floor dpm

Alternatively int. wall dpc and floor dpm could be linked at this level – dpc then stepped down in length of wall to gen.
dpc level for suspended floor dpc
dpc
150 min ground level

(3) External wall

Diagram D3.4/60 *Sections through suspended timber and solid ground floors illustrating methods of achieving adequate ventilation, covering the ground (under the suspended floor) and the correct use of dpcs to prevent direct contact between timber and porous building materials. All dimensions are in millimetres*

Table T3.4/2   *Comparison of materials used as dpcs*

Material (with British Standard relevant to material as such)	Relevant British Standard for use as dpc	Minimum thickness (mm)	Minimum weight (kg/m²)	Resistance to compressive loading	Resistance to lateral loading (tests on this property are difficult to interpret)	Freedom from extrusion	Durability — Covered	Durability — Exposed	Ease of handling on site, and ease of storage	Workability (Ability to be made into shapes) — High temperature	Low temperature	Ability to keep shape without restraint	Ease in forming corners and junctions	Resistance to damage (at 20°C) — Extensibility	Tearing (resistance to 'first tear', not continuation; and taken in weakest direction of material)	Puncture	Preforming readily available	Ease of forming and sealing laps
Rigid materials — Brick (to BS 3921)	BS743	Two courses; 150 mm	–	c	c	c	c	c	a	Only suitable for rising damp function, in carefully selected situations								
Slate (to BS 3798)	BS 743	Two courses; slate 4 mm thick	–	b	b	c	c	c	a	See note above								
Malleable metals — Lead (to BS 1178) Lead must be coated with bitumen	BS 743	1.80 (code no. 4)	19.50	b	c	c	c	c	a	c	c	c	b	a	c	c	No	a
Copper to BS 743	BS 743	0.25 (if exposed, must be thicker see BS 743)	2.28	c	b	c	c	c	a	c	c	c	b	a	c	c	No	a
Zinc (to BS 849) Zinc may require coating with bitumen	None (but see CP 143, Part 5)	0.81 (14 g)	5.40	c	b	c	c	b	a	b	b	c	a	a	c	c	Yes	a
Aluminium (to BS 1470)	None (but see CP 143, Part 15)	0.91 (20 g)	2.80	c	b	c	c	b	a	b	b	c	a	a	c	c	Yes	a
Zinc/lead (Proprietary product)	None (but see zinc, above)	0.60	4.30	c	c	c	c	b	a	b	b	c	a	a	c	c	Yes	a
Pressed steel[5]	None		22.24 g	–	c	a	c	b[6]	b[6]	a	Only preformed		c	b	Nil	c	c	Yes	b
Bitumen-based products — Bitumen/hessian base*	BS 743, type A	2.5[4]	3.80	a	b	a	b	a	a	b	a	a	b	a	b	a	No	c
Bitumen/fibre base*	BS 743, type B	2.5[4]	3.30	a	b	a	b	a	a	b	a	a	a	a	a	a	No	a
Bitumen/asbestos base	BS 743, type C	2.5[4]	3.80	a	b	a	b	a	a	b	a	a	a	a	a	a	No	b
Bitumen/hessian base/lead*	BS 743, type D	3.0[4]	4.40	a	b	a	c	b	a	b	a	b	b	a	b	b	No	c
Bitumen/fibre base/lead*	BS 743, type E	3.0[4]	4.40	a	b	a	c	b	a	b	a	b	a	a	a	a	No	b
Bitumen/asbestos base/lead	BS 743, type F	3.0[4]	4.90	a	b	a	c	b	a	b	a	b	a	a	a	a	No	b
Bitumen/hessian base/aluminium	None (but superior to type A above)	3.0[4]	4.10	a	b	a	c	b	a	b	a	b	b	a	b	b	No	c
Bitumen/fibre base/aluminium	None (but superior to type B above)	3.0[4]	4.90	a	b	a	c	b	a	b	a	b	a	a	a	a	No	b

Material (with British Standard relevant to material as such)	Relevant British Standard for use as dpc	Minimum thickness (mm)	Minimum weight (kg/m²)	Resistance to compressive loading	Resistance to lateral loading (tests on this property are difficult to interpret)	Freedom from extrusion	Durability — Covered	Durability — Exposed	Ease of handling on site, and ease of storage	Workability — High temperature	Workability — Low temperature	Ability to keep shape without restraint	Ease in forming corners and junctions	Resistance to damage — Extensibility	Tearing (resistance to 'first tear', not continuation: and taken in weakest direction of material)	Puncture	Preforming readily available	Ease of forming and sealing laps
Polymer-based² Polythene*	BS 743	0.46	0.48	c	b	c	b	a	c	b	b	a	a	c	b	a	No	a
Polypropylene (proprietary product)	None	0.8 and 1.5	–	c	b	c	b	a	a	b	b	a	b	b	b	b	Yes	a
Pitch polymer*	None	1.27	1.50	c	c	c	b	a	c	c	c	a	c	b	b	b	Yes	c
Bitumen polymer (proprietary product)	None	1.25	1.60	c	c	c	b	b	c	c	c	a	c	b	b	a	No	c
In-situ coatings Mastic asphalt (to BS 1097 & BS 1418)	BS 743	12.0	–	a	b	a	b	a	a	c	c	c	b	Virtually nil	N/a	a³	N/a	N/a
Epoxy resin/sand	None	6.0	–	b	c	c	c	a	b	a	c	c	c	As above	N/a	a	N/a	N/a
Bitumen/rubber glass fibre	None	1.5	–	a	a	a	b	a	b	c	c	c	c	a	N/a	b	N/a	N/a
Pitch epoxy/glass fibre	None	1.5	–	b	c	c	c	b	b	c	c	c	c	a	N/a	b	N/a	N/a

a Below average.   b Average.   c Above average   N/a Not applicable.   * Denotes types most frequently used.

Notes:
General: properties of materials may differ according to individual manufacturer. The alternatives should be carefully examined physically: some manufacturers produce materials to higher specifications.
1. Inherently durable in themselves but susceptible to fracture under building movement.
2. A number of proprietary products in this group have Agrément certificates.
3. Resistance to puncture is not applicable, so resistance to fracture is assessed.
4. Approximate thickness: not part of specification.
5. Pressed steel is fairly rigid; malleable skirt or apron of lead is often attached. Steel is available galvanized or stainless.
6. Stainless versions would be above average.

From Duell, J. and Lawson, F., *Damp Proof Course Detailing*, 2nd edn, The Architecural Press, London, 1983.

(c) Walls

(1) Methods of exclusion
Modern walls achieve exclusion of water rising by capillary action from the ground by means of a dpc, which comprises a *continuous* impermeable membrane the width of the wall. Historically, before suitable materials became available at a reasonable cost, the principle of controlled penetration was relied upon. The extent of problems with rising damp in the walls of old buildings indicates the inadequacy of this method.

(2) Range of materials
There is a wide range of materials suitable for forming dpcs in walls, given in BS 743 and summarized in Table T3.4/2. Criteria for the selection of dpc materials include durability, flexibility, resistance to extrusion and sliding, ease of handling and cost.

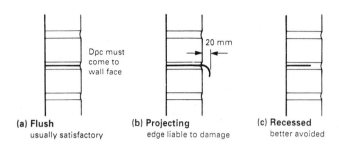

(a) **Flush**
usually satisfactory

Dpc must come to wall face

(b) **Projecting**
edge liable to damage

20 mm

(c) **Recessed**
better avoided

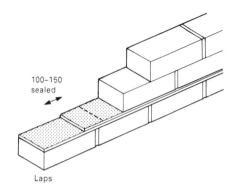

100–150 sealed

Laps

Diagram D3.4/61 *Basic rules for dpcs in masonry walls*

(3) Rules and precautions
Continuity: The dpc should cover the full thickness of the wall and should not be set back from the wall face for pointing, or be bridged by rendering (Diagram D3.4/61). Flexible dpcs should be lapped and sealed at joints with a lap not less than 100 mm. The type of seal depends on the dpc material (for example, hot bitumen may be used for bituminous dpcs and special adhesives for polymer-based dpcs). The dpc must be continuous with the dpm in adjacent solid floors and, where necessary, a vertical dpc should be provided at the inner face of the wall (see Diagram D3.4/56).

Height above ground level: At the base of a wall a dpc should not be less than 150 mm above the level of the ground. If the ground level varies, the dpc must step accordingly, and this means that a flexible dpc has to be used. Bridging of the dpc often takes place by earth or leaves banked against the wall. Most of the effects of rain splashing up from adjoining paved or planted ground occurs within a 150 mm zone.

Mortar bed: All flexible dpcs should be provided with a mortar bed which is free from projections liable to cause damage.

(4) Providing dpcs in existing walls
With so much of the building stock of the UK of an age when dpcs were not provided, and rising damp a widespread problem as a result, it is necessary to briefly consider the provision of dpcs in existing walls.
   There are basically two techniques, insertion and injection:

- *Insertion*: Where a cavity tray is absent (for the definition of cavity trays see 'Junctions' below), cutting out short lengths of brickwork at a time and inserting pieces of dpc or special fittings is unavoidable. At one time it was necessary to use a similar technique for inserting simple dpcs. However, for simple dpcs in walls it is much more common today to use injection methods.
- *Injection*: In this technique a number of holes are drilled at regular intervals along the base of a wall, into which water-repellent substances are injected under gravity or pressure. Solutions used include silicones, usually in conjunction with water-displacing fluids to counter the existing dampness of the wall, an aqueous siliconate/latex mixture, hot wax or setting resins. Precautions that must be taken are, first, to repair brickwork and pointing so that the wall can actually hold the solutions and, second, to ensure that existing plaster is removed and replaced by waterproof render, in order to prevent the damp-proof 'zone' being bridged internally.

5 Junctions

As discussed in the section on junctions under 'Principles of exclusion' above, the task of achieving exclusion at joints is done by reconciling the objective of **continuity** in the face of the constraint of **separate lives**. In practice, the range of conceivable junctions verges on the infinite: all that is attempted here is to set out some general observations and illustrate in diagrams how exclusion is achieved between the main elements and components of buildings which work using the different methods of exclusion already described.

In the same way as at joints, seals against water and air are required at junctions. At junctions where a limited amount of movement is anticipated, such as a window-to-masonry wall junction, sealants may be used. More generally, a seal against water is achieved by using a geometrical seal which works on the lapping principle. The lapping may be of an entire element over another, such as a pitched roof overhanging a wall, or it may be provided by using an intermediate strip of impervious material known as a flashing. Flashings are commonly made of lead, since it meets several performance requirements outstandingly well – particularly ease and adaptability of application, and durability – although these requirements can be met adequately by other materials in some circumstances.

Whereas flashings are exposed to the elements, their counterpart within the thickness of the element (in the case of porous materials) are dpcs and damp-proof trays (otherwise known as cavity trays). The objective of dpcs and damp-proof trays is to achieve a clear and continuous wet/dry boundary, and in the case of trays, to drain water to the exterior.

Diagram D3.4/62 illustrates the strategy for excluding rainwater at the junctions between the primary building elements. The different exclusion methods used by each element are considered in turn. The numbering key used is set out in Tables T3.4/3 and T3.4/4. Note that only the exclusion of rainwater and ground moisture is covered: continuity of the method adopted to control vapour movement and condensation must also be provided, whether it be the use of a vapour control layer or ventilation. Similarly, although continuity of other performance requirements, such as thermal insulation, is fundamental, diagrams do not attempt to illustrate these. Solutions to unusual conditions can often be found by thinking of the adjoining pieces *as if* they were one of the standard elements.

Sources

Books
Anderson, J. M. and Gill, J. R., *Rainscreen cladding: a guide to design principles and practice*, CIRIA and Butterworths, London, 1983
Brookes, A. J., *Cladding of Buildings*, Construction Press, London, 1983
BRE, *Principles of Modern Building*, Vol. 1, 3rd edn, HMSO, London, 1959 and Vol. 2, 1st edn, 1961
Briggs Amasco Ltd, *Flat roofing: a guide to good practice*, 1982
Duell, J. and Lawson, F., *Damp proof course detailing* 2nd edn, The Architectural Press, London, 1988
Fitzmaurice, R., *Principles of Modern Buildings, Volume 1: Walls, Partitions and Chimneys*, HMSO, London, 1938
Property Services Agency, *Technical guide to flat roofing*, Vols 1 and 2, 1987
Rostron, R. M., *Light Cladding of Buildings*, The Architectural Press, London, 1964

Table T3.4/3   *Pictorial key to the junctions between elements described in Diagram D3.4/62. Read from left to right (e.g. the group of junctions of walls to roofs are prefixed '5')*

Element	Roof	Wall	Ground	Opening	Projection
Roof	1.1.2	2.1		3.1	4.1
	1.2.2	2.2		3.2	4.2
Wall	5	6	7	8	9

Table T3.4/4   *Numerical key to the junctions between elements described in Diagram D3.4/62*

Element		Roof		Wall			Ground	Opening	Projection
	Exclusion method	Complete imperm.	Multiple layers	Complete imperm.	Controlled penet.	Multiple layers			
Roof	Complete impermeability	1.1.1	1.1.2	2.1.1	2.1.2	2.1.3		3.1	4.1
	Multiple layers		1.2.2	2.2.1	2.2.2	2.2.3		3.2	4.2
Wall	Complete impermeability	5.1.1	5.1.2	6.1.1	6.1.2	6.1.3	7.1	8.1	9.1
	Controlled penetration	5.2.1	5.2.2		6.2.2	6.2.3	7.2	8.2	9.2
	Multiple layers	5.3.1	5.3.2			6.3.3	7.3	8.3	9.3

Roofs
1 Roof to roof

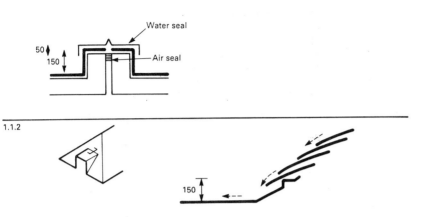

1.1.2

1.2.2

Lead roll
A

Lead soakers

Hip tiles
A

Close mitred hip
B

Lead valley
B

Swept valley
B

2 Roof to wall

2.1

Roof of 'complete impermeability' type

2.1.1

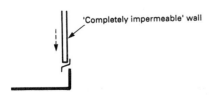

'Completely impermeable' wall

Diagram D3.4/62 *The strategic exclusion of moisture at junctions between building elements. For key to numbers see Tables T3.4/3 and T3.4/4*

2.1.2

'Controlled penetration' wall

Flashing

Risk of rain penetration
Not recommended
without careful consideration of
materials and exposure

2.1.3

'Muliple layers' wall

Flashing

2.2

Roof of 'multiple layers' type

2.2.1

Render

Cover flashing

Soakers

Soakers are L-shaped
pieces of lead
interweaved between
the units at the
abutment

2.2.2

'Controlled penetration' wall

Risk of
rain penetration

2.2.3

'Multiple layers' wall

Weepholes

Damp-proof tray (stepped)

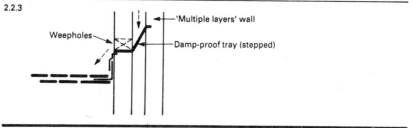

3 Openings in roofs

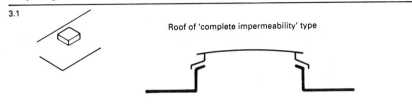

3.1

Roof of 'complete impermeability' type

3.2

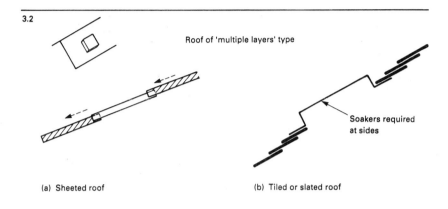

Roof of 'multiple layers' type

Soakers required at sides

(a) Sheeted roof   (b) Tiled or slated roof

4 Projections through roofs

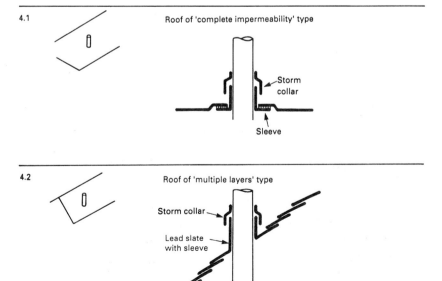

4.1

Roof of 'complete impermeability' type

Storm collar

Sleeve

4.2

Roof of 'multiple layers' type

Storm collar

Lead slate with sleeve

Walls

5 Wall to roof

5.1.1

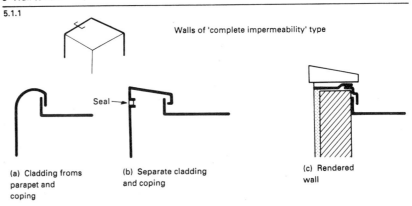

Walls of 'complete impermeability' type

(a) Cladding froms parapet and coping

(b) Separate cladding and coping

Seal →

(c) Rendered wall

5.1.2

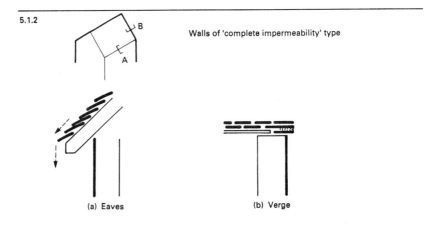

Walls of 'complete impermeability' type

(a) Eaves

(b) Verge

5.2.1

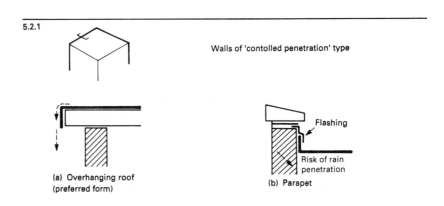

Walls of 'contolled penetration' type

(a) Overhanging roof (preferred form)

Flashing

Risk of rain penetration

(b) Parapet

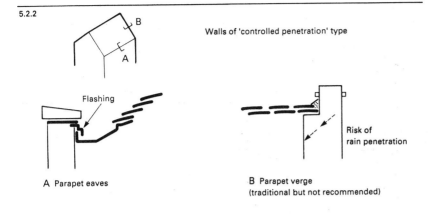

5.2.2

Walls of 'controlled penetration' type

Flashing

Risk of
rain penetration

A Parapet eaves

B Parapet verge
(traditional but not recommended)

(Note: arrangements shown in 5.1.2 also apply)

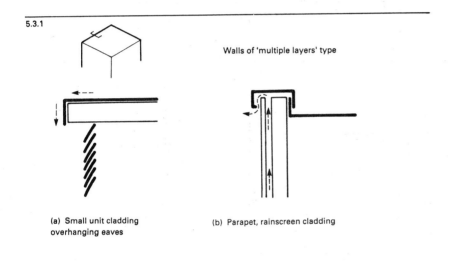

5.3.1

Walls of 'multiple layers' type

(a) Small unit cladding
overhanging eaves

(b) Parapet, rainscreen cladding

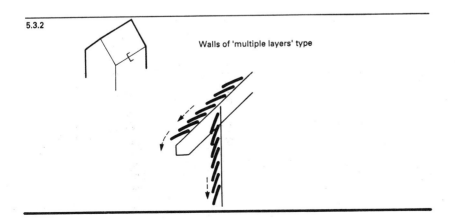

5.3.2

Walls of 'multiple layers' type

6 Wall to wall

6.1.1

Walls of 'controlled impermeability' type

A

B

6.1.2

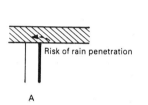

Risk of rain penetration

A

B1

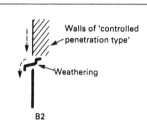

Walls of 'controlled penetration type'

Weathering

B2

6.1.3

A

B1

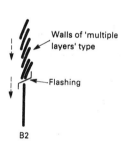

Walls of 'multiple layers' type

Flashing

B2

6.2.2

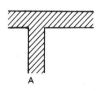

A

6.2.3

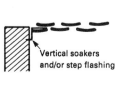

Vertical soakers and/or step flashing

A

B1

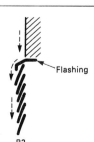

Flashing

B2

6.3.3

Soakers

Special corner
tiles may be used

A

7 Wall to ground

7.1

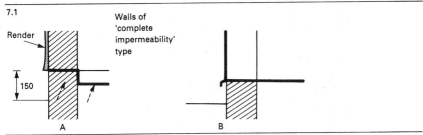

Walls of
'complete
impermeability'
type

Render

150

A   B

7.2

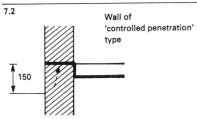

Wall of
'controlled penetration'
type

150

7.3

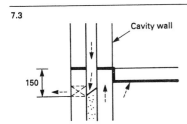

Cavity wall

150

8 Openings in walls

8.1

Walls of 'complete impermeability' type

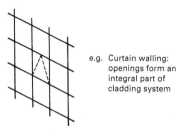

e.g. Curtain walling:
openings form an
integral part of
cladding system

8.2 Openings in 'controlled penetration' type walls

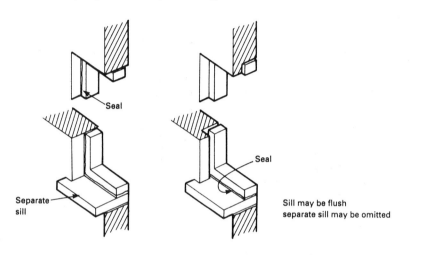

Seal

Seal

Separate
sill

Sill may be flush
separate sill may be omitted

Openings in 'multiple layers' type walls

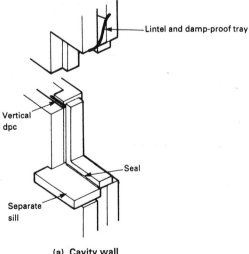

Lintel and damp-proof tray

Vertical
dpc

Seal

Separate
sill

(a) Cavity wall

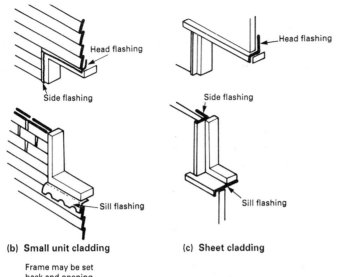

(b) Small unit cladding

(c) Sheet cladding

Frame may be set
back and opening
lined with timber,
or have tiled returns

9 Projections through walls

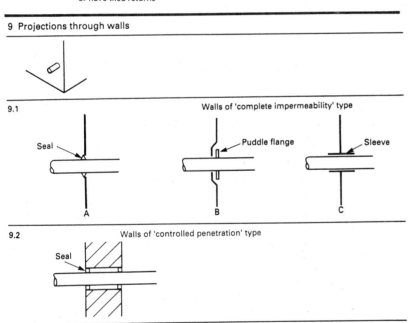

9.1   Walls of 'complete impermeability' type

9.2   Walls of 'controlled penetration' type

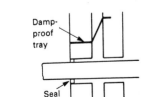

9.3   Walls of 'multiple layers' type

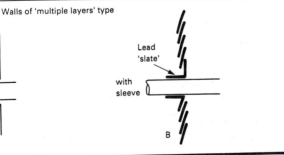

Simmons, J, *The pattern of English building,* (Clifton-Taylor, A, ed)
Faber and Faber, London, 1987
See also standard textbooks of building construction such as *McKay's Building Construction*, Longman, Harlow, *Michell's Building Construction*, Batsford, London

BRE

National Building Studies Research Paper No. 23, *Condensation in sheeted roofs*, Pratt, A. W.

Current paper
CP86/74, *Window to wall joints*

Information papers
IP 4/83, *Window to wall jointing*, May 1983
IP 13/87, *Ventilating cold deck flat roofs*
IP 2/88, *Rain penetration of cavity walls: report of a survey of properties in England and Wales*
IP 2/88, *Summer condensation on vapour checks*

Defect action sheets
1, *Slated or tiled pitched roofs: ventilation to outside air,* May 1982
3, *Slated or tiled pitched roofs: restricting the entry of water vapour from the house,* June 1982
6, *External walls: reducing the risk of interstitial condensation,* July 1982
9, *Pitched roofs: sarking felt underlay – drainage from roof,* November 1982
12, *Cavity trays in external walls: preventing water penetration,* December 1982
15, *Wood windows: rain penetration at perimeter joints,* January 1983
17, *External masonry walls insulated with mineral fibre cavity width batts: resisting rain penetration,* February 1983
19, *External masonry walls: wall ties – selection and specification,* February 1983
22, *Ground floors: replacing suspended timber with solid concrete – dpc's and dpm's,* March 1983
34, *Flat roofs: built-up bitumen felt – remedying rain penetration at abutments and upstands,* August 1983
35, *Substructure: dpc's and dpm's – specification,* September 1983
36, *Substructure: dpc's and dpm's – installation,* September 1983
37, *External walls: rendering – resisting rain penetration,* October 1983
68, *External walls: joints with windows and doors – detailing for sealants,* December 1985
69, *External walls: joints with windows and doors – application of sealants,* December 1985
79, *External masonry walls: partial fill cavity insulation: resisting rain penetration,* June 1986
85, *Brick walls – injected dpc's,* August 1986
94, *Masonry chimneys: DPC's and flashings – location,* February 1987
95, *Masonry chimneys: DPC's and flashings – installation,* February 1987
98, *Windows: resisting rain penetration at perimeter joints,* April 1987
106, *Cavity parapets – avoiding rain penetration,* August 1987

Digests
54, *Damp-proofing solid floors,* 1971
77, *Damp-proof courses,* 1971
110, *Condensation*
157, *Calcium silicate (sandlime, flintlime) brickwork,* 1981
160, *Mortars for bricklaying,* December 1973
180, *Condensation in roofs*
196, *External rendered finishes,* December 1976
236, *Cavity insulation,* April 1980

245, *Rising damp in walls: diagnosis and treatment*, 1981
273, *Perforated clay bricks*, May 1983
270, *Condensation in insulated domestic roofs*, February 1983
277, *Built-in cavity insulation for housing*, September 1983
Report BRE/BBA/NFBTE/NHBC, *Cavity insulation of masonry walls – dampness risks and how to minimise them*, 1983

BSI
BS 743: 1970, *Specification for materials for damp proof courses*
BS 849: 1939, *Code of practice for plain sheet zinc roofing*
BS 4315, *Methods of test for resistance to air and water penetration*, Part 1 1968: *Windows and gasket glazing systems*, Part 2 1970: *Permeable walling constructions*
BS 5247: 1975, *Code of practice for sheet roof and wall coverings*, Part 14 *Corrugated asbestos cement*
BS 5385: Part 2: 1978, *Code of practice for external ceramic wall tiling and mosaics*
BS 5390: 1976, *Code of practice for stone masonry*
BS 5534: Part 1: 1985, *Code of practice for slating and tiling: design*
BS 5628: Part 3: 1985, *Code of practice for use of masonry. Materials and components: design and workmanship*
BS 6213: 1982, *Guide to selection of constructional sealants*
BS 6229: 1982, *Code of practice for flat roofs with continuously supported coverings*
BS 6398: 1983, *Specification for bitumen dpc's for masonry*
BS 6477, *Specification for water repellants for masonry surfaces*
BS 6515: 1984, *Specification for polyethylene dpc's for masonry*
BS 6576: 1985, *Code of practice for installation of chemical damp-proof courses*
BS 6915, *Design and construction of fully supported lead sheet roof and wall coverings*
BS 8102: 1973, *Code of practice for protection of structures against water from the ground*
CP 143, *Code of practice for sheet roof and wall coverings*
   Part 1: *Aluminium, corrugated and troughed*
   Part 5: *Zinc*
   Part 10: *Galvanised corrugated steel*
   Part 11: *Lead*
   Part 12: *Copper*
   Part 15: *Aluminium*
CP 144: Part 3: 1970, *Code of practice for roof coverings: built-up bitumen felt*
CP 144: Part 4: 1970, *Code of practice for roof coverings: mastic asphalt*

Journals
Architects' Journal, Darby, K., 'Thatch as a modern building material', 13 September 1986
Architects' Journal, Falconer, P., 'Metal industrial roofs moisture problems', 7 May 1986

National Research Council of Canada, Department of Building Research
Canadian Building Digest, **40**, 'Rain penetration and its control', April 1973
Canadian Building Digest, **130**, 'Wetting and drying of porous materials', October 1970

Trade organization
Lead Development Association, *Lead sheet in building: a guide to good practice*, 1978

Efflorescence

Background

1 The phenomenon

Deposits of soluble salts at or near the surface of a porous material as the result of evaporation of the water in which the salts are dissolved is usually known by the term *efflorescence*. On evaporation the concentration of salts in solution increases until they finally crystallize out. In general, efflorescence is harmless, the superficial white deposits often contributing more to a temporary change of appearance of the surface on which it forms. Butterworth* has aptly described efflorescence as *usually* being 'a skin trouble and not a deep-seated disease'. However, it is the exceptions which require, if not repay, a deeper consideration, not only of the physical processes involved but also of the derivation of the soluble salts (Photograph P3.5/1).

2 Scope

Under certain conditions the soluble salts may be responsible for deleterious effects on materials. These effects are included in *3.6 Chemical attack*, p. 385 ff. The main purpose of this chapter is, in addition to discussing efflorescence, to outline the sources, behaviour and *basic* effects of soluble salts. For purposes of explanation it is more convenient to sub-divide the various aspects of the whole problem. However, in practice, it should be remembered, one or more of the separate aspects usually operate together.

In order to cover efflorescence in particular and soluble salts in general the following headings are used:

1. General considerations, which will include an outline of the whole problem so that the significance of the individual aspects dealt with separately under the remaining headings may be better appreciated;
2. Nature of soluble salts;
3. Sources of soluble salts;
4. Physical processes involved in the formation of efflorescence;
5. Lime staining of the surface of brickwork, which is often mistaken for efflorescence. Unlike efflorescence, the white deposit is insoluble. The phenomenon is therefore better considered separately although some of the mechanisms involved are similar in many respects to those associated with efflorescence;
6. Precautions. Certain aspects included here are introduced or explained in *3.1 General considerations*, *3.2 Exposure*, *3.3 Moisture content* and *3.4 Exclusion*. However, the coverage here is made as self-contained as possible, giving in outline those aspects previously covered in detail.

Bricks and Modern Research, by B. Butterworth, Crosby Lockwood, London, 1948. The classic work on the subject.

(a)

(b)

Photograph P3.5/1 *Efflorescence causes more commonly superficial white deposits on (mostly brickwork) surfaces, causing a temporary change of appearance (a); occasionally the soluble salts can give rise to deleterious effects (b)*

General considerations

1 Effects of soluble salts

Depending on their nature, distribution and quantity in materials and complete constructions in which they occur, soluble salts may produce one or more of the following effects.

(a) Temporary and unsightly efflorescence

A temporary, although unsightly, white efflorescence on the exposed external or internal surfaces of materials (Photograph P3.5/2), the particular surface on which they occur being related to the mode and direction of drying out.

(b) Lime staining

The salts involved are initially soluble but later become insoluble. The staining results from free lime leached from cement-based products, such as mortars and concrete, and should not be confused with the reaction of certain compounds in Portland cement that can give rise to sulphate attack (see (f) below).

(c) Surface disruption

When the salts crystallize within the pores at or near the surface of a material the expansion caused during the crystallization is sometimes capable of disrupting the surface of the material. Expansion may, however, take place as a result of a change of state of the salt deposited below the surface. Efflorescence within a material is referred to as 'crypto-florescence'.

(d) Coloured stains

Some efflorescences may give rise to more or less permanent coloured stains (see *3.6 Chemical attack*, pp. 402–5).

(a)

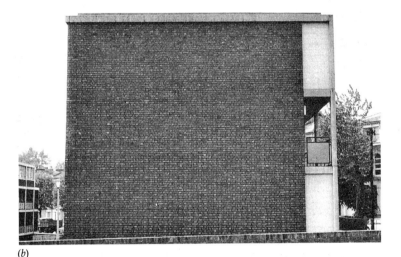

(b)

Photograph P3.5/2 *Although transient in nature, efflorescence may take several years to disappear through natural weathering. (a) Photograph taken a few months after completion with much efflorescence in evidence and (b) after one year when the efflorescence had disappeared. Efflorescence had still not reappeared after a further 20 years*

(e) Hard glassy skin

The formation of a hard glassy skin on the surface of a material usually causes deterioration of the surface due to blistering or exfoliation (see *3.6 Chemical attack*, p. 386 ff).

(f) Sulphate attack

When soluble salts react with certain compounds in Portland cement or hydraulic limes (as in mortars, renders or concrete) there is expansion, ultimate softening and possible disintegration of the material due to what is known as 'sulphate attack' (see *3.6 Chemical attack*, p. 405 ff).

(g) Loss of adhesion

Finishes such as plasters, renders and paint coatings may lose their adhesion to the background as a result of the formation of efflorescence after the finish has been applied or to the presence of efflorescence before the finish is applied (see *3.6 Chemical attack*, p. 383).

2 Variable factors

It is understandable that the phenomenon of efflorescence, together with the behaviour and effects of soluble salts in solution, often present a baffling problem. As an example, consider superficial efflorescence. The precise position of its occurrence is no clear indication of its origin. In many cases, salts derived from what can best be termed 'external sources' are usually significant. On the other hand, only very small quantities of salts are required to form even a prominent efflorescence. In brickwork the small quantities required usually occur in the bricks and mortar.

Among other factors which must be taken into account, two are closely interrelated:

1. The porosity of the materials concerned, particularly the difference in behaviour between fine- and coarse-pored materials as these relate to the drying-out pattern in which the wetting and drying cycle is also important (see particularly *3.3 Moisture content*, under 'General influences, 1 Capillarity, p. 230). This raises the significance of the problems associated with what Schaffer has called 'selective decay', and often occurs when a material is not homogeneous or when materials with different physical properties are used with one another.
2. The precise distribution of the salts, rather than their quantity, within a material. The significance of this may be understood if it is realized that, to be dissolved, the salts require water; to be transferred, they need water as the 'transfer' medium. Consequently, the salts will remain in a sense innocuous *provided* water cannot reach them. Although obvious, cognisance of this particular aspect can help greatly in an understanding of the precautions which should be taken when trying to prevent the effects of soluble salts. This may be extremely relevant when trying to prevent their more deleterious effects.

3 Tests

(a) Effects of variations

Variations in the properties (porosity and soluble salt content are notable) of materials that are commonly affected, in one way or another, by soluble salts in solution, also contribute to the difficulties often encountered in dealing effectively with the problem.

(b) Interpretation of results

It is possible to test samples of materials either for their soluble salt content or for their liability to effloresce. Despite this, none of these tests, within the context of the materials involved, is conclusive evidence that there will be no efflorescence or other deleterious effects, when *other* samples of the same material are used in a building. Among other things, the precise relationship between the material under review with other materials in the construction and the factors which influence the behaviour of the salts is important, and these are difficult to simulate for test purposes.

In brickwork, for example, there have been no deleterious effects when bricks with a high sulphate content (according to tests carried out on samples) have been used, while the converse has been true with bricks of low sulphate content. Examples such as these do not, however, necessarily invalidate the usefulness of tests. They only help to reinforce two things: first, the importance of correlating test results with actual experience of the particular material in practice; second, the necessity for an understanding of the behaviour of soluble salts in solution, the importance of the context of the *whole of a construction* and the advisability of taking, wherever practicable, reasonable precautions. Many, though not all, of these are associated with preventing a construction from either becoming excessively damp (*3.3 Moisture content* is relevant) or remaining damp for excessively long periods of time (*3.4 Exclusion*).

(c) BS requirements

(1) Soluble salt categories
BS 3291: 1985, *British Standard Specification for Clay bricks* classifies bricks according to their salt content into two categories:

1. *Low (L)*, in which the percentage of mass soluble ions shall not exceed the following:
   Calcium   0.30
   Magnesium   0.03
   Potassium   0.03
   Sodium   0.03
   Sulphate   0.50
2. *Normal (N)* – no limit on soluble salt content.

Previously, the classification was by quality (i.e. internal, ordinary and special).

(2) Efflorescence categories
The Standard categorizes bricks according to the area of efflorescence found on the surface of a test specimen as follows:

Nil. No perceptible deposits of salts.
Slight. Up to 10% of the area of the face covered with a deposit of salts but unaccompanied by powdering or flaking of the surface.
Moderate. More than 10% but not more than 50% of the area of the face covered with a deposit of salts but unaccompanied by powdering or flaking of the surface.
Heavy. More than 50% of the face covered by a deposit of salts and/or powdering or flaking of the surface.

(3) Soluble salt content tests
Tests of the soluble content of a complete load of bricks would produce a normal distribution curve. Clearly, it is impracticable to test bricks on this scale. On the other hand, the result of a single brick would be meaningless, as it must lie anywhere within the normal distribution curve. The Standard reduces to an acceptable level the chances of unrepresentative sampling by requiring the sample for soluble salt analysis to be prepared from ten bricks from a load of not more than 15 000. The Standard describes two sampling procedures, random and representative.

It should be noted that most brick manufacturers are able to give guidance on the soluble salt content of their bricks. Based on the efflorescence tests, bricks with a high concentration of water-soluble salts are effectively excluded in bricks in the normal category.

4 Apparent transient nature of efflorescence

Another aspect of efflorescence, which often encourages what may be termed a certain degree of cynicism, is its apparent transient nature in a great many cases. For some, it would appear to be a phenomenon which must be associated only with new buildings (brick, unfairly at that) on which it usually occurs, sometimes to the temporary annoyance of owners, architects and builders, soon after completion, but usually disappears within a comparatively short period of time as the result of the combined effects of wind and rain. In other cases, the efflorescences may persist for a little longer, when it seems to be a seasonal phenomenon; appearing prominently in the spring and summer but disappearing by the autumn. The cycle may persist for a number of years, but usually (unless there is a perpetual supply of salts from external sources) the amount of the deposit recurs with decreasing intensity, until it finally disappears after a few seasons.

Despite these 'fortunate' occurrences, which probably take place in the majority of cases, it is still possible, under certain conditions, for the efflorescences to remain hidden behind the surface, probably causing slow but eventual damage. If, on the other hand, the salts remain in solution for long enough within the material, damage due to chemical attack could also be taking place slowly. Consequently, there may be a considerable time lapse from the completion of a building until the deleterious effects of the soluble salts are 'suddenly' made apparent.

It is obviously extremely easy to over-estimate or exaggerate the effects of soluble salts, while many of the precautions, which could be taken to try to prevent some of its effects, are sometimes seen to be tedious. Nevertheless, it may be noted that, even when every precaution has been taken, it is still possible for some isolated damage or disfigurement to occur. As in most cases in building practice, the effects which result when no precautions at all have been taken are usually extremely pronounced, and sometimes regretted.

5 Materials commonly affected

The problems of efflorescences and salts in solution can take place in any porous material in which water transfer is possible. However, in the past, brickwork and stonework, including the constituents of the relevant jointing and/or surfacing materials used, have received the greatest attention. In fact, the major part of research, carried out in all parts of the world, has been devoted to these two materials. The effects of soluble salts may equally well be produced on or within other porous materials such as tiles, faience and concrete, while they may also damage finishes such as paint or weaken the bond of adhesives.

In many ways the effects of soluble salts on some materials, hitherto regarded as 'immune', may not become pronounced until the material is used in new ways or, as important, it is more widely used. As will be seen later, the increased use of Portland cement in mortars, for example, is now seen as one of the more important sources of salts in brickwork.

6 Importance of sources of salts

It cannot be emphasized too often that salts, other than those initially present or inherent in a particular material, derived from external sources and transferred through a construction or from one part of a construction to another, are of fundamental importance in all considerations. An analysis of any particular situation aimed at assessing the necessity for any precautions should, therefore, always take these external sources into account. It may be as well to note here that, although fairly obvious, it is surprising how often, particularly in stone plinths, the soil (including hardcore under any adjacent paving) has been overlooked as a source of salts. Illustrations of these kinds of examples have been included in earlier chapters; some others are included later in this one.

Nature of soluble salts*

1 Composition

The composition of the soluble salts which may be encountered is, to a large extent, determined by their sources ('Sources of soluble salts', p. 356). As the latter are numerous, so are the compositions of the salts. However, soluble salts more commonly associated with the materials, including the jointing and surfacing materials, making up a construction, are the sulphates and carbonates of calcium, magnesium, sodium and potassium.

Some common salts are: calcium sulphate (gypsum); magnesium sulphate (Epsom salt); sodium sulphate (Glauber's salt); potassium sulphate (sulphate of potash); ferrous sulphate (sulphate of iron or copperas); calcium carbonate (lime); magnesium carbonate; sodium carbonate; and potassium carbonate.

The presence of nitrates and chlorides is usually indicative of salts derived from groundwater, although sulphates and carbonates may also be present. Chlorides may also originate from sea spray, although they may also be present in some polluted atmospheres.

2 Solubility

In general, the sulphates constitute the more commonly found salts in most materials. (Any carbonates present can be converted into sulphates.) All, except calcium sulphate, are very soluble in water. Although generally regarded as an insoluble compound, calcium sulphate does dissolve in water to an appreciable extent (soluble to the extent of 1 part in 500). Its solubility cannot therefore be entirely overlooked. The important point is that the other sulphates are soluble to an even greater extent.

3 Significance and effects

Only the sulphates, for the reason given in 2 above and because they are generally the most aggressive of the salts, are included here.

*A general discussion of *solutions* is given in *Materials for Building*, Vol. 1, *1.06 Solutions*, particularly 'True solutions' and 'Solids in liquids', and of *salts* in Vol. 1, *1.10 Chemical reactions*.

(a) Calcium sulphate

Calcium sulphate, mainly because of its relatively low solubility, does not normally give rise to efflorescence. However, its innocuous character only persists provided (1) there are no appreciable quantities of other soluble matter present, *or* (2) that materials are not maintained in a damp condition for long periods, *or* (3) a hard glassy skin is not formed. Solutions of calcium sulphate may give rise to sulphate attack if dampness persists for long enough. Calcium sulphate may form a hard glassy skin on sheltered areas of a building not freely washed by rainwater. The skin is not protective and may lead to disruption of the surface of the material on which it forms.

The solubility of calcium sulphate may be increased if potassium sulphate is present. Calcium and potassium sulphate combine to form a double salt which is more soluble, and appreciable amounts may be found in efflorescences.

Calcium sulphate may occur in stonework as a result of decomposition. In brickwork its presence is, as explained later (see 'Sources of soluble salts', p. 356), a consequence of manufacture. Consequently, calcium sulphate is unavoidably present and over one-third of all bricks produced in the UK contain upwards of 3.0% of it. However, for the reasons already given, such quantities may not necessarily be detrimental to the brick.

(b) Magnesium sulphate

This salt is probably responsible for most of the failures caused by soluble salts, despite the fact that it is rarely present in appreciable quantities (in brickwork the amounts present seldom exceed 0.5%). In stonework magnesium sulphate results from the decomposition of magnesium carbonate exposed to the atmosphere. The failures caused by magnesium sulphate may take various forms. However, once it has gained a foothold and suitable conditions persist, its destructive effects may be serious – some building stones are particularly susceptible to attack.

The salt may crystallize out at or near the surface of a material. On external surfaces, because of its solubility, the magnesium sulphate is readily washed away. This does, to a large extent, minimize the possibility of destruction of materials. Thus sheltered parts of a building are more vulnerable.

When magnesium sulphate appears on the surface of plaster, decoration may be impracticable or disintegration of the surface may occur. Under certain conditions, when lime is present in the plaster, crystal growth at the interface of the background and the plaster may exert sufficient pressure to separate the plaster (and sometimes the surface of the brick as well – see *3.6 Chemical attack*, p. 383).

(c) Potassium sulphate and sodium sulphate

During crystallization potassium sulphate tends to assume a hard, glassy form, while sodium sulphate gives a fluffy deposit. Decay is usually likely to be most severe where the salts concerned can exist in more than one state of hydration. Externally, the transition temperature between the two states is within the range of outdoor temperatures. Sodium sulphate, for example, at temperatures below 32.5°C crystallizes with ten molecules of water of crystallization (Na_2SO_4 10 H_2O). Known as sodium sulphate decahydrate, this compound changes into the anhydrous salt (Na_2SO_4) at 32.5°C even under water. In air it begins to lose water below this temperature. Heat from the sun can change it to the anhydrous form, but a subsequent shower of rain may convert this to the decahydrate which will occupy four times the volume. Consequently, if the salt is deposited in the pores of a material the dimensional changes can lead to destructive pressures in the pores and powdering of the surface.

(d) Ferrous sulphate and vanadium

Rusty stains, particularly on mortar joints in brickwork, are usually due to iron salts which are changed to ferric oxide by oxygen and lime. Yellowish-green stains are due to an efflorescence of a coloured salt which contains vanadium.

Sources of soluble salts

The sources from which salts may be derived can, for convenience, be grouped into three classes, namely, (1) salts originally present in a material before its incorporation into a building; (2) salts derived from the decomposition of a material exposed to the atmosphere after its incorporation into a building; and (3) salts derived from external sources. Each of these will be discussed separately, including their own sub-divisions, but, as pointed out earlier, it is as well to remember that, within the context of building practice, all sources are usually closely interrelated.

1 Salts originally present in a material

In this class the salts which may be present in a material before its incorporation into a building are chiefly those which result from what may best be termed the manufacture (either natural or man-made) of a particular material. Solid-state materials, e.g. fired clay products, quarried and dressed stone or concrete, are implied. Any salts which may find their way into a particular material after manufacture are regarded, for the purposes of this book, as being derived from external sources.

It is convenient to consider groups of materials separately.

(a) Clay bricks

The origin of salts is attributed to *three* main sources, namely, the clay itself (including the tempering water), pyrites, and the action of sulphur from the fuels used in firing. The relative importance of any of these three sources has not been established conclusively. Each of them is dealt with separately. As firing temperature is significant, particularly as regards the quantity and nature of salt which may be present, it is also discussed. Table T3.5/1 sets out the results of soluble salt analyses and efflorescence tests (after Butterworth). The results, it should be noted, were obtained before the publication of BS 3921.

(1) Clay
Although not considered to be a major origin of salts, when salt is present it is usually gypsum. The reason for its presence is usually due to the fact that it may not be in a form that can be easily separated from the clay before it is used for making bricks. Where, however, the

Table T3.5/1   *Results of soluble salt analyses and efflorescence tests on clay building bricks*

Type of brick	Fletton		London stock		Colliery shale Stiff-plastic-pressed common		Keuper marl wirecut common	
	Sample 1	Sample 2	Sample 1	Sample 2	Sample 1	Sample 2	Sample 1	Sample 2
Total soluble (%wt)	3.70	2.39	0.20	0.77	0.96	0.29	0.86	5.70
Composition:								
R_2O_3[a]	0.06	0.02	Nil	0.02	0.11	0.02	Nil	Nil
Ca^{++} (calcium)	0.97	0.62	0.06	0.18	0.06	0.03	0.21	1.40
Mg^{++} (magnesium)	0.01	<0.01	<0.01	<0.01	0.03	0.03	0.03	0.04
Na^+ (sodium)	0.02	0.06	0.01	0.02	0.01	0.01	0.04	0.08
K^+ (potassium)	0.02	0.02	0.01	0.05	0.03	0.03	0.05	0.17
SO_4'' (sulphate)	2.13	1.42	0.11	0.42	0.29	0.08	0.49	3.38
SiO_3'' (silicate)	0.04	0.06	Nil	0.05	0.01	0.02	0.04	0.03
Liability to efflorescence	Nil	Slight	Nil	Slight	Heavy, iron and potassium sulphates	Moderate, magnesium and alkali sulphates	Slight	Variable slight to heavy alkali and calcium sulphates

[a] R_2O_3 is a conventional symbol for the sesquioxides of iron and aluminium (Fe_2O_3 and Al_2O_3) which are precipitated together in the course of analysis, and are not separated when there is too little to justify the trouble involved.

From Butterworth, B., *Clay building bricks*, National Building Studies, Bulletin No. 1, HMSO, London, 1948, Table 5, p. 18.

Notes:
1. Two analyses are given for each type of brick in order to show that there may be a variation, sometimes a wide one, between different samples from the same district or of the same variety.
2. Because of the variations likely, it is unwise to attach too much importance to *small* differences in analytical results. This is not a reflection on the accuracy of the analytic methods used but rather an emphasis on the variability of the material. The results of periodic analyses would enable some idea of the usual variation to be obtained. This would make comparison of different makes of brick valid.
3. The results given in the table have been based on the analytical method given in BS 1257: 1945, which is now superseded. The methods of anlysis now included in BS 3921: 1965 are aimed to improve the extraction of the more soluble constituents (magnesium, potassium and sodium). These, although the minor constitutents, are the most aggressive. Analyses made by the old method should not be used in conjunction with the limits specified for soluble salts for special quality bricks in the new standard. For the present purposes, and bearing in mind the allowable tolerance (see note 2), the analyses given in the table are considered adequate.
4. The analyses do not report the presence of salts, such as calcium sulphate or sodium silicate, but record instead the bases and acids of which they are composed. This is due to the necessities of the analytical method.
5. The accuracy of extraction is limited by the fact that the samples used contain both powdered brick and the salts. The former must first be removed before the latter can be analysed. Both the old and the new British Standards specify a method of extraction which enables analysts 'to steer a course between inefficient extraction and too violent a treatment'. It is wise to check that the results of analyses have been obtained by the standard method, particularly if they show phenomenally low results.
6. A wide range of data is given in Bonnell, D. G. R. and Butterworth, B., *The Properties of Clay Building Bricks made in the United Kingdom*, HMSO, London, 1950.

gypsum does occur in massive form, separation is not difficult. Butterworth has noted that there is at least one works in England where gypsum for plaster and clay for bricks are dug from the same quarry, and yet the soluble salts content of the bricks is in no way abnormal.

(2) Pyrites (iron sulphide)
This may be present in certain clays. The salts are formed during heating of the clay in air when the pyrites take up oxygen, giving gaseous oxides of sulphur which further react with bases in the clay to form sulphates.

(3) Sulphur action

Sulphur in fuels (coal and oil (see *3.2 Exposure*, 'Sulphur dioxide', p. 174) used for firing the bricks are the most general source of salts. During firing, sulphur burns, forming oxides capable of reacting with the clay. In the early stages of firing there is an increase in the amount of salts, but these begin to decompose again at the temperatures usual in the firing of bricks. Decomposition proceeds further as the firing temperature is raised.

(4) Firing temperature

In part, the soluble content of particular types of brick varies over such wide limits, because certain salts tend to be decomposed and expelled from bricks by hard firing. For example, magnesium sulphate is more easily decomposed in the firing of bricks than is calcium sulphate. A firing temperature of between 1000° and 1500°C is sufficient to get rid of magnesium sulphate. However, magnesium sulphate is often absent, even from lightly fired bricks, and as many bricks fired at temperatures below 1000°C appear quite satisfactorily hard, it is possible for a manufacturer to be taken unawares if there is a local change in the composition of the clay (isolated seam or pocket of clay in a pit containing either magnesium sulphate or carbonate). This sort of experience underlines the importance of regular analyses at works. After the first discovery of the change in the composition of the clay, for example, a higher firing temperature should be used. This will guard against any further local changes that may occur, but will also mean a change in colour in the bricks, as lighter-coloured bricks are fired at a lower temperature.

Variations in the soluble content of bricks of a particular type may, then, be found in those kilns where close temperature control is not exercised. Higher soluble contents will generally occur in those bricks coming from the cooler parts of the kiln. For the reasons already outlined under 'General considerations', apart possibly from magnesium sulphate, the soluble content of any particular type of brick is not necessarily indicative that either efflorescence and/or chemical attack will result. However, some *under-fired* bricks may be badly attacked and these should therefore be carefully considered before use in building.*

Although ferrous sulphate, like magnesium sulphate, is not found in the majority of bricks, when it does occur it is present in very small amounts. However, the conditions required for the formation of ferrous sulphate, namely, a reducing atmosphere and a high firing temperature, are also those necessary to the production of some of the most attractive colours in hand-made facing bricks. The same conditions are also used in some kilns firing ordinary bricks from colliery shale. However, the presence of ferrous sulphate is unlikely to be important, as far as rust staining is concerned, in the work in which ordinary bricks from colliery shale are likely to be used. When bricks of this type are plastered with gypsum plasters some staining may appear. (See 'Physical processes, 6 The effects of precipitation', p. 368.)

Even relatively low firing temperatures appear to expel sodium chloride (common salt). Consequently, this salt is not found among the deposits of soluble salts, even though it may be present (sometimes specifically added to the clay) before firing. Although the decrease in sulphate content is not great enough to warrant its general addition to clay, if sodium chloride is present in the early stages of firing the sulphate content of the fully fired brick may be less than it would have otherwise been. Butterworth has suggested, however, that the addition of sodium chloride is a possibility that might be examined by a brickmaker who is troubled with efflorescence and cannot conveniently raise the firing temperature.

*Under-firing may also lead to considerable expansion of the brick (see *3.3 Moisture content*, p. 235).

(b) Sandlime (calcium silicate) and concrete bricks and blocks

In general, it is very rare for sandlime bricks to contain soluble salts. The same is usually true of concrete bricks and blocks. Efflorescence on or within the surface is more likely due to salts derived from external sources.

(c) Natural stones

The natural building stones which are of significance here are the sedimentary stones. Despite the fact that it would be expected that these stones should have certain amounts of salts due to their method of formation – originally laid under water, the majority under sea water – the actual amounts from this source are negligible. Such salts as may have been present have been largely leached out by the action of rainwater.

The traces of chlorides and sulphates found in some specimens of stone are not considered to constitute a common cause of efflorescence in building stones. However, it should be noted that contamination of stone by sea salt may occur if the stone has at some time been exposed to sea water. Sometimes this has been accidental, but there is at least one known case (another is suspected) in which a salt was applied in the mistaken belief that salt would serve as a preservative.

Apart from isolated cases, stone 'obtains' its soluble salt content after it has been incorporated into a building as a result of decomposition of calcareous matter in it (see '2 Decomposition', p. 361) or from other external sources (backing materials are often significant).

(d) Aggregates

Aggregates, such as sand, clinker, crushed stone and gravels, may be used for mortars, plasters, renders and concretes. Some of these and others such as colliery shale may also be used as hardcore in foundations, under ground-floor slabs or under paving. It is in the general sense, i.e. to include all uses, that the term 'aggregates' is being used.

The soluble salt content of the various types of aggregate available is liable to wide variation. In addition to sulphur compounds including sulphates (gypsum is often notable), soluble alkalis may also be present. Coal dust, which can have deleterious effects, may be found in aggregates from certain districts. In general, however, clean river, pit or crushed natural stone sands, and crushed stone or gravel aggregates are substantially free from soluble salts. The soluble salt content of *crushed brick* aggregate will depend on the soluble content of the brick used, although it is reasonable to assume that in its crushed state the brick is more likely to 'expose' its salt content. However, in general, crushed brick, if clean, is a good aggregate; if contaminated with old plaster and other rubbish, it may be very bad.

There are some aggregates which do require special consideration if they are to be used. These are as follows.

(1) Sea sand
It is perhaps all too obvious that sea sand will contain salt in it. Nevertheless, there is often a temptation to use unwashed sea sand for mortars or plastering or concrete in districts near the coast or tidal estuaries. In addition to the possibility of causing efflorescence, the presence of sea salt, which is hygroscopic and likely to absorb moisture from the atmosphere in humid weather, results in work appearing

damp. If sea sand is to be used for building work it should always be *washed*.

(2) Ashes and clinker

Furnace ashes and clinker are sometimes employed as mortar aggregates; clinker may be used for lightweight concrete. They may also be used as hardcore. Both may contain sulphur compounds including sulphates. For mortars and concrete, materials of doubtful origin should not be used. Well-burnt clinker from large industrial furnaces is, in general, likely to have a reasonably constant composition.

For use as hardcore, the soluble salt content is only of importance in so far as it influences the precautions, usually in terms of dpcs or dpms, which should be taken to prevent water transferred through the hardcore coming into contact with or subsequently being transferred through other porous materials.

(3) Colliery shale

This materials, a waste seen in unsightly heaps near coal mines, is often used as a hardcore. Initially the material consisted of shale with a proportion of carbonaceous materials which had burnt slowly and had fired the shale to a varying extent, ranging from well fired to very under-fired. Consequently, the physical condition and the chemical composition of the burnt shale are both very variable, and the material may contain appreciable amounts of soluble salts, particularly sulphates. In the absence of adequate precautions, this material has been known to give rise to severe sulphate attack of concrete ground-floor slabs laid on it.

(e) Cementing agents

Whatever the soluble salt content of the aggregate used, soluble salts may be derived from the cementing agents used. Both lime and Portland cements may be responsible, with their soluble alkali content being important.

(1) Lime

Mortars or plasters made with pure high-calcium lime will be practically free from soluble salts. However, limes made by burning limestone containing clay, namely semi-hydraulic and hydraulic lime, usually contain the soluble alkalis of soda and potash. These may appear as efflorescences either as carbonates or sulphates, or a mixture of both. Sulphates will be formed if there is calcium sulphate present. As the amount of soluble alkali increases with the hydraulicity of the lime (roughly proportional to the amount of clay present in the limestone prior to burning), eminently hydraulic limes may contain a maximum of nearly 1.0% of potash and 0.4% of soda, which corresponds to about 2.5% of soluble sulphates.

(2) Portland cements

Although Portland cements contain appreciably less soluble alkali than the maximum found in hydraulic limes, they need special consideration, particularly if the mixing water is allowed to migrate to other building units, from mortar to bricks for example, *before* the cement has set.

In addition to the alkalis of potash and soda, Portland cements also contain calcium sulphate which is added in manufacture to control the set of the cement. On being mixed with water, the potash, soda and sulphate go into solution. After a few hours, however, the sulphate is fixed in an insoluble form as calcium sulpho-aluminate, leaving the alkalis as hydroxides, which are gradually converted into carbonates by carbon dioxide in the atmosphere. If the sulphate, however, is removed with the mixing water (usually by capillarity in an adjoining porous building unit) from the environment in which it would normally be rendered insoluble the quantity of soluble salts in the material into which it has been transferred will be increased.

2 Salts derived from decomposition of a material

Salts derived from this source are usually those which result when a material is exposed after its inclusion in a building and mainly as a result of the effects of a polluted atmosphere. Calcareous sedimentary stones, mainly the limestones, are the chief sources of salts. Lime leached from the exposed surfaces (often of poor quality) may also, under certain conditions, be relevant (see 'Lime staining', p. 371, for leaching from within cement-based products).

Some of the basic effects of the action of acid gases in the atmosphere (carbon dioxide and sulphur dioxide) on the carbonates of either calcium or magnesium have already been outlined in *3.1 General considerations*, 'Weathering, 2 Chemical weathering', p. 145, and *3.2 Exposure*, 'Sulphur dioxide', p. 174. For the present purpose it is considered advisable to discuss, in slightly more detail, the formation of salts which result from the action of carbon dioxide and/or sulphur dioxide in the atmosphere as it applies specifically to calcium and magnesium carbonate in the calcareous sedimentary stones. The chemical reactions described will, in some cases, also be applicable in principle to other materials in which or on which carbonates are present. Concrete is a notable example.

(a) Action of carbon dioxide

Although carbon dioxide is a normal constituent of the atmosphere, the normal proportions may be exceeded, particularly in atmospheres polluted as a result of fuel burning. Although only slightly soluble in pure water, calcium carbonate is more readily dissolved in water containing carbon dioxide.

The result of this reaction is the formation of a solution of calcium bicarbonate, which is unstable. Boiling or evaporation of the solution results in calcium carbonate being redeposited. When this happens on the exposed face of a building the calcareous matter is removed from the stone. Either the calcium carbonate is *carried away* to another part of the building *or* it remains behind in the material and will, on evaporation of the water, be *deposited* again.

However, it is important to note that, whereas the calcium carbonate in building stones is present in a crystalline form, thus offering a small area to attack, the calcium carbonate redeposited from solution, though probably still crystalline, is usually a finely divided incoherent powder, thus offering a large area to attack, particularly by sulphur gases. The gradual erosion of stonework by the removal of small quantities of carbonate, as described, cannot normally be regarded as being damaging to the stonework, but the effect of the action of sulphur gases is far more serious.

It may be noted that in the case of concrete, calcium carbonate removed from the surface by the solvent action of carbonic acid is replaced by diffusion of calcium hydroxide from the interior. Carbon dioxide changes the hydroxide to calcium carbonate. The depth to which the calcium hydroxide is changed depends on the porosity of the concrete, but does not normally exceed 12 mm, even for porous concrete.

(b) Action of sulphur dioxide

Sulphur dioxide in the atmosphere may form either sulphurous or sulphuric acid, depending on conditions previously described (see *3.2 Exposure*, p. 174). The reaction of sulphurous acid and calcium carbonate results in a relatively insoluble compound, calcium sulphite ($CaSO_3$), which combines with oxygen to form calcium sulphate ($CaSO_4$). Sulphuric acid, being stronger, reacts with the calcium carbonate to form calcium sulphate.

The action of the sulphur gases on the calcium carbonate is, it should be noted, to liberate carbon dioxide, which can react with further quantities of calcium carbonate as already described under (a). In practice, it is, of course, difficult to separate, the combined effects of both acid gases, as the reactions and processes involved are complex.

When sulphur gases react with *magnesium limestones* both magnesium and calcium sulphate are formed. Magnesium sulphate, as already noted, in addition to being aggressive is easily dissolved in water and hence readily washed away; calcium sulphate, though basically less aggressive, is not as soluble.

3 Salts derived from external sources

In this class some external sources may be difficult to define. In practice, each circumstance must be considered on its merits. For the present purposes, external sources are deemed to imply those other than the particular material under consideration. Thus in brickwork, for example, the brick is regarded as the particular material under consideration, so that jointing becomes an external source. There may, of course, be other external sources. Mortar in stonework would also be an external source, although if a backing of brickwork were used, both the bricks and the mortar would constitute an external source, relative to the stone.

External sources may, therefore, be grouped as follows:

1. Jointing materials;
2. Backing materials;
3. Decomposed materials;
4. The soil, including any hardcore used under paving and ground-floor slabs;
5. The atmosphere;
6. User requirements; and
7. Cleansing and other maintenance methods.

(a) Jointing materials

The basic sources of salts in either the aggregate or the cementing agents have already been outlined under '1 Salts originally present in materials, (d) and (e)'. It only remains to discuss the contribution of mortars in *brickwork*. As regards stonework, in which the mortar joint is, in ashlar work at any rate, extremely narrow, the principles are somewhat the same, although staining by the mortar is more usual (see *3.6 Chemical attack*, p. 402). For convenience a note on 'plasticizers' is added.

In *brickwork* three points arise: first, the relative volume of mortar as against that of the bricks, second, the contribution of lime as compared to Portland cement and, third, the transfer of the mixing water from the mortar to the bricks. The mortar is usually about one-sixth of the brickwork, while the cementing agents form only one-third to one-quarter of the mortar. Taken in isolation, and as described previously, the percentages of soluble alkalis in hydraulic limes and cements are much higher than those usually found in bricks. However, taking the smaller volume of mortar into account, the actual quantity of the more soluble salts introduced into the brickwork by the cementing agents may not, in practice, therefore, greatly exceed that contributed by clay bricks.

Limes do not appear to contribute as much to soluble salt content as do the Portland cements. This is probably due to the wide variations in the soluble alkali content of limes in general, with non-hydraulic limes in particular rarely contributing to soluble salt content. However, there does seem to have been an increase in efflorescence since the increased use of cement for mortars. In support of this, research in recent years has shown that the amount of efflorescence in brickwork built with any given brick tends to increase as the proportion of Portland cement rises. This, among other things, further supports the case for using cement:lime:sand mixes that are no stronger than necessary.

Finally, there is the transfer of mixing water to the bricks by which the sulphate is extracted from its setting environment. This will occur if the bricks are very dry at the time of laying. However, excessive wetting of the bricks prior to laying increases not only the risk of greater efflorescence on drying out but also the possibility of loss of adhesion between the bricks and the mortar. The right **balance** must therefore be sought. (See *3.4 Exclusion*, pp. 315–16.)

When considering the use of either sandlime or concrete bricks (including blocks), the mortar will naturally contribute far more to soluble salt content, as these bricks, particularly sandlimes, do not normally contain any significant amounts of salts themselves.

As mortar *plasticizers* are now often used in Portland cement mortars it may be noted that there has, as yet, been no evidence to suggest that these contribute significantly to efflorescence. This is probably due to the fact that they are added in such small quantities. As regards undercoats to plaster, it has been found that a plasticized mortar undercoat is more effective than a cement:lime:sand undercoat in preventing salts in the body of the wall from reaching the surface. Although domestic detergents may be effective as mortar plasticizers, they should not be used, as many of them contain sodium sulphate which could contribute to efflorescence.

(b) Backing materials

It is with *stonework* that backing materials are chiefly an external source of soluble salts. In developments of ashlar work it has become common practice to fix rather thin slabs of stone to either a brick or concrete background.* The possibility of the transference of soluble salts from the backing to the stone is likely when the porous sedimentary stones are used as the facing. With the more impervious types of stone, such as granite, marble or slate, the joints, subject to their composition and porosity, may be affected.

Although the sedimentary stones appear to be the materials most commonly affected by the soluble salt content of backing materials, other porous materials (bricks or tiles or concrete) and some paint films or adhesives may also be affected. In all cases, the soluble salt content of the backing can vary between wide limits.

*More recent developments include the use of even thinner slabs of stone as facings to precast concrete panels. The long-term performance of such panels in relation to efflorescence, chemical attack or frost action is, as yet, unknown.

Photograph P3.5/3 *Moisture rise in the absence of adequate damp-proofing and the subsequent efflorescence, staining and decay to limestone due to soluble salts derived from the soil. Decay first observed about three years after completion*

(c) Decomposed materials

Decomposition of materials exposed to weathering as described earlier can, if the decomposed matter is transferred by rainwater to other materials in the same construction, provide an external source of soluble salts (see Photograph P3.2/1, p. 211). In brickwork (many Victorian examples are notable) limestone dressings can be significant. Nowadays, lime leached from the exposed surface of concrete may also be included (see Photograph 3.2/2, p. 211).

(d) The soil

The reasons for the soluble salt content to be found in soils and in groundwater have already been given both in *3.1 General considerations*, particularly under '2 Weathering', p. 144 ff and *3.2 Exposure*, p. 196 ff. Suffice it to emphasize here that, in general, the soil provides one of the most prolific external sources of salts, which, among other things, can almost be regarded as perpetual (see Photograph P3.5/3).

Hardcore used under pavings or ground floor slabs should, if it consists of such materials as ashes, clinker or colliery shale, be regarded in the same way as soils, particularly in so far as precautions that may be taken to prevent the transference of the soluble salts.

The principles outlined previously in *3.4 Exclusion*, p. 325 ff are applicable. For convenience, a traditional example illustrating moisture rise and subsequent efflorescence, staining or decay as related to stone plinths is included in Diagram D3.5/1.

Diagram D3.5/1

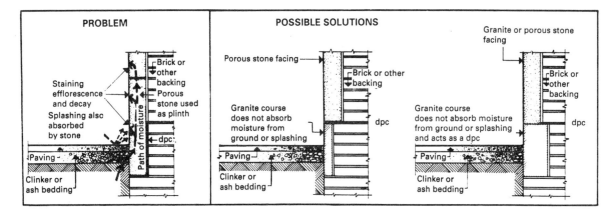

(e) The atmosphere

Account has already been taken of the effects of the two main acid gases, namely, carbon dioxide and sulphur dioxide, in the atmosphere on calcareous material (see earlier) and the products which result from the action. In addition to these, chlorides and other salts, either derived from fuel burning or from sea spray, may be significant. It is necessary to note that chlorides have often been found in considerable amounts in areas well away from the sea. An indication of the distribution of chlorides in country areas well away from large towns, for example, is given in the map in *3.2 Exposure*, p. 176, while concentrations in various types of districts are given in the chart form (p. 177). Local measurements or experience may often repay investigation in particular cases, although it may be taken as axiomatic that facades exposed directly to sea spray are potentially vulnerable.

(f) User requirements

Many industrial processes may be potential sources of salts which could be transferred to porous materials in solution or with the water vapour in the atmosphere. As regards the latter, there would appear to be greater risk, if there is a possibility of the polluted water vapour condensing within the thickness of a construction. Industrial processes which are likely to become potential sources of soluble salts are both numerous and variable. It is, therefore, important that each case is considered on its merits. In addition to industrial processes, the chemical nature of any materials which may be stored in buildings of any kind should be taken into account. Fertilizers, among other materials, are potential sources of soluble salts. A rather fortuitous yet, in certain cases, significant source of soluble salts, applicable to *all types of buildings*, is leakage from waste pipes, particularly those from kitchen sinks. Leakage from services is also discussed in *3.2 Exposure*, p. 212.

(g) Cleansing and other maintenance methods

Certain chemical compositions used for daily or regular cleansing of buildings or during maintenance may constitute accidental sources of soluble salts. In the case of stone, the BRE has drawn attention to 'even purposeful (though misguided) contamination of stone' through the use of certain detergents for cleaning or 'preservatives' for preservation.

Among cleansing agents which should be regarded as potential sources of soluble salts or decay are caustic soda, washing soda, soda ash, household scouring powders and some of the modern organic detergents. The last contain a high proportion of sodium sulphate. However, it may be noted that there are also some detergents which consist solely of organic compounds and are free from sodium sulphate or other inorganic salt. Although there is no reason to suppose that these will have any detrimental effect, in the case of stonework, organic detergents may not assist in the cleaning process very materially.

Physical processes

1 Background

(a) Need for understanding

The formation of efflorescence depends on a number of physical processes, and both the salts and the transfer of water in and out of porous materials are involved in these. Specific conditions dictate the extent to which any of the many processes take place. Thus efflorescence can be a baffling phenomenon. An understanding of the physical processes involved should help to explain the apparent anomalies.

(b) Scope

All the relevant aspects previously discussed or explained are brought together and applied specifically to the physical processes involved in the formation of efflorescence.* Chemical effects are covered in *3.6 Chemical attack*.

Brickwork is used as the point of departure for porous materials. The principles explained are also applicable generally to other porous materials. Differences in detail are noted where relevant. For purposes of explanation, the different factors involved are dealt with separately. In practice, such separation seldom occurs and, importantly, all the factors are interrelated to some extent, thereby creating a complex 'mechanism'.

2 The effects of solubility

A solution of a soluble salt in water is an example of a solution of a solid in a liquid. Saturation of the solution occurs when no more salt can be dissolved in the water. Factors which limit the weight of a given salt that can be dissolved in a fixed weight of water are the properties of the salt (variable) and temperature (variable). Although there are exceptions, warm water will usually dissolve more of a salt than cold water. As an example, consider sodium sulphate. One hundred parts of water will dissolve nine parts of sodium sulphate crystals at 10°C and 19.4 parts at 20°C. Once saturated, cooling of a given solution will result in some of the salt separating out in solid form. On the other hand, without any change in temperature, loss of water through evaporation will result in some of the salt being deposited.

Whether or not a porous material is likely to have salts deposited when it *begins* to dry out will depend on whether the salts it contains are either saturated or near to saturation before drying out commences. For example, if a brick, containing only a fraction of 1% of sodium sulphate, absorbs, say, 25% of water, the relationship between the amount of sodium sulphate and water is insufficient to cause a saturated solution. In effect, there is too much water present, and consequently there is no reason why any salt should come out as an efflorescence when the bricks start to dry. (Some salt could eventually be deposited if water continued to evaporate from the solution after saturation had been reached.) On the other hand, if a second brick, containing the same amount of salt as the first, only absorbs 2% of water, the solution would be nearer saturation, and efflorescence would be more likely to appear on drying out. In the last case, efflorescence may also appear if, instead of drying out, there is a drop of a few degrees in temperature. In this, the difference in temperature (as between summer and winter) is notable.

3 The effects of pore structure

The pore structure of the porous material containing a saturated solution will influence drying out and the position in which salts may be deposited on evaporation of the water. Evaporation of water can

*For moisture transfer of water see *3.3 Moisture content*, p. 230. The nature and sources of salts are covered earlier in this chapter, p. 354 ff.

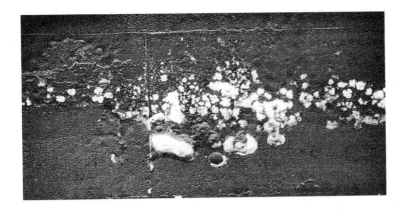

(a)

(b)

Photograph P3.5/4 *An example of crypto-florescence (i.e. efflorescence forming at or near the surface) of limestone. The soluble salts are visible because of the exfoliation of the surface of the stone*

only occur when a free water surface is presented to an environment in which evaporation can take place.

For purposes of explanation, two distinctly different types of pore structure, namely, *fine-pored* and *coarse-pored*, are considered. In the case of the fine-pored structure (close-textured brick) capillary forces will tend to draw the solution so that the free water surfaces in the pores are near the face exposed to the evaporating environment. Visible deposits of salt crystals will take place as evaporation, and thus crystallization, will tend to occur at the exposed face. With a coarse-pored structure (open-textured brick) the larger dimensions of the pores will cause the free water surfaces to fall below the exposed surface. Nevertheless, evaporation by diffusion and by pore ventilation could take place, resulting in crystallization occurring below the exposed surface, i.e. crypto-florescence (see Photograph 3.5/4). Although the pore structure of a material as a whole will influence the rate of water transfer (fast with fine pores and slower with coarse ones) the pore structure at the surface is particularly significant as regards the position in which salts will be deposited.

Some apparent anomalies may now be explained. Efflorescence tends to appear more readily on many dense bricks (facings are notable) than on common bricks, mainly because the former are fine-pored and the latter coarse-pored. Sometimes, however, efflorescence may only appear on certain areas of materials. This may be explained by the fact that these areas are probably finer-pored than the remainder in the material. Thus, in addition to the factors already outlined, the salt solution would be drawn to the finer-pored areas, due to the stronger capillary action which they are capable of exerting (see Photograph P3.5/5).

The manner in which pore structure may be distributed in various parts of the same unit of material, different units of the same material or different materials used in a composite construction will also influence, other things being equal, the position in which salts may be deposited. Mortar and bricks provide a good example for explanation. If either the *mortar* or the *pointing* (the latter is extremely significant, as it represents the drying surface) is dense, drying out at the mortar joints will be either prevented or restricted. Drying out of the wall, therefore, takes place mostly through the bricks. Consequently, soluble salts will tend to concentrate on the bricks. The lessons to be learnt from this are twofold. First, pointing should not be denser than the mortar and, second, the mortar should, ideally, match the porosity of the units it is joining. This is yet another reason for avoiding, wherever possible, the

Photograph P3.5/5 *Variation in the pore structure of materials (individual units, the jointing medium or an element as a whole) often accounts for the rather haphazard distribution of efflorescence. In (a) the efflorescence is mainly on the surface and in (b) mostly in the mortar joints*

use of mortars which are stronger than the units being joined. It should be noted that, where the building unit is impermeable, salts may be deposited on the mortar joint if this is porous and permeable.

4 The effects of crystal form of the salts

The readiness with which efflorecence forms depends on the shape of the crystals of the salt or mixture of salts that is deposited from solution. If the first small crystals that form in the surface pores are deposited in such a pattern that still finer capillary passages are left between them through which solution can be drawn, then new crystals can form on top of the old. Salts will, however, be discouraged from growing outwards if the angles between the faces of the salt crystals are such that adjacent crystals fit together and leave no spaces.

Although this complex and difficult field has not been fully explored, it is known that sodium sulphate forms efflorescences much more readily than potassium sulphate. The latter may be more readily formed in the presence of other salts, while even small amounts of sodium sulphate will effloresce readily. Magnesium sulphate also effloresces almost as readily as sodium sulphate.

5 The effects of hygroscopicity of salts

The alkalis (sodium hydroxide and potassium hydroxide) left in solution when Portland cement is mixed with water are not so likely to form efflorescence because they are hygroscopic. Their greater affinity for water means that they can pick up water from the air. Consequently, they are unlikely to be dried to the point of appearing as efflorescence at normal outdoor temperature. However, after carbonation, sodium carbonate may appear in efflorescences, but even this is generally unlikely, as it is usually mixed with potassium carbonate, which is also hygroscopic.

6 The effects of precipitation

Although slightly outside the scope of this present discussion, it may be noted here that staining on plaster caused by soluble iron salts in the bricks does not take place if the iron is precipitated. Undercoats containing lime and/or Portland cement precipitate the iron and do not allow it to pass through to the surface with any water there may be in the wall as the latter dries. The same is not, however, true of gypsum plasters, as these do not have alkaline constituents capable of precipitating the iron. In addition, the pore structure of gypsum plasters favours the passage of salts to the surface.

7 The effects of the distribution of salts in a material

Analyses of salt content are more commonly, though not entirely, carried out on bricks rather than other porous materials. Such analyses usually report on the amount of salt present in an average sample of powder obtained by taking drillings from a number of bricks, or by crushing pieces cut from them. Such analyses are adequate for most purposes, particularly as with most, though not all, bricks the salts are fairly uniformly distributed. Nevertheless, it is the exceptions which prove the rule, and sometimes salts may be concentrated close to the surface. In cases such as these, an average analysis would not show the quantity of salts that is *readily* available to form efflorescences.

Bricks likely to have high concentrations of salts near the surface are usually those made from a white – or yellow – burning calcareous clay. Even with bricks made from this type of clay only those (usually only a few, in any case) fired near the grates or charging shafts of a kiln, where they may absorb sulphur gases from the burning coal, are likely to be affected.

8 The effects of the transfer of water in a material

Although pore structure, including the manner in which the pores are distributed in a material, influence the manner in which, or the rate at which, water is transferred in and out of a porous material, a discussion of some of the basic principles of the transfer of water without referring to the precise pore structure can explain the reasons why efflorescence is so common in new work and also why it may recur. Account must, of course, also be taken of the other factors which have already been outlined. For the present purposes a brick wall is considered, but reference is also made, for convenience, to 'special' effects of the wetting and drying cycle on roofing tiles and sandstones.

(a) Absorption during laying

Efflorescence which forms round the edges of bricks suggests that, during laying, the brick has absorbed water from a mortar containing Portland cement, thus also absorbing alkali sulphates before they can be converted into insoluble compounds, as previously explained (see 'Sources of salts, 1 (e) (2) Portland cement', p. 360).

(b) Exposure to rainwater

Assuming there has been no absorption of water from a Portland cement-based mortar by the bricks during actual laying, the alkali sulphates may still be transferred to the bricks if an unfinished wall is left uncovered during heavy rain. Rainwater filtering down through the top of the wall will effect the transfer. 'Evidence' of heavy rain on an unprotected wall may be recognized, many weeks later, by a band of efflorescence on the affected courses (see Photograph P3.5/6).

(c) Construction water

Apart from local saturation as described in (b) above, sufficient water may be introduced during construction (see, particularly, *3.2 Exposure*, p. 200 ff), to bring out an efflorescence. Initially, the water is uniformly distributed through the wall. Drying out, however, takes place mainly through the most favourably exposed face. (This is usually the external face, but it may sometimes be the internal face.) Consequently, there is a general movement of water, carrying in solution (not necessarily saturated) any salts that may be present towards the exposed face of the wall. This explains why efflorescence is most likely to occur during the early life of a building.

(d) Wetting and drying cycles

(1) Temporary disappearance
The fact that deposits of salts have disappeared from the surface of a wall does not necessarily mean that they have been washed away by rain (some may have been removed by wind), but rather that they have

Photograph P3.5/6 *The areas of concentrate efflorescence are indicative of those parts of the top of the external leaf that were not suitably protected during construction*

been washed back into the wall, where they may remain or from which they may be subsequently (usually months after being absorbed by the wall) 'extracted' to reappear as efflorescence.

(2) Re-absorption of salts

Rain falling on a wall covered with efflorescence soon dissolves the soluble salts which will be re-absorbed into the wall. The dissolved salts will continue to penetrate further into the wall as long as the rain persists. However, whereas the initial solution of dissolved salts will be highly concentrated, subsequent solutions which are absorbed by the wall will have progressively lower concentrations of salt. As the rainwater continues to flow over the exposed surface of the wall there is less salt for it to dissolve. Consequently, a stage is reached, probably soon after the initial fall of rain, when the salt solution is followed by 'pure' rainwater. (The term '(pure' is used in a relative sense to differentiate it from the salt solution. It is more conveniently referred to hereafter as water.) In the narrow pores of the bricks, mixing of the salt solution and the water that follows it will be very slow. Once the rain has ceased and the wall commences to dry out, it is quite probable that drying will reach the stage where the surface pores, which contained the water rather than the solution, are empty before any salts have been brought back near the surface. This produces a state of equilibrium in which the washing of salts into the wall by rain normally compensates for any tendency that they may have to come out again in dry weather.

(3) Disturbance of equilibrium and reappearance

The fact that efflorescence may reappear in the second spring of a building's life may be explined partly by two factors which favour efflorescence, namely, the lower temperature and slower evaporation of spring as compared with summer and partly by a disturbance of the equilibrium already described. Disturbance of the equilibrium can occur during the winter, when the surface of wall may remain wet for long enough to allow any salts previously washed into the wall to diffuse back towards the surface. Fortunately, any efflorescence formed during the second spring will be less prominent than the first outbreak. This is due to the fact that the wall will usually be drier, and consequently there is less water in it to move towards the surface, carrying salts with it.

(4) Clay roofing tiles

Clay roofing tiles present a special problem. In a tiled roof, for example, the tail ends are exposed (wet) while the heads, which are fixed to the roof structure, are sheltered (dry). Consequently, there is a tendency for water to creep up the tiles and evaporate at the heads. Salts will gradually be transferred to the heads where they crystallize out (**high>low**).

(5) Sandstones and 'contour scaling'

Some kinds of sandstones (occasionally the less durable limestones) develop what has been termed 'contour scaling' – the face of the stone breaks away at a depth of perhaps 6 mm or more, following the contours of the surface, plane or curved, in sills, mouldings, etc. This is attributed to repeated wetting and drying of the stone by rain, known as the 'moisture rhythm' in which soluble matter is transferred from the body of the stone to its surface. During the process the exposed surfaces remain outwardly sound while a narrow zone at a more or less uniform distance behind the face becomes soft and friable until the face eventually breaks off.

This form of weathering is explained as follows. Rainwater falling on the surface penetrates to a limited but more or less constant depth. A small fraction of the constituents of the stone within the depth of penetration is dissolved by the rainwater. Any soluble matter in solution is deposited at or near the evaporating surfaces as the stone dries out. Repetition of the cycle tends to consolidate the outer zone at the expense of cementitious matter in the underlying zone, which eventually loses cohesion. Although even silica can be dissolved and transported as described, calcium sulphate, which is commonly present in the scales and in the friable zone, also plays a part. The presence of calcium sulphate, even when there are no obvious external sources, underlines the importance of decomposition of materials due to the action of acid gases in the atmosphere.

Lime staining

1 Background

Hitherto it has not been uncommon, as described earlier, for lime to be leached from the exposed surfaces of limestones or concrete and then deposited on brick and other surfaces below. The white deposit is the result of chemical actions that cause the decomposition of the surface (see p. 261–2). Today it has become increasingly common for a white deposit to form on brickwork, initially at the mortar joints and in severe cases later spreading to cover the bricks as well (see Photographs P3.5/7 and P3.5/8). This phenomenon is often mistaken for efflorescence. It shares some of the processes involved in efflorescence, the effects of the transfer of water in a material in particular (p. 365 ff), but differs in that free lime, leached from cement- and lime-based products within a construction is brought to the exposed surface, where it is converted to an insoluble form of lime. Unlike efflorescence, it is not temporary and, importantly, is difficult to remove. Another manifestation of the phenomenon occurs in reinforced concrete slabs where the lime leaches out at exposed cracks or day joints. The saturated lime usually forms small stalactites (see Photograph P3.5/9), the dripping from which stains surfaces below them.

2 Mechanism

(a) Reactions

The mechanism involved is relatively simple. All cement-based products contain free lime that is soluble in water. Water that finds its way through mortars or concrete takes free lime with it as calcium hydroxide to the exposed surface, where carbon dioxide in the atmosphere reacts with it to form insoluble calcium carbonate.*

(b) Transfer of water

Lime staining only occurs when water is transferred through the cement-based product. The basic mechanism is the same for all products (see *3.3 Moisture content*, p. 230). However, transfer through brickwork, claddings and concrete needs separate consideration.

(1) Brickwork

By its nature and function, the outer leaf of brickwork in a cavity wall is inherently prone to water leaking through the mortar joints to the

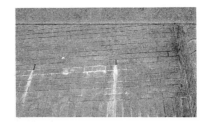

Photograph P3.5/7 *An example of lime staining of new brickwork (external leaf of cavity wall) soon after completion. This looks like efflorescence but closer examination revealed otherwise. Note how lime has found its way preferentially at the weepholes. Compare with Photograph P3.5/8*

Photograph P3.5/8 *Another example of lime staining. In this case it is intense and confined mostly to the surface of the (hard) bricks. The path of the lime is through the softer mortar*

*The changing of calcium hydroxide to calcium carbonate by carbonic acid (i.e. carbon dioxide dissolved in water) also occurs in concrete exposed to the atmosphere. This process, known as carbonation, proceeds from the exterior to the interior, and is relevant in problems associated with the corrosion of steel reinforcement (see *3.7 Corrosion*, 'Protection, 6 Concrete', p. 524).

Photograph P3.5/9 *Regular dripping of lime from reinforced concrete roof slab causing stalactites to form. Excess dripping caused lime staining of vitreous enamel mosaic cladding below (not in photograph)*

cavity side. Water may, of course, also find its way down the perpends onto the bed joints. The amount of water that may travel by either route depends basically on the absorption of the brick (or other masonry unit), apart, that is, from the exposure of the masonry to rainwater. Generally, the less absorbent the unit, the greater the amount of water that will travel through the joints. Perforated bricks, by their nature, have built-in reservoirs in which water may collect. The base of these reservoirs is the bed joint into which water may subsequently be transferred. It appears, therefore, that perforated bricks of low absorption have a greater potential for causing lime staining.

(2) Claddings

Water may also be transferred through joints in brick or stone claddings to concrete backgrounds, as described for brickwork above. In addition, water finding its way through the joints may then be transferred to any mortar backing that may be used or onto the concrete background itself. The greatest amount of staining occurs if the backing mortar is of poor quality because water is transferred through it more easily.

(3) Concrete

Examples of water transfer through concrete are more commonly associated with roofs or balconies where there has been a breakdown of the waterproofing and/or weatherproofing (notably perimeter skirtings). Water that penetrates the water-/weatherproofing then travels along the surface of the concrete, finding its way out at cracks or dayjoints.

3 Precautions

(1) Brickwork

The most intense lime staining occurs during the drying-out period, when the tops of unfinished brickwork have not been properly protected. The primary aim of any precautions should therefore be the protection of work from rainwater penetration during construction. All water-excluding details must ensure that the brickwork cannot become saturated during its life. The mortar joints should be properly filled and the mortar should be of good quality.

Removal of lime staining is not an easy matter. Some form of chemical treatment is necessary, but this should not be carried out without first testing a small area of the brickwork (see 'Removal of efflorescence', p. 375).

(2) Cladding

Protection of the tops of unfinished work is an essential precaution during construction. All water-excluding details must ensure that the cladding and its background cannot become saturated during its life. The coating of the concrete background with a waterproofing compound should help to reduce the amount of staining significantly. It will not prevent staining arising from water flowing through or out of the joints.

(3) Concrete

All water-excluding details must ensure that water cannot penetrate to the top surface of the concrete. The amount of staining may be reduced when such penetration occurs, provided the concrete is well compacted and no cracks develop.

Precautions

1 Generally

The primary aim of precautions in relation to efflorescence is to reduce the amount of the deposit and/or the deleterious effects of soluble salts. In the light of all the variable and interactive factors involved, it is seldom possible in practice to avoid efflorescence completely. To reduce the risk of the phenomenon occurring, the precautions that should be taken are similar or closely related to those concerning moisture movements (3.3 *Moisture content*), chemical actions (3.6 *Chemical attack*), corrosion (3.7 *Corrosion*) and frost action (3.8 *Frost action*).* As in all building work, the interrelationship and interaction of what may appear to be self-contained aspects must not be overlooked.

Although the precautions necessary to reduce the risk of efflorescence occurring may differ in detail from the other aspects described above, all share one fundamental and important principle, namely, that materials and certain parts of constructions must be prevented from becoming damp or saturated for excessively long periods of time. In relation to efflorescence specifically, the precautions that should be taken are related to: the selection of materials; details of design; handling, storage and protection of materials; drying out before the application of finishes; and removal of efflorescence.

2 Selection of materials

Care in the selection of materials regarding either their soluble salt content or their liability to form efflorescence should, in the first instance, be related to all the component parts of a construction. Thus in brickwork, for example, both the bricks and the mortar need consideration; as regards the mortar (plaster, renders or concretes are also relevant), the aggregates, cementing agents and the water must be included.

The choice of suitable materials for a particular condition must take into account conditions of exposure. Once this has been established, either by measurement or experience, the limits of soluble salt content or liability to effloresce may be established. This is particularly true of *clay bricks* which would, in any case, contain some salts. Under severe conditions of exposure, materials with a high soluble-salts content should be avoided. In this context, the amounts of magnesium sulphate, particularly in some types of clay bricks, are significant.

Most useful information on the performance of particular materials under particular conditions (both are important) may usually be obtained by careful inquiry. Evidence of quality should be sought from either manufacturers or suppliers. In the case of bricks and stone their reputation in the district from which they originate or in which they are predominantly used may prove to be useful sources of information. However, account should be taken of any probable effects of changes in environment or methods of use. In the case of clay bricks, *Clay Bricks of the United Kingdom*, by D. R. G. Bonnell and B. Butterworth, National Brick Advisory Council Paper Five, HMSO, London, 1950, is an invaluable reference, despite its age. Tests to BS 3921 could provide helpful results.

As regards the *sedimentary stones* it is useful to note that, in general, most of these are unlikely to contain significant, if any, soluble salts. Some types, particularly some qualities, may be the exception. With *limestones*, their susceptibility to deterioration due to crystallization of soluble salts is the property which is most important in distinguishing a durable stone.

*The relevant precautions are given under the specific material covered in each chapter. Hence no references to the pages.

Specification of *mortar* mixes requires care. Their most important property is that they should be the weakest that conditions allow (see *3.8 Frost action*, p. 553). In this, account should be taken of the fact that mortars containing high-calcium limes are unlikely to contribute to soluble salt content of a construction. The soluble salt content is, however, likely to rise as the hydraulicity of the lime increases, but more important, Portland cement in the mortar is likely to be the most significant contributor of soluble salts. Similar considerations are required with plaster and renders. Tests of various mortar mixes for their liability to cause efflorescence or staining may often be advisable with stonework.

3 Details of design

Details of design are primarily concerned with methods which aim to exclude as much water as possible (dpcs and other water-excluding features are significant). However, an equally important consideration is the relationship of different materials in the same construction.

As regards exclusion of water, special care is required with those parts which are likely to be more heavily exposed than others.* It is important that the body of a construction is prevented from becoming saturated by water percolating through horizontal features. *Parapets* are notable as points of entry. Accordingly, these should incorporate appropriate dpcs or flashings. Other projecting horizontal surfaces may require weatherings and flashings, while even rainwater outlets need suitable detailing. It is probably axiomatic that care in the design of dpcs and dpms in walls and floors is necessary so as to limit the rise of salts from the soil to a predetermined (i.e. designed) level.

It is always as well to remember the *weakness of joints* between units. These, as illustrated in *3.1 General considerations* and *3.4 Exclusion*, are notorious as entry points for water into the body of a construction. With large masses of materials (concrete is notable) cracks which may develop may become troublesome.

The relationship of different materials in the same construction may occur in basically two different ways, namely, when one is used as a backing for the other or where they are combined in the same facade in juxtaposition or at different levels.

Problems with backing materials occur mainly, though not entirely, with *sedimentary stones*. A common practice is to coat either the surface of the backing material or the concealed surface of the stone or both with a waterproofing compound. Although generally successful, it should be noted that water and hence salts (or other matter) may penetrate at the joints. A similar consideration needs to be applied to impermeable units having porous, permeable joints. Limestone or cast stone dressings, string courses, etc., used with certain types of clay bricks (usually the soft porous varieties), should be avoided unless experience has shown that the bricks are resistant to damage from crystallization. However, it should be noted that changes in appearance, rather than damage as such, from lime washings can also take place on other materials.

4 Handling, storage and protection of materials

Contamination of materials can take place very easily either during handling or storage, particularly on a building site, although the works or the yards of manufacturers and builders' merchants cannot be

*These locations are similar to those for possible frost action – see Diagram D3.8/8, p. 551.

exempt. Care should always be taken to ensure that incompatible materials are not stored in contact with one another, particularly under damp conditions. In general, materials should not be stacked or stored in contact with the ground. Some form of raised platform of dpm material (polythene sheet has proved invaluable for this purpose) should preferably be used.

Materials should be protected from becoming saturated by rainwater while stacked or stored on the site and later when they have been incorporated into a construction.

5 Drying out

Some applications, such as adhesives and paints, cannot be successfully applied either on damp surfaces or on surfaces which have deposits of soluble salts. The latter should always be removed (see '6 Removal of efflorescence') prior to the use of applications which rely on adhesion (see *2.2 Strength and use of materials*, 'Jointing, fixing and sealing', p. 95 ff). Apparently dry surfaces should be checked. (See *3.1 General considerations* under 'Dampness', p. 152.) If there is still sufficient water left in the body of the construction, efflorescence may still form. However, the salts are likely to be deposited under the adhesive or paint film causing loss of adhesion. On the other hand, salts in solution may attack the film. Some paints, for example, are particularly susceptible to alkaline attack – the alkalis usually derived from damp backgrounds containing Portland cement. Although some compositions are more resistant than others, the degree of dampness of the background is nevertheless important. (See *3.6 Chemical attack*, p. 392–3, for detailed discussion.)

6 Removal of efflorescence

The way efflorescence may be removed from a surface depends on the type of efflorescence and the nature of the surface.

(a) Brickwork

(1) White powdery deposit
There is a simple rule for the removal of white deposits from the surface of brickwork: *chemical methods should never be used*. On new brickwork the efflorescence should first be allowed to weather away naturally during the warmer, drier months.

The residual deposit may be removed by brushing with a stiff-bristle brush. The deposit should be collected as it is being brushed away, care being taken to ensure that it does not enter the brickwork at lower levels. Any deposit still remaining may be removed or reduced by using clean cold water. The washing down may cause the salts to be re-absorbed by the brickwork, as the salt is water soluble. The amount of re-absorption may be minimized by using a clean, damp sponge, rinsed frequently in clean, cold water. Where a deposit appears on old brickwork the cause of the efflorescence should be investigated and remedied before removing it as described above.

(2) Yellow or green efflorescence
Washing the brickwork with inorganic acid solutions often produces a dark-coloured stain. In the first instance the efflorescence should be allowed to weather away naturally.

If the weathering fails to remove the efflorescence or if it is pronounced, chemical methods should be tried but only after a small area has been tested. One of three solutions may be used: tetra-sodium salt of diaminoethane tetra-acetic acid, EDTA (50 g/l) and washing down afterwards; oxalic acid solution (20–35 g/l) followed by washing soda solution (12 g/l) after the stain has been bleached; and sodium hypochlorite or household bleach in concentrated solution followed by washing soda solution (12 g/l) after the stain has been bleached. In the last two the washing soda solution is left on the wall.

(3) Iron staining

This type of staining occurs normally on the surface of the joint. Where severe staining takes place on the brickwork it is best left to weather away naturally. Removal of the stain from the mortar joint is best achieved by scraping or rubbing with a round file or carborundum slip while the mortar is relatively weak. Where this is ineffective, trials are needed to test which of five alternative chemical treatments will be the most effective.

(4) Manganese staining

Generally, as for iron staining above.

(b) Plasters and renders

Plasters and renders to receive a paint finish should be allowed to dry out and any deposit removed by brushing only.

(c) Stonework

The care which should be exercised in the cleaning of stonework may be noted here. In addition to the normal soot and grime deposits found on many stone buildings in polluted atmospheres, the sedimentary stones (usually limestones) may have areas of a hard, impervious skin (usually forms in sheltered areas). This skin on the stone, particularly when it is covered with soot or grime (as often happens), increases the difficulties of cleaning and sometimes encourages methods (mainly chemicals and detergents) which are likely to do more harm than good to the stone after it has been cleaned. As already noted, earlier efflorescence or staining may result. However laborious the process may be, the use of clean water, either as steam or water spray, is still regarded as the most effective method of cleaning stonework. Because of the solubility of the encrusted zones (usually calcium sulphate or calcium carbonate) water soon softens the deposits sufficiently for removal by light brushing. Vigorous brushing of delicate mouldings can be damaging. Effective cleaning of stonework is a specialized job and should, therefore, only be undertaken by established and reputable firms. Their experience will enable them to use methods which will not be deterimental to the stone. Above all, most experienced firms are prepared to guarantee that no caustic soda, soda ash or other harmful chemical will be used.

Sources

Butterworth, B., *Bricks and modern research*, Crosby Lockwood, 1948
Lea, F. M., *The Chemistry of Cement and Concrete*, revised edition (of Lea and Desch), Edward Arnold, London, 1956

BRE
Principles of Modern Building, Vol. 1, 3rd edn, HMSO, London, 1959
Butterworth, B., 'Efflorescence and the staining of brickwork', Article reprinted from the *Brick Bulletin* and amended December 1962

Current papers
Eldridge, H. J., 'Concrete floors on shale hardcore', Building Research Current Paper, Design series 30, 1964

Digests
(1st series) 20 and 21, *The weathering, preservation and maintenance of natural stone* (Parts I and II), July 1950 (revised March 1965) and August 1950
160, *Mortars for bricklaying*, December 1973
164 and 165, *Clay brickwork: 1 and 2*, Minor revisions 1980 and May 1974
269, *The selection of natural building stone*, January 1983
273, *Perforated clay bricks*, May 1983
276, *Hardcore*, August 1983
280, *Cleaning external surfaces of buildings*, December 1983

National Building Studies
'Some common defects in brickwork', National Building Studies, Bulletin No. 9, HMSO, London, 1950

Reports
Schaffer, R., *The weathering of natural building stones*, Building Research Special Report No. 18, HMSO, London, 1932 (facsimile reprint 1972 with a new Appendix II on the cleaning of external building surfaces and colourless treatments for masonry)

Brick Development Association
Harding, J. R. and Smith, R. A., *Cleaning of brickwork*, BDA Building Note 2, by R. G. D. Brown (ed.), September 1982
Harding, J. R. and Smith, R. A., *Brickwork durability*, BDA Design Note 7, August 1983

BSI
BS 3921: 1985, *British Standard Specification for clay bricks*
BS 5628: 1985, *British Standard code of practice for use of masonry*, Part 3, *Materials and components, design and workmanship*

Chemical attack

Background

1 Scope

(a) Generally

Chemical actions play an important part in two different phases of a building: first, during its construction in which off-site manufacturing processes should be included and, second, throughout its life. During construction, the chemical actions are necessary to enable forms of construction to be achieved; in the life of a building they are almost always destructive. It is the latter aspect of chemical actions that are considered here, hence the term 'chemical attack'.

(b) Primary agencies for attack

It is usually helpful to have in mind the more important or primary agencies responsible for the relevant chemical reactions, and many are involved. Some are of the conventional kind involving recognizable chemicals;* others concern biological sources whose 'reactions' are different (if they have reactions at all); and a third group relates to a more or less invisible source – solar radiation. For convenience, all, except corrosion,† are included here under the following three groups:

1. Chemical;
2. Biological; and
3. Ultraviolet radiation.

The first two are characterized by the need for moisture to be present for chemical reactions to take place or to facilitate the reactions; the third by the need for exposure to sunlight.

(c) Coverage

The three primary agencies listed above are included separately in this chapter but under the following subdivisions:

Chemical
- Unsound materials (i.e. inherently unstable in the presence of moisture)
- Attack by *acids*
- Attack by *alkalis*
- Attack by *sulphates*

Biological
- Fungi and insects
- Wood-rots

*Later the term 'chemical' is used for convenience to signify this group of primary agencies.

†Corrosion, purposely excluded from the primary agencies, is dealt with in detail in *3.7 Corrosion*.

*Ultraviolet radiation**

In each case consideration is given to the basic chemical mechanisms involved, the related influencing factors and the precautions that should be taken. In almost all cases, reference is made to closely related aspects in *3.2 Exposure*, *3.3 Moisture content* and *3.4 Exclusion*.

2 Effects of changes in materials and constructions

(a) Shift in emphasis

For the reasons given earlier in the *Introduction*, 'Changes and their effects' (p. 4), the problems associated with chemical attack have shifted in emphasis. Obviously, the nature of the chemical reactions or mechanisms involved have not changed. Some of the traditional problems such as those associated with unsound materials and, to some extent, acid attack have diminished in importance in that they appear not to occur as frequently as they did. Nevertheless, these problems still need to be addressed.

(b) Changes in exposure

Adding significant amounts of insulation in constructions has created many physical problems, as discussed in the earlier parts of this book. As importantly, their inclusion in forms of constructions that are markedly different from those encountered in practice hitherto has led to greater exposure of materials in constructions to moisture from surface and interstitial condensation, the latter in particular. Some materials, such as concrete, are used far more often, so problems associated with them become more common. Looked at in another way, concrete is more widely exposed. New materials, such as the plastics, present chemical problems that take on a new significance.

(c) Examples of changes

The following examples should assist in focusing on the effects of some of the changes that have taken place over the last ten years or so:

- Timber used in roof constructions were mostly not at much risk of decay in the past. Now that risk can be significant. There is therefore a greater need for an understanding of preservative processes, among other factors.
- Changes in the procurement and seasoning of timber and in joinery practice has meant that timber windows and doors are much at risk of decay.
- Changes in forms of construction, forms of heating and living habits in domestic buildings gave rise to what could be called an epidemic of mould growth during the 1950s and 1960s, due mainly to the combined effects of intense surface condensation and high internal air humidities.
- The chemistry associated with alkali-reactive aggregates has been known for a long time – at least since 1956. Then it was thought to be a problem associated only with some aggregates found in America, Australia and New Zealand. Common British aggregates did not appear to have alkali-reactive constituents.† With the greater use of concrete, aggregates from sources other than those that were common have been used. Some of these have contained alkali-reactive constituents, and concrete in which they were used has been

*Ultraviolet radiation is covered in detail in *Materials for Building*, Vol. 4, pp. 148–150. For completeness, that coverage is reproduced later.

†Alkali-reactive aggregates were included in *Materials for Building*, Vol. 3, p. 25, in a single paragraph. There mention was made of a survey of common British aggregates quoted by Lea which failed to reveal any containing alkali-reactive constituents.

attacked, usually in the form of cracking – referred to sometimes as 'concrete cancer'. Earlier knowledge on the subject has had to be reread and rehearsed.

- The greater use of cement in preference to lime in mortars for brickwork has helped to focus more attention of the risks of sulphate attack, the mechanisms of which have long been known. The same attention has also been focused on concrete, a cement-based product, because of its greater use in building.
- Plastics in various forms have brought with them problems of site application, mostly concerning adhesion. After their application there have been difficulties associated with polymer migration and the deleterious effects of exposure to ultraviolet radiation.*

(d) Emerging problems and knowledge

Added to these known problems are apparently emerging difficulties associated with traditional materials such as stone that is now being used in new ways, in thin sections in particular. The chemical performance of stone as used in thick sections is well documented. What is likely to happen chemically or physically to thin sections is therefore unknown. The same may apply to other traditional materials used in new ways. Accordingly, there is a need for interpretation of present knowledge of the **creative pessimism** kind. This chapter explains that knowledge in relation to chemical attack.

3 Variations and their effects

(a) Influencing factors generally

The agencies responsible for chemical attack vary from material to material and from one location to another in a construction or in a particular part of the country. Whether attack is likely to occur depends on the chemical properties of a material, its exposure to dampness or sunlight and the nature and composition of the particular agency within or adjacent to a material. The source of the agency is obviously important. In some cases it will be derived from user requirements (e.g. in factories using chemicals or in maintenance employing certain cleaning compounds) and in others from what might be termed naturally occurring sources. In all cases the aggressiveness of the agencies will be important and the rate at which the reactions take place will also be influential. In this, the effects of physical changes (such as movements and frost action) that may result in cracking or other chemical changes (such as carbonation of concrete) may initiate or increase chemical attack.

(b) Rate of attack

It is important to note that chemical reactions do not necessarily take place instantaneously, while the rate at which reactions proceed depends on the concentration of the 'chemicals' involved, their continued 'supply' and the length of time during which the relative constituents are 'in contact'. In building, chemical reactions, including those of a biological kind, that have a destructive effect take place comparatively slowly. Thus, other things being equal, short periods of exposure to dampness or sunlight generally have little effect as far as chemical attack is concerned.

(c) Chemical agencies

A distinction can be drawn between the basic difference in the consequences of chemical agencies on porous and non-porous

*In addition, that is, to problems associated with the environment and the 'greenhouse effect'.

materials. In the latter, as in corrosion for example, the attack is first confined to the surface of a material because the agency for the attack is external to the material. It may, of course, proceed towards the interior but progress is normally fairly slow. This means the use of protective coatings to control the attack (apart, that is, from using a material that is in any case less prone to the attack). (Surface coatings and related matters are discussed in detail in *3.7 Corrosion*, p. 508 ff.)

In the case of porous materials the source of the agency may be external to the material or within it, or both. Attack may therefore be confined to the surface or may occur within a material. This means that the control of the attack does not rely entirely on the use of protective coatings, although in some cases these may be invaluable. It also means that full account must be taken of the porosity of any given material and all the aspects that this influences (e.g. absorption, degree of permeability, transfer of moisture and wetting/drying cycles – see *3.3 Moisture content*, 'General influences', p. 230 ff). In nearly all cases, circumstances that are likely to lead to prolonged wetting (i.e. at or near saturation) should be avoided. If they cannot, as might occur in exposed parts of a construction (e.g. certain parts of walls or earth-retaining structures), then precautions must be taken according-ly. These may include the use of alternative materials.

(d) Biological agencies

Apart from insect attack (included here for convenience), the fungi involved share the same life cycle (and other characteristics). They require moisture (among other ingredients) to start and (mostly) to continue growing. Mould growth is characterized by being confined to surfaces (of building materials and most contents of buildings); timber decay by starting at the surface and then progressing inwards. The spores for both are 'freely' available in the atmosphere, simply waiting for the appropriate conditions to exist for growth.

The most common causes for mould growth are condensation or excessively high internal air humidity for long periods of time (or combinations of both). It is these closely related problems that must be addressed in design. Fungicides applied to surfaces likely to be affected may help to prevent growth in mild cases of dampness but may only delay growth in severe cases. The problem has to be tackled at source (i.e. by controlling the amount of condensation and/or high internal air humidities).

In principle, the moisture content of timber should be kept below 20% if the risk of its decay is to be reduced to a minimum. To achieve this involves ensuring that the environment surrounding the timber is maintained to obtain this moisture equilibrium or, in the case of joinery, checking through design that water is not allowed to become entrapped in joints. In practice it may not always be possible to achieve the ideal conditions for a sufficiently short period of time. Consequent-ly, **creative pessimism**, if not practical reality, cautions the use of protective measures in the form of preservative treatments and/or protective coatings. For different reasons, preservative treatment of timber is recommended where insect attack is likely.

(e) Ultraviolet radiation

The deleterious effects of sunlight on organic materials, such as liquid applied coatings (paint), plastics, sealants and roofing materials but excluding wood, are included. Emphasis is given to the influencing factors rather than the precautions.

Unsound materials

1 Meaning and scope

(a) Meaning

The use of the term 'unsound' in the present content relates specifically to those materials which are unstable when exposed to damp or humid atmospheric conditions. The defects which arise when such materials are exposed, and included here, are due to expansion that accompanies the chemical reaction when an expansive particle and moisture combine. The fact that both must be present for a defect to occur is important when considering either prevention or cure. The defects which do occur are mainly, though not entirely, associated with unslaked lime (quicklime, i.e. calcium oxide) used in plaster or mortars or found in some clay bricks. Other materials which may be unsound include sand (may contain certain impurities) but more particularly ashes and clinker used either for mortars or for making clinker concrete which may contain unburnt particles of coal or other impurities. The effects of these impurities, together with others which may be derived from failure to ensure cleanliness of all materials used, are similar to those produced by quicklime.

(b) Scope

Emphasis is placed here on the *delayed hydration* of quicklime, as it probably has been the more common cause of failures.* However, brief references are made to other relevant factors.

2 Basic mechanisms

Lime used in building practice is derived essentially from calcium carbonate ($CaCO_3$) – either limestone or chalk may be the raw material – which is calcined (i.e. heated) to produce calcium oxide (CaO), commonly known as quicklime. During calcination the following chemical reaction takes place:

$$CaCO_3 \rightarrow CaO + CO_2$$

(calcium carbonate) $\rightarrow$ (calcium oxide (quicklime))   (carbon dioxide)

Before the quicklime can be used it must first be slaked (a process of hydration) to produce calcium hydroxide ($Ca(OH)_2$) as follows:

$$CaO + H_2O \rightleftharpoons Ca(OH)_2$$

(calcium oxide (quicklime)) + (water) $\rightleftharpoons$ (calcium hydroxide)

Within the present context the calcium hydroxide is the sound material and the quicklime the unsound one. During the slaking process, expansion takes place in addition to the evolution of heat.

In use, if slaked lime contains particles of quicklime, and if those particles are able to hydrate in the set or solid material containing them, the resultant expansion may cause defects. Generally, delayed hydration of quicklime in materials in the plastic state is unlikely to be deleterious, as the expansion can be accommodated before the material sets or hardens. It may be noted here that delayed hydration of quicklime can, and does, take place with moisture absorbed from the atmosphere. This type of slaking is known as 'air-slaking'. Generally, there is greater danger of deleterious effects of expansion caused by air slaking than by immersion.

*Some of the techniques described later (such as the site slaking of lime) are not as common as they were. They are included nevertheless for completeness.

3 Plasters and mortars

Although the manifestations of defects due to the presence of unsound materials are not the same in plasters and mortars, it is convenient to group them together, as they share some of the basic causes of defects, chief of which is the delayed hydration of quicklime.

(a) Slaking of quicklime

Quicklime may be run to a putty in a pit or container. This is an operation commonly carried out on a building site. On the other hand, the quicklime may be treated by a steaming process in a factory, in which case a dry-hydrated lime is made available ready for use. In general, the latter is less likely to result in defects due to the delayed hydration of quicklime.

Among *precautions* which should be taken to minimize the risk of subsequent expansion of quicklime is the avoidance of the use of the bottom layer of material in the pit in which quicklime has been run to a putty. This layer contains the larger particles which are most likely to be unsound. When dry-hydrated lime is used it is still recommended that it should be soaked overnight before use. The main object of the soaking treatment is to improve the working properties of the lime, but there is the added advantage that the soaking will reduce the possibility of the expansion of any unhydrated particles which may be present.

(b) Impurities in aggregates

- *Sand* may contain coal dust from soft lignite and some bituminous coals, or iron pyrites, both of which may oxidize. The oxidation results in expansion.
- *Ashes and clinker* may contain unburnt particles or calcium oxide, both of which may lead to expansion when moisture is absorbed. Plaster failures have, in some cases, been traced to the presence of calcium oxide in the clinker used in either concrete blocks or *in-situ* concrete over which the plaster was applied.
- *Dirt, dust and other impurities* which may be present due to failure to ensure cleanliness may also result in expansions large enough to cause failures.

(c) Manifestations of defects: plasters

The terms 'blowing', 'popping' and 'pitting' are used when describing the appearance of a particular type of defect in an internal plaster finish due to the delayed hydration of quicklime (see Photographs P3.6/1 and P3.6/2), but see (e) later for the effects of magnesium sulphate. Conical holes usually appear in the finished work, varying in diameter according to the position in which the material responsible for the defect occurs. When the material responsible for the defect is confined to the finishing coat, the individual blemishes are usually not larger than 13 mm in diameter, often considerably less, but these may increase to as much as 50–75 mm if unsound material occurs in the undercoat. The formation of the conical holes invariably follows the following sequence of events: a bulge first appears on the plaster face, a fine crack tending to be circular in form then develops, the plaster inside the circle lifts, and finally drops out leaving the conical hole.

Photograph P3.6/1 *Blowing of a finishing coat of plaster (scale three-quarters full size). (From BRS Digest No. 26 (First series), Blowing, popping or pitting of internal plaster, January 1951, Fig. 1, Building Research Establishment, Crown Copyright)*

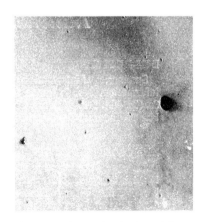

Photograph P3.6/2 *Blowing of plaster caused by unsound material in the background. (From BRS Digest No. 26, Fig. 1, Building Research Establishment, Crown Copyright)*

(d) Manifestations of defects: mortars

Defects may show one or more of the following symptoms:

1. Small pits, with nodules of friable material at their bases, form in the mortar joint.
2. Strong pointing mortars are displaced, and pits develop in the weaker bedding mortar.
3. General expansion occurs, with deformation and consequent cracking of the brickwork, accompanied by disintegration of the mortar. This effect is similar to that caused by sulphate attack.

(e) Presence of magnesium sulphate

A type of failure which is associated with lime-based, i.e. alkaline, plasters (neutral or acid plasters, such as calcium sulphate, are not affected) and which, fortunately does not occur, often is the loss of adhesion of plaster due to the formation of needle-like crystals of magnesium sulphate *below* the surface of the plaster. The formation of the crystals is associated with efflorescence, but the reason why the efflorescence occurs below the surface is attributed to the reaction of magnesium sulphate and the calcium hydroxide in the wet plaster. The explanation for this may be summarized as follows.

Magnesium sulphate in the bricks over which the lime-based plaster is applied dissolves in the water introduced during construction, and reacts chemically with slaked lime (calcium hydroxide) from the plaster, forming a gelatinous deposit of magnesium hydroxide:

$$MgSO_4 + Ca(OH)_2 \rightarrow CaSO_4 + Mg(OH)_2$$

The deposit forms below the surface of the brick because the surface pores of the latter are impregnated with lime during the application of the plaster. In addition, the deposit acts as a semi-permeable membrane, and so water can get through but not the salts. As the wall dries out, the salt accumulates against the membrane.

The reason the same type of failure does not occur with neutral or acid plasters, such as gypsum plaster, is because magnesium hydroxide is not precipitated. There is no lime in these plasters and consequently efflorescence forms in the normal way on the surface of the plaster. It follows, therefore, that bricks which are likely to cause loss of adhesion of lime-based plasters should preferably be plastered with calcium sulphate or other neutral or acid plasters. It may be noted that once the 'disease' has occurred, it is necessary to strip the defective plaster completely, while the wall must be plugged and battened before replastering (that is, the new plaster must not come into contact with the brickwork).

4 Bricks

Bricks made from calcareous clays may contain calcium oxide and have been observed to blow during the erection of brickwork. Experience has shown that hydration of the quicklime will usually occur at this stage rather than later, and is, therefore, unlikely to be the cause of blowing of plaster after it has been applied. Hydration does more commonly result in splitting of the bricks themselves.

The presence of calcium oxide in bricks occurs as a result of the conversion, during the burning process, of calcium carbonate which is intimately mixed with the clay. Failures, on the other hand, are more commonly associated with the presence of large lumps of calcium oxide which may be derived from limestone pebbles or fossils in the clay. If the lumps of quicklime are allowed to slake by absorbing moisture from the air they are liable to expand and split the brick.

Solutions aimed at preventing the bricks from splitting include:

1. Grinding the clay finely prior to the making of the bricks.
2. Immersing the bricks in water after burning, so as to slake the quicklime to a plastic mass that cannot exert any pressure on the brick. It should be emphasized that air slaking does not normally slake the quicklime to a plastic mass, and consequently results in splitting.*
3. By incorporating a small proportion (¼–1%) of common salt into the clay to promote chemical combination of the lime with the clay during burning.

Sources

BRS Digest (1st series) 26, *Blowing, popping or pitting of internal plaster,* January 1951
National Building Studies, Bulletin No. 9, *Some common defects in brickwork,* HMSO, London, 1950
Ministry of Works Advisory leaflet No. 15, *Sands for plasters, mortars and renderings,* HMSO, London, 1960
Butterworth, B., 'Efflorescence and staining of brickwork', reprint from the *Brick Bulletin,* amended December 1962, published by the National Federation of Clay Industries

Acids

1 Scope

The extent to which acids may have a deleterious effect on materials varies considerably. This is due not only to the variation in resistance which given materials may offer to attack but also to the variability of the concentration, and, as important, the sources of acids. It has, therefore, been considered advisable to deal primarily with those materials which have been commonly affected by atmospheric gases, i.e. stone, brick and cement-based products. These are discussed separately. In the case of cement-based products some consideration is given to the presence of certain acids which arise out of industrial processes.

2 Acid gases and stonework

The basic effects of the two more important acid gases, carbon dioxide and sulphur dioxide, on the carbonates present in the calcareous sedimentary stones (limestones and some sandstones are relevant), are outlined in *3.5 Efflorescence,* '2 Salts derived from decomposition of a material' (p. 361). In this section the significance of these effects on the more commonly used sedimentary stones is discussed, while a note on the weathering of slate, a metamorphic stone, is also included, as calcite, one of the minerals which occurs in slate, is readily attacked by atmospheric acids.

(a) Solubility

In principle, the fact that limestones are slightly soluble in water, and even more so in water containing carbon dioxide and sulphur, means that the surface of the stone, particularly in polluted atmospheres, is

---

*Bricks known to contain quicklime which may lead to splitting should preferably be immersed as soon as practicable after burning. If, however, this is not done, then it may be necessary to undertake the slaking on-site prior to the bricks being incorporated into a construction. This would constitute one of the exceptional cases when complete wetting of bricks may be justified. Nevertheless, it is important that they are allowed to dry out sufficiently before being incorporated into a construction to avoid problems associated with shrinkage and loss of adhesion.

kept clean, because the surface is *gradually* eroded, and so unsightly effects produced by sulphur fumes are washed away while soot (and other sticky deposits) is unable to gain a foothold. The washing effects are, therefore, *basically* beneficial. However, in practice such washing effects only take place on surfaces which are freely exposed to rain.

This, in addition to the variations which may be encountered in the physical structure of the stone, the non-uniformity of the surface and the effects of building design on the flow of water, results in the erosion being far from uniform (see *3.9 Flow and changes in appearance*, Photographs P3.9/1–6) and contributes to the so-called 'soot and whitewash' effect commonly seen at one time in towns and cities but gradually disappearing as the Clean Air Act bites (see *3.2 Exposure*, 'Atmosphere', p. 168 ff).

The basic effect of solubility on the fine-textured limestones with a uniform structure is for the surface to remain smooth; on the coarse-textured limestones the fossil fragments, which are more resistant to erosive effects than the surrounding material, stand out in low relief after a time, as a result of the differential erosion (see Photograph P3.6/3). The latter may also result in the exposure of the bedding of the stone, if the stone has hard and soft beds; this only happens occasionally. Carvings and sculpture, unless of good, frost-resistant limestone, tend in time to lose their finer detail and may assume a weather-torn appearance. This may also occur in plain sections (see Photograph P3.6/4).

In sheltered positions where rainwater is unable to erode the surface of the stone, sooty material not only accumulates on the surface but is also firmly bound to the surface with calcium sulphate and calcium carbonate (see '(b) Skin formation' below). Consequently, a marked contrast develops between the washed and unwashed areas – 'the soot and whitewash' effect. However, in addition, hard black encrustations may form on the surface. Water flowing over the surface of the stone contains calcium carbonate and calcium sulphate in solution with sooty material in suspension; on evaporation of the water, these are deposited as the hard black encrustations.

It may be noted here that although stones which are insoluble or virtually insoluble in water, such as granites and some of the sandstones (constituent grains of the latter bonded with siliceous or ferruginous matter), also tend to collect surface deposits, these are usually much harder and more insoluble than those which collect on limestones, and are therefore much more difficult to remove. Consequently, a more or less continuous black film collects on the insoluble stones with little differentiation in weathering between sheltered and exposed surfaces. The deposits on limestone, though hard and intractable in the dry condition, can be softened and removed with water.

(b) Skin formation

(1) Limit of beneficial effect
The beneficial effect of the gradual erosion of the surface of limestones only occurs on those areas which are freely washed by rainwater. In sheltered areas, on the other hand, calcium sulphate, one of the products of the action of the acid atmospheric gases, particularly sulphur dioxide, tends to form a hard, glassy and impermeable surface skin which is harmful rather than protective, as it tends to blister and exfoliate. The fact that this skin formation is found in sheltered areas is mainly due to the relative insolubility of calcium sulphate. Although the skin may form on relatively new stone, it later becomes obscured with soot and dirt.

Photograph P3.6/3 *An example of the solubility of limestone due to acid attack. Fossil fragments, that are more resistant to erosive effects than the surrounding material, stand out in low relief after a time as a result of the differential erosion. This is characteristic of the coarse-textured limestones. Carvings and sculpture, unless of good, frost-resistance limestone, tend in time to lose their finer detail and may assume a 'weather-torn' appearance*

Photograph P3.6/4 *The 'weather-torn' appearance shown in Photograph P3.6/3 may also occur in plain sections, as can be seen in this sill, particularly the right-hand side*

(2) Reasons for harmful effects

The reasons why the sulphate skin is harmful have not yet been fully understood. Among other things, the impermeable nature of the skin may result in the formation of crypto-florescence which may tend to push off the skin plus any stone to which the skin may be adhering. On the other hand, the skin has properties different from those of the surface of the stone. Differences in moisture and thermal movements, for example, are significant, and could account for cracking or exfoliation (see Photograph P3.6/5). The deleterious effects of exfoliation are generally more severe when the sulphate skin forms on layers of friable stone, as these layers are weak and thus comparatively easily pulled away from the remainder of the stone with the skin. Finally, the actual crystallization of the calcium sulphate on the surface may cause blistering.

(a)

(b)

Photograph P3.6/5 *Two manifestations of the formation of sulphate skins on stonework. (a) Blistering and scaling of the surfaces with cracks at the arrises and active centres of decay where the surfaces are broken; (b) scabby protuberances that may appear even on well-washed surfaces*

(3) Susceptibility of sandstones

Some stones, it may be noted, are more susceptible to the deleterious effects of sulphate skins than others. In fact, some varieties of stone offer very good resistance. Although limestones have been referred to in particular, it should be noted that some of the calcareous sandstones may also suffer damage from sulphate skins.

(4) Manifestations summarized

The manifestations of the formation of sulphate skins may be summarized as follows:

- Spontaneous blistering and scaling of the surfaces;
- Cracks develop at the arrises and active centres of decay make themselves apparent where the surfaces are broken;
- Hard surfaces may be found to overlie friable layers of stone, and sometimes scabby protuberances appear, even on well-exposed surfaces.

(c) Erosion

Some types of magnesium limestone develop a 'cavernous' type of decay in which deep holes are eaten into the stone due to the presence of magnesium sulphate (see Photograph P3.6/6). This is caused by the detachment of successive scales behind which crystals of the active salt (magnesium sulphate) are usually apparent. The effects of this type of decay caused by the aggressive magnesium sulphate, one of the products of the action of an acid atmosphere on magnesium limestones (calcium sulphate is also produced, as previously explained in *3.5 Efflorescence*, p. 362) which is extremely soluble and hence readily washed away, is more commonly found in the shelter of projecting features, with a very marked contrast in the condition of sheltered and exposed surfaces. However, the decay may also be found in an otherwise well-exposed surface, as the harmful effects of magnesium sulphate will continue after it has managed to gain a foothold in small depressions. Such depressions may be due to the surface characteristics of the stone or to accidental damage (bomb splinters, for example, have been seen to initiate this type of decay).

(d) Adjacent stone

The soluble products derived from limestones may cause the development of an unsightly type of decay – usually a form of blistering or scaling, probably due to the calcium sulphate crystallizing – on some sandstones. Although the latter vary in their resistance to this type of decay, it is generally wise to avoid the association of both sandstone and limestone in the same facade. The deleterious effects of the washings from limestone buildings can often be seen in sandstone steps and paving.

(e) Weathering of slates

Such weathering of slates that does take place is largely due to the action of atmospheric gases. Slate is a metamorphic rock and its mineral complexity and degree of crystallinity are governed by the conditions to which the original material (usually, but not always, clay) was exposed. However, of the minerals which occur in slate, only calcite is readily attacked by the acids commonly found in the atmosphere. Consequently, only slates which contain significant amounts of calcite are likely to prove of poor weathering quality. On the other hand, atmospheric

Photograph P3.6/6 *Two examples of the 'cavernous' type of decay which develops is some types of magnesium limestone. Both examples are of very thick magnesium limestone that has been in service for about 600 years. Such service life may not necessarily be obtained from very thin sections, as more commonly used nowadays*

sulphation is not necessarily the sole cause of slate decay, but calcium sulphate can be formed from calcite and pyrite present in the slate by repeated wetting and drying, while the presence of carbonaceous matter aids this reaction.*

In practice, the weathering action often occurs where one slate overlaps another, for it is here that moisture tends to be held by capillarity. The upper exposed surface is usually unaffected as the washing action of rainwater tends to prevent decay.

Sources

Schaffer, R., *The weathering of natural building stones*, Building Research Special Report No. 18, HMSO, London, 1932 (facsimile reprint 1972)
BRS Digest (First series) 20, *The weathering, preservation and maintenance of natural stone* (Part 1), July 1950 (revised March 1965)
BRE Digest 177, *Decay and conservation of stone masonry*, minor revisions 1984
BRE Digest 269, *The selection of natural building stone*, January 1983
BRE Digest 280, *Cleaning external surfaces of buildings*, December 1983

3 Acid gases and brickwork

(a) Mechanisms

In general, bricks are highly resistant to acid gases in the atmosphere. However, when facing bricks are used with limestone dressings they sometimes undergo a form of decay associated with calcium sulphate

*The durability of slates can be determined by subjecting samples to the action of sulphuric acid, as required by BS 680, for example. Microscopic examination of thin sections may also be of value.

Photograph P3.6/7 *An example of blistering of bricks as a result of soluble material from limestone washed over them. In general, bricks are resistant to the effects of acid gases*

caused by and characteristic of the weathering of limestone in polluted atmospheres. The formation of blisters is characteristic of the decay, due, it is thought, to the crystallization of calcium sulphate on the surface of the brick. The blisters are usually sooty in appearance. The calcium sulphate is derived from the limestone, so attack of the bricks only takes place if soluble matter from the limestone is washed over the bricks, as can be seen in Photograph P3.6/7. Soft bricks, although otherwise resistant, are particularly susceptible to decay from calcium sulphate action.

4 Acids and cement-based products

Acids of various kinds may attack concrete and other cement-based products. The various types of acids, particularly their main sources, are more conveniently discussed separately.

(a) Organic acids

A number of organic acids may come into contact with concrete or other cement-based products mainly in buildings of an industrial nature. Two groups of acids may be identified, namely those of relatively low molecular weight, such as lactic acid and butyric acid (result from souring of milk and butter – dairies significant), acetic acid (vinegar – food pickling and other industries significant), oxalic and tartaric acid (fruit juices significant), and those of relatively high molecular weight, such as oleic, stearic and palmitic acids, encountered as constituents of various oils and fats.

Generally, most types of cement are attacked by the acids mentioned, although some types may be more resistant than others, with Portland cement usually the least resistant. Special acid-resisting cements are available. Depending on the concentrations of acid which may be present, materials which are resistant to acids should be used as linings to protect the concrete. (In some cases acid-resistant paints may be sufficient.) It is also important to note that acid attack may be encouraged if concrete or a cement-based product (mortar in a tiled floor, for example) is damaged by abrasion or weakened by alkaline detergents.

(b) Sewage

In general, concrete and concrete pipes are not attacked appreciably by normal sewage, but certain sewage conditions arising mainly from industrial processes may cause severe attack. Normal sewage has an alkaline reaction, but effluents arising from industrial processes may sometimes be acid, unless neutralized immediately by discharge into an excess of normal sewage. As acidic effluents are hazardous if they are allowed to discharge into concrete pipes, or into any cement-jointed pipe, the effluents should always be neutralized before discharging into the drainage system.

Concrete and mortars have been known to be attacked indirectly by hydrogen sulphide, which under some conditions may be evolved from normal sewage action of anaerobic bacteria on organic sulphur compounds and on sulphates and other inorganic sulphur compounds. Serious trouble has mainly occurred in, for example, North and South Africa, Australia and the USA.

(c) Acid gases

Sulphur dioxide and carbon dioxide are the main gases which normally attack concrete and cement-based products, but then usually only under moist conditions. In general, trouble is more commonly associated with the sulphur products of combustion of fuel. Thus chimneys, both industrial and domestic, of concrete or of brick with cement mortar may be attacked, particularly if condensation occurs within the chimney. Attack may, of course, also take place in chemical factories.

With industrial chimneys methods may be devised to limit the amount of sulphur compounds which arise from combustion of fuel, although other protective measures within the flue may also be required. The treatment of domestic chimneys in order to avoid attack is considered under 'Sulphates, 7 Domestic chimneys', p. 421.

It may be noted that there are likely to be high concentrations of sulphur dioxide in the atmosphere in the vicinity of large power stations when flue gases are emitted without treatment, and under these conditions Portland cement concrete may suffer surface attack (high-alumina cement is usually more resistant).

Sources

Lea, F. M., *The Chemistry of Cement and Concrete*, revised edition (of Lea and Desch), Edward Arnold, London, 1956

Alkalis

1 Scope

The action of alkalis is associated mainly, though not entirely, with cement-based products, particularly those containing Portland cement. The latter is the main source of the alkali, although in some cases it could be lime. The action of alkalis may manifest itself in a number of ways, depending on the other materials involved. Those included here are:

1. The saponification of certain paint films (that is, the conversion of oil to soap which is dissolved in water) applied to surfaces of plaster or renderings containing cement (or, in some cases, lime), asbestos (or asbestos-free) cement sheets and concrete;
2. Expansion and cracking of concrete as a result of alkali-reactive aggregates;
3. The staining of new limestone masonry as the result of the reaction of the alkali content of mortars and traces of organic matter present in the stone;
4. The staining of brickwork due to the action of iron compounds in bricks and alkali in the mortar and of lime leached from concrete;
5. Effects of alkaline solutions on glass.

All of the above are self-contained problems and are, therefore, dealt with separately. For convenience, staining due to *causes other than the presence of alkalis are* included in (3) and (4). Where relevant, sources of alkali other than cement or lime are noted.

2 Paint films

(a) Scope

Emphasis is given here to the alkali attack of paint films applied to cement-based products where the Portland cement provides the main (though not necessarily the only) source of the alkalis. Attack of the kind described here can take place in other products (notably plasters) that may contain certain amounts of lime or be capable of having lime transferred through them and provide the surface for the paint. Equally important is the possibility of alkali attack from strong cleansing agents (soda and caustic soda) or alkaline paint removers.

(b) Initiation and intensity of attack

As with all chemical reactions, alkali attack can only take place if the following conditions exist simultaneously:

- Presence of an alkali;
- Alkali-sensitive base in the paint film; and,
- Presence of moisture.

The degree of the breakdown of the paint film depends therefore on:

- The strength of the alkali present;
- The sensitivity of the paint film in contact with the alkali; and
- The duration of the wet conditions to maintain the activity of the alkali.

(c) Mechanisms

(1) Saponification
'Saponification' takes place when a paint film is attacked by alkalis. It is the splitting of oils into fatty acids and glycerol (a form of soap) and leads to sticky, soft, brownish films, water blisters and oily brown droplets and runs. It is usually the oil medium in the paint film that is attacked, linseed oil being the most vulnerable with certain alkyds following close behind it. Synthetic resins are highly resistant.

(2) Effects
When first attacked by alkali, paint films become soft and sticky and partially water soluble. If the attack is severe *and* moist conditions prevail (e.g. externally from rain and internally from condensation or from the substrate), sticky runs may develop of a yellow or yellowish-coloured liquid, or water sticky blisters may be present. The stickiness may disappear upon drying out but peeling, cracking and flaking may then occur, as the film will have lost binder and have become brittle. Gloss paints lose their gloss and, as with flat oil paints and the like, show brownish discoloration when saponified by alkali.

(3) Source of the lime
Any caustic alkali will saponify an alkali-sensitive oil base in the paint. Calcium hydroxide is a caustic alkali and is formed when lime is mixed with water (see 'Unsound materials', earlier, p. 382). It also exists as 'free lime' in cement and remains so until it has been carbonated (i.e. combined with atmospheric carbon dioxide) to form calcium carbonate that is harmless in the present context.* However, carbonation of the free lime takes place at the *surface* of the concrete (or other cement-based product), being transferred there during the drying-out process. Unless the material has matured fully and dried out completely, further free lime is available to be brought to the surface if the interior of the material subsequently becomes wet (see 'General influences' in *3.3 Moisture content*, p. 230 ff). If a paint film covers the exposed surface, then the caustic alkali in the lime is available for attack of the film.

*For the measurement and quantities of alkalis in cement, see '3 Reactive aggregates', (d) later, p. 398.

Alkalis may, in addition, be derived from sands used in concrete, mortar, plaster or renders. They may also be derived from an external source and then migrate to and then possibly through the material (in much the same way as salts are transferred (see *3.5 Efflorescence*, p. 369)).

(d) Resistance of paints to attack

A number of different paint formulations are available for application to various surfaces, including those of cement-based products. Developments in paint technology aimed at improving specific properties are constantly taking place. Not all types of paint are attacked by alkalis; there is a wide variation in the resistance to alkali attack that paints of different types will provide, as can be seen from Table T3.6/1. The table also gives the permeance of each paint, and this can be considered as a guide to the paint's 'breathability'. It should be noted that any given brand of paint may have alkali resistance or other properties different from those given in the table.

Not all of the paints in the table are suitable as finishing paints. Many of the bituminous types are better as 'isolators'. As a guide, paints and media suitable as finishes for painting of alkaline surfaces such as lime plasters, concrete, new brickwork, cement renderings and asbestos-free cement are as follows:

Type of paint	Medium
Emulsion	Vinyl, acrylic or styrene/acrylic copolymers
Solvent-type paints	Styrene/acrylic copolymers
	Plasticized chlorinated rubber
	Alkyd
	Two-pack epoxy
Cement paints	Portland cement

(e) Precautions

(1) Generally

The precautions that should be taken to minimize the risk of alkali attack of the paint film are associated with:

- Selection of a suitable type of paint or painting system (e.g. primers, undercoats and top or finishing coats) as related to;
- The type of background to which the paint is to be applied (e.g. degree of alkalinity of the background); and
- The presence of moisture *either* at the time of painting *or* subsequently during the life of the film.

It is important to recognize that each of these will vary. In addition, account should be taken of the possible effects of the properties of the paint or painting system such as resistance to attack and 'breathability'. The factors that should be taken into account and aspects related to them are summarized below.

(2) Moisture content

There can, of course, be no attack of an alkali-sensitive paint film if there is no moisture present, whatever the alkali content of the background on which the paint has been applied. This means that, at the time of painting, it is essential that the condition of the substrate for the paint is dry enough. Assessment of this condition should be made on the basis of measurements rather than touch.* As a guide, Table T3.6/2 gives the BRE guidance on wall moisture content with choice of paint for early decoration.

*See *3.1 General considerations*, 'Dampness', p. 152 ff, for methods of measurement available and *3.3 Moisture content*, p. 230 ff, for the factors that are likely to influence surface dryness.

Table T3.6/1   *Alkali resistance and vapour permeance of paints and coatings for interior and exterior use*

Type of paint	Alkali resistance	Vapour permeance[a]
Cement paints	High	High but will reduce rain penetration of porous materials
Emulsion paint: General-purpose good quality	Low	Medium
Matt or 'contract' lower quality	Medium	High
Exterior, smooth	Should be high	Medium
Exterior sand-textured or fibre filled	High	Medium
'Vinyl' silk, satin or gloss	Low or medium	Low or medium
Bituminous emulsions	High	Low or very low
Textured 'plastic' paint	Low	Medium
Textured or coarse mineral aggregate-filled emulsion coatings (including 'organic' renderings)	High	Medium
Gloss paint and enamels (mainly alkyd resin based)	Low	Low
Polyurethane (one-pack)	Medium	Low
Flat oil	Low	Medium or low
Imitation stone paints	Low	Medium or low
Chlorinated rubber (masonry) paints	Very high	Very low
Epoxy ester paints (one-pack)	High	Low
Textured coatings, sprayed or roller applied (*not* emulsion)	Low	Medium
Mineral aggregate coatings (ready-mix or aggregate applied to an adhesive base coat)	Medium	Medium
Chemically cured paints (epoxy- and polyurethane-based 'aliphatic' type for good colour retention)	High (some paints are low)	Very low
Multi-colour paints (emulsion-based examples also available)	Medium	Low
Bitumen paints (solution type) Tar paints solution type)	High	Very low

[a] Approximate permeances for temperate conditions (BS 2972) for typical film thicknesses (one or two coats as recommended by manufacturer) are in the following ranges:

Range	Water vapour permeance		Permeance/g
	g/m^2/24 h	g/sMN	
High	30–40	0.145–0.2	5–7
Medium	13–30	0.067–0.145	7–15
Low	5–13	0.025–0.067	15–40
Very low	Less than 5	Less than 0.025	Greater than 40

From BRE Digest 197, *Painting walls Part 1: choice of paint*, new edition 1982, Table 1. Building Research Establishment, Crown Copyright.

The painting of boards or corrugated sheets containing cement (such as asbestos cement or, nowadays, asbestos-free cement) needs special attention. In service, moisture may be transferred through the back of the board or sheet (i.e. the back or unpainted surface). Condensation on that surface (or high levels of internal humidity) will enable moisture to migrate through the material, taking free lime with it. This passage of moisture should be prevented by the application of a suitable paint system (alkali-resistant primer or colourless paint) to the back and edges of the material. Apart from preventing the alkali attack of the paint, the board will be **balanced** and therefore unlikely to bow or

Table T3.6/2   *Wall moisture content with choice of paint for early decoration*

Relative humididity in equilibrium with surface (%)	Wall condition	Electical meter indication	Recommendation	Suitable paint types	
				On alkaline surfaces	On neutral surfaces
100	Moisture visible	Red zone	Preferably postpone decoration and dry further; if treatment essential dry the surface before painting	Cement-based paint; possibly water-thinned epoxy paint; bituminous emulsion paint	Cement-based paint but not on gypsum plaster; possibly water-thinned epoxy paint; bituminous emulsion paint
90–100	Wet or damp patches, no obvious moisture on surface	Red zone	Preferably postpone; painting may be possible but with high risk of failure	As above; possibly emulsion-based paints	As above; emulsion paint internally
90–75	Drying, doubtful visual indication	Amber zone	Decoration possible with limitations and some risk at higher levels of moisture	Some emulsion paints; masonry paints (not chlorinated rubber); possibly epoxy paints	Most emulsion paints (except glossy); masonry paints; plaster primer; flat oil paints
Below 75	Dry	Green zone	No restriction	All paints on alkali-resisting primers; chlorinated rubber paint; epoxy and polyurethane paints; acrylic and acrylic copolymer solution paints; some emulsion paints; (alkali resistance only as a precaution against future dampness)	Oil or emulsion paints (flat, semi-gloss and gloss); masonry paints; acrylic and acrylic copolymer solution paints; epoxy or polyurethane paints, one- or two-pack types

From BRE Digest 198, *Painting Walls Part 2: Failures and remedies,* minor revisions 1984, Table 2, Building Research Establishment, Crown Copyright.

otherwise deform. As to the possibility of the moisture content of a substrate being increased during service, account must be taken of the effectiveness of the method of exclusion used (see *3.4 Exclusion*).

(3) Degree of alkalinity

The degree of alkalinity at the surface of a cement-based product is likely to vary considerably, for the reasons explained earlier in '(b) Mechanisms'. It might seem possible to expose a cement-based product for a certain amount of time so that adequate carbonation of the free lime at or near the surface has taken place. This is not usually practical, except possibly with thin materials such as asbestos-free boards and sheets. An alternative would be to neutralize the alkalis with solutions such as hydrochloric acid, zinc sulphate and other chemicals. These treatments are not always effective, and in any event only neutralize the alkali at or near the surface. They need care in their use and, apart from causing other damage or deterioration, they may interfere with the proper bond of the paint film to the substrate. **Creative pessimism** would caution that special paint treatment is the appropriate and less risky alternative.

(4) Type of paint

It is generally recommended that all surfaces of all substrates likely to contain either Portland cement or lime should be painted with alkali-resistant materials. However, where this is not possible it is essential that at least one or preferably two coats of alkali-resistant primer is first applied. As these make any paint system impermeable to moisture the substrate must be dry before the primer is applied.

Sources

Hamburg, H. R. and Morgans, W. M. (edited and revised by), *Hess's Paint Film Defects*, 3rd edn, Chapman and Hall, London, 1979
Lea, F. M., *The Chemistry of Cement and Concrete* (revised edition of Lea and Desch), Edward Arnold, London, 1956
BRE Digest 197 and 198, *Painting walls, Part 1: Choice of paint* and *Part 2: Failures and Remedies*, 1982 (new edition) and 1984 (minor revisions)
BRS Digest (1st series) 38, *Painting asbestos cement*, January 1952

3 Reactive aggregates

(a) Scope

Reactive minerals in certain types of aggregate can deteriorate the concrete in which they are used as a result of an interaction between the minerals and alkaline pore fluids (the latter principally originating from the Portland cement). The deterioration of the concrete manifests itself as cracking, usually some years after the concrete has hardened. The mechanism of deterioration is known as alkali aggregate reaction (AAR), and can occur in a number of forms, the most common being alkali silica reaction (ASR). Alkali carbonate reaction (ACR) is another form, but this is less common worldwide. No harmful cases of it have been observed in the UK and it is therefore not covered here.

As explained in (b) below, AAR in general and ASR in particular are phenomena new to the UK. Much work has already been carried out by BRE, among others, on how to identify ASR as the cause of cracking in concrete structures and what to do to reduce the risk of it occurring in new concrete structures. Knowledge of the subject is in its infancy in the UK, at any rate, and experience in other countries has been shown by the BRE not to apply in all respects to the UK. Coverage here is restricted, therefore, to an explanation of the relevant factors. Detailed advice should be obtained from experienced specialists in the field.

(b) Background and risk

ASR was identified in the USA in 1940 and has occurred in other countries such as New Zealand and Australia.* In the UK, the first case of the reaction was found to be the cause of cracking in a concrete dam in Jersey in 1971. Later in 1976 it was identified as the cause of the cracking in the concrete bases of electricity substations in south-west England. By 1988 well over 180 structures, only a few of which were in buildings, had been affected, mainly in the south-west and Midlands of England. These represent a very small proportion of all concrete structures in use.

Although the problem should not be of great cause for concern for concrete structures in general and in buildings in particular, it certainly cannot be ignored. Experience has shown that ASR has often come to light as a result of cracking of concrete from other causes (e.g. corrosion of the reinforcement). Like all other chemical reactions defined in this section, it requires the presence of moisture. The latter has to reach the reactive minerals but these have to be present in sufficient amounts and in contact with water to cause cracking.

Thus the presence of reactive aggregates in concrete does not, by itself, signify that there will be cracking of the concrete in which they have been included. The cement/aggregate combinations in use in the UK will only react detrimentally under adverse combinations of:†

- High cement content;
- High alkali content in the cement; and
- Exposure to water.

(c) Mechanism and effects

(1) Gel formation
The interaction between the alkaline pore fluid and siliceous minerals in some aggregates results in the formation of a calcium alkali silicate gel. This gel absorbs water, thereby producing a volume expansion which disrupts the concrete. The reaction appears to happen slowly, as the disruptive effects of the reaction take at least five years to manifest themselves.

(2) Cracking
In concrete that has been subjected to ASR cracks the mode of cracking depends on whether the expansive forces have been restrained:

- *In unrestrained concrete*, the cracks have a characteristic random distribution often referred to as map cracking (Photograph P3.6/8). This type of cracking, also associated with concrete shrinkage (see *3.3 Moisture content*, p. 262) or frost action,‡ has a network of fine, barely visible cracks bounded by a few larger ones. If the map cracking is due to shrinkage it will probably have taken place during the first year, whereas frost action is likely to have caused spalling of the surface as well (see *3.8 Frost action*, p. 549).¶
- *In reinforced concrete* the expansive forces due to ASR are restrained by the reinforcement and the cracking is thereby modified, with the cracks now running parallel to the main reinforcing bars or prestressing tendons (see Photograph P3.6/9). The cracks may be bordered by gel if the ASR is particularly active. This may be mixed

Photograph P3.6/8 *Typical cracking patterns in unrestrained concrete containing alkali-reactive aggregate. (From BRE Digest 330,* Alkali aggregate reactions in concrete, *March 1988, Fig. 1, Building Research Establishment, Crown Copyright)*

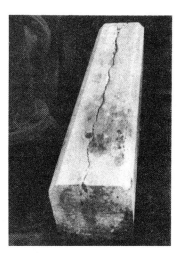

Photograph P3.6/9 *Cracking of concrete containing alkali-reactive aggregate where the expansive forces have been restrained by prestressing. (From BRE Digest 330, Fig. 2, Crown Copyright)*

*As reported by Lea and mentioned in *Materials for Building*, Vol. 3, p. 25.

†It should be noted that the list refers only to cement/aggregate combinations. The conditions necessary for the reactions to occur are listed in (d) later, p. 398.

‡It is also associated with sulphate attack of rendering (see 'Sulphates' later, p. 420 ff).

¶Surface pop-outs can also be caused by reactive aggregate particles, but few cases of such pop-outs as a result of ASR have been observed in the UK.

with lime which normally leaks from cracked concrete. Both turn to a whitish carbonate on exposure to the atmosphere.

(3) Identification
Importantly, the only certain evidence of ASR is provided by microscopical examination of the interior of concrete for the positive identification of the presence of the gel and of the aggregate particles that have reacted.

(4) Long-term durability
The long-term durability of concrete in respect of reinforcement corrosion and frost action may be affected by the cracking induced by ASR. Conversely, ASR may lead to reinforcement corrosion or frost action or both.

(d) Sources of alkalis

(1) Measurement
For the present purposes, alkali means the alkali metals of sodium and potassium expressed as their oxides. The alkali is generally expressed as the equivalent of Na_2O and it is the sum of the Na_2O molecular equivalents of the actual Na_2O and K_2O:

$$\% \text{equiv. } Na_2O = \%Na_2O + \%K_2O \times 0.658$$

(2) In cement
The alkalis in ordinary Portland cement (OPC) are derived from the raw materials. The levels of alkali in OPC manufactured in the UK are normally in the range 0.4–1.0%.* OPC to BS 12 with *a guaranteed alkali level* of 0.6% or less is *not obtainable* in the UK. However, cement manufacturers in the UK can supply *sulphate-resisting* Portland cement to BS 4027 having a guaranteed alkali content of not more than 0.6% Na_2O equivalent, *provided* this guarantee is requested at the time of ordering.

(3) Slags and ash
So far as is known, ground granulated blastfurnace slag (ggbfs) and pulverized fuel ash (pfa) to BS 3891: Part 1, *inhibit* alkali reactions, although the mechanism involved is not understood. Both have comparatively high total alkali levels but most of this alkali is combined in the glassy structure of the material and is released relatively slowly in the concrete as hydrates. It is assumed, therefore, that ggbfs and pfa contribute no reactive alkali to the concrete. Accordingly, the alkali content of concrete or the cementitious material may be reduced by replacing part of the OPC by ggbfs or pfa.

(e) Conditions necessary for the reaction

Deterioration of concrete only happens when three conditions occur simultaneously:

- A sufficiently alkaline solution in the pore structure of the concrete;
- An aggregate combination susceptible to attack by this alkaline solution; and
- A sufficient supply of water.

*As measured in accordance with BS 4550: Part 2, clause 16.

Each is explained separately below.

(f) Alkalinity of the pore solution

(1) Generally

The alkalinity of the pore solution (the hydroxyl ion concentration) depends inherently on:

- The equivalent sodium oxide in the Portland cement (see (c) above); and
- The cement content of the mix.

However, the alkalinity may be increased:

- By other alkali metals added at the mixing stage; or
- From external sources such as sea water or de-icing salts.*

(2) Limiting the hydroxyl ion concentration

Experience in the USA and elsewhere has suggested that Portland cements with an equivalent Na_2O of less than 0.6% are unlikely to suffer from ASR even in combination with a reactive aggregate. (See (d)(2) above for alkali contents of cements.)

(3) Limiting the cement content

In addition to the alkaline level in the cement, account must be taken of cement content in the mix, as the alkalinity of the concrete is also dependent on this. This means that the *total amount of alkali* in the concrete mix can be restricted by another specification method.

From investigations of structures and laboratory experiments BRE has suggested that a maximum limit of 3.0 kg of Na_2O equivalent per cubic metre of concrete (the 3.0 kg limit, for convenience) should be appropriate except in some particularly vulnerable structures. In calculating this limit *all sources of alkalis* are included.

Another limiting factor is the cement content. An alkaline level of 1.0% equivalent Na_2O can be used without further protection against ASR with the 3.0 kg limit in a mix containing *no* cement replacement materials and with *salt-free* aggregate *provided* the cement content of the mix does not exceed 300 kg/m³. The cement content can be increased to 500 kg/m³ with an alkali level of only 0.6% Na_2O.

(4) Particularly vulnerable structures

- Concrete slabs sunk into wet ground but with the upper surface exposed to drying;
- Structures frequently saturated or in areas of high humidity such as highways and multi-storey car parks, and water-retaining or buried structures.

For all of these a lower total alkali content must be specified or an alternative specification route must be followed: for example, use of a low-alkali (i.e. less than 0.6/Na_2O) cement, of ggbfs or pfa (see (d)(3) earlier) or of a non-reactive aggregate combination.

(g) Susceptible aggregate combinations

(1) Siliceous aggregates

Siliceous aggregates are the most common type of aggregate susceptible to alkali attack worldwide. It is the only type definitely identified as being susceptible in the UK. The reactivity of minerals of silica depends on the quality of the ordering of their crystal structure.

Opal is the most reactive of the different silica minerals because it has a very disordered structure; *unrestrained quartz* is normally unreactive because its structure is well ordered. In between there are many other

*The contribution of these salts is not well understood. British aggregates usually contain very small amounts of soluble alkali, apart from salt, and have not been found to have a significant influence on the alkalinity of concrete in the UK. This is not necessarily the case with aggregates from other countries.

silica minerals with varying orders of their structure. These include microcrystalline and cryptocrystalline quartz and strained quartz. Opaline quartz has not been found in significant amounts in aggregates from the UK mainland. The reactive aggregates in the mainland, from both land-based and sea-dredged sand and gravel, have been found to contain minerals of microcrystalline and cryptocrystalline quartz and chalcedony in cherts and flint and strained quartz present in some quartzites.

(2) Tests

There are, as yet, no British Standard tests for the susceptibility of aggregates to attack by alkalis. The American ASTM tests have been found, according to BRE, not to predict accurately the reactivity of UK aggregates. A method of test has been developed at BRE that shows a good correlation with field evidence of which aggregates are susceptible.*

(3) Resistance to attack

From BRE's experience, aggregates composed of rock or artificial material types listed below are very unlikely to be susceptible to damaging attack by cement alkalis:

Basalt
Diorite
Gabbro
Gneiss
Granite
Limestone
Marble quartz (discrete and not highly strained)
Micro granite
Quartz (discrete and not highly strained)
Syenite
Trachyte
Tuff
Air-cooled blastfurnace slag (BS 1047)
Expanded clay/shale/slate
Sintered pfa

Aggregates of these rocks can only be considered non-reactive *if*:

- The rock is not contaminated with forms of silica such as chert (which may occur in limestone) or chalcedony (which may be present in igneous or other rocks) in amounts sufficient to cause ASR; and
- Other aggregates containing significant amounts of reactive silica (e.g. in sand used in concrete mixes) are not added in amounts significant to make the proportion of silica in the whole aggregate a damaging one.

(4) Proportion of reactive silica

The proportion of reactive silica in the combined aggregate has a very important influence on its reactivity. The worst expansion occurs at what is known as the 'pessimum' proportion and is related to the reactivity of the aggregate. For example, the pessimum for a highly reactive opaline material is as little as 2–4%. Beyond this, there will be less expansion, and a point may be reached where there is no significant expansion. The pessimum for less reactive aggregates is much higher, in some cases as much as 100%.

Although flints and cherts have proved to be the source of reactive silica in the UK, it is considered that a combination of fine and coarse

*A British Standard based on the BRE work is, it is understood, in the course of preparation.

aggregate which contains more than 60% by mass of them is unlikely to cause ASR damage. Another factor to be taken into account is the porosity of the coarse aggregate. The use of reactive fines with a dense, low-porosity coarse aggregate is potentially more damaging than the same fines with an inert but porous coarse aggregate.

(h) Quantity of water

Sufficient water must be present for the damaging expansion to take place. Although theoretically there is enough mixing water in mass concrete to cause expansion, in practice, expansion has only occurred in structures affected by ASR in the UK after they have been exposed to an external source of water. In bridges, for example, severe condensation on the underside of the deck may be sufficient.

Consequently, saturation of the concrete in use is the factor that must be considered, and avoiding saturation of the concrete is, in most cases, the most important precaution that should be taken. In this, the relative impermeability of the concrete is an important factor (see *3.3 Moisture content*, p. 262 ff).

(i) Precautions

In the light of the present state of knowledge the precautions that should be taken cannot be as definitive as might be preferred. A question-and-answer procedure seems to be the most appropriate for general purposes.

(1) Questions
On the basis of present experience, concrete should be safe from ASR if the answer to *any* of the following questions is 'no':

- Will the concrete be exposed to an external source of moisture?*
- Will the total alkali content of the concrete mix exceed 3.0 kg of equivalent Na_2O per cubic metre, or will the concrete be exposed to an external source of alkali or alkali concentration by moisture migration?
- Is the aggregate combination thought to be potentially reactive?

(2) Response to answers
If the answer to *all three* questions is 'yes', ASR is a possibility and *some changes in design* should be made. When making any changes, care should be taken that none of them impairs the strength or other aspects of the durability of the concrete.

(3) Changes in design
Changes in design that might be made include any one or a combination of the following precautions:

- Keep the concrete dry;
- Use a Portland cement with a lower alkali level such that the equivalent Na_2O in the cement is 0.6% or less or the total alkali level in the concrete is 3.0 kg/m^3 or less;
- Substitute sufficient of the Portland cement with ggbfs of pfa (at least 25%) to reduce the alkali level of the concrete to 3.0 kg/m^3 or less. Include one-half of the acid-soluble alkali in the ggbfs or one-sixth of the total alkali in the pfa in the calculation. Alternatively, use a blend of 50% or more ggbfs with OPC such that the acid-soluble alkali content of the blend does not exceed 1.1% Na_2O equivalent.
- Change to a non-expansive aggregate combination.

*In mass concrete the mixing water may be sufficient to cause degradation.

Sources

Lea, F. M., *The Chemistry of Cement and Concrete* (revised edition of Lea and Desch), Edward Arnold, London, 1956
Minimising the risk of alkali-silica reaction, Guidance notes, Report of a working party of C&CA, September 1983
BRE Digest 330, *Alkali aggregate reactions in concrete*, March 1988
Levitt, M., *Precast Concrete, materials, manufacture, properties and usage*, Applied Science, London, 1982

4 Staining of limestone

(a) Generally

Disfiguring brownish or yellowish stains, which are unsightly, often develop on the surface of limestone masonry, usually near the joints, soon after it has been incorporated into a building. In sheltered positions the stains may persist for years, but generally they soon disappear or are masked by superficial deposits. Sometimes brownish stains also develop on newly cleaned limestone previously covered with soot. The stains which occur on new limestone are attributable to a reaction between the alkali compounds of a mortar and organic matter in the stone; those that are found on newly cleaned limestone are attributable to soluble matter absorbed from soot deposits into the stone and brought to the surface as the stone dries out after cleaning.

(b) New limestone

The mechanism involved in the staining of new limestone is similar to that involved in the production of soluble salts from mortars (see *3.5 Efflorescence*, p. 360). Alkali compounds are leached from the mortar as hydroxides and carbonates into the stone and there react with organic matter which is present in most limestones to form a soluble organic salt that is deposited on the surface of the stone and appears as the stain (see Photograph P3.6/10). Mortar used for both jointing and grouting may 'supply' the alkali compounds. Both the presence and availability of the alkalis may vary. Waterproofing the back of the stone does, in some measure, help to reduce the transference of the alkali compounds from the grout behind the stone; the limitation of the this treatment is the weakness at the joint.)

Other treatments, such as applying a lime slurry to the back of the stone, are reported to have given satisfactory results, even when the stone was backed with concrete.

If the risk of staining is to be minimized, even if it is only from the jointing, then it is probably better to test different mortars in respect of their staining potential.

A simple test for the production of the stain has been devised.* A cube of stone is allowed to stand on a cube of mortar with a layer of clean washed blotting paper between them. The cube of mortar is placed in a shallow dish or tray containing distilled water, and the water rises by capillarity through the mortar and into the stone, where it evaporates at the surface. Thus the relative tendency for staining of any particular combinations of stone and mortar can be observed.

No rule can be laid down as to the relative merits of different mortars in respect of staining, for the effect depends not only on the amount of alkali present but also on its availability. For example, bad stains on Portland stone have been caused by both ordinary and rapid-hardening Portland cement (the grey cements) in this type of test, while there has

*See *The weathering of natural building stones*, by R. Schaffer, Building Research Special Report 18, HMSO, London, 1932 (facsimile edition 1972).

been considerably less staining when a white cement has been used. The least amount of staining has been produced by a specially prepared alkali-free cement. From the tests which have been carried out it would seem that mortars of high-calcium (fat) limes and similar limes gauged with a moderate amount of cement (preferably, white cement) will cause less staining than straight cement mortars.

(c) Newly cleaned limestone

The brownish stains mentioned above which sometimes occur on newly cleaned limestone are generally fortuitous and are not the fault of the operator. Their development depends, to some extent, on prevailing weather conditions. Slow drying would encourage staining; rapid drying discourages it. The stains often disappear quite quickly; sometimes they may persist for a considerable time. Although not always effective, repeated rinsing with water is helpful.

Photograph P3.6/10 *The disfiguring browning or yellowish stains that often develop on the surface of limestone masonry as a result of a reaction between alkali compounds in the mortar and organic compounds in the stone. The stains develop soon after being incorporated into the wall construction and usually disappear after a comparatively short period of exposure except in sheltered positions, where they may persist for several years*

Sources

Schaffer, R., *The weathering of natural building stones*, Building Research Special Report 18, HMSO, 1932 (Facsimile edition, 1977)
Principles of Modern Building, Vol. 1, 3rd edn, HMSO, London, 1959
BRS Digests (First series) 20 and 21, *The weathering, preservation and maintenance of natural stone* (Parts I and II), July 1950 (revised March 1965) and August 1950
BRE Digest 269, *The selection of natural building stone*, January 1983
BRE Digest 280, *Cleaning external surfaces of buildings*, December 1983

5 Staining of brickwork

Two types of staining are included here, namely, rusty stains from iron compounds in either the bricks or mortar and white stains from concrete. However, it may be noted that stains may also result from embedded ironwork, copper and bronze (see *3.7 Corrosion*, p. 482) and from vegetation.

(a) Rusty stains

Rusty stains on mortar joints are usually due to iron salts, especially ferrous sulphate, which form efflorescences when they first come to the surface. However, they react chemically with the lime in the mortar and are subsequently washed over the mortar joints by rain.* If bricks known to give rusty stains due to the presence of iron salts are to be

*The staining which may occur on plasters of gypsum applied over bricks which contain iron salts is discussed in *3.5 Efflorescence*, 'Physical processes, 6 Effect of precipitation', (p. 368).

used it is wise to allow the stain to develop harmlessly on the bed joints and then to cover it by pointing after the work is complete. It is unlikely that the staining will recur. On the other hand, if this method cannot be adopted, the amount of staining can be considerably reduced, if, as Butterworth advises,* the unfinished walls are kept covered to prevent rain percolating through the brickwork until the mortar has aged for at least a week.

Mortar sands that contain particles of ironstone may result in rusty stains, which generally flow from particular grains in the surface of the mortar. Repointing is the only cure.

(b) White stains

White stains which do not disappear when the brickwork is washed by rainwater, and thus distinguishable from efflorescence, may be caused by water percolating through poor-quality concrete in adjoining elements such as lintels, copings, etc., particularly when the requisite dpcs have been omitted. The stains consist of calcium carbonate which is formed when atmospheric carbon dioxide reacts with the free lime, calcium hydroxide, leached from the concrete, and may, under favourable conditions, build up to an appreciable thickness of stalactite-like material.

Removal of these stains can be carried out as follows. The brickwork must first be well wetted, and then brushed over with *diluted* hydrochloric acid (spirits of salts) to dissolve the deposits. Finally, the work must be thoroughly washed down with clean water. In order to avoid a recurrence, attention should be given to taking the necessary steps to prevent further percolation of water.

6 Glass

All glasses are inherently water- and alkali-sensitive. The effects of alkalis on glass must now be considered for two different products: windows, doors and cladding (for convenience, referred to as 'sheet glass') and as fibres for reinforcement of cement.

(a) Sheet glass

The durability of sheet glass in buildings is generally excellent under ordinary conditions, but deterioration is more rapid is the presence of alkalis. These may be derived from paint removers or from rainwater running off new concrete or a cement-rendered wall. In either case the alkali needs to be in contact with the glass for some time before the surface degrades.

(b) In GRC

Glass fibre-reinforced cement (grc) is a relatively new composite building material that has been used in buildings in various ways. The inherent alkali sensitivity of silicate glass fibres in a highly alkaline environment of hydrated cements has been reduced initially by the inclusion of a fairly large proportion of zirconium dioxide and subsequently by the modification of the surface of the fibres. Modification of the cement used can also improve alkali resistance. In 1971 a new glass fibre, made of a new glass and trade-marked Cem-FIL, was introduced, and Cem-FIL 2 appeared in 1979. The latter differs from the earlier version in that its surface has been modified to increase further its alkali resistance. (See BRE Digest 331, *GRC*, April 1988.)

*Butterworth B., 'Efflorescence and the staining of brickwork', Article reprinted from the *Brick Bulletin* and amended December 1962.

Sulphates

1 Background

(a) Meaning and occurrence generally

The chemical reactions of sulphates and certain constituents of cement and hydraulic lime in cement-based products such as concrete, mortar and renderings cause considerable expansion that leads to the gradual softening and/or disintegration in the form of cracking of the product. 'Sulphate attack' is the term used to describe the results of these reactions. The attack is more commonly associated with cement-based products in which OPC has been used, although other types of cement (including other types of Portland) may be attacked. It is more likely to occur in constructions that are persistently damp.

Problems associated with sulphate attack have been around for many years. There are a large number of variables involved, but research into these and other factors has meant that the phenomenon is now better understood. The precautions that should be taken are much clearer, albeit still qualified in certain cases.

(b) Scope

As sulphate attack involves soluble salts, the explanations of the way these may be transmitted as explained in *3.5 Efflorescence* (p. 365) are relevant. Additionally there are those factors that influence exposure (*3.2 Exposure*), moisture content (*3.3 Moisture content*) and exclusion (*3.4 Exclusion*), to which reference is made as appropriate. In this section the aim is, first, to explain those factors that are common to sulphate attack in all cement-based products and then to deal with the detailed implications of and precautions for particular products or elements of construction, such as concrete, brickwork, rendering and domestic chimneys.

2 General considerations

Although there are likely to be detailed and important variations with given cement-based products under given conditions, it is possible to identify the existence of some fundamental criteria common in all cases of sulphate attack. Apart from the basic mechanisms involved, which can conveniently be considered separately, these are associated with the nature and derivation of the soluble salts likely to cause attack, the constituents of the cement involved, and the significance of the characteristics of the materials in which the attack takes place together with the effects of exposure to water.

(a) Basic requirements

Three things are basically necessary for sulphate attack to take place:

- Sulphates;
- Tricalcium aluminate* (a normal constituent of cements); and
- Water.

(b) Sulphates

The sulphates which most commonly attack *set* cements are those of sodium, potassium, calcium and magnesium. It is important to note

*This compound has been included in the basic requirements as it appears to be generally regarded as the most significant in sulphate attack, particularly when Portland cements are involved, and secondary to free lime (calcium hydroxide), which may also react with sulphates. The significance of these reactions together with basic reasons for the particularly high resistance to attack provided by high-alumina cement (rich in aluminates, an apparent paradox) are noted in '3 Basic mechanisms' later.

that it is in fact the set cement which is attacked. In general, mortars and concretes attacked by calcium or sodium sulphate become eventually reduced to a soft mush, but when magnesium sulphate, which is by far the most aggressive, is the main destructive agent the mortar or concrete remains hard, although it becomes much expanded and the disrupted mass usually consists of hard granular particles.

The nature, sources and behaviour of sulphates are included, in detail, in *3.5 Efflorescence* (p. 354–71) and are therefore not repeated here, apart from emphasizing the importance of solubility. In general, two sulphates commonly encountered, those of sodium and magnesium, are highly soluble, and their presence is therefore far more significant than that of calcium sulphate, which is, by comparison, less soluble. An aspect not previously covered in detail, namely, the variability of soluble salt content of soils, is now included for convenience, in '4 Concrete', p. 410.

(c) Rate of attack

The rate at which sulphate attack may proceed is governed by a number of interrelated factors, which may be summarized as follows:

1. *The quantity and type of sulphate present.* Generally, magnesium sulphate has the most far-reaching effects.
2. *The quantity of tricalcium aluminate present in the cement used.* With Portland cements, those with a low tricalcium aluminate content are more resistant to sulphate attack than those with high tricalcium aluminate content.
3. *The permeability of the cement-based product.* This is one of the factors governing the extent to which water may 'bring together' the soluble sulphates and tricalcium aluminate (or other relevant constituent in the cement used – see '3 Basic mechanisms' for the exceptions) so that the requisite chemical reaction may proceed. In general, highly impermeable products will only suffer surface attack, and then usually by sulphates from external sources; with highly permeable products, attack may proceed simultaneously throughout the mass of the product and by sulphates either present in the material or from external sources, or both.

    Permeability of cement-based products, which is dependent on mix design and control of quality on-site or in the factory, plays an extremely important part in the resistance which a product will offer to sulphate attack. Thus in the case of concrete, for example, precast units are likely to be more resistant than concrete cast *in situ*. The latter also suffers the added disadvantage of being more susceptible to attack if exposed to sulphate sources during its 'green' state.
4. *The permeability of adjacent materials.* This includes permeability of soils adjacent to and in contact with the cement-based product, which may provide possible sources of sulphates. The more impermeable these surrounding materials, the less likely are sulphates to be transferred to the cement-based product.
5. *The availability and quantity of water.* Without water there can be no chemical reaction, while certain minimum quantities of water are necessary for the reaction to proceed. For any given condition, the rate of attack increases as the duration of the presence of water increases. Time, it should be noted, is extremely important.
6. *Pressure differences.* Closely associated with the permeability of the cement-based product is whether pressure differences will tend to force a sulphate solution through the material. In this connection, it may be noted here that the most severe condition occurs when a cement-based product, such as a concrete retaining wall, is exposed

to water pressure on one side and to air on the other, in which case evaporation is promoted, thus tending to increase the volume of sulphates in solution drawn through the material. In addition, evaporation at the exposed face leaves the sulphate salts behind. This then increases the concentration of those salts in the water in the concrete. The very soluble magnesium and sodium sulphates would constitute a greater danger from this effect than would the more insoluble calcium sulphate. A similar, though less severe, condition occurs in a partly buried concrete mass, when water is drawn up by capillary forces from the portion below ground and evaporates at the exposed surface.

3 Basic mechanisms

(a) Scope

The chemistry associated with sulphate attack on set cement is extremely complex, with a number of different compounds involved in what may best be termed a series of reactions, some of which are interrelated. However, tricalcium aluminate appears to be the most important compound in set Portland cement, and governs the resistance a given cement will offer to attack, although the reactions of other compounds, such as calcium hydroxide (free lime), must also be taken into account. On the other hand, the type of sulphate involved in any reaction has a marked effect on the rate at which attack may proceed and on the deleterious results in the cement-based product.

The purpose of this section is mainly to outline, in general terms, the fundamental chemistry involved in sulphate attack; to note the effects of magnesium sulphate; to consider the influence of sea water; and to compare the resistance of various cements to sulphate attack.

(b) Fundamental chemistry

There are two basic reactions which account for expansion and disruption of mortars and concrete attack by sulphates:

1. The formation of calcium sulphate* (gypsum) by the conversion of calcium hydroxide (free lime). Various sulphates, including sodium, potassium, and magnesium may react with calcium hydroxide. The molecular volume of calcium hydroxide ($Ca(OH)_2$) is 33.2 ml, while that of gypsum ($CaSO_4 . 2H_2O$) is 74.3 ml. Thus the formation of gypsum from the free lime results in more than a doubling of the solid volume.
2. The formation of the more insoluble calcium sulphoaluminate from the combination of hydrated tricalcium aluminate and gypsum. This, too, is accompanied by an increase in solid volume. The increase is rather smaller than that described in (1) above, but is nevertheless significant.

(c) Effects of magnesium sulphate

In addition to reacting with the aluminates and calcium hydroxide as do the other sulphates, magnesium sulphate is particularly aggressive because it decomposes the hydrated calcium silicates (usually unaffected by the other bases), and, by continued action, also decomposes calcium sulphoaluminate.

Initially, magnesium sulphate has a similar action to that of the other sulphates on the hydrated calcium aluminates, and calcium sulphoalu-

*It is extremely important to note that under no circumstances should calcium sulphate be added to Portland cements (or hydraulic limes) in order to alter the setting or other properties of the cement (or lime). In the British Standard Specifications, as in most specifications in other countries, the amount of calcium sulphate which may be present in the finished cement (i.e. ex-works) is strictly limited to a few per cent, with the amount added during manufacture varying between 1% and 3%. The limited amount of calcium sulphate allowed is valuable in the control it affords of the setting time and in its favourable effect on strength. On the other hand, the presence of large quantities of calcium sulphate leads to slow expansion in the *set* cement, and this accounts for its stringent limitation.

minate is formed together with magnesium hydroxide. However, calcium sulphoaluminate is itself unstable in the presence of a magnesium sulphate solution. Continued action of the magnesium solution results in the decomposition of the calcium sulphoaluminate to form gypsum, hydrated alumina and magnesium hydroxide. In this, the length of exposure to the magnesium sulphate solution is significant. Thus, after a fairly long period of action, the external skin of a mortar or concrete is relatively free of calcium sulphoaluminate while gypsum is present in a large amount. However, in the interior of the mortar or concrete, where access to the solution has been much slower, both calcium sulphoaluminate and gypsum are found.

Another characteristic of the action of magnesium sulphate on mortars and concrete is the formation of a hard glassy skin on the surface which tends to hinder the penetration of the solution. This skin is formed by the deposition of magnesium hydroxide in the pores of the cement-based product, and accounts for the fact that the disrupted mass of a product attacked by magnesium sulphate more often consists of hard granular particles. This may be contrasted with attack by either sodium or calcium sulphate when the cement-based product becomes soft and incoherent.

(d) The effects of sea water

Although the predominant constituents of sea water are the chlorides, there are, nevertheless, significant quantities of sulphates present. The percentage of magnesium sulphate, for example, may be in the region of 5% of the total (see *3.1 General considerations*, Table T3.1/1, p. 132) and the chemical action of sea water on concrete, is, in fact, due to the action of magnesium sulphate. However, and this is important, the presence of chlorides in sea water retards and inhibits the expansion of concrete by sulphate solutions, although the degree of reaction remains unchanged.

The chemical action of sea water is better considered as one of a series of reactions proceeding concurrently. Calcium hydroxide and calcium sulphate are considerably more soluble in sea water than in plain water and the leaching actions which occur (accelerated, to some extent, by wave motion) remove lime and calcium sulphate from the concrete. At the same time, the reaction with magnesium sulphate leads to the formation of calcium sulphoaluminate (as described previously in (c)) which may cause expansion. This then renders the concrete open to further attack and leaching, and so the cycle may be repeated. The relative contribution to deterioration which may be attributed to either expansion or leaching is largely dependent on conditions. The rate of chemical attack is increased by temperature, while both the rate and its effects are influenced by the type of cement.

(e) Increasing the resistance of set cement

(1) Cements compared
The resistance of various types of cement to sulphate attack are compared in Chart C3.6/1. It will be seen that most types of Portland cement have low resistance, except low-heat (medium) and sulphate resisting (high). The resistance increases with other types of cement, notably supersulphate (i.e. cement containing blastfurnace slag) and high-alumina cement. Pozzolanic cement (a naturally occurring cement) offers high resistance. Partial replacement of Portland cement with either pozzolanas or pulverized fuel ash (pfa), a material covered in 'Alkalis' under '3 Reactive aggregates' (p. 398), will also increase the set cement's resistance to attack.

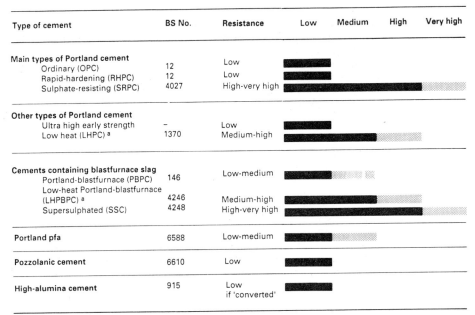

Type of cement	BS No.	Resistance	Low	Medium	High	Very high
Main types of Portland cement						
Ordinary (OPC)	12	Low				
Rapid-hardening (RHPC)	12	Low				
Sulphate-resisting (SRPC)	4027	High-very high				
Other types of Portland cement						
Ultra high early strength	–	Low				
Low heat (LHPC) [a]	1370	Medium-high				
Cements containing blastfurnace slag						
Portland-blastfurnace (PBPC)	146	Low-medium				
Low-heat Portland-blastfurnace (LHPBPC) [a]	4246	Medium-high				
Supersulphated (SSC)	4248	High-very high				
Portland pfa	6588	Low-medium				
Pozzolanic cement	6610	Low				
High-alumina cement	915	Low if 'converted'				

[a] Normally available in the UK to special order.
[b] For general guidance only.

Chart C3.6/1 *Illustration of the resistance to sulphates of various types of cement. (Read with Table T3.6/3.) (From BRE Digest 325,* Concrete, Part 1: Materials, *October 1987 Building Research Establishment, Crown Copyright)*

(2) Reducing the tricalcium aluminate
Sulphate-resisting cements differ from ordinary cements in that they have a low calculated tricalcium aluminate content *and* a carefully controlled iron oxide-to-alumina ratio. Sulphate-resisting Portland cements have been manufactured and used in the UK since 1949. Initially, these cements conformed to BS 12 for OPC with the composition so adjusted to give a cement containing not more than a few per cent of tricalcium aluminate. Now these cements are covered by BS 4027 that limits the amount of tricalcium aluminate to 3.5% by weight. Although a limit is not placed on the ratio of alumina to iron oxide the Standard gives no reason why it should not be reduced below the value of 0.66 specified in BS 12.

(3) Resistance of other cements and replacement materials
Among reasons offered for the high resistance to sulphate attack of *pozzolanic cement* and the very high resistance of both *supersulphate* and *high alumina* cements (and to sea water) is the absence of limitation of calcium hydroxide in the set cement. This apparently is not the only reason, because high-alumina cement, for example, contains a large proportion of aluminates (approximately 40%). With all these cements it would appear that a protective film forms over the vulnerable compounds and thereby provides a high resistance to sulphate attack. Another factor may be the increased impermeability of the set concrete.

Pozzolanas or pfa are used as partial replacement of Portland cement in concrete, the set concrete regarded as having improved resistance to sulphate attack. Such concrete is made when OPC is mixed with pfa to BS 3892: Part 1: 1982. The degree of resistance offered by the concrete depends on the amount of pfa mixed, a pfa content exceeding 25% providing the optimum resistance (see '4 Concrete', p. 410).

(4) Steam curing

Steam curing of concrete adds to the inherent chemical resistance of the set cement to attack. This is because various transformations of the set cement occur during the curing. Another contributory factor is the greater impermeability of the set cement. Among the chemical transformations that take place is the removal of calcium hydroxide, which increases the resistance to sodium sulphate, although this gives less protection against magnesium sulphate that can attack the hydrated calcium silicates as outlined in 3(c) earlier (p. 407).

4 Concrete

(a) Scope

Emphasis is given here to sulphate attack on concrete in *sulphate-bearing clays and groundwaters** and on the measures that should be taken to prevent serious deterioration of the concrete. However, it should be noted that sulphate attack of concrete may also occur if it is exposed to industrial wastes that contain sulphates. Such cases need special consideration in the selection of cements and the adoption of precautions, much on the lines discussed later. It should be noted in particular that special measures are necessary when both sulphates and acids are present, as may occur, for example, in soils near colliery waste tips.

Shale used as hardcore for concrete floors has given rise to a number of failures due to sulphate attack. The related precautionary measures are dealt with separately at the end of this sub-section.

(b) Factors influencing attack

The factors that influence sulphate attack generally explained earlier (p. 405) are equally applicable to concrete. For convenience, those related to concrete in sulphate-bearing clays and groundwaters are:

- The amount and nature of the sulphate present;
- The level of the water table and its seasonal variation;
- The flow of groundwater and soil porosity;
- The form of construction; and
- The quality of the concrete.

It should be noted that if sulphates cannot be prevented from reaching the structure, the only defence against attack lies in the control of the quality of the concrete, which must be fully compacted and of low permeability.

(c) Sulphates commonly encountered

(1) Naturally occurring

Sulphates occur mainly in strata of London clay, Lower lias, Oxford clay, Kimmeridge clay and Keuper marl (see *3.1 General considerations*, 'Assessment of severity of conditions', p. 162). Salts occurring most abundantly are

- Calcium sulphate (gypsum)†
- Magnesium sulphate (Epsom salt)†
- Sodium sulphate (Glauber's salt)†

Sulphuric acid and sulphates in acid solution occur much less often but might be found where pyrite in the soil is being slowly oxidized (for example, near colliery tips and on marshy land).

*Much of the concrete used in buildings is exposed to attack by sulphate-bearing clays and groundwaters. Hence the emphasis given here.

†As both magnesium and sodium sulphate are highly soluble in water, and so much more soluble than calcium sulphate, the amount of sulphur trioxide in the groundwater in which they may occur may be much greater. Also the concentration in groundwater of either magnesium or sodium sulphate can vary much more widely than that of calcium sulphate, and for this reason they must be regarded as potentially more dangerous. However, it is to be remembered that, in addition to the nature of the sulphate salts involved, their amount is also important.

(2) 'Man-made'
Sulphates are sometimes present in materials such as:

- *Colliery shales* used as fill beneath concrete ground-floor slabs (see p. 415);
- Brick rubble, particularly with adhering plaster; and
- Some industrial and mine wastes.

(d) Variability of sulphate content

(1) Generally
In the UK the distribution of sulphate salts in a clay is often very irregular and may vary much between points only 15 m apart. As a result of constant leaching by rainwater, there is also a considerable variation in sulphate salt distribution with depth, and consequently the top 0.6 m or 0.9 m of soil are often, though not always, relatively free of sulphate salts. On the other hand, considerable amounts are more usually found at depths from 0.9 m to 1.8 m, or even more, below the surface.

These factors, particularly the wide variations in salt content that can occur between points quite close together, are important when making any assessment of the precautionary measures which may be advisable in any given conditions. However, some caution is also necessary when analysing the results of any field tests which may be undertaken to determine the sulphate content of a soil or groundwater (i.e. the degree of exposure), because local conditions, such as the weather (at the time of the test), the prevailing wetness or dryness of a soil, and drainage facilities and fluctuations in the water table are extremely significant.

(2) Wetness and dryness of the soil
Weather conditions at the time of making tests are important, because, in extreme cases, the concentration found in water in dry weather may be several times as large as that found in wet weather. A soil with a relatively high sulphate content but which is dry for long periods, or is so drained that the solutions formed are removed from the vicinity of the concrete will be *much less* destructive than one with a lower sulphate content and where the physical conditions are more favourable, that is, wet for long periods or inadequately drained. Generally, soils in which wetting and drying occur fairly frequently are likely to be particularly destructive. It may also be noted here that the salt content of a soil is only important in so far as it represents the reserve supplies available for maintaining, or increasing, the salt content of the waters contained in it or draining from it.

(3) Transfer
The rate with which the sulphates may be replenished under 'service' conditions is dependent on the rate water is transferred through the soil, and this, in turn, is dependent on either drainage or fluctuations in the level of the water table. 'Assisted' drainage may occur, for example, when pipelines are laid at a gradient. The effect of this is to allow the sulphate-bearing waters to flow along the lines. In so doing, sulphates may then be transferred from sulphate-bearing clay, at a high level, to non-sulphate-bearing clay at a lower one, thus increasing the severity of exposure of the concrete. The same may also occur, but in a more restricted way, in shorter excavations. In general, foundations and concrete piles do not form channels along which the flow of groundwater is likely to occur.

(e) Precautions

(1) **Creative pessimism**

As explained earlier, the factors associated with the attack of concrete by sulphates are variable. The precautions that are described below should therefore be treated as guides. Whenever there is doubt about any of the influencing factors it is wiser to assume the worst and take precautions accordingly.

(2) *Scope*

Table T3.6/3 is used as the reference point for the precautions that should be taken. What follows are notes on the more important aspects of the information contained in or related to the table. These commence with the relevance of the form of construction and type and quality of the concrete. They are followed by an explanation of the classification of soil conditions and related measurements.

(3) *Form of construction*

The degree of exposure of concrete is, in addition to the sulphate content of the soil, dependent on the form of construction in which it is used. The recommendations as to the type and quality of concrete given in Table T3.6/3 should, in this respect, be assumed to apply to the least severe condition of exposure. Such a condition is likely to apply to concrete which is *completely buried* under conditions such that the excavation does not form a channel along which a flow of groundwater is likely to occur. Foundations to buildings will usually be this class.

The *most severe* condition of exposure is likely to be found when concrete is subjected to one-sided hydrostatic pressure, as in retaining walls. A similar, though less severe, condition occurs in a partly buried concrete mass when water is drawn up by capillary forces from the portion below ground and evaporates at the exposed surface (e.g. ground-floor slabs). Severe conditions of exposure can also occur with thin sections or sections with thin bars and therefore smaller cover. In all these cases, consideration should be given to a further *reduction* of the water/cement ratio, and, if necessary, an *increase* in cement content to ensure adequate workability.

(4) *Type and quality of concrete*

Whatever cement may be used, the principal requirement for concrete is that it should be *dense* and *impermeable*, as the attack of concrete proceeds inwards from the surface, while the rate of attack is dependent on the ease with which water can penetrate into the concrete. In order to achieve dense and impermeable concrete it is necessary that particular attention be given to those factors which influence the quality of concrete, namely, richness of the mix (ratio of cement to aggregate), water content (water/cement ratio), quality and grading of the aggregate, methods of mixing, placing, and curing, and, also the age of the concrete. As regards the last, the vulnerability of cast-*in-situ* concrete to attack when it is still in the green state is worth noting. In some cases, additional protective measures may be required.

The use of *admixtures* needs a special note. Air-entraining or water-reducing agents may give some *limited* improvement in sulphate resistance in that a better-quality concrete is achieved. Those containing workability aids improve compaction and allow the use of lower water/cement ratios, but those with calcium chloride are *not* recommended. Admixtures should *not* be used with high-alumina cement without consulting the cement manufacturers.

Table T3.6/3 *Requirements for concrete exposed to sulphate attack (Note: Recommendations for concrete in near-neutral groundwater only)*

Class	Concentrations of sulphates expressed as SO_3			Type of cement		Requirements for dense fully compacted concrete made with aggregates meeting the requirements of BS 882 or 1047	
	In soil		In groundwater (g/l)			Minimum cement content[a] (kg/m³)	Maximum free water cement[a] ratio
	Total SO_3 (%)	SO_3 in 2:1 water:soil extract (g/l)					
1	Less than 0.2	Less than 1.0	Less than 0.3	Ordinary Portland cement (OPC) or rapid-hardening Portland cement (RHPC) or combinations of either cement with slag[c] or pfa[d]	Plain concrete[b]	275	0.65
					Reinforced concrete	300	0.60
				Portland blastfurnace cement (PBFC)			
2	0.2–0.5	1.0–1.9	0.3–1.2	OPC or RHPC or combinations of either cement with slag or pfa PBFC		330	0.50
				OPC or RHPC, combined with minimum 70% or maximum 90% slag[e] OPC or RHPC, combined with minimum 25% or maximum 40% pfa[f]		310	0.55
				Sulphate-resisting Portland cement (SRPC)		280	0.55
3	0.5–1.0	1.9–3.1	1.2–2.5	OPC or RHPC, combined with minimum 70% or maximum 90% slag OPC or RHPC, combined with minimum 25% or maximum 40% pfa		380	0.45
				SRPC		330	0.50
4	1.0–2.0	3.1–5.6	2.5–5.0	SRPC		370	0.45
5	Over 2	Over 5.6	Over 5.0	SRPC+ protective coating[g]		370	0.45

[a] Inclusive of content of pfa or slag. These cement contents relate to 20 mm nominal maximum size aggregate. In order to maintain the cement content of the mortar fraction at similar values, the minimum cement contents given should be increased by 50 kg/m³ for 10 mm nominal maximum size aggregate and may be decreased by 40 kg/m³ for 40 mm nominal maximum size aggregate.
[b] When using strip foundations and trench fill for low-rise buildings in Class 1 sulphate conditions further relaxation to 220 kg/m³ is permissible in the cement content for C20 grade concrete.
[c] Ground granulated blastfurnace slag. A new BS is in preparation.
[d] Pulverized-fuel ash to BS 3892: Part 1: 1982.
[e] Per cent by weight of slag/cement mixture.
[f] Per cent by weight of pfa/cement mixture.
[g] See BS 8102: 1990: *Protection of structures against water from the ground.*

From BRE Digest 250, *Concrete in sulphate-bearing soils and groundwaters*, minor revisons, 1986, Building Research Establishment, Crown Copyright.

In order to ensure minimum permeability and a hard surface when any type of concrete uses supersulphated cement it is important to give particular care during the initial curing period.

The recommendations for the type and quality of concrete given in Table T3.6/3 for each class of soil condition specify the *minimum* cement content in kg/m^3 of aggregate. This allows for some modification of the ratio of fine to coarse aggregate. It should be noted that the cement contents quoted are *minima*. In certain cases, additional amounts of cement may be required if workability is high, or aggregates are harsh, or if hand compaction cannot be avoided. As far as the water/cement ratios are concerned, it is important to note that those specified are *maxima* and based on the 'free' water present in the mix, and should therefore not be exceeded. 'Free' water, it is important to note, is the total weight of water in the concrete less that absorbed by the aggregate.

(5) Classification of sites
BRE have *five* classes of site for the UK, given in the table below in ascending order of severity (i.e. Class 1 is the least severe and Class 5 the most). Each class is based on the sulphate contents of the soil and groundwater. The table also gives the recommended type of cement (see 3 '(e) Increasing resistance of set cement', p. 408) and the corresponding *minimum* cement content for each of the classes. BRE emphasize that the divisions between the classes are somewhat *arbitrarily drawn* and that the recommendations are based on their judgements of present knowledge. This means, apply **creative pessimism**!

(6) Sampling and analysis
In the table each class has concentrations of sulphates expressed as SO$_3$ in soils (two methods) and in groundwater (one method). The procedure for sampling and testing are covered in BS 1377: 1975, *Methods of test for soil for civil engineering* and BS 5930: 1981, *Site investigations*. Methods of extraction of sulphates in soils and analysis of groundwaters for sulphates have been devised by BRE.* Apart from the care needed in carrying out the tests, account needs to be taken of the variations that are likely to occur on-site (see (d) p. 411) and the relative amounts of the potentially less dangerous calcium sulphate as opposed to the potentially more damaging sodium and magnesium sulphates.

With soils, the classification used for a particular site based on the results of sampling needs to be adjusted, depending on whether the 2:1 or 1:1 method was used. BRE gives the following criteria for each method:

	Concentration of SO$_3$ (g/l)	
Site classification	in 2:1 extract	in 1:1 extract
1	0–1.0	0– 1.0
2	1.0–1.9	1.0– 2.5
3	1.9–3.1	2.5– 5.0
4	3.1–5.6	5.0–10.0
5	Over 5.6	Over 10.0

With groundwater there is the difficulty of obtaining samples that are not diluted with surface water. As noted earlier (see p. 411), the three

*See BRE Current Paper 2/79.

most abundant salts (and expressed as grams of sulphur trioxide (SO₃) per litre of solution) differ considerably in their solubilities. That of calcium sulphate is only 1.2 (SO₃) per litre compared with about 150 and 200 times this value for magnesium and sodium sulphates, respectively. The SO₃ content of groundwaters therefore gives an indication whether one (or both) of the latter sulphates is present and the division of column 4 of Table T3.6/3 between Classes 2 and 3 directly relates to this. The BRE method refers to sulphate contents as SO₃ (sulphur trioxide).*

On the *interpretation* of analytical results BRE advises classification based on:

- *Solely the analysis of groundwater*: The classification should correspond to the *highest* sulphate concentration recorded.
- *Analysis of soil samples – small in number and vary widely*: Worth taking further samples for analysis.
- *A large number of soil samples*: The *highest* 20% of results. The selection should be more stringent when the soil samples have been combined before analysis.

Photograph P3.6/11 *Distortion and cracking of a thin concrete ground-floor slab caused by sulphate attack. (From BRE Digest 276, Hardcore, August 1983, Fig. 1, Building Research Establishment, Crown Copyright)*

(7) Protective coatings

A concrete on a Class 5 site not only requires sulphate-resisting Portland cement but also a protective coating. A coating of this kind would be impermeable and may also be used where, for example, it is impracticable to use a more sulphate-resistant cement, or if it is required to provide temporary protection to the concrete while it is curing. For the latter condition, the use of bitumized paper, hessian or similar materials may be sufficient for the least severe exposure. When, however, a very high degree of durability is required of the coating, then consideration should be given to the use of asphalt or bituminous felts, often using tanking techniques.

(f) Concrete floors in shale hardcore

(1) Background/scope

Mention is made earlier (see (c), p. 410) and in *3.5 Efflorescence*, 'Soluble salts', p. 359, that colliery shale may contain appreciable amounts of soluble salts, particularly sulphates. The soluble salt content of the shale is liable to wide variations, and some tips (the unsightly heaps of waste material near coal mines) contain as much as 5%, whereas others contain less than 1%. Shale has been used as hardcore for concrete ground-floor slabs in a similar way to hardcore of other materials, namely, by being tipped into the foundation area and compacted, using water as necessary. The concrete floor slab has, in turn, been cast on top of the hardcore without a dpm between the hardcore and the concrete. Following investigation by BRE of cases of failure of concrete slabs laid on colliery shale hardcore, it is clear that certain precautionary measures are necessary if the material has to be used. Before outlining these measures, it is considered advisable to describe the nature and causes of the failures which have occurred.

(2) Failures

Failures of concrete slabs laid on shale hardcore in houses have taken various forms, but usually the slab lifts and cracks. Considerable cracking of the concrete occurs in the later stages. In extreme cases, on very wet sites, the underside may become mushy (see Photograph P3.6/11).

*Results may be reported in other ways for which the following conversions apply:

- Sulphate ion (SO₄) *multiply* by a factor of 0.833 to convert to SO₃;
- Concentrations of SO₃ in groundwater given in parts per million (ppm) *divide* by 1000 to convert to grams per litre.

In some cases the upward movement of the slab has been accompanied by an outward movement, resulting in the outer walls being pushed out of place. In cavity construction the inner leaf may be pushed towards the outer leaf without necessarily moving the latter. However, if the cavity ties are stiff enough the movement may be transferred to the outer leaf. When the latter is of brickwork a crack usually forms somewhere near the corner of the building; but with an outer leaf of precast concrete units, there may be a displacement of the units, causing gaps between them.

Two reasons have been advanced for the lifting of the concrete slabs: (1) swelling of the shale as it gets wet, thus pushing up the concrete, and (2) transference of salts from the shale into the concrete, causing the sulphate attack of the concrete, which is accompanied by expansion. Although the swelling of the shale would appear to be insignificant, the possibility of it occurring, particularly on permanently very wet sites, should be considered with certain types of shale.

The most severe condition of exposure is likely to occur on poorly drained sites, as there will then be sufficient water present to allow transference of salts from the shale to the concrete. The chance of a sufficient quantity of salts being transferred in this way is considerably reduced on well-drained sites and on those where the water table is nearly always low.

(3) Precautions

Sulphate attack is unlikely to occur if the colliery shale to be used has a low content of soluble salts (see '(e) Precautions' and (5) Classification of sites, p. 414). However, due to the variability of the soluble salt content which can be expected in shale (even if it is from the same tip), it would be necessary to test several samples. Even this procedure may not necessarily guarantee that all the shale to be used would have a low sulphate content. Consequently, it is advisable to prevent the salts from being transferred from the shale to the concrete. This may be done by laying a sheet of polythene (at least 0.2 mm thick) as a dpm on the hardcore before the concrete is placed. The dpm should be taken up the sides of the brickwork or concrete footings to link up with the wall dpc (see *3.4 Exclusion*, p. 319).

On sites where the ground is permanently very wet, it is wise, if the shale is of a type which is likely to swell, to place an extra polythene layer on the subsoil before placing the hardcore. Care is needed to ensure that the polythene or other material is not unduly perforated by the hardcore. Good drainage would help to keep water away.

Sources

Note: Almost all the pre-1980 publications are included for background to the problems.
BRE Digest 250, Concrete in sulphate-bearing soils and groundwater, minor revisions, 1986
BRE Digest 276, *Hardcore*, August 1983
BRE Digest (First series) 31, *Concrete in sulphate-bearing clays and groundwater*, June 1951
Eldridge, H. J., *Concrete floors on shale hardcore*, Building Research Current Paper, Design series 30, 1964
Lea, F. M., *The Chemistry of Cement and Concrete* (revised edition of Lea and Desch), Edward Arnold, London, 1956
Principles of Modern Building, Vol. 1, 3rd edn, HMSO, London, 1959
Levitt, M., *Precast Concrete: materials, manufacture, properties and usage*, Applied Science, London, 1982

Photograph P3.6/12 *Typical pattern of cracking in mortar joints of brickwork due to the expansive forces characteristic of sulphate attack. (From BRE, Defect Action Sheet 113,* Brickwork: prevention of sulphate attack, *January 1988, Fig. 1, Building Research Establishment, Crown Copyright)*

5 Brickwork

Sulphate attack of fairfaced brickwork is more commonly associated with clay bricks, and these types are implied throughout. Rendered brickwork is covered under '6 Rendering'.

(a) Manifestations

Basically it is the mortar, particularly in the bed joints, in brickwork which suffers as a result of sulphate attack sometimes as early as two years after construction. The mortar affected generally has a whitish appearance; the mortar in close contact with the bricks will often be whiter than in the centre of the joint. The ways in which the attack may manifest itself may be summarized as follows:

1. Expansion of the mortar, leading to deformation and cracking of the brickwork (see Photograph P3.6/12). In this the mortar in the bed joints, particularly in the early stages of attack, remain rigid although cracked, with considerable increase in the height of the wall taking place. Height increases of as much as 0.2% in the external leaf of a cavity wall to a two-storeyed dwelling have been reported. Bowing of parapets may occur.
2. The edges of the individual bricks may spall.
3. The mortar deteriorates. Generally, the affected mortar cracks along the length of the joint (i.e. laminates) while the surface may fall off. The surface effects are very similar to those caused by frost action. However, frost action usually shows less extensive cracking and more surface spalling. When much water is present the affected mortar may be reduced to a soft mush.
4. Serious disintegration of the brickwork generally, which is the ultimate result following the lamination of the mortar joints outlined in (3) above (see Photograph P3.6/13).

Photograph P3.6/13 *Disintegration of bricks due to sulphate attack of the mortar. (From BRE Digest 89,* Sulphate attack on brickwork, *minor revisions 1971, Fig. 4, p. 2, Building Research Establishment, Crown Copyright)*

(b) Causes

In principle, the causes for the attack of the mortar are similar to those already discussed under '1 General considerations' and '2 Basic mechanisms'. However, it is necessary to note here the more common source of sulphates and exposure to water.

The chief source of sulphates is the bricks, which are likely to vary considerably in their sulphate content, as explained in *3.5 Efflorescence*, '1 Salts originally present in a material, (a) Clay bricks'. In general, no correlation has been found between sulphate content of bricks and sulphate expansion. For example, brickwork does not expand if the bricks have a low sulphate content; there is marked expansion with bricks containing a high sulphate content; but with those with intermediate amounts of sulphate, expansion has varied widely. In this, variations in tricalcium alumiante content of the cement used in a mortar and the presence (or absence) of water are significant.

Normally, the vertical surfaces of walls do not remain wet for long enough for attack to take place, although soluble salts in the bricks may have been transferred during the wet period. In some of the wetter parts of the UK (see *3.1 General considerations*, p. 140 ff *3.2 Exposure*, p. 180 ff) this assumption may have to be questioned. Sulphate attack is therefore usually, but not always, confined to those areas of walls where water has been allowed to percolate into the body of the brickwork containing sulphates, such as may occur through the joints of copings or sills. Areas which are likely to remain *persistently damp* or damp for prolonged periods must, however, also be included. Such areas would include: (1) highly exposed parts of a wall, such as parapets; (2) from foundation level to dpc level; (3) completely free-standing walls, such as boundary walls; and (4) earth-retaining walls (extra source of sulphates, i.e. the soil or sulphate-bearing groundwaters, together with only one evaporating surface are significant). It may also be noted that defective service pipes may also be an important source of water (and sometimes of sulphates as well).

(c) Precautions

In general, the precautions which could be taken to minimize the risk of sulphate attack taking place in brickwork (as in other cement-based materials) involve the control of one of the three variables – sulphates, tricalcium-aluminate or water.

So far as the sulphates are concerned, BS 3921: 1985 now classifies bricks into two categories: Low (L) or Normal (N). Normal bricks have no limit on soluble salt content. For Low bricks, the percentage of mass soluble ions should not exceed the following:

Calcium	0.30
Magnesium	0.03
Potassium	0.03
Sodium	0.03
Sulphate	0.50

By classifying bricks in this way the Standard recognizes that many will be used in situations where there is little or no risk of sulphate attack (e.g. in internal walls and partitions). It is therefore important that the sulphate content of a brick for a particular use and exposure is checked beforehand.

Despite the limitations of BS 3921, bricks of low sulphate content complying with the requirements of the standard should be used. This is not always possible, and so extra precautions namely, *either* to increase the resistance of the mortars to sulphate attack *or* to limit the extent to which the brickwork becomes and remains wet should be taken. In exceptional cases it may be necessary to employ both methods.

The sulphate resistance of mortars can be increased *either* by specifying richer mixes (for example, 1:½:4–4½ or 1:5–6 with plasticizer in place of lime) thereby making the mortar less porous, *or* by using sulphate-resisting, supersulphate or high-alumina cements. It is important to note that lime additions should not be used with high-alumina cements (ground limestone may be used).

Precautions may be summarized as follows:

1. Generally ensure that the brickwork is kept dry in service – attention to details at eaves, verges, sills, parapets (including chimneys) and copings to prevent the ingress of water (see *3.4 Exclusion*, Diagram D3.4/46, p. 318) is important. Here it may be emphasized that sulphate attack is unlikely to occur if brickwork does not become and remain unduly wet.

2. Ideally, parapets and free-standing walls should be avoided. Where this is not possible it is essential to:

- Use low-sulphate bricks;
- Design copings with generous overhang and adequate drip with dpcs under them.
- Provide dpcs at bases of free-standing walls that are above expected soil levels and at roof level in parapets;
- Provide expansion joints not more than 12 m apart;
- Specify sulphate-resisting mortar mixes.

3. Although it may be generally assumed that no special precautions are necessary in vertical walls between dpc and roof level (in this the height of any parapets are excluded), the use of sulphate-resisting Portland cement is a worthwhile precaution in the wetter parts of the UK.

4. In retaining walls, manholes, brickwork below dpc level, and in other situations where wet conditions for relatively long periods are likely to be expected, only bricks having a low sulphate content should generally be used. Such bricks should be employed in conjunction with sulphate-resisting mortar mixes (e.g. 1:½:4½ or stronger, and preferably containing a sulphate-resisting cement). In this an alternative to clay bricks would be sand-lime or concrete bricks as these seldom have sufficient amounts of sulphates (see *3.5 Efflorescence*, p. 359).

   Where, however, in earth-retaining walls it is necessary to use bricks containing appreciable amounts of sulphates, then the retaining wall should be built of *in-situ* concrete, without a batter, with the facing brick layer separated from it by a cavity. Adequate copings and expansion joints should be provided as for parapet walls while weepholes at the base of the cavity to ensure drainage but not venting of the cavity should be incorporated.

5. Brickwork exposed to sea spray or other saltbearing water should preferably be built of dense bricks and cement mortar.

6. Gypsum plaster should never be added to mortars containing Portland cement or hydraulic lime.

Sources

BRE Digest 89, *Sulphate attack on brickwork*, minor revisions 1971

BRE Digest 160, *Mortars for bricklaying*, December 1973

BRE Digests 164 and 165, *Clay brickwork: 1 and 2*, minor revisions 1980 and May 1974

BRE Direct Action Sheet 113, *Brickwork: prevention of sulphate attack* January 1988

Some common defects in brickwork, National Building Studies, Bulletin 9, HMSO, London, 1950

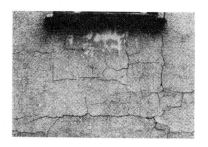

Photograph P3.6/14 *Typical 'map' or 'crocodile' pattern of cracking associated with sulphate attack of rendering. (From BRE Digest 89, Fig. 89. p. 2, Building Research Establishment, Crown Copyright)*

6 Rendering

It is important to note that sulphate attack of rendered brickwork (clay bricks are again implied) is more commonly associated with dense cement renderings (1:2 or 1:3, cement:sand).

(a) Manifestations

With rendered brickwork there are two component parts which may be attacked, namely, the mortar joints as previously discussed under '5 Brickwork' and the rendering itself. The amount of expansion vertically may be as much as 2%, that is, ten times as much as in facing brickwork. The ways in which the attack may manifest itself on rendered brickwork may be summarized as follows:

1. The 'map' or 'crocodile' pattern of cracking which the rendering may develop as a result of drying out shrinkage of the rendering is accentuated, and may later be followed by horizontal cracks formed along the lines of the mortar joints (see Photograph P3.6/14).
2. Wide horizontal and vertical cracks may appear in the renderings, while there may be some outward curling of the rendering at the cracks.
3. The adhesion of the rendering to the brickwork may fail, resulting in the rendering falling off, either from individual bricks or in fairly large sheets, depending on the extent to which its undersurface has also been attacked by sulphates. Often the brickwork thus exposed will show white efflorescence.

(b) Causes

The causes of sulphate attack are generally the same as those previously outlined under '5 Brickwork'. However, there are some detailed considerations which must be taken into account.

Under similar conditions, the likelihood of the failure, due to sulphate attack, of brickwork rendered with dense cement renders is much greater than that of unrendered brickwork. Two factors are mainly responsible for this. First, there is a much greater surface area on which attack may take place, while, in addition, there are the mortar joints which may also be attacked. Second, dense renders are more or less impervious, thus encouraging far longer periods of dampness in the brickwork should water gain access. In this there is the tendency of dense renders to develop cracks, particularly during the drying-out period; these cracks may then provide a further means of ingress for water (that is, in addition to the other common defects which allow water to percolate into brickwork). Once the water has gained entry, it cannot subsequently readily evaporate.

(c) Precautions

1. Particular attention is required to details at eaves, verges, sills, parapets and copings (see *3.4 Exclusion*, p. 317) to prevent the ingress of water. In parapets it is wise to ensure that one face is left unrendered (the 'roof' face, for example). This then provides one face from which any water that may have gained access may evaporate.
2. Dense cement renderings (1:2 or 1:3, cement:sand) should not be applied to bricks containing large amounts of sulphates, due to the tendency of such renderings to form shrinkage cracks which allow

water to enter the brickwork. This water cannot readily evaporate through the dense rendering. Dense renderings may not be required, even if they do not develop cracks, except possibly in the more heavily exposed parts of the UK. Consequently, consideration should be given generally, but not exclusively, to the use of more porous and hence permeable renderings, containing cement gauged with lime, such as 1:1:6, cement:lime:sand, which permit evaporation from the brickwork more readily than the dense mixes; they may still afford adequate protection against rain penetration and are less liable to cracking. Thus the permeable renderings are less likely to accentuate sulphate attack.

3. In those areas in which sulphate attack is common, or where exposure to driving rain is heavy, and if rendering is essential it is advisable, in addition to choosing bricks with a low sulphate content, to use a cement which is more resistant to sulphate attack than OPC for both the mortar and the rendering. Suitable cements are sulphate-resisting Portland cement, supersulphated cement or high-alumina cement.

4. Gypsum plaster should never be added to mortars or renderings containing Portland cement or hydraulic lime.

Sources

BRE Digest 89, *Sulphate attack on brickwork*, minor revisions 1971
BRE Digests 164 and 165, *Clay brickwork: 1 and 2*, minor revisions 1980 and May 1974
BRE Digest 196, *External rendered finishes*, December 1976 (summarizes information and recommendations in BS 5262: 1976)
Some common defects in brickwork, National Building Studies, Bulletin 9, HMSO, London, 1950

7 Domestic chimneys

Sulphate attack of domestic chimneys mainly occurs when the condensation risk with domestic boilers and slow combustion stoves and cookers has been overlooked. The risk of condensation, it may be noted here, generally increases as the efficiency of the boiler increases, while the flue gases themselves may be destructive. (See also 'Acid action', p. 389.)

In addition to the signs, mentioned previously for brickwork (see Photograph P3.6/15), both unrendered and rendered, sulphate attack in chimney stacks may cause a gradual curvature or leaning of the stack.

In principle, precautions aimed at preventing attack from taking place (that is, in addition to those which should be taken for exposure to rain, as previously described) are concerned with either reducing or eliminating the condensation by increasing the thermal insulation of the chimney, or by introducing relatively dry air into the flue through a ventilator just below the ceiling. Alternatively, an impermeable flue lining which prevents any moisture that condenses within the flue from entering the brickwork may be provided. For the modern high-efficiency boiler it is, however, advisable, not only to ensure adequate thermal insulation (together with the correct flue size) but also to incorporate a flue lining. The use of a flue lining, together with adequate insulation, particularly when existing flues are to be used, is essential. For convenience, flue linings which may be used in both new and existing chimneys are set out below. Insulation requirements are included.

Photograph P3.6/15 *Distintegration of brickwork to an unlined domestic flue due to sulphate attack of the mortar*

(a) *Flue linings*

Lining materials suitable for use in both new and existing flues, with spigot and socket or rebated joints, include:

1. *Impervious clay pipes*, often salt-glazed but may include certain classes of ceramic-glazed and unglazed pipes.
2. *Dense concrete pipes*, which may require internal surface treatment with an acid-resisting compound.

For new work, proprietary *precast concrete flue blocks* are available for construction of external flues.

In addition to the above, the following materials are available for use in *existing flues*:

3. *Refractory concrete pipes*, made of concrete from crushed firebrick aggregate and high-alumina cement which is, to some extent, acid resisting. The precast sections have rebated joints that should prevent seepage.
4. *Flexible and jointless metal linings* intended primarily for use with gas- and oil-fired boilers, circular in cross section and may be of stainless steel or aluminium, or made with an outer leaf of aluminium separated from an aluminium or lead foil inner leaf by corrugated paper.
5. *Lightweight concrete infills*, formed by a method that is basically different from those described above, namely, by 'casting' the infill between the existing chimney and an inflated rubber 'core'. Experience with this method is limited, and, due mainly to the fact that the flue surface is not impermeable to flue gases, there appears to be a substantial risk that the gases may diffuse into the voids in the concrete, subsequently condense, and start attacking the cement.

(b) *Staining*

In the absence of precautions which will reduce the condensation risk in chimneys, the sooty and tarry deposits arising from combustion of fuel may cause ugly stains on ceilings and walls (usually the chimney breast). These stains are difficult to remove and consequently remedial measures involve the complete removal of plaster or of applying a new lining. Before replastering, care must be taken to ensure that the stains have also been removed from the background.

Sources

Some common defects in brickwork, National Building Studies, Bulletin 9, HMSO, London, 1950

BRS Digest (Second Series) 60, *Chimney design for domestic boilers*, July 1965

Fungi and insects

1 Background

(a) *Scope*

The characteristics, behaviour and effects fungi and insects found in buildings are explained here. The only similarity between the two is that they both need food, suitable types of which are commonly found in buildings and usually cause much damage to the materials on or in which they grow or live. It is for this reason, and for convenience as well, that they are included here. So far as the fungi are concerned, this section provides the basis for a detailed consideration of the effects of wood-rotting fungi, and the precautions that should be taken to reduce the risk of rot occurring are covered in the next section (p. 431).

The aim throughout is to provide the fundamental basis for an understanding of the problems and their solutions during the design and construction of buildings. How problems in existing buildings should be investigated, the relevant fungi or insects identified or remedial solutions devised is a matter for expert attention and advice.

(b) Context

Damage to wood by fungi (the wood-rots in particular) and insects has a long history in buildings, as have the methods to increase the resistance of timber to decay or attack by means of 'preservatives'. Mould growth, also a fungus, lay dormant, so to speak, until about 25 years ago, when conditions of moisture in certain buildings (notably dwellings) allowed the mould to grow in an intense and disturbing way (see Photograph P3.6/16). Apart from, or in addition to, high conditions of humidity within a dwelling, condensation has been shown to be an important contributing factor to the widespread outbreaks of mould growth. The fungi that cause mould growth, in common with most other fungi that cause the decay of wood, can be a health hazard for people with respiratory problems.

In the light of recent experience the problems associated with fungi, whether mould growth or wood decay, should be considered as one, because all the building fungi share the same life cycle and roughly the same conditions, while both need moisture to start growing and, in most cases, to continuing growing/spreading. Whereas not so long ago it was appropriate to consider the avoidance of those design and construction features that would result in conditions sufficiently damp to lead to wood decay, consideration must now be given to the likelihood of conditions of dampness from all sources of moisture everywhere on, along or within constructions (see *3.2 Exposure*).

Photograph P3.6/16 *A general view of mould growth in a corner due to surface condensation*

2 Fungi

(a) Characteristics

Fungi that cause mould growth or destroy wood belong to a large group of very simple plants that includes edible mushrooms and toadstools. There are many species of fungus, each with a Latin name. The life cycle of a fungus is illustrated schematically in Diagram D3.6/1, and is similar to the life cycle of the higher plants. Fungi take the form of a fine web of microscopic root-like threads which grow over and into the materials from which they extract the food they need. The reproductive parts are contained in the fruiting body (encircled in the diagram) that are equivalent to the flowers and fruits of higher plants. The spores (seeds in the higher plants) are released into the air by the fruiting body, usually in very large numbers. This accounts in part for the ability of fungi, mould fungi in particular, to spread rapidly.

Germination of a spore commences when it has found the appropriate conditions, and growth continues while those conditions persist. The growth consists of the development of exceedingly fine hollow threads, known as *hyphae*, that grow in length by elongation of their tips. The *hyphae* may be arranged loosely or bunched closely, to form soft cushions. Dense skins, sheets, lumps or long strings are the manifestation of the closely interwoven tubes known as the *mycelium*

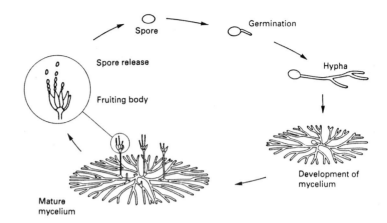

Diagram D3.6/1 *Schematic illustration of the life cycle of a typical fungus. (From Bravery, A. F., Mould and its control, BRE IP 11/85, June 1985, Fig. 1)*

(see Table T3.6/4 and Photograph P3.6/17). A mature *mycelium* produces fruiting bodies (Photograph P3.6/18). The colour of a fungus may be derived by the colour of the *hyphae* or the spores or a combination of both. In any event, each fungus has its own identifying colour(s). With moulds, the spores give the powdery appearance associated with them.

(b) Conditions for growth

(1) Main requirements
The main requirements for the growth of fungi are:

- A source of infection;
- Food;
- Water;
- Oxygen; and
- Suitable temperature.

(2) Water – the limiting factor
Of these, only water is normally the limiting factor in most buildings. This is because:

- The source of the infection (i.e. the spores) is always present in buildings to a greater or lesser extent;
- Food, often only required in small quantities, is normally available;*
- Aeration levels always provide sufficient oxygen;† and
- Fungi have a wide tolerance of temperature (0–60°C), growth increasing rapidly as the temperature rises above freezing until the optimum is reached.‡

(c) Sources of moisture

One or more of the following are sources of moisture for fungal growth:

- Water vapour in the atmosphere (usually expressed as % rh);
- Condensation;
- Rain penetration;
- Rising damp; and
- Faulty services.

All of these are covered in *3.2 Exposure* but see also *3.3 Moisture content* ('Timber', p. 239 ff).

*For moulds: normal levels of dust and other deposits even in clean and well-maintained houses are sufficient; for wood-rots: cellulose and lignin in wood or other materials with a similar composition such as paper and straw.

†It should be noted that there is no air in saturated timber because the cells are filled with water. This is why such timber is quite immune from attack and why wooden galleys submerged since Roman times or the timber from lake dwellings have been found intact.

‡In the case of the cellar fungus, for example, growth is four times as fast at 24°C as it is at 10°C.

Photograph P3.6/17 *An example of sheets of dry rot mycelium exposed by removal of timber wall panelling. Dry rot is characteristically different from other types of wood-rot – see text. (From BRE Digest 299, Dry rot: its recognition and control, July 1985, Fig. 2, p. 3, Building Research Establishment, Crown Copyright)*

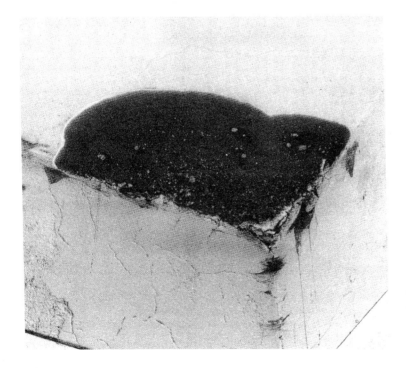

Photograph P3.6/18 *Dry rot fruit-body with spores trapped on the cobwebs and deposited on horizontal surfaces. (From BRE Digest 299, Fig. 3, p. 6, Building Research Establishment, Crown Copyright)*

Table T3.6/4 *Types of fungi and their occurrence, effects and characteristics*

Rot type	Fungus	Occurrence	Effect on wood	Mycelium	Strands	Fruit-body
Dry	*Serpula lacrymans* (brown rot) Dry rot fungus	Mostly softwoods. Major building decay fungus often causing extensive damage. A brown rot which typically occurs on or in wood embedded in or in contact with wet brickwork. Sensitive to high temperatures (over 25°C) and drying and therefore rarely found on exposed timbers or situations where fluctuating conditions are likely (e.g. well-ventilated sub-floors, roofing timbers). More likely to be found in humid, unventilated places where the air is still and the woodwork damp but not saturated. Able to grow through bricks and mortar, although cannot feed on these. Importantly, strands can transport moisture from damp area, allowing spread of fungs to dry wood in unventilated conditions. Appearance of fruit-body may be first indication of outbreak	Decayed wood brown in colour, typically with deep cracks along and across the grain, light in weight and crumbles between fingers. No skin of sound wood	Silky-white sheets or cottonwool-like cushions with patches of lemon yellow or lilac tinges where exposed to light. In less humid conditions forms thin, felted grey skin. During active growth the advancing hyphal edge forms a silky fringe	White to grey, branching, sometimes as thick as a pencil, brittle when dry	Usually on wood/wall joint, rare on exterior of building. Tough, fleshy, pancake or bracket-shaped. Centre is yellow-ochre when young, darkening to rusty red when mature due to spore production, covered with shallow pores or folds; margin white or grey. *Spores*: profuse and may settle as fine layer of reddish-brown dust on horizontal surfaces
Wet	*Coniophora puteana* (brown rot) The cellar fungus	Softwoods and hardwoods. Commonest cause of decay of woodwork which has become *soaked* by water	Wood darkens with cracks along and across the grain but usually less deep than those caused by *Serpula lacrymans*. Where conditions cause drying of the wood surface an apparently sound skin of timber often occurs, which may crack longitudinally as the decay progresses beneath. Freshly colonized wood usually shows a yellow coloration	Only present in conditions of *high humidity*. Yellow to brownish in colour, off-white under impervious floor coverings. May spread superficially over damp plaster or brickwork	Thin, usually brown or black, though yellowish when young	Rare in buildings. Thin, lying flat on substrate and with small irregular lumps. Olive-green to olive-brown with cream margin, paler when young

Type	Fungus	Where found	Effect on wood	Mycelium	Strands	Fruit-body
Wet	'Poria' species including *Amyloporia xantha*, *Fibroporia vaillantii*, *Poria placenta* (brown rots) (*Fibroporia vaillantii* also referred to as white pore or mine fungus)	Softwoods. Together this group are a common cause of rot of *damp woodwork in buildings, frequently in areas of higher temperature*, and can lead to extensive damage (particularly *Fibroporia vaillantii*)	Resembles *Serpula lacrymans* – wood breaks up into cuboidal pieces but decayed wood is lighter in colour and cracks are not as deep as those caused by *Serpula lacrymans*	White or cream sheets or fern-like growths. May discolour brown on contact with iron.	Seldom thicker than twine, white to cream, remain flexible when dry (only well developed in *Fibroporia vaillantii*)	Irregular lumpy sheets or plates, white or cream to pale yellow. Spore-bearing surface white to pale yellow, occasionally with pink patches (*Poria placenta* only), with numerous minute pores
Wet	*Phellinus contiguus* (white rot)	Softwoods and hardwoods. Common on external joinery	Wood bleaches and eventually develops stringy, fibrous appearance. No cuboidal cracking; does not crumble	Tawny-brown tufts – may be found around fruit-body or in crevices in wood	None	Occasionally found. Tough, elongated, ochre to dark brown in colour, covered in minute pores
Wet	*Donkioporia expansa* (white rot)	Hardwoods, particularly oak, though may spread to adjacent softwood. Common on timbers where there has been *persistent water leakage*. Can cause more extensive damage to oak than any other fungus. Often found at ends of beams embedded in damp walls. Damage may be confined to interior of beam and not noticed until fruit-body appears. Often associated with death-watch beetle attack	Wood becomes bleached and is reduced to lint-like consistency leaving stringy white fibres. Decayed wood easily crushed but does not crumble	Yellow to red-brown thick felted growth, often shaped to contours of wood, exudes drops of yellowish-brown liquid	None	Thin, leathery, plate or bracket shaped, or thick, hard and woody. Brown or buff coloured. Spore-bearing surface cinnamon-brown or fawn with numerous minute pores; often several pore layers present
Wet	*Pleurotus ostreatus* (white rot). Oyster fungus	Generally found on broad-leaved trees. Occasionally found in buildings, usually associated with decay of *panel products*	Board lightens in colour; in particleboards, chips tend to separate	Whitish, woolly mat.	None	Grey or fawn mushroom with whitish, plate-like gills beneath, with or without stalk; stalk, if present, not central
Wet	*Asterostroma* spp. (white rot)	Softwoods. Usually found on joinery (e.g. skirting boards), often limited in extent of spread	Wood becomes bleached and develops stringy, fibrous appearance. No cuboidal cracking; does not crumble	White, cream or buff sheets, not always present	Thin, white, with rough appearance. Remains flexible when dry. Sometimes can cross masonry over long distances	Thin, sheet-like, hardly distinguishable from mycelium. No pores
Wet	*Paxillus panuoides* (brown rot)	Softwoods. Prefers very damp situations	Initially a vivid yellow discoloration appears wherever mycelium is present. In an advanced stage the decayed wood becomes soft and cheesy, and on drying, deep longitudinal fissures and some fine cross-cracks appear	Soft, hairy or woolly, dull yellow with occasional tinges of violet	Thin, colour as mycelium, do not darken with age	Dingy-yellow, fan or funnel-shaped, without stalk. Gills yellow and branch frequently

From Bravery, A. F., Berry, R. W., Carey, J. K. and Cooper, D. E., *Recognising wood rot and insect damage in buildings*, BRE Report, 1987, pp. 15–35, Building Research Establishment, Crown Copyright.

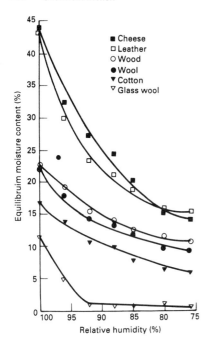

Diagram D3.6/2 *The equilibrium moisture contents of several materials at 20°C (85°F) and different relative humidities (Block, 1953). (From Bravery, Fig. 5, Building Research Establishment, Crown Copyright)*

(d) Food and moisture content

The food and moisture content requirements of moulds and wood-rotting fungi differ in detail.

(1) Mould growth
The spores will germinate on a number of materials, including dust and other deposits, cheese, leather, wool, cotton and glass wool. However, normal levels of dust and other deposits, even in well-cleaned and maintained interiors, are sufficient for germination. Abnormal deposits of food encourage mould, but they are not an essential prerequisite.

The key to the susceptibility of materials in buildings to mould is mainly in their moisture content. As explained in *3.3 Moisture content* (see 'Equilibrium moisture content', p. 225), different materials have different equilibrium moisture contents. Diagram D3.6/2 illustrates graphically the emcs of materials normally associated with mould growth. These have different emcs for the same condition of relative humidity. The emc does not give the amount of moisture in the material available to the fungus. This is why not all materials are equally susceptible to mould, even under the same conditions of relative humidity. For example, mould develops on leather at 76% rh but does not do so on wood below 85% rh nor on glass below 96% rh. On brick and painted surfaces BRE has found that mould was negligible below 88% rh but increased markedly above 95% rh.

Individual species of fungi vary in their tolerance to moisture conditions. Those commonly found in dwellings can survive under very low levels of available water; others have moderate tolerance. High water availability is required by those species of fungi commonly found on foods such as bread or cheese. Temperature is also another influencing factor.

In terms of precautions, there needs to be a rule that will take into account the complexity of the interaction between temperature, humidity and the properties of the food in materials and the varying tolerances of different mould fungi. BRE suggest that, as *a general rule,* marginal conditions occur when a relatively humidity is consistently close to 70% rh or if it fluctuates markedly above or below this level at different times. BRE regard this level as a significant threshold, because above it the probability of condensation occurring at some time on cold surfaces increases noticeably.

For convenience, the factors that affect condensation and mould growth in dwellings are briefly:

● Moisture generation;
● Ventilation;
● Thermal insulation; and
● Surface absorption.

All except the last are referred to in *3.3 Moisture content*, p. 230 ff. Absorption by surfaces of condensation may reduce the risk of mould growth occurring if the water that is absorbed can be evaporated relatively soon after the condensation has stopped.*

(2) Wood-rots
Wood-rots present a simpler picture as to both food and moisture content. This is probably due to greater experience with decay by wood-rots.

The food picture is simpler because the rots feed on either the cellulose or the lignin (or both) in the wood. For this purpose, the fungi can be divided into two types: brown and white rots (see Table T3.6/4):

*BRE is conducting research into this with a view to advising on surface-absorption properties to reduce the incidence of mould growth.

- *Brown rots* feed mainly on the cellulose;
- *White rots* feed on the cellulose and the lignin but to a varying extent, depending on the particular fungus.

What may be done by way of an added precaution is to 'sterilize' the food supply by what is known as 'preservation' (covered later under 'Wood-rots', p. 434).

On moisture content, species of fungus do vary as to the moisture content of the wood at which their growth is initiated and sustained. For safety, it is wise to adopt a maximum moisture content of wood of 20%. This is about the level at which dry rot, the most virulent and damaging of the wood-rots, will start to colonize wood. Although high humidities for relatively long periods should not be overlooked as the cause of high equilibrium moisture contents in wood, experience so far has shown that such contents are invariably due to dampness from rainwater penetration and the like (see (c) earlier).

3 Insects

(a) Characteristics and life cycle

Like fungi, insects come in many species, all Latin-named. Those, like beetles, that are able to use wood as a source of food damage building timbers with small holes 'drilled' into their interior and some insects can cause serious harm. The insects have a fairly similar life cycle (see Diagram D3.6/3). This follows the stages in the development of moths and butterflies, namely, from egg to caterpillar (or *larva*) followed by a resting period (*pupation*) as a chrysalis until the moth or butterfly emerges. In insect attack on wood the activity of the larvae is damaging and therefore significant.

The cycle of a typical wood-boring insect is as follows:

- Eggs are laid by the adults on the surface of the wood in splits or in bark.

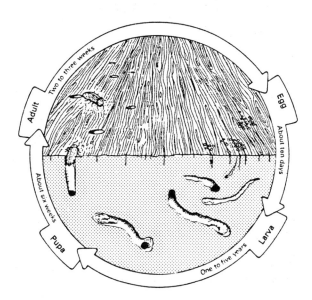

Diagram D3.6/3 *Schematic illustration of the life cycle of a typical wood-boring insect. (From BRE Digest 307,* Identifying damage by wood-boring insects, *March 1986, Fig. 1,* Building Research Establishment, *Crown Copyright)*

- Within about two to three weeks the eggs hatch into active, grub-like larvae which eat their way *into* the wood, thereby creating tunnels. The damage caused to timber results largely from the feeding and tunnelling of the larvae,* which take between one and five years to grow fully.
- When full grown they undergo a transformation, first as a pupa and then as an adult insect. The adults emerge from the infested wood, leaving behind the characteristic holes.
- The adults do not themselves cause further damage but, after mating, lay further eggs, thereby spreading the infestation.†

(b) Moisture content

While the larvae of insects can cause their characteristic damage without the aid of moisture, the presence of moisture assists them in their job. In many cases, insect attack is found in conjunction with or as a result of investigations into attack by wood-rots.

(c) Precautions

There is no shortage of recommendations on the investigation of and remedial treatments for insect attack. There is little or no specific advice from authoritative sources on what might be done initially to prevent the attack occurring. It is therefore necessary to apply the lessons to be learnt in the eradication of infestations. These include:

1. While insects likely to cause damage are not necessarily as widespread or as concealed as fungi, the likelihood of attack occurring in a particular area should be checked first.
2. Whenever in doubt, ensure that the timber is treated with an appropriate preservative treatment. Treatments that also 'sterilize' the food needed by wood-rots are obviously an advantage.
3. As in the case of wood-rots, ensure that timbers are designed, detailed and constructed so as to remain dry during their service life.

Sources

Desch, H. E. (revised by Dinwoodie, J. M.), *Timber, its Structure, Properties and Utilisation*, 6th edn, Macmillan, London, 1981
Richardson, Barry A., *Remedial Treatments of Buildings*, The Construction Press, London, 1980

BRE
Digests
139, *Control of lichens, moulds and similar growth*, new edition 1982
238, *Reducing the risk of pest infestations: design recommendations and literature review*, June 1980
297, *Surface condensation and mould growth in traditionally-built dwellings*, May 1985
299, *Dry rot: its recognition and control*, July 1985
327, *Insecticidal treatments against wood-boring insects*, December 1987
307, *Identifying damage by wood-boring insects*, March 1986

Information papers
15/83, Berry, R. W. and Orsler, R. J., *Emulsion-based formulations for remedial treatments against woodworm*, October 1983
11/85 Bravery, A. F., *Mould and its control*, June 1985

Report
Bravery, A. F., Berry, R. W., Carey, J. K. and Cooper, D. E., *Recognising wood rot and insect damage in buildings*, 1987

*What is known as 'bore dust' are wood pellets excreted by the larvae, and this fills the tunnels. The size and cross section of the tunnels and, to a lesser extent, the characteristics of the bore dust depend on the type of insect involved.

†Identification of adult insects and their larvae from those that may be found on (or in) infested wood is difficult without experience and reference to detailed literature. The characteristics of the damaged wood itself is therefore the more usual basis for identification by experts.

Wood-rots

1 Scope

This section builds on characteristics, behaviour and effects in general of the fungi explained in 'Fungi and insects' earlier (p. 422 ff). Attention is now focused on the wood-rot fungi for which detailed consideration is given to the types of rots commonly found in building timbers, the resistance of timber to decay by these fungi, commonly occurring problems and their resolution in principle and the precautions that should be taken to reduce the risk of rot or decay occurring in timber.

2 Types of rot

(a) White and brown

Table 3.6/4 gives the occurrence, characteristics and effects of wood-rots that cause decay of wood. It will be seen that there are only two types of rot: wet and dry. The fungi are also divided into two types, brown and white rots. These differ mainly in the food they prefer, as explained earlier (see (2)(d)(2), p. 428). Another distinguishing characteristic is the effect each has on the wood it has decayed. The brown rots cause the wood to crack, with the cracks generally forming a cuboidal pattern. With the white rots there is generally no cracking; instead, the wood becomes bleached or stringy and may be crushed with relative ease.

(b) Dry and wet

The terms *dry rot* and *wet rot* can be misleading if they are taken to mean the conditions required for the active growth of fungi. For reasons given earlier, 20% moisture content is a *critical minimum* for active growth of all fungi. A distinction has to be made between active and optimum growth. For the latter most fungi require high moisture contents of the wood to be between 35% and 50%, with some preferring even wetter conditions. To add to the complexity, the minimum moisture content required for spores to germinate is higher than that for the actively growing mycelium.

The terms become more meaningful in relation to the state of the timber in the final stages of the activity of the fungus. In the case of *dry rot* the wood may be dry and friable whereas with *wet rot* the fungi require the wood to be comparatively wet for them to initiate or continue attack. As *a general guide*, bearing in mind that there is only one dry rot, it is useful to note that wet rots need a higher moisture content of wood for germination and activity of the fungi, remain wet after the activity has stopped and, importantly, cease activity where drier conditions prevail. *Dry rot*, on the other hand, apart from requiring a lower moisture content of the wood for germination and activity, may *continue to be active* even when the moisture content is lowered because of the ability of the fungus to conduct water from wetter to drier parts of the timber. The fungus can also spread over the surfaces of materials or through weak jointing between units. It is for these reasons that dry rot is extremely virulent and often difficult to eradicate. Prevention of this type of decay, representing the worst case in that it requires the lowest moisture content of the fungi for germination and activity, is therefore very important.

(c) Dry rot

As mentioned above, prevention of dry rot is important. Its special characteristics and effects merit special consideration. Its strands or strings may vary in thickness from thin threads to hard strands as thick as a pencil lead and are able to pass across inert substances such as bricks or metals. Consequently, the fungus can be transported from timber in one area of a building to that in another. In addition, the strands are capable of passing through the joints in walls, so the fungus can spread from room to room.

The most important characteristic of the strands is that the special hyphae they contain are capable of carrying water from damp to dry areas. This enables the fungus to attack relatively dry parts of a building.* The strands also contain a reserve supply of food materials. This means that growth can be renewed after the replacement of infected timber if the surrounding walls have not been sterilized.

A characteristic of the fruit-bodies is that they produce rusty red spores in very large numbers, many millions being shed from a single fruit-body.† It is therefore comparatively easy for the infestation to become widespread.

Wood thoroughly decayed by dry rot has the appearance of charred wood. It is friable, light and dry, and falls to powder under the fingers. The wood is also broken up in more less cubical pieces due to cracks running both along and across the grain. The cracks are usually deep.‡

3 Resistance of timbers

(a) Generally

When considering the resistance which any particular species of wood may have to fungal attack account must be taken of the fact that there is *no risk of attack while the timber remains perfectly dry*. For reasons given earlier, this means that, for practical purposes and safety, the timber must remain at a moisture content of 20%, except for short periods of time. Otherwise, the consideration of the resistance of wood in general may be approached from two directions: first, the natural resistance that different species of wood offer and, second, the resistance of the different species of wood when treated with preservatives. As regards the latter it is important to note that different species of wood respond differently to preservative treatments (see below).

(b) Natural resistance

(1) Variations and sapwood
As is to be expected, there is a very large variation in the resistance to fungal attack provided by different species of wood. In this the structure and rate of growth of a given species is influential. However, far more important is the difference in resistance between the *sapwood* and *heartwood* of all the species. As a *general rule*, the sapwood of all species has little or no resistance to fungal attack. For practical purposes it should be regarded as being perishable under conditions of moisture content that are likely to lead to such attack.

(2) Resistance of heartwood
The heartwood is more durable than the sapwood because in the growing tree it contains extractives¶ that are toxic to fungi. The variation that occurs in the natural resistance between species depends on the type, quantity and distribution of these extractives.

*During active growth the fungus often produces drops of moisture – hence the name *lacrymans*, the Latin for weeping.

†The spores are capable of being blown about with the slightest draught and can be carried by insects and vermin. In many cases it is the red 'dust' that is the first indication that there is dry rot. Another might be the distinctive 'mushroomy' smell of the mycelium.

‡This appearance is characteristic of brown rots in general. However, they are usually more pronounced in dry rot. The differences between the two might not be apparent to the uninitiated, and is another reason for the identification of the fungus causing an infestation to be undertaken by an expert in the field.

¶These are naturally occurring chemicals that can be extracted with solvents.

(3) Grading durability

Since 1932, BRE* have been conducting field and laboratory tests to determine the natural resistance of *heartwoods* of different species to fungal attack. In essence, these have been carried out using stakes with a cross section of 50 × 50 mm in ground contact. From the results of these tests BRE have grouped timbers into five grades of durability related, it should be stressed again, to their *heartwood*. The grades are as follows.

Durability grade	Average life of 50 × 50 mm stakes in ground contact (years)
Perishable	Less than 5
Non-durable	5–10
Moderately durable	10–15
Durable	15–25
Very durable	Over 25

In those cases where timber contains a large proportion of sapwood such timber will have a shorter life than its durability grade suggests, because it is perishable or non-durable.

BRE has classified the heartwood of about 200 species of timber. The breakdown, with some typical species commonly used, is as follows:

Perishable (29)
Abura
Alder
Ash, European
Beech, European
Birch, yellow
Poplar, black and Italian
Ramin

Non-durable (55)
Beech, silver
Elm
Fir, excluding American
Gaboon
Hemlock, western
Maple
Pine (excluding American, maritime and Caribbean pitch)
Spruce

Moderately durable (39)
Cedar, Port Orford and western red (British grown)
Fir, Douglas (American only)
Gurjun, Indian
Keruing, Larch
Mahogany, African
Oak (excluding American and European)
Sapele
Walnut
Yellowwood

Durable (40)
Agba

*At the Princes Risborough Laboratory, now transferred to the BRE site at Garston.

Cedar (excluding Port Orford and western red, British grown)
'Cypress'
Mahogany, American
Oak, European
Utile

Very durable (37)
Afromosia
Ebony, African
Greenheart
Iroke
Jarrah
Opepe
Purpleheart
Teak (including Rhodesian)

From this selective list it will be seen that species such as beech and poplar are perishable. They may be reduced to a powder in a few months by fungal attack. Others such as greenheart and teak (very durable) can resist attack for many years. The choice of timbers, on which reliance is placed on their natural resistance to attack, is limited, and many of the more resistant timbers are hardwoods. In specifications, a proper description of the timber required is important.*

(c) Preservative treatments

(1) Aim and choice
The aim of any treatment of timber is to increase its natural resistance to attack by wood-rots by making toxic the food required by the fungi. There are a number of chemicals with different degrees of toxicity that will achieve this and different ways in which the chemicals may be impregnated. Timbers vary in their resistance to impregnation, and the choice of an appropriate preservative is therefore not simple. What follows is a summary of the more important factors that need consideration.

(2) Types of preservatives
There are three groups of preservative:

- Tar oil and distillates from coal tar;
- Organic solvent solutions; and
- Water-borne and inorganic solvents dissolved in water.

The types, their ingredients, treatment methods used and preservative properties are given in Table T3.6/5 and the treatments suitable for various uses in Table T3.6/6.

(3) Methods of application
Summarized, the methods of application available for preservatives are as follows:

- *Brushing and spraying*: The penetration obtained is normally shallow. Retreatment is usually required every few years, in which case all surfaces must be accessible to the preservative. Brushing and spraying are normally more appropriate for protection against insect attack.
- *Deluging*: Organic solvent preservative is sprayed or flooded over the timber as it is fed through a tunnel on a conveyor system.
- *Immersion*: Timber is submerged in a bath of preservative liquid for periods ranging from a few seconds or several minutes (usually called 'dipping') to several hours or days ('soaking' or 'steeping').

*It is important not to confuse heartwood and hardwood – heartwood applies to all species; hardwood to specific ones. In specifications it is also important that descriptions such as 'all timber to be hardwood' should be avoided and the name of particular species given instead.

Table T3.6/5 *Classification of wood preservatives*

Preservative types and ingredients		Treatment methods	Preservative properties
Tar oil	1 Coal tar creosote to BS 144	Pressure Open tank	Resist leaching and are particularly suitable for external work. Have a characteristic odour, and can stain adjacent materials.
Distillate from coal tar	2 Coal tar oil to BS 3051	Pressure Open tank Immersion Brush	Non-corrosive to metals, and treated timber presents no special fire hazard after a few months' drying. Not suitable normally for timber that is to be painted. Impart a degree of water repellency to timber which helps in retarding dimensional movement.
Organic solvents to BS 5707: Part 1. (Solutions of one or more organic fungicides/insecticides in organic solvents, usually petroleum oil distillate)	1 Copper naphthenate 2 Zinc naphthenate 3 Pentachlorophenol and derivatives 4 Tributyltin oxide 5 gamma-HCH 6 Dieldrin	Double vacuum Immersion Delunging Brush/spray	Most of the preservatives are resistant to leaching but some are subject to loss by evaporation: suitable for exterior and interior use. Not generally corrosive to metals, and non-staining. Treated timber clean in appearance and, when solvent has dried off, can usually be painted and glued satisfactorily. Treatment does not cause swelling of the timber and these preservatives can be employed on accurately machined wood and components without trouble from movement or distortion. Solvents are readily flammable, but once they have evaporated the treated timber presents no fire hazard. Water-repellent additives can be included to retard timber moisture changes in service.
Water-borne Inorganic salts dissolved in water	1 Copper/chrome/ arsenic to BS 4072 2 Copper/chrome to BS 3452	Pressure	(1) and (2) undergo chemical changes within the wood and become resistant to leaching: suitable for exterior and interior use. (3) is leachable to some extent, but can be used outside if the wood is painted. Generally non-staining and non-flammable. Copper-based preservatives may induce some metal corrosion in severe environments. Timber must be re-dried after treatment: clean (but sometimes coloured) in appearance: can be painted and glued satisfactorily.
	3 Disodium octaborate	Diffusion	

From BRE Digest 201, *Wood preservatives; pretreatment application methods,* minor revisions 1984, Table 1, Building Research Establishment, Crown Copyright.

Table T3.3/6 *Wood preservative treatments for various uses*

Use	Pressure[a]	Open tank[b]	Double-vacuum[c]	Diffusion[d]	Immersion[e]
Sea or fresh water	√				
In contact with the ground	√	√			
External, but not in ground contact:					
unpainted	√[f]	√	√		√[f]
painted	√[f]		√	√	
Interior timbers in high humidity or subject to heavy condensation	√[f]		√	√	
Internal timbers, against insects or with low risk of decay	√[f]		√	√	√[f]

[a] Full-cell process: creosote or water-borne. Empty-cell process: creosote.
[b] Usually creosote.
[c] Usually organic solvent.
[d] Usually boron.
[e] Usually organic solvent; sometimes creosote.
[f] Not creosote if timber is to be painted, or where staining of adjacent materials may occur.

From BRE Digest 201, Table 2, Building Research Establishment, Crown Copyright.

The degree of treatment depends on the length of time during immersion, and that required to increase the amount of protection is not linear. For example, the level of treatment after 3 minutes is double that achieved in 10 seconds, but to double the degree of protection achieved in 3 minutes, immersion for 30 minutes is required.

- *Double-vacuum*: Timber is flooded after a partial vacuum has been created in a treatment cylinder. Once flooded, the vacuum is released and the timber allowed to remain in the preservative fluid for up to an hour either under atmospheric pressure or at a pressure of up to 3 bar. The preservative is then emptied from the cylinder and a final vacuum is applied to the timber. This recovers some of the solution from the wood and provides it with a dry surface.
- *Hot and cold open tank*: With this method the timber is submerged in cold preservative (usually creosote), which is then heated to 85–94°C and held at that temperature for 1–3 hours, so that much of the air in the wood is expelled. The preservative is then cooled to ambient air temperature with the timber in it, which may take many hours. During the cooling, the residual air in the wood contracts, thereby drawing liquid into the wood.
- *Pressure*: This is defined as one in which timber is placed in a closed cylinder and preservative fluid is forced into the wood by artificially applied high pressure, generally 10–14 bar. An important characteristic of this method is the difference in the degree of penetration of the fluid into the sapwood (i.e. the virtually non-resistant part of the timber) and the heartwood (the more resistant part of the timber). The process ensures that the sapwood is penetrated completely but with less deep penetration into the heartwood, depending on the species of wood. There are two types of process: the full cell and the empty cell.
- *Diffusion*: Whereas all the methods described above are for use with seasoned timber, diffusion can only be used on green timber. The method consists of applying a concentrated solution of a water-borne preservative to the surface of the wood and then stacking the timber for several weeks under cover, which restricts drying of the wood. With this method it is possible to obtain complete penetration of the preservative, even in timbers which are otherwise very difficult to impregnate.

(4) Resistance of timber to impregnation

As with other aspects, timber species vary greatly in their response to treatment, and the effectiveness of any treatment depends, of course, on the amount of preservative that the timber can be made to absorb and the depth of penetration. Some species can be impregnated rapidly and fully by pressure treatment; others absorb little preservative, even after a similar treatment lasting several hours. The extent to which a particular species will resist treatment is related to the anatomical structure of the wood (i.e. the size, shape and number of the individual wood cells and their ability to conduct liquids).

In the growing tree the sapwood conducts liquids, therefore it usually treats more readily, irrespective of the heartwood characteristics. The species of the wood and the method of application of the preservative will influence the effectiveness of the treatment. With sapwood, wood laid down towards the end of the growing season may allow preservative treatment more readily. In some hardwoods only particular types of cell can take up the preservative.

For many years BRE have used a standard test to assess the treatability of timbers. Heartwood specimens (50 × 50 mm cross section) from various positions in different logs (and, where possible, sapwood as well) are tested for their uptake of preservative fluid. BRE have devised an index of the extent to which various timbers can be impregnated, divided into the following four arbitrary groups:

Group 1 (permeable)
Amenable to treatment and can be penetrated completely under pressure without difficulty.

Group 2 (moderately resistant)
Fairly amenable to treatment; 6–8 mm lateral penetration in softwoods after 2 or 3 hours under pressure (or the penetration of a large proportion of vessels in hardwood) is usual.

Group 3 (resistant)
Difficult to treat, requiring lengthy periods of treatment. It is often very difficult to obtain more than 3–6 mm lateral penetration.

Group 4 (extremely resistant)
Virtually impervious to treatment. Absorbs little preservative even after lengthy treatment. Both lateral and longitudinal penetration minimal.

Most sapwood can be treated fairly easily. As to the heartwoods, BRE have assessed many of the perishable ones to be permeable (Group 1); most others are moderately resistant (Group 2); none are extremely resistant. As *a general rule*, the extent to which heartwoods can be treated reduces as their durability increases. With very durable species almost all are extremely resistant (Group 4).

(d) Resistance of timber and other organic products

(1) Plywood
Two factors influence the durability of plywood to decay. These are:

- The resistance of the wood to decay; and
- The resistance of the adhesive to moisture and to decomposition by microorganisms.

The *exterior grades* of plywood do incorporate synthetic adhesives which are resistant to moisture and microbial decomposition. However, the veneers themselves are not usually treated with wood preservative, and thus the plywood as a whole is not immune from dry rot, although it will not delaminate even after prolonged exposure to damp conditions. The resistance to decay of plywood of this type may therefore be regarded as similar to that of the wood from which it was made.

Certain caution is required in the use of 'interior' grade plywood. The term 'interior' should *not* be regarded as one which allows plywood of this kind to be used under any internal climate. In fact, this type of plywood should not be used in any situation where it is likely to remain damp for any length of time.

(2) Other organic products
In general, wall boards made from defibrated wood, sugar cane, bagasse and similar cellulose materials are readily attacked by dry rot if exposed to persistently damp conditions. However, the incorporation of a preservative in some brands of wall boards renders these resistant to decay (and attack by white ants). The resistance which hardboards may offer to attack varies with their porosity, although all types will rot under bad conditions – the soft, more porous types are generally less resistant.

Woodwool slabs, though resistant to decay, are not impervious to the strands of the dry rot fungus. Saturation with water may lead to slow disintegration as a result of fungal decay. Under the influence of moisture alone, some types of slab incorporating gypsum with the cement may lose their cohesion. Resistance of the slabs to decay is not necessarily increased by preserving the woodwool, it being thought that the cement renders inactive the preservatives used.

4 Precautions

(a) Generally

(1) Scope
Precautions may be divided in those which are necessary (1) in the selection and storage of timber and (2) under the conditions in which timber may be used.

(2) Two simple rules
In principle, control of fungal decay in buildings can be achieved by following two simple rules: first, by using only sound, well-seasoned timber, free from any incipient decay and, second, by maintaining the timber in a *perfectly dry* state.

(3) Dry state
Precautions should be related to the worst condition, namely attack by the dry rot fungus. Perfectly dry implies a moisture content of not more than 20%. Although the choice of the correct quality of timber should not, in practice, present insurmountable difficulties, the maintenance of the dry state may not be easy to comply with, and to insist on strict adherence to this rule is, in practical terms, unrealistic, particularly as the resistance of any species of timber, including the weaker sapwood, may be adequately increased by the use of preservative treatments (see 3(a) above). However, it is still wise to ensure that even the most resistant timbers, or those made more resistant by treatment with preservatives, are not allowed to become unduly damp for prolonged periods. The reason for this is really quite simple. The more resistant timbers, or those treated with preservatives, are not necessarily completely immune from attack under severe conditions of exposure to moisture.

(4) Controlling fluctuations in moisture content
In general, the dry rot safety line, that is, a moisture content of not greater than 20%, can be fairly easily achieved with most air-seasoned timber in the UK (see Diagram D3.3/6), while in a building heated intermittently the service moisture content may drop to 11% or 15%. However, in a centrally heated building the moisture content of the timber may fall to 8% or 10%. Moisture contents below 17% cannot be achieved without kiln seasoning. At the same time, in order to avoid problems associated with dimensional fluctuations in timber, as described in detail in *3.3 Moisture content* under 'Timber', p. 239 ff, it is necessary to ensure that when timber is installed in a building its moisture content is not significantly different from that to be expected in service. The margin between installed and average moisture content in service is, in general, less critical for timber used for constructional purposes than for joinery and flooring, mainly because the average service moisture content of constructional timber is generally higher than that for joinery (see Chart C3.3/8, p. 251). Control of dimensional

fluctuations in timber is related to fungal attack in so far as the fluctuations may cause *joints* in joinery, for example, *to open up*, and in exposed positions, allow the ingress of water which, in turn, may become trapped and hence may cause sufficient dampness for decay to take place (wet rot usually significant). Thus for a variety of different reasons it is necessary to ensure that adequate precautions are taken which will *not allow* timber to undergo excessive fluctuations in moisture content.

(5) Preservative treatment and conditions likely to lead to dampness
Creative pessimism might advise that all timber used in a building should be treated with preservative. This might be unnecessary or too costly in some circumstances. Choosing timber naturally resistant to fungal decay might be an alternative. However, in the first instance a thorough analysis should be made as to the likelihood of conditions of dampness occurring during the drying-out period of a building *and* in service.

Conditions which are likely to give rise to dampness are covered variously in *3.1 General considerations*, *3.2 Exposure*, and, in particular, *3.3 Moisture content* (this has special and detailed reference to timber), and are, therefore, not repeated here except in outline as relevant later. In a great many cases, water, and hence dampness, may be excluded by paying proper attention to constructional details (subsequent maintenance is also extremely significant), as covered in detail in *3.4 Exclusion*. It is important to note that *attention to details* of construction so as to exclude water from reaching timber is one of the most important precautions to be taken in preventing fungal attack.

From the considerations given in previous sections it will be seen that, in general, timber which is likely to be subjected to fungal attack is, to some degree, *concealed* and/or in *contact with* other porous materials which may transfer moisture to the timber (moisture 'released' during the drying out of a building is most significant). In the case of constructional work, including roof timbers and decking, the whole of the timber is concealed and partly in contact with other (usually, but not always, porous) materials; in the case of floor boards, skirtings and joinery such as frames (windows, doors, cupboards, etc.) only the back or side(s), except for timber floor boards covered with other finishes such as carpet, linoleum, vinyl, etc., are concealed and usually the concealed face(s) are in contact with other (usually, but not always, porous) materials. In all these cases, *interstitial condensation* might be the source of dampness.

(b) Selection of timber

(1) Seasoning and yard storage
Only sound, well-seasoned timber should be used, free from any incipient decay. Wherever possible, it is wise to ascertain the conditions under which the timber has been stored. No timber which originates from a yard in which careless stacking or unclean conditions are tolerated should be allowed on a building site.

(2) Joinery – sap-/heartwood and soft-/hardwood
In the case of softwoods for joinery, particularly mass-produced joinery, it may be noted that the species normally used are chosen more for their economy and ease of machining rather than for their resistance to decay. Baltic redwood (also known as red or yellow deal, red pine or fir) is a commonly used softwood; its heartwood being more resistant to decay than its sapwood (see 3(b) earlier, p. 432 ff). The heartwood of

Douglas fir is also used. Subject to the conditions of service, the timber may have to be preserved. Among hardwoods which have proved acceptable as regards their resistance to decay in highly exposed positions (external sills, for example), are English oak (traditional hardwood), teak, utile, gurgun and agba. All these should, as mentioned earlier, be specified by name, as merely specifying 'hardwood' could lead to the use of unsuitable timbers such as abura, beech, obeche and ramin, all of which have poor resistance to decay. If it is required that the perishable sapwood of an otherwise decay-resistant timber should *not* be included, then 'heartwood only' should be specified.

(c) Storage of timber

(1) Generally
Precautions necessary in the storage of timber should be aimed to ensure that, at best, the timber does not undergo significant moisture content fluctuations (see *3.3 Moisture content*, 'Timber 6 Precautions', p. 253 ff and 'Timber based products, Precautions for individual products'), or, at worst, does not become damp. Care in storage is not only required at the timber yard, or in the joinery works, but also on the building site. It is perhaps only too axiomatic that care taken at the timber yard or in the joinery works is of little value unless matched by similar precautions on a building site.

(2) Constructional timber
All precautions taken in connection with constructional timber should aim to ensure that the moisture content of the timber does not exceed 24% (see *3.3 Moisture content*, p. 250).

The precautions described briefly below are only those which should be taken on a *building site*.

- Timber should preferably be stored within a building. If this cannot be achieved it should be stacked on bearers to avoid close contact with the ground.
- Dry, well-seasoned timber should not be allowed to absorb excessive amounts of atmospheric moisture. To achieve this, it should be closely stacked and wholly covered with tarpaulins or polythene sheeting around and under the stack as well as over it.
- Timber which arrives on the site in a moist condition due to inadequate seasoning and drying, exposure to rain, or impregnation with water-borne preservatives should be open-piled in such a way that air can circulate round each piece to promote further drying. A top cover only should be given to the open-piled stacks; the sides must be left open so as not to hinder air flow through the stack. If the stacks are stored in a building to dry out, the windows should be left open. It should be noted that timber that has been treated with water-borne preservatives will not have been redried in a kiln, unless this was especially ordered. Consequently, it is normally necessary for such treated timber to be open-piled and air-dried.

(3) Joinery
Whereas it may be possible to allow timber used for constructional purposes, particularly that which is not prefabricated, to dry out after delivery to the site as described in (2) above, the same is not true of joinery. Consequently, *extra care* is required. Taking into account the importance of the limitation of dimensional fluctuations, precautions for joinery may be summarized as follows:

1. Careful storage after manufacture.
2. Protection during delivery to the site, avoiding exposure to the weather.
3. Careful storage on-site, before and after fixing, avoiding conditions of exposure which are materially different from those expected in service. Before use, all timber should be adequately protected from exposure to the weather (use of tarpaulins and polythene sheeting as described (see '(2) Constructional timber' above), and stacked clear of the ground.
4. Where adequate protection of joinery on site cannot be ensured, delay the delivery of the joinery to the site until it is needed.
5. Timber which is to receive paint may be primed* immediately after manufacture. Single-coated 'pink' shop primers vary in quality and will delay the ingress of water for only a comparatively short period of time, and cannot, therefore, be relied upon as a means of protecting timber. Storage under cover is always preferable, but if this is not possible then it is necessary to specify a better primer. A water-repellent preservative† plus a good-quality primer, or two coats of aluminium primer‡ would be suitable. In some cases it may be practicable to apply the complete painting system before delivery on-site. However, it is important to note that any protective coating must completely envelope all faces of the timber.

(d) Timber floors

The risk of fungal attack of timber floors differs according to the position and construction of the floor. There are three different categories, as described below.

(1) Unventilated ground floors
An important characteristic of floors of this type is that the finish is 'fixed' to a solid concrete base in contact with the ground. Consequently, maintaining the timber in a dry state depends on the extent to which, and the positions in which, dpcs may be necessary. (See *3.4 Exclusion*, p. 328.)

Two types of floors, namely those of wood block, and those of board or strip, may be identified, each of which requires different detailed precautions. These are considered separately.

Wood block floors. Table T3.6/7 gives the resistance of floor finishes to ground moisture. Wood blocks laid in *hot-applied* bitumen or pitch adhesives (Group B in the table) require no further protection against rising damp, although it is important to note that the adhesive in which the blocks are dipped should form a continuous layer and keep them out of contact with the concrete. Blocks laid with cold bitumen adhesives (Group D in the table) require additional protection from rising damp, and this means the inclusion of a dpm as described in *3.4 Exclusion* (see p. 328). If mastic asphalt underlay is used, it is important to ensure that the adhesive is compatible with the asphalt.

Board or strip flooring. Board and strip flooring may be nailed to battens fixed to the concrete sub-floor either with floor clips or embedded in a screed as shown in Diagram D3.6/4. Floors of this type must be protected by a dpm which may, as shown in the diagram, be either in sandwich form (that is, between concrete bed and screed) or on the surface of the screed. The choice of position for the dpm will depend on circumstances; one important point to be borne in mind is residual moisture, which may be in the screed at the time the membrane is laid. A surface membrane will protect the floor not only against rising damp but also against the effects of residual moisture in

*The primers may give joinery a uniform appearance. However, after short exposure primers become weak or powdery and unfit to take further coats of paint. Many paint failures can be traced back to poor-quality primers.

†If a water-repellent type of preservative is specified it should be of a paintable type and the preservative supplier and joinery manufacturer must ensure that it will be compatible with the paint.

‡It should be noted that, in addition to keeping water out, aluminium primers will also prevent moisture within the timber from getting out. Consequently, before priming, the timber should be correctly seasoned, or the moisture or solvent from preservative treatment allowed to dry out. Failure to do this may allow blistering to occur.

Table T3.6/7   *Resistance of floor finishes to ground moisture*

Group	Material	Properties
A Finish and dpm combined	Pitch mastic flooring Mastic asphalt flooring	Capable of resisting rising dampness without dimensional or material failure
B Finishes that may be used without extra damp protection	Concrete Terrazzo Concrete or clay tiles	Capable of transmitting rising dampness without dimensional, material or adhesion failure
	Cement/latex Cement/bitumen	Capable of partially transmitting rising dampness without dimensional or material failure and generally without adhesion failure
	Wood composition blocks (laid in cement mortar) Wood blocks (dipped and laid in hot pitch or bitumen)	Capable of partially transmitting rising dampness without material failure and generally without dimensional or adhesion failure. *Only in exceptional conditions of site dampness is there risk of dimensional instability*
C Finishes that are not necessarily trouble-free without damp protection	Thermoplastic flooring tiles (BS 2592) PVC (vinyl) asbestos floor tiles (BS 3260)	Capable of partially transmitting rising dampness through the joints without dimensional failure and generally without adhesion or material failure. Water penetration at the joints may result in decay at the edges in some conditions when groundwater contains dissolved salts or alkalis
D Finishes which require reliable protection against damp	Magnesite	Capable of transmitting rising dampness but adversely affected by water (softens and disintegrates under wet conditions)
	Flexible PVC flooring in sheet or tile form (BS 3261)	Impervious, but the flooring adhesive is sensitive to moisture (may expand under damp conditions)
	PVA Emulsion cement	Impervious, but dimensionally sensitive to moisture (adhesive for tiles also sensitive to moisture)
	Rubber	Impervious, but prone to adhesion failure mainly through sensitivity of its adhesive (may expand under damp conditions)
	Linoleum	Sensitive to alkaline moisture attack through breakdown of bond and adhesive film (may expand under damp conditions)
	Cork (carpet and tile) Wood (block in cold adhesives and strip and board flooring) Chipboard	Acutely sensitive to moisture with dimensional or material failure

Important: With floor warming a damp-proof sandwich membrane is necessary whatever the floor finish.

Notes: The degree of protection to be provided by the dpm will be dependent on:
1. Wetness of the site.
2. Temperature gradient through the floor.
3. Contamination of aggregates, hardcore and fill with sea salts.

Order of protective value of sandwich membranes (see text):
1. Mastic asphalt.
2. Bitumen sheeting.
3. Hot applied pitch or bitumen.
4. Cold applied bitumen solution and coal tar pitch/rubber emulsion or bitumen/rubber emulsion.
5. Polythene or polyethylene film sheeting.

From CP 102, *Protection of buildings against water from the ground*, 1963, Tables 1 and 2, pp. 49–50: BRS Digest (2nd Series) No. 54. *Damp-proofing solid floors*, Tables 1 and 2, pp. 2–3, HMSO, London, January, 1965.

Note: CP 102 is being revised as BS 8102. A revised version of Tables 1 and 2 should be included in a forthcoming part of BS 8102.

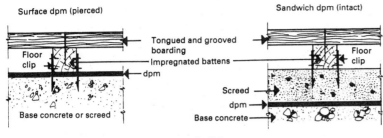

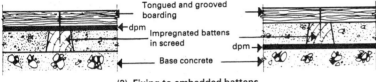

Diagram D3.6/4 *Alternative methods of fixing boards and batten flooring to ground floors using surface or sandwich dpms*

the screed, although it will be pierced by the legs of the flooring clips or by the flooring nails. A sandwich membrane, on the other hand, need not be pierced, but it is only satisfactory when sufficient time can be allowed to dry the screed above the membrane *before* the boards are fixed in position. In practice, the drying-out time may be so long as to make the sandwich method less useful. It must be noted that failure to allow the floor screed to dry out thoroughly before the boards are laid may lead to decay; the risk of decay will be greatly increased if, in addition, the floors are covered prematurely with impervious floor coverings, such as rubber, linoleum, plastics, etc. (notably interstitial condensation).

The battens, however they are used, should be impregnated with preservative. A recommended additional precaution is the brush application of a preservative to the underside of the boards. The selected preservative should not stain or otherwise disfigure the exposed surface of the timber.

Finally, it is recommended that only tongued and grooved boarding should be used in these floors. This is a precaution aimed to prevent water used for washing, or which may be accidentally spilled, from passing readily through the joints, because any water reaching the underside of the boarding will be slow to evaporate, and dangerous conditions may, therefore, persist for a considerable time.

(2) Ventilated ground floors
Ground floors of timber boarding on joists which are totally suspended have a relatively large air space beneath them and if this space is efficiently ventilated it is unnecessary to treat the timber with preservative. There is, however, also a need to provide a dpc in all sleeper walls and other supports in actual capillary contact with the ground. Methods by which adequate ventilation may be ensured and the correct use of dpcs are described in *3.4 Exclusion*, p. 328 and Diagram D3.4/60, p. 331. For convenience, the need for taking these precautions is summarized below.

The dense oversite concrete, which should not be less than 102 mm thick, is required in order to minimize the evaporation of water from the ground into the air space below the floor. It is essential that the oversite concrete covers the whole area under the suspended floor. An alternative to the oversite concrete would be a damp-resisting coating to BS 2832 laid on a well-compacted base of hardcore, blinded with ashes to form a level surface free from fissures. The surface of either the concrete or the damp-resisting coating should not be lower than the level of the surrounding ground. Where it is not possible to meet this requirement then it is necessary to ensure that the site is suitably drained so that inundation of the area cannot occur. It may be noted here that care should be taken to ensure that there are no shavings, pieces of wood employed during excavation or from wooden forms for concrete work, soil or debris left behind on the surface of the oversite concrete or damp-resisting layer after the floor has been laid. All these should be removed systematically during the course of the work on-site.

Ventilation of the underfloor space is necessary so as to ensure that moisture-laden air is replaced with drier air. As explained earlier, it is of the utmost importance that the air in the space below the floor is not allowed to become saturated. In this it should be realized that even when the utmost care has been taken with the oversite concrete or damp-resisting layer, some moisture will probably still find its way to the surface. In order to ensure that adequate and efficient cross-ventilation is maintained it is necessary to comply, as described in *3.4 Exclusion*, p. 328, with the following:

- The clear depth between the underside of the joints and the top of the site covering must not be less than 150 mm;
- Air bricks in the external walls should be provided so as to give at least 3200 mm^2 *open area* per metre run of external wall, and should be placed as high as possible on opposite walls;
- All sleeper walls should be built honeycombed;
- All cross-walls should be provided with vent holes;
- Care should be taken to avoid unventilated air pockets such as may occur near bay windows; and
- Ducts should be formed under hearths and solid floors wherever these might interrupt cross ventilation.

The efficient ventilation of the underfloor space implies a risk of draughts and a reduction in thermal insulation. In order to minimize the risk of objectionable draughts it is advisable to use tongue and grooved boarding rather than plain edge or sheet materials. An improvement in thermal insulation can be gained by providing a layer of insulating material below the boarding. Provided the floor is properly ventilated, the provision of the insulating layer should not increase the risk of fungal attack.

The importance of the provision of dpcs in sleeper walls and other supports in capillary contact with the ground need not be repeated here. However, it may be noted that in order to ensure that there is a clear space (that is, no contact) between the ends of the joints and walls, the former should be cut back.

(3) Upper floors
In general, fungal attack of timbers in upper floors has been rare. This may not always be the case. Although upper floors are not exposed to moisture rising from the ground, the air above and below them, in certain buildings and under certain conditions, may have humidities sufficiently high to give rise to fungal attack in general or mould growth in particular. Other circumstances that should be considered are:

- *Suspended timber floors* of boarding on joists, being essentially a 'dry' construction, do not need special precautions, except (and this is important) that any timbers built into walls that might become damp should be given a suitable preservative treatment.

- The types of floors with *boards and battens*, as outlined previously under '(1) Unventilated floors', may, in the case of upper floors, be used on various forms of structural base, as, for example *in-situ* or precast reinforced concrete, hollow beams, or similar constructions. In these cases the dpm would be omitted. However, the structural base or any screed laid on the base may contain considerable quantities of residual moisture (lightweight screeds are particularly notable in this respect). Consequently, the flooring should not be laid until drying out is well advanced, while the battens should be given protection during the drying-out period. Temporary protection may be obtained by brush application of a preservative; the same treatment may, for safety, also be extended to the underside of the boards. Although pressure impregnation is not essential, it may be considered for the battens, because not only does it give the protection required during drying-out but it also continues to protect the battens from decay as a result of accidental spillage, etc.

- *Wood block* or *parquet flooring* laid on concrete, hollow tile or similar structural upper floors do not require special protection against fungal attack. However, the concrete and screeds should be dry before the timber flooring is laid.

(e) Timber roof construction

All types of roofs classified as 'cold deck designs' in pitched, sloping or flat forms having constructional timbers are inherently at risk of interstitial condensation occurring (see *3.4 Exclusion*, p. 301 and 310) and therefore to fungal attack. With these roofs, proper ventilation of the roof space or cavity (depending on the form of the roof), is essential. The amount of ventilation that may be adequate in particular cases (i.e. the rules for this ventilation) is likely to change as more experience is gained (see *3.4 Exclusion*, p. 310 ff, for the current rules). However, and importantly, the use of timber treated with preservative is recommended in any case, because at times external environmental conditions (notably wind) or internal environmental conditions (notably excessive moisture production) may cause the ventilation to be intermittently inadequate.

(f) Timber framing and cladding

Timber framing may be used simply as framing for a number of different lightweight claddings or as framing/cladding in what are known as timber-framed buildings. In either type the moisture content of the timber framing and/or the timber cladding (where used) may be sufficiently high to cause fungal attack of the timber or perhaps simply mould growth. The correct use of vapour control layers (see *3.4 Exclusion*, p. 300) is essential, as are details that will ensure that there is no penetration of rainwater. It is wise to ensure that all timber used is treated with an appropriate preservative.

(g) Internal joinery

In general, internal joinery is not at risk of fungal decay once the surfaces with which it is in contact have dried out. The same is certainly not true while the building is drying out. It is therefore important that no joinery is fixed until the surfaces with which it is to be in contact are sufficiently dry. This condition should always be checked using one of the techniques described in *3.1 General considerations*, 'Dampness', p. 152 ff. Alternatively, the concealed surfaces of the timber (i.e. those in contact with porous materials or non-materials that are likely to have damp surfaces temporarily) may be coated with an appropriate impervious paint.

(h) External joinery

Detailed consideration is given here to windows. The same principles apply to external doors.

(1) Background
The type of decay associated with timber windows is invariably wet rot (brown and white rots). In the past such decay has only occurred sporadically, but experience over the past 15–20 years has shown that the decay has been more widespread, in newly built houses in particular. In many cases it has occurred within five or six years from construction and in all parts of the UK, although it has tended to take place sooner in the wetter areas. In any event, decay has been found to be most pronounced in those windows which are heavily exposed to water wherever such exposure may occur (i.e. both externally, the lower floors in particular, and internally, the latter usually because of high humidities such as occur in bathrooms and kitchens). Significantly, usually the lower parts of the window have been particularly susceptible to attack.

Once infected, the decay of the timber develops gradually, so it may be several years before there is any detectable softening of the wood. In many cases decay starts in the centre of a joint and works to the surface. It is therefore only detectable when it has broken through the external surface (and the paint film on it). By this time the decay is in an advanced state (see Photograph P3.6/19).

The interacting factors, some of which have been included earlier, that influence the service life of joinery are:

- Selection of suitable timber – see 3(b) earlier;
- Selection of designs that limit moisture uptake – see (2) later;
- Application of appropriate preservative treatments – see 3(c) earlier;
- Care of joinery prior to installation – see (c)(1) earlier;
- Regular maintenance of protective coatings – see (j); and
- Prompt attention to any necessary repairs.

(2) Design and fabrication
Joints are the most likely places for water to enter. Water was excluded (or better excluded) in traditional wedged, mortice and tenon joints because these were primed prior to assembly and were less prone to

Photograph P3.6/19 *Decay of external timber window. The decay seen in (a) shows only as slight shrinkage of the wood causing surface cupping but when the wood is prodded with screwdriver the decay revealed is extensive as seen in (b). (From BRE Digest 304,* Preventing decay in external joinery, *December 1985, Fig. 1, p. 1, Building Research Establishment, Crown Copyright)*

(a)

(b)

opening up. In contrast, many modern joint designs are more easily loosened as a result of moisture movements, thereby allowing water to enter the joint. In addition, the glue used to provide stiffness does not seal the joint surfaces in the same way as priming, so water can still be absorbed by the timber (see Photograph P3.6/20).

The precautions that should be taken to reduce the entry of water at joints include:

- Joints should be tight fitting and designed to protect the end grain from wetting;
- The end grain of all joints should be sealed with an appropriate end-grain sealer – see Diagram D3.6/5;
- The avoidance of voids in which water can collect – see Photograph P3.6/21 and Diagram D3.6/6;
- Glazing beads must be compatible with the finishing system;
- Ensure that horizontal surfaces slope properly and/or do not provide surfaces on which water can collect for long periods of time – see Diagram D3.6/7 (note also that multipiece sills are particularly vulnerable if water can collect on the exposed surface);
- Dowels (in doors, notably) should be made from durable or preserved timber, the latter before assembly; and
- Ensure that plywood in panels is of an exterior grade that will resist attack (see p. 437 earlier) and that the end grains are sealed.

(3) Painting

Good primers are essential for success of the paint system. It is important not to over-estimate the ability of a good-quality primer to prevent the ingress of water – poor-quality primers usually lead to many paint failures. Nevertheless, off-site primed joinery should be stored under cover and the finishing coats applied as soon as possible after installation.

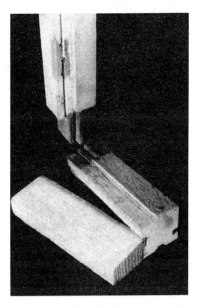

Photograph P3.6/20 *The complexity of joint design in a modern window frame and sill means that joints expose a large surface area to the risk of moisture penetration. (From BRE Digest 304, Fig. 1, p. 2, Building Research Establishment, Crown Copyright)*

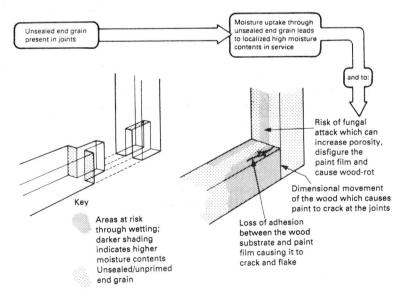

Diagram D3.6/5 *Effects of moisture uptake on exterior joinery with unsealed end grain. (From: Miller, E. R., Boxall, J. and Carey, J. K., External joinery: end grain sealers and moisture control, BRE IP 20/87, December 1987, Fig. 1, Building Research Establishment, Crown Copyright)*

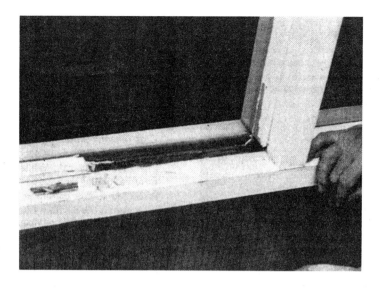

Photograph P3.6/21 *Groove machined for outer glazing bead running right through mullion/transom joint and outer glazing bead nailed not glued or primed. This has led to the entry of moisture and subsequent decay. (From BRE Digest 304, Fig. 2, p. 3, Crown Copyright)*

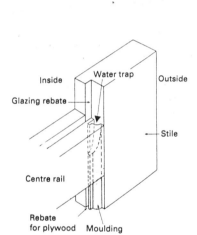

Diagram D3.6/6 *Illustration of a void present in the centre rail joint of a door.* Note: *The inside face is shown without moulding to simplify the diagram. (From BRE Digest 304, Fig. 3, Building Research Establishment, Crown Copyright)*

(4) BRE recommendations
- Consider whether the requirements of specifications and codes of practice are adequate to ensure satisfactory durability of the timber in the particular circumstances under consideration;
- Consider design details carefully to ensure that they do not encourage water penetration;
- Specify preservative treatment, preferably by a double-vacuum process to ensure adequate service from European redwood and other timbers of low decay resistance;
- Specify all timbers by name and insist on the exclusion of sapwood if a durable species is to be used without treatment;
- Protect joinery from the weather during delivery to the site and on-site before installation;
- Specify a good-quality BS primer and a finishing system specifically designed for exterior use; and
- Inspect for evidence of decay as part of pre-painting maintenance inspections.

(i) Painting woodwork

(1) Influencing factors generally
Paint behaviour on timber is influenced by the cellular structure of the wood. The size, distribution and orientation of the cells is dependent on the species and the way the log was converted. Hardwoods have a

Diagram D3.6/7 *Section through the bottom rail of an opening light showing horizontal collecting surfaces when closed and water traps when open. (From BRE Digest, 304, Fig. 4, Building Research Establishment, Crown Copyright)*

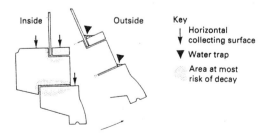

greater variety of cell type than do softwoods. Another influencing factor is moisture movement (see *3.3 Moisture content*, p. 239 ff). Movements across the grain are the largest, and these can cause splitting of the paint film on drying: differential swelling of early and late wood bands creates raised grain and strains in paint films. While the paint film reduces the range of moisture movements, excessive moisture content can cause blistering and loss of adhesion.

The properties of the wood and paint system and related precautions that influence the durability of the paint film are summarized below.

(2) Penetration/adhesion
The penetration of paints (in contrast to preservatives generally) is restricted by their high viscosity and pigment content. Although paint penetration, to some extent, is important (the amount depending on the paint), deep penetration is not essential for adhesion. Some penetration to produce a mechanical key is helpful.

(3) End-grain uptake
The uptake by the end grain is rapid. This means that sealing it, with a generous application of adhesive or primer, prior to the assembly of joints in joinery is essential.

(4) Surface smoothness
A smooth, planed surface is essential. Planing and sawing cut and expose cavities within the cell fibres, thereby allowing the paint to penetrate better. However, it should be noted that stain treatments of the penetrating type generally give better results on rough-sawn wood.

(5) Extractives
Many timbers, including pine, larch, gurjun, keruing and agba, contain resins which sometimes exude and impair the painted surface. The resin may be activated by treatment with solvent-borne preservatives. Other woods, such as teak and certain types of cedar, contain oil which interferes with the drying, hardening and adhesion of paint. Swabbing with solvents and the use of aluminium primers are the usual remedies. Water-soluble extractives, when present in large quantities (for example, in oak, chestnut and walnut), can also interfere with the drying and hardening of priming paints. They may also discolour water-thinned paints without necessarily affecting their durability.

(6) Plywood
The veneer providing the surface for the paint system influences the durability of the paint. Moisture movement of the veneer causes fine cracks (checks) on the surface, and checking varies according to the species of the wood. It is particularly marked with Douglas fir and birch; less so with African mahogany and gaboon. A resin-treated paper overlay provides a more effective base than a primer, a coat of which is not required. Exterior plywood may not necessarily have a durable veneer. Where this is the case, the plywood should be treated with a preservative to increase the life of the paint.

(7) Preservative treatment
Treated timbers may have excessive moisture in them, in which case the life span of the paint will be shortened. In addition, further efforts should be made to exclude it in service by, for example, end-grain sealing.

(8) Stain and mould

Blue-stain fungi and surface moulds may disfigure the paintwork but do not reduce the strength or the durability of the timber. Preservation of the wood has little effect on either. Where mould is particularly common or anticipated, a mould-resistant paint may be needed.

(9) Knotting

Knotting is used to reduce staining by resin in knots or streaks in the timber. Preservative treatments may reduce the viscosity of the resin, thereby making it more likely to exude. Exudation is increased by exposure to sunlight. Dark colours increase the temperature, and timbers with fewer knots are better suited to exposure to sunlight.

(10) Priming and primer

A good priming coat is the essential foundation for a durable paint system. A primer should not, however, be expected to protect wood on-site for several months. High moisture contents resulting from poor-quality primers (or good-quality primers allowed to degrade) can cause blistering and peeling in the subsequent coats of the paint. Primed joinery should be properly stacked and protected before its installation. There are a number of primers, such as lead-based, low-lead aklyd-based, aluminium-based, emulsion and water-repellent preservative primers.

(11) Undercoats

Undercoats should match a finishing coat and be compatible with the priming coats. Differences of opinion exist as to whether a four-coat paint system should consist of primer, two undercoats and one top coat; or primer, undercoat and two top coats. In principle, better protection of external joinery is afforded by two top coats.

(12) Top or finishing coat

Alkyd gloss paints with appropriate undercoats still form the major type of exterior finish on woodwork. Their initial good elasticity to accommodate movements is lost on ageing. Lead-based paints offer good durability but their toxicity is now considered an undesirable feature. Polyurethane (one-pack) paints are not necessarily as durable outdoors as the conventional alkyd type; they are particularly useful indoors for hard wear and resistance to washing with soap, detergents and bleach. Sanded and textured paints based on emulsions, oleo-resinous or epoxy binders are used mainly on timber panel products, and vary considerably in their durability and water permeability. Exterior stains are not as durable as paints but their renewal is easier and cheaper.

In recent years paint systems with high water vapour permeability have been made increasingly available. While such paints obviously assist in reducing paint failures due to high moisture contents in wood, their 'breathability' does mean that the timber is subjected to greater moisture movements than would be the case with more impermeable paint films. Further experience with the breathing paints is necessary to determine the **balance** that is actually achieved.

Sources

Desch, H. E. (revised by Dinwoodie, J. M), *Timber, its Structure, Properties and Utilisation*, 6th edn, Macmillan, London, 1981

BRE

Digests

201, *Wood preservatives: pretreatment application methods,* minor revisions 1984

261, *Painting woodwork,* May 1982

296, *Timbers: their natural durability and resistance to preservative treatment,* April 1986

297, *Surface condensation and mould growth in traditionally-built dwellings,* May 1985

299, *Dry rot: its recognition and control,* July 1985

304, *Preventing decay in external joinery,* December 1985

340, *Choosing wood adhesives,* January 1989

Information papers

11/85, Bravery, A. F., *Mould and its control,* June 1985

16/87, Boxall, J. and Smith, G. A., *Maintaining paintwork on exterior timber,* November 1987

17/87, Dearing, T. B. and Miller, E. R., *Factory-applied priming paints for exterior joinery,* December 1987

20/87, Miller, E. R., Boxall, J. and Carey, J. K., *External joinery: end-grain sealers and moisture control,* December 1987

Report

Bravery, A. F., Berry, R. W., Carey, J. K. and Cooper, D. E., *Recognising wood rot and insect damage in buildings,* 1987

Ultraviolet radiation

1 Background

(a) Scope

The sun, referred to here simply as sunlight, is the source of ultraviolet radiation that degrades organic materials. The chemistry involved is complex, and water is not essential in the reactions, though it can be influential in certain circumstances. Unlike the other factors described earlier in this chapter, chemical and physical changes are related. The precautions that may be taken also need to be considered differently. Among other things, experience with the materials commonly affected such as liquid coatings (paints, plastics, sealants and roofing materials, the plastics-based ones in particular), is comparatively limited. Despite all these differences, ultraviolet radiation is included here for completeness.

(b) Nature of compounds

Organic compounds are mainly carbon based while the possible number of compounds of carbon are almost limitless. For building purposes, organic compounds require molecules of large size, the large molecules generally being polymers. However, the size of molecule may have to be altered to suit particular applications, while other 'additives' may be incorporated to aid manufacture or application or to increase durability. At this stage the problem of degradation is best treated in a general way, mainly because insufficient experience is available for conclusive evidence of the efficacy of most organic compounds, the chemical composition of which may undergo changes during manufacture in order to improve durability. It may be noted that, whereas in some cases it is possible to predict life expectancies of about 20 years, in others the period may be as low as two years.

(c) Inorganic/organic compounds

In thermal terms, organic compounds differ from most inorganic compounds in two important respects, namely, by having low melting and boiling points. Most large organic molecules will melt or decompose at temperatures of 300–400°C. At elevated temperatures and in the presence of air, organic compounds are seldom as stable as inorganic materials, mainly because the carbon can be oxidized to carbon dioxide while the hydrogen which is usually present can be oxidized to water.

2 General considerations

(a) Forms of breakdown

Surface degradation which is induced by weathering includes fading, yellowing or degradation, erosion and sometimes the accumulation of dirt. In translucent or transparent materials these effects may be characterized by loss of light transmission.

Loss of plasticizers or degradation of the basic material may lead to the material becoming brittle.

(b) Influencing factors

The ultraviolet component of sunlight is the most important single factor responsible for the breakdown of organic compounds, while it is also responsible for producing changes in colour. Of equal importance is the fact that the ultraviolet radiation initiates many of the chemical reactions by which organic compounds are oxidized and degraded. The reactions that take place are often chain reactions, which are accelerated by favourable conditions of warmth, oxygen and moisture.

3 Exposure

Detailed consideration of solar radiation, including the extent to which it is modified before it reaches the earth and the significance of orientation, all of which are relevant here, is included in *Materials for Building*, Vol. 4, *4.02 Exposure* (see pp. 38–41). For the present purposes it is necessary to consider the significance of ultraviolet radiation and orientation and location.

(a) Ultraviolet radiation

Organic compounds are characterized by the fact that the primary bonds between the molecules are chemical, their strength depending upon the elements involved. These bonds can be broken by sources of energy that exceed the attractive forces between the atoms, thus disrupting the molecules. In radiation terms, the level of energy required to cause disruption is reached at approximately 350 nm for the carbon–carbon bond and this is well within the range of the shorter, i.e. near-ultraviolet (wavelength range 300–400 nm), radiation component of solar energy received at sea level. It is important to note that, although the sun emits radiation down to 200 nm, the middle ultraviolet (wavelength range 200–300 nm) is absorbed by water vapour and ozone before it reaches the earth (see *Materials for Building*, Vol. 4, *4.02 Exposure*, p. 40) so that only near-ultraviolet radiation is received at ground level.* Equally important is the greater reduction in energy that

*The significance of the absorption of the middle ultraviolet is related to one of the characteristics of radiation, namely, that the shorter the wavelength, the higher the energy content.

takes place when the sun is not at the zenith because of time of day or year or latitude, while more scattering of shorter wavelength radiation occurs when the sun is at low angles. Consequently, the proportion of (the destructive) shorter wavelength is small, and there are also marked differences in the proportions received during the summer (more) and the winter (less). In addition to the geometric effects, both the total radiation and the intensity of ultraviolet are reduced by clouds and smoke.

(b) Orientation and location

Although the ultraviolet component of sunlight is the most significant factor responsible for the degradation of organic compounds, heat and moisture must also be taken into account. Consequently, the rate of deterioration will depend on how and where the materials are exposed to all these agencies. The following generalizations should be helpful in making *basic* assessments:

1. Vertically exposed materials weather better than horizontally exposed ones, primarily because the latter are exposed to sunlight for longer periods and because they may take longer to dry.
2. The time taken for vertically exposed materials to dry may not be significant particularly if the surfaces are not exposed to sunlight for long periods. For example, drying after rain takes longer on the north face of a building, but then this face is subjected to less sunlight than the south face, where drying would be more rapid.
3. Degradation is likely to be more rapid in areas with clean atmospheres, as the intensity of ultraviolet radiation is reduced in industrially polluted atmospheres while surfaces may be further protected by a film of dirt.
4. The severest conditions of exposure are likely in coastal areas due to the combined effects of the moist salt-laden atmosphere and high ultraviolet radiation intensity.

4 Actions and effects

The main purpose here is to discuss in more detail the actions and effects of those factors that are influential in the degradation of organic compounds. As mentioned above, the single most important factor is the ultraviolet component of radiation.

(a) Action of radiation

Absorbed ultraviolet radiation may lead to degradation, because it contains sufficient energy to cause a chemical reaction at the excited atom. A similar action does not occur when longer-wave radiation (i.e. visible light and infrared) is absorbed because, although a molecule may be raised to an excited state (usually at one particular atom) by the absorbed radiation, it may return to its unexcited state by dissipating the absorbed energy by reradiation of fluorescence, phosphorescence or heat.

Fundamental to the actions that may take place is the extent to which a material actually absorbs the radiation falling on it. Furthermore, the material may be selective in the bands of radiation that it absorbs, the selectivity depending largely on molecular arrangements. For example, in the case of visible light, if none of the groups of molecules that absorb visible light are present, the material appears colourless (or white if light is reflected). Lack of absorption or transparency to a particular wavelength signifies that the radiation passes through the material *without effect*. Thus materials that transmit all the ultraviolet down to 300 nm will not be degraded on exposure to sunlight. One of the reasons for the excellent exterior durability of *acrylics* such as polymethylmethacrylate is that they do not absorb until well down the middle ultraviolet range.* In contrast, polysytrene is affected by exterior exposure (usually turning yellow and losing some of its mechanical properties) because it is made from molecular groups that absorb at the lower end of the near-ultraviolet range.

It is important to note that the transparency of the basic polymer may be affected if impurities that cause absorption are not removed during the manufacturing process. In this, even small quantities of absorbers can be extremely significant, because absorption frequently produces chemical groups that absorb additional ultraviolet and so the reaction accelerates with time.

(b) Chemical changes due to radiation

Degradation due to chemical changes caused by ultraviolet radiation may take two forms.

1. The reverse of the polymerization reaction that originally produced the large molecules, in which the polymer may be broken down into isolated locations (chain scission) or it may completely revert to small molecules ('unzipping' of the polymer). The latter occurs very slowly when radiation is the only factor.
2. Increase in cross linking of molecules than was originally present, thus causing the material to become harder and more brittle. This occurs because the smaller molecules produced by chain scission (see (1) above) or reactive sites on large molecules react with other chains.

Two visible defects may occur. First, if the ultraviolet alters the resin's internal structures to those that absorb blue visible light, the reflected light appears yellow. Second, if the colorant used in the material is affected by ultraviolet, even if the polymer itself is not, fading results.

The only chemical effect of visible and infrared radiation is to speed up the rate of reactions that may be occurring from other causes. As a guide, an increase of 10°C doubles the rates of chemical reactions.

(c) Physical changes due to radiation

Physical changes due to radiation result from the chemical reactions that have taken place. Reduction in molecular size due to either chain scission or 'unzipping' results in changes in the physical characteristics for which large molecules are required. When cross linking occurs the material is affected because of the relationship between cross-link density and physical properties. For example, if flexibility is required for the building material to perform its function (coatings and sealants are notable), the additional hardness and brittleness caused by cross linking results in the material becoming less extensible, thus leading to cracking or loss of adhesion as the material can no longer accommodate movements.

Cross linking contributes more to the degradation of organic building materials exposed to ultraviolet radiation than does the depolymerization reaction. In some cases initial radiation products are coloured and absorb subsequent ultraviolet radiation. Deeper penetration of the radiation is thus prevented, and as the outer layer only becomes cross

*The benefits of transparency to materials used in bulk can sometimes be a liability if the background to which a clear (thin) coating is applied is affected by radiation. The failure of clear acrylic coatings on timber results from the degradation of the outer layer of wood which, in turn, provides insufficient support for the coating which then peels off.

linked, cracking is restricted to the surface. Materials with a thick cross section will have surface cracking or crazing. In contrast, the cracking in coatings which are relatively thin may be deep enough to penetrate the background. If, however, pigmentation restricts excess cross linking to a very thin layer and if the cracks are microscopic they will result in chalking from the gradual erosion of this layer.

(d) Thermal conditions

Temperature can affect organic compounds, both physically and chemically. Among important aspects are thermal shock resulting from rapid changes from hot to cold or vice versa. Temperature can have a marked effect on mechanical properties, and here the molecular structure of the polymer can be significant.

The low rate of heat transmission of most organic materials is influential in the internal stressing from thermal shock that can be experienced in plastics of thick cross section. Either surface or interior cracking can result, depending on conditions (see *2.3 Cracking*, p. 54 ff for general principles involved). Of equal significance is the fact that repeated thermal cycling can cause exudation of some of the less efficient plasticizers, resulting in the formation of a whitish layer on the surface, similar to that which results from fading caused by ultraviolet radiation.

The selection of the proper grade of asphalts and tars as roofing materials requires special care, as these organic materials have demands placed on them from temperature extremes. For example, a composition that is sufficiently hard for it not to flow in the summer may become brittle and crack badly in the winter. In contrast, those that do not become hard in the winter will flow in the summer. Being black, the materials are good absorbers of radiation and so may reach high temperatures in the summer, particularly if backed by insulation. The application of reflective treatments is necessary not only in order to reduce the temperature of the material but also to reduce degradation. The use of white or light-coloured roofing membranes based on liquid- or film-applied synthetic resins is relatively recent and their efficacy has yet to be proved. So far, these materials exhibit undesirable dirt collection or chalking.

The effect of temperature on chemical changes is, as already noted, to change the rate of reactions (see (b) previously). Oxidation, for example, takes place far more rapidly at elevated than at room temperature.

(e) Combined effects

1. *Radiation and water*. In general terms, most organic compounds are little affected by moisture, although glass-reinforced plastics can be damaged if the fibres are too close to the surface. Although radiation and water tend to be present at different times, materials can be subjected to radiation while still wet (i.e. after rain or from overnight condensation). The action of the combined elements can follow several different forms, with radiation speeding up the effect of water or vice versa. One important effect of radiation is to raise the temperature of a material to a level where solution or hydrolysis can take place. In general terms, the action of water is to remove plasticizers or low-molecular fragments. It may be noted here that the greying of exposed timber is due to the leaching by water of irradiated lignin.

2. *Radiation and oxygen*. The combined effect of oxygen and radiation, known as photo-oxidation, is greater than that of radiation and water. But, again, degradation of materials subjected to oxygen is much faster in the presence of radiation. Because of the reinforcing action of radiation and oxygen, anti-oxidants as well as ultraviolet absorbers are generally added to plastics (to reduce discoloration, for example) designed for exterior use. It is, however, important that the anti-oxidant does not absorb ultraviolet.

3. *Radiation and heat*. As infrared radiation is converted into heat, radiation referred to here is the ultraviolet component only. Under the influence of ultraviolet, degradation reactions may occur at much lower temperatures than would be expected to be found on natural exposure. Plastics that have been held at high temperatures for too long during moulding are more susceptible to ultraviolet degradation upon exposure.

5 Increasing durability

Three basic approaches can be taken to reduce the effects of radiation. Summarized, these are:

1. *Use of non-absorbent polymers*: Useful for transparent materials but complete transparency is difficult to achieve except on the laboratory scale.

2. *Use of polymers with high bond strengths*: In this method it is necessary to make polymers from combinations of elements whose bond strengths exceed the energy available in solar radiation. Only a few polymers can be created in this way but many of these are unfortunately readily decomposed by water and oxygen. The silicones are the best-known polymers of this type.

3. *Preventing absorption*: The most common approach is to prevent the polymer from absorbing ultraviolet radiation, and this can be easily accomplished if the material does not have to be transparent by the incorporation of pigments that reflect radiation or absorb it preferentially. Materials that incorporate reflective pigments usually have some degradation at the surface, frequently resulting in the loss of gloss on exposure, because radiation has to pass through the top layers twice. Black pigments allow complete absorption and are very effective, as demonstrated by the increase in polyethylene from one to 20 years with the addition of 1% carbon black. Clear polymers present far more difficulties if radiation absorption is to be prevented. Apart from the fact that absorbers need to be carefully selected, they are themselves slowly degraded, so do not last indefinitely.

Corrosion

Introduction

1 The scale of the problem

The destruction of metals by corrosion is a widespread problem. It is not confined to buildings and is therefore the concern of everybody. The annual cost of corrosion is considerable in all the developed countries, consequently it is not surprising that a great deal of study and research has been, and continues to be, devoted to the causes and remedies. Part of the answer is seen to lie in the use of alternative materials such as plastics.

In buildings there has been a significant increase in the use of metals for a number of different reasons (for example, the greater use of steel in reinforced concrete and in steel structures and of metals in wall and roof claddings as well as in environmental services). Plastics and composites not using metals are not always appropriate as alternatives to metals in many situations. Having an understanding of the problems associated with corrosion remains important.

2 Inevitability and complexity

Corrosion of the ferrous metals, that is, the rusting of iron and steelwork, is most commonly encountered, with the result that the problems of corrosion are often, but mistakenly, associated with this group of metals alone. An extremely important aspect of corrosion, which in simple terms is the chemical interaction of a metal with its surroundings, is that it is inherently inevitable. This means that all metals, particularly those used in buildings, will, given the right environmental conditions, corrode. However, whether or not corrosion is deleterious depends on the type of metal and the type of corrosion. In this, a distinction has to be drawn between surface films, essentially the product of gaseous corrosion, which are protective (the green patina which develops on copper roofs, for example), and other forms of corrosion which occur in the presence of water, such as rusting of iron, where the corrosion products are not usually protective, thus allowing the metal to be progressively eaten away. In general, it is the non-ferrous metals which develop protective surface films, but even these vary in the degree of protection they may provide. Given the right environmental conditions, the non-ferrous metals will also corrode, sometimes very rapidly. This chapter is concerned with the deleterious effects of corrosion. One of the environmental factors necessary for attack to take place is the *presence of water*.

The resistance which any metal, or alloy, will provide to corrosion is, environmental conditions apart, largely dependent on the degree to which the metal, as used, varies from the metal in the natural state. The natural, or original, state represents the more stable condition and, however processed, it is natural for metals to tend to return to the state from which they came. Thus, the noble metals platinum, silver and gold and the 'near' noble metal copper are found in the metallic state in nature and are very resistant to corrosion. Iron and steel as used, on the other hand, are not found in the metallic state and are correspondingly less resistant to corrosion. With most metals, considerable conversion of the original state, as found in the ores from which metals are extracted, is necessary, At the same time, it may be noted that the corrosion products of iron, for example, are all found in the natural state, thus illustrating the *tendency of a metal to return to its natural state* (an example of *reversion*, p. 14).

In all considerations of corrosion the relationship between a given metal and given environmental conditions is extremely important. In this it is also important to note that there is a large number of metals, including alloys, available, and these, may, in practice, be subjected to a wide range of environmental conditions. Many of the latter when related to given metals may result in intensive attack by corrosion. It is not surprising, therefore, that the problems associated with corrosion are complex.

3 Minimizing the effects of corrosion

In building, as in other spheres, the fact that corrosion will take place is not of itself vitally important. The *rate of attack* and the fact that attack is more commonly *localized* (uniform attack can be convenienctly dealt with in terms of adequate thickness of metal relative to service life) are basically far more important. At the same time, the possible effects of corrosion need careful consideration. In general, corrosion is a form of waste. In buildings, such wastage is not necessarily confined to the destruction of the metal, which may have to be replaced, but also the destruction (or, at best, changes in appearance) of other materials which may be caused during, or as a result of, corrosion. The corrosion of metals in services (pipes, tanks, boilers, radiators, etc.) may lead not only to interruption of the services but also to accidental leakage of water into the fabric of a building (see *3.2 Exposure*, 'Faulty services' p. 212). The corrosion of embedded metal work (the steel in reinforced concrete, for example) may lead to spalling and cracking of the surrounding material.

If the deleterious effects of corrosion are to be minimized, then attention should be paid to three main requirements:

1. Care in design, that is, avoiding conditions which are likely to give rise to corrosion;
2. Care in the selection of metals for given conditions; and
3. Care in providing the necessary protection.

In the final analysis, the successful use of metals depends on the combined efforts of *manufacturer* (manufacturing details, particularly of components for services), *architect/designer* (building details), the *builder* (ensuring work is carried out properly) and the *user* (ensuring that there is adequate care and maintenance).

4 Importance of durability and maintenance

The use of metals requires, perhaps more than with other materials, that special consideration be given to the relationship of durability and maintenance. In this, it may be noted that, in a great number of cases, metals are selected for reasons other than their resistance to corrosion.

Strength properties, paticularly when these can be obtained economically (first cost, that is), are often more important, as is often the case with structural iron or steelwork, for example. Strength may, of course, be applicable to situations other than the main structure of a building. However, in order to maintain 'economical' strength properties, the metal has to be protected from corrosion, and preservation generally implies continual, but periodic, maintenance, the period depending on the durability of the kind of protection employed. Exceptions would include those situations in which the metal may be inaccessible – fixings, steel in concrete, etc. Although it is possible to use metals which have equal, or near-equal, strength properties but with higher resistance to corrosion (less maintenance), such metals make first cost high. Strength requirements are subject to such wide variations that it is always advisable, in any given circumstance, to evaluate cost implications (first cost + maintenance cost = final cost) of the possible range of metals.* In this it is significant to note the increased use of the more corrosion-resistant metals such as stainless steel or copper alloys for a wider range of applications than hitherto. At the same time, there have been notable advances in the development of methods of protection.

5 Initiation and rate of attack

As already mentioned, the problems associated with corrosion of metals are complex. In practice, corrosion may take various forms, while there are many variable factors that influence its initiation and rate. In a great many cases the *time* element is important. For example, corrosion may be initiated because of the existence of one set of conditions but may subsequently progress because of another. The conditions prevailing at any given time may influence the rate of attack. In buildings, environmental conditions which may result in corrosion are dynamic rather than static. Importantly, design details may often encourage conditions that will increase the rate of attack.

6 Scope

To recognize the need for the special care that should be taken in the use of metals in buildings requires an understanding of the causes of and remedies for corrosion. However, initially it is important to grasp the interrelationship of the various factors involved. Accordingly, general principles are considered first. These should also help to illustrate the complexities involved. Having considered the general principles, attention is then focused on the effects of corrosion, the exposure conditions under which corrosion is likely to occur, the resistance to corrosion provided by various metals commonly used in buildings and the methods of protection available. Special consideration is given to the precautions that might be taken in the selection, design and maintenance of metals.

General considerations

1 Scope

The main purpose of this section is to outline the basic principles of corrosion generally and to note the more important factors involved, so as to form a background for the sections that follow dealing specifically with the corrosion of metals used in building practice.

*Guidance on the life expectancy of metals and protective coatings is now increasingly available from BSI in particular. Notes on these are included later.

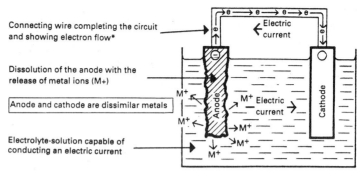

Connecting wire completing the circuit
and showing electron flow*

Dissolution of the anode with the
release of metal ions (M+)

Anode and cathode are dissimilar metals

Electrolyte-solution capable of
conducting an electric current

Diagram D3.7/1 *Principle of a simple battery*

*Note: Electrons are current carriers but, by convention, they
flow in the opposite direction to the current.

Although various theories have been advanced to account for the anomalies of corrosion, it is now generally agreed that there is a relationship between corrosion and electrolysis. This has given rise to the electrochemical theory, that is, that a corroding metal behaves as if it is part of a kind of electric cell or wet battery. It has been shown that corrosion is accompanied by the setting up of *small* electric currents. Electric currents are produced in a battery, for example, when two different metals are suspended in a chemical solution. When the circuit is completed (the exposed portions of the metals connected by a wire), one metal, known as the *anode*, dissolves, while an electric current flows through the solution from this corroding metal to the other, known as the *cathode* (see Diagram D3.7/1).

In practice, different metals may be in contact with moisture so as to behave like a small battery (galvanic cell, as it is often known in corrosion theory). However, currents (and thus corrosion) may also be set up in single metals when one part becomes the anode and another the cathode. In both cases the moisture involved may contain air or other dissolved chemical substances, which conducts electricity.

In some ways the basic principles of corrosion based on the electrochemical theory are fairly straightforward. However, the study, or understanding of corrosion is complicated by the fact that there are so many variable factors involved that are likely to influence the initiation, course, rate and final result (i.e. form) of corrosion. Some of these factors include the purity of the metal, the composition and interrelation of all the substances with which it comes into contact, the presence of bacteria and the possibility of minute externally produced electric currents being present. It is also necessary to remember that the time element may be significant. Corrosion may well be initiated by one set of conditions, but over a period of time other conditions may exist which enable it to proceed.

2 Definition

Corrosion may be simply defined as the destructive chemical attack of a metal by agents with which it comes into contact. In fact, destruction occurs as a result of the interaction of a metal with its environment. It is important that a distinction is made between corrosion and erosion. The latter is the destruction of materials by mechanical agencies. However, it may be noted that corrosion and erosion often occur together.

Although the chemical changes of a metal that take place during corrosion (here more conveniently referred to as 'corrosion changes') are generally taken to imply wearing or eating away of the metal, this is not strictly the case, as corrosion changes may be divided into two classes, namely, those which produce a solid film and those which do not. The former usually performs a protective function, that is, once the film has formed, corrosion of the underlying metal is stifled (but see also '3 Surface films', later), while in the absence of a film formation corrosion generally proceeds until a reactant has been exhausted. It is the latter, known as *electrochemical corrosion*, that is emphasized in this section.

3 Surface films

(a) Formation

The formation of surface films on metals is regarded as gaseous corrosion in which oxidation takes place. The most significant fact about film-forming reactions is that they are usually very rapid in the opening stages, but become increasingly slow as the film thickens, thereby isolating the metal and air from one another. Furthermore, the fact that such reactions often choke themselves is important, as it does help to explain why metals exposed to the atmosphere usually escape destruction, particularly when the atmosphere is dry.

(b) Protection provided

The degree of protection which surface films may give to the underlying metal when exposed to environments other than gases depends not only on the resistance of the film itself but also on the extent to which it adheres to the underlying metal. The poor resistance of the surface film formed on iron exposed to the atmosphere may therefore be explained by the fact that the film (iron oxide) is porous and only adheres loosely to the underlying metal. On the other hand, the high resistance of the surface film formed on copper (among other non-ferrous metals) may be attributed to the fact that the film is solid and adheres firmly to the underlying metal. Two other factors which will influence the resistance a surface film will provide to subsequent corrosion is its uniformity and its continuity. Non-uniformity, that is, varying thickness, and discontinuity, that is, a break in the film, may often account for localized corrosion.

(c) Direct oxidation

It is also important to note that surface film formation, which is the result of direct oxidation, takes place at the point where the oxidizing impinges on the exposed metal. This, as discussed in '4 Basic mechanisms' below, is not generally the case with electrochemical corrosion, as the metal goes into solution in *one* place, with oxygen taken up at a *second* place, and with the oxide or hydroxide formed at a *third*. It is because the solid corrosion product is formed at a distance from the point of attack that it cannot stifle the action. Thus, direct oxidation is generally less dangerous than electrochemical corrosion.

4 Basic mechanisms

It is now generally held that the corrosion of metals is basically *electrochemical in nature* – see p. 460 and Diagram D3.7/1. Consequently, the attack is basically a chemical reaction accompanied by the passage of an electric current. The flow of electricity occurs between certain areas of a metal surface, known as *anodes* and *cathodes*, through a solution capable of conducting an electric current, known as an *electrolyte*. However, the flow of electricity can only occur when a potential difference exists between the anode and the cathode. During the flow of electricity *destruction of the anode* occurs.

(a) Electrolyte

An electrolyte is a liquid that contains ions, which are positively or negatively charged atoms or groups of atoms in an aqueous solution. In equilibrium, the negative and positive charges are in electrical balance. Pure water, for example, contains an equal number of hydrogens (H^+) and hydroxyl (OH^-) ions. Electrical current is conveyed through the solution by ionic migration. Thus electrolytes conduct electricity but are decomposed by it.

Acids, alkalis and many other solutions (those containing salts are significant) are considerably more ionized than water, and therefore act as good electrolytes.

(b) Simple corrosion cell

During corrosion there is a flow of electricity resulting in *simultaneous processes* taking place at the anodes and cathodes. A number of reactions are possible at the cathode, while the formation of corrosion products occurs as a result of the interaction of cathodic and anodic processes. The basic mechanisms involved may be illustrated with a simple corrosion cell.

(1) Flow of electricity

A simple corrosion cell consisting of a metal immersed in an electrolyte is illustrated in Diagram D3.7/2. At the anode, which is the region at the lower potential, the atoms dissolve to form ions, leaving behind electrons and giving, in the case of ions, the following reaction:

$$Fe \rightarrow Fe^{++} + 2e(\text{electrons})$$

The freed electrons travel through the metal to the cathode, that is, that part of the metal at the higher potential, where they are utilized in the reduction of either ions or oxygen. A two-way ionic migration also occurs in the solution in which positively charged cations migrate to the cathode and negatively charged anions to the anode. (Examples of cations and anions in water are H^+ and OH^-; in sodium chloride, Na^+ and Cl^-; and in sodium sulphate Na^+ and SO_4^{--}.) Thus a complete

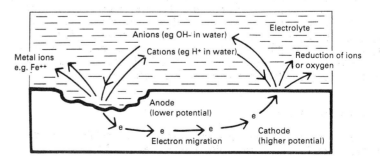

Diagram D3.7/2 *Simple corrosion cell – compare with Diagram D3.7/1*

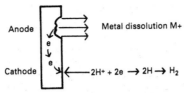

(i) Hydrogen reduction – in acids

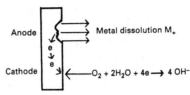

(ii) Oxygen reduction – in natural waters
(i.e. only slightly acidic or alkaline)

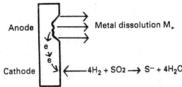

(iii) Reduction of sulphate
(with the aid bacteria
- anaerobic solid notable)

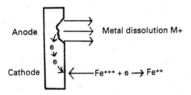

(iv) Reduction of metal ions – acid waters

Diagram D3.7/3 *The four basic reactions possible at the cathode and the electrolytes in which they occur. In all cases the reaction at the anode is the dissociation of the metal*

circuit is formed by the metal and solution, and the passage of a current through the circuit. The electrons are the current carriers in the metal (it is important to note that, by convention, they flow in the opposite direction to the current), while in the solution the current is carried by the ions.

(2) Cathodic reactions
The nature of the electrolyte will govern the reaction that will take place at the cathode. However, four basic reactions may be identified, as shown in Diagram D3.7/3 and which may be summarized as follows:

(i)   Reduction of hydrogen:

$$2H^+ + 2e \rightarrow 2H \rightarrow H_2$$

This is the main cathodic process in acid solutions, and the metal dissolves with the simultaneous evolution of hydrogen gas. (See (iv) for an alternative cathodic process in acid waters.)

(ii) Reduction of oxygen:

$$O_2 + 2H_2O + 4e \rightarrow 4OH-$$

This reaction is generally responsible for corrosion of metals in natural waters which have an approximately neutral reaction, that is, they are only slightly acidic or alkaline.

(iii) Reduction of sulphate (with the aid of bacteria):

$$4H_2 + SO_4^{--} \rightarrow S^{--} + 4H_2O$$

This reaction takes place when there is an absence of oxygen, as may occur in underground conditions and requires the presence of adequate dissolved sulphate and the bacteria *Disulpho-vibrio disulphuricans*. The bacteria use cathodic hydrogen in their living process and bring about the reduction of sulphate to form sulphide. A similar cathodic reaction can also take place in aerated solutions that occur beneath any impervious corrosion product which prevents oxygen from gaining access to the surface of the metal.

(iv) Reduction of metal ions:

$$Fe^{+++} + e \rightarrow Fe^{++}$$

This is an alternative cathodic process to that outlined in (i) for acids and occurs with metals which have two valencies, such as copper and iron, and can exist in solution as cupric and cuprous, or ferric and ferrous ions, respectively. This type of cathodic reaction is usually considered to be the cause of the corrosion of iron in acid mine waters, when the ferrous iron formed at the cathode is subsequently oxidized back to the ferric form by dissolved oxygen.

Although the cathodic reactions are quoted separately, this does not preclude the possibility of more than one of the four participating in the overall cathodic process in any given circumstance. In this, the time element may be significant. Time apart, reductions of metal ions (reaction (iv)) and hydrogen ions (reaction (i)) can occur in acid solutions, while a certain amount of hydrogen is evolved even when oxygen is present in solutions which are only slightly acidic or alkaline. On the other hand, when all oxygen (reaction (ii)) has been removed during corrosion in a closed vessel, the sulphate-reducing bacteria (reaction (iii)) can take over, thus allowing the corrosion to proceed.

(3) Formation of corrosion product

The formation of the corrosion product, as in the case of rust shown in Diagram D3.7/4, results from the interaction between anodic and cathodic products. The metal ions (Fe^{++}) dissolved from the anode and the hydroxyl ions ($2OH^-$) from the cathodic reactions move in opposite directions through the electrolyte because of their positive and negative charges. When they encounter each other, they react to form ferrous hydroxide, which precipitates to form a visible corrosion product:

$$Fe^{++} + 2OH^- \rightarrow Fe\,(OH)_2$$

The ferrous hydroxide is a white product, but in oxygenated conditions this will rapidly oxidize to form, first, ferric hydroxide:

$$4Fe(OH)_2 + O_2 + 2H_2O \rightarrow 4Fe(OH)_3$$

As the ferric hydroxide is unstable it subsequently loses water to form hydrated ferric oxide, $FeO(OH)$, or Fe_2O_3 (red rust):

$$Fe(OH)_3 \rightarrow FeO(OH) + H_2O$$

It is significant to note that the products of both the cathode and the anode are soluble bodies and will not stifle attack. At the same time, the

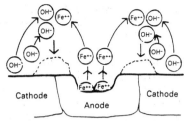

Movement through electrolyte of metal ions (Fe⁺⁺)
from anode and hydroxyl ions (2OH⁻) from cathode
to form FERROUS HYDROXIDE which precipitates
(visible white corrosion products)

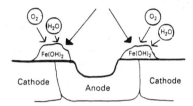

Dissolved oxygen oxidizes the ferrous hydroxide
and with water forms unstable ferric hydroxide
which loses water to form FERRIC OXIDE

as

FeO(OH) or Fe_2O_2 (red rust)

Diagram D3.7/4 *The formation of corrosion products on iron. (Read from top to bottom)*

solid substances formed when they meet also cannot stifle further attack of the anode, as the corrosion product forms at a distance from the point of attack.

An important characteristic of most of the solid compounds formed by *corrosion* is that they occupy a *larger volume* than the metal destroyed in producing them. Rust, for example, normally occupies a larger volume than the iron contained in it. This aspect is particularly important when metals are connected or embedded, as the expansion accompanying corrosion can lead to the development of forces strong enough to cause breakage. In a known case of the failure of a line of rivets holding two steel plates together, the rust which formed between the plates acted in the same way as if a wedge has been driven between them. In building terms, examples include the cracking of brick, stone or concrete in which embedded iron or steel have corroded.

(c) Dissimilar metals in contact

The example of the simple corrosion cell given in (b) (p. 462) is based on a single metal immersed in an electrolyte. Apart from the elecrolyte, the factors which give rise to variations in potential between one part and

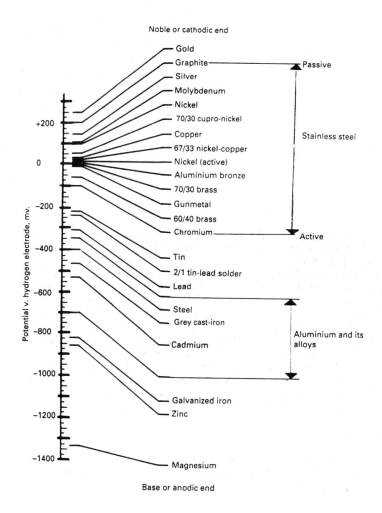

Diagram D3.7/5 *Corrosion cell formed by two dissimilar metals – compare with Diagram D3.7/1*

another, and hence the flow of electricity, are variable, but are generally due to some form heterogeneity of the metal. (See '6 Initiation of attack', p. 470).

When different metals or alloys are immersed in an electrolyte each, after a time, attains a potential which is characteristic for that metal and the electrolyte (for each metal the potentials will vary with the nature of the electrolyte). Thus, if two metals with different potentials (one higher than the other) are connected together, there will be a flow of current from the metal with the *higher* potential to that with the *lower* potential, as shown in Diagram D3.7/5. The metal with the higher potential is said to be *electropositive*, or *cathodic* to the metal with the lower potential. This has given rise to a series known as the *galvanic series* (sometimes also the electrochemical series) as shown in Chart C3.7/1.

The value of the galvanic series is that it does enable the behaviour of metals when connected together to be predicted in principle. Two common examples found in building practice may help to illustrate this. If copper and zinc are in contact with one another (copper piping and a galvanized steel tank, for example), then the copper will be

Chart C3.7/1 *Practical galvanic series of metals and alloys*

cathodic to the zinc, and the zinc, being anodic, will corrode. If brass and aluminium are in contact (brass hinges to aluminium window frames, for example), then the brass will be cathodic to the aluminium, and the aluminium, being anodic, will corrode. In addition, the galvanic series enables some prediction to be made of the rate of corrosion of the anodic metal. In this, the potential difference between the cathodic and anodic metals is significant. Thus, in the two examples given, and assuming a similar electrolyte, there is likely to be a faster rate of attack of the zinc than of the aluminium – the potential difference between copper and zinc is greater than between brass and aluminium.

5 Corrosion classification

Classification of corrosion depends on the environment to which a metal is exposed. In general, four classes may be identified namely, gaseous, atmospheric, immersed or underground. Gaseous corrosion is essentially associated with the formation of surface films, as briefly discussed in '3 Surface films' previously and is, therefore, excluded here. An essential feature of the remaining three is that they all require the *presence of water*, as the corrosion is electrochemical in nature. It is convenient to discuss these three classes of corrosion under separate headings.

(a) Atmospheric corrosion

Metals freely exposed to the atmosphere receive an unlimited supply of oxygen. Attack depends, therefore, on the presence of water and impurities dissolved in it. The dissolved impurities which would commonly include sulphur dioxide (polluted industrial atmospheres) or salt (marine atmospheres) form efficient electrolytes (see 3.2 *Exposure*, p. 168 ff), thus promoting corrosion.

The maintenance of dampness, and hence electrolytes, is often assisted by hygroscopic bodies which adhere to the surface of the metal. Such hygroscopic bodies would include soot, rust and magnesium chloride, the last originating from marine spray.

In atmospheric corrosion *relative humidity* is a useful guide for predicting both the likelihood and rate of corrosion. It has been shown that there is a sudden rise in the rate of corrosion above a certain *critical humidity*. Above this humidity atmospheric pollution becomes the decisive factor. In general, serious corrosion is unlikely to take place at relative humidities *below* 70%. For convenience, Diagram D3.7/6 illustrates the annual variation in relative humidity for some places in the UK. It will be seen that relative humidity only falls below the critical value of 70% for comparatively short periods during the year. When making use of information such as that included in Diagram D3.7/6 it is important to remember that ambient temperature has an effect, although it is the diurnal fluctuations in temperature which determine the incidence and duration of condensation, which are more significant than the average value. In addition, the presence of deliquescent particles, as already noted, can be highly injurious, and corrosion can, therefore, take place below the critical value of 70% relative humidity.

An interesting example of what may best be described as *two-stage atmospheric corrosion* occurs in the formation of the green patina commonly seen on copper roofs. In the early stages the copper darkens. This is due to a dark deposit of sulphide, oxide and soot. The subsequent formation of the final green patina is due partly to the action of sulphuric acid in the soot and partly to the oxidation of the

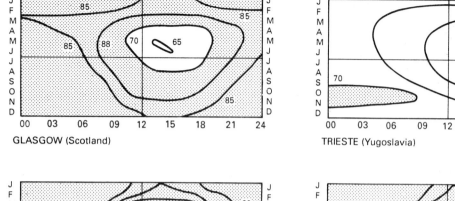

GLASGOW (Scotland)

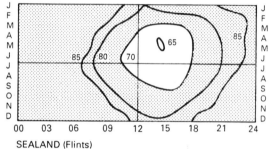

TRIESTE (Yugoslavia)

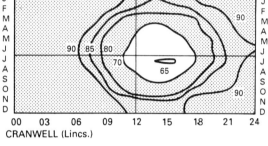

CRANWELL (Lincs.)

SEALAND (Flints)

Diagram D3.7/6 *Comparison of annual variations in the relative humidity of four locations in the UK (Glasgow, Cranwell, Sealand and Kew) and in Trieste. (Data from* Averages of humidity for the British Isles, *HMSO, London, 1962 (for the UK); and* Atmospheric corrosion resistance of zinc, *ZDA and AZ11, 1965, Fig. 13 (for Trieste)*

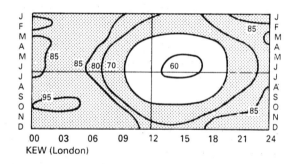

KEW (London)

copper sulphide. The green patina serves to protect the underlying metal from attack. In this example it should be noted that some gaseous corrosion is also involved.

(b) Immersed corrosion

Whereas atmospheric corrosion is mainly controlled by moisture, corrosion in totally immersed conditions is regulated mainly by the availability of oxygen. The amount of dissolved oxygen which may occur in waters is liable to wide variations. However, even when the oxygen supply is severely depleted, corrosion can proceed with the aid of sulphate-reducing bacteria. The composition of the water is important in that it affects electrical conductivity. In certain cases, calcareous deposits on metals, formed from hard waters, can have a protective value. The corrosiveness of water is also dependent on temperature. Generally, corrosion proceeds more rapidly as the temperature rises. Finally, under conditions of flow, the rate of water flow often increases the rate of corrosion for two reasons. First, the supply of oxygen is promoted and, second, the adhesion of protective corrosion products may be prevented.

(c) Underground corrosion

Corrosion of metals buried underground may take place due to the action of three processes, namely, basic electrochemical action, sulphate-reducing bacteria, or stray currents. Although none of these processes takes place exclusively (in any given circumstance, one or more may be operative – the time element is important), it is more convenient to consider them separately.

(1) Basic electrochemical action

Conditions existing in the soil may vary between something comparable to atmospheric exposure and what is almost equivalent to complete immersion. The nature of the soil is an important governing factor. Soils such as sand and chalk, because of their permeable nature, have a plentiful supply of atmospheric oxygen. Such soils are said to be *aerobic*. Consequently, metals buried in open aerobic soils have considerable portions of their surface exposed to oxygen, which is sufficient to ensure that ferric products are formed close to the metal. The rust which may soon appear on the surface once it is wetted stifles further attack. In soils which are completely waterlogged and deficient of free oxygen (known as *anaerobic* soils) corrosion is usually slow, unless sulphate-reducing bacteria are present.

Corrosion is likely to be localized, and intense, when a metal is buried in soils of an intermediate character. In this, the presence of air pockets are significant as differential aeration currents (see 6(b)) may flow, when oxygen is taken up at the air pockets and attack is directed on the places where the soil presses on the metallic surface. Air pockets may be a natural feature of the soil or, as important, may be produced artificially when soil is thrown back into a trench after pipe laying, so that spaces are left between the individual spadefuls. Disturbance of the soil, and hence the subsequent formation of air pockets, can occur not only with trenches for pipe or cable laying but also with holes dug to receive a steel stanchion. It is also important to note that backfilling creates an entirely different environment from that found in undisturbed soil only a few feet away. Such differences, and there are others which may occur naturally (the boundary between horizontal soil strata, for example), may lead to the formation of a current.

The chemical nature of the soil, and in particular different constituents, may also account for the setting up of corrosion currents. Thus the acidity or alkalinity of the soil will also affect corrosion. *Made-up ground* containing *ashes and clinker* in which steel pipes have been buried requires special consideration, as under these conditions the steel is rapidly corroded. The reason for this is not due to the acidity of the ashes and clinker (they tend to the alkaline) but to their content of water-soluble matter, which yields electrolytes of low resistivity with the soil water. The presence of any unburnt carbonaceous matter may also promote corrosion, as such matter will act as the cathode of a corrosion cell and, being hygroscopic, will retain moisture in contact with the metal.

(2) Corrosion by sulphate-reducing bacteria

Sulphate-reducing bacteria, previously outlined in 4 (b)(2) – reaction (iv) – can flourish only in anaerobic soils such as waterlogged clays

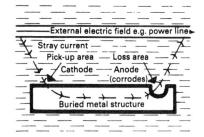

Diagram D3.7/7 *Corrosion cell formed by stray currents from an external electric field*

containing sulphates and organic matter. Whereas soils with no free oxygen are usually non-corrosive, if sterile, similar soils containing the bacteria will be highly corrosive, as the organisms present enable sulphates to act as hydrogen acceptors, with reduction to sulphides. The actions of the sulphate-reducing bacteria are even more dangerous than the purely electrochemical types of attack, particularly as the organisms can continue to multiply.

The final corrosion product is a mixture of rust and black iron sulphide. Freshly exposed adjacent soil will be seen to be discoloured and it (or the crust of corrosion product) will evolve hydrogen sulphide when wetted with hydrochloric acid.

(3) Corrosion by stray currents

Metal pipes and structures buried in the soil may act as conductors and pick up stray currents from such sources as tramlines (now generally non-existent in the UK), power and telephone cables, thus giving rise to serious risk of corrosion by stray currents. Part of the current (up to 15–20%) from the main power line would stray to enter the buried metal and then leave it some distance away to rejoin the main power line. Corrosion generally occurs at the 'loss areas' (Diagram D3.7/7). In the case of steel, attack does occur at the 'loss areas' (the anodes); but with lead sheathing (to cables) attack sometimes occurs at both the 'loss areas' (the anodes) and the 'pick-up areas' (the cathodes), as in the presence of salt the cathodic reaction would produce alkali, which can attack lead. Methods of protecting buried metals from corrosion by stray currents include: insulating the joints, electrical drainage and sacrificial anodes.

Steelwork embedded in concrete and used in chemical works where electrochemical cells are installed in the buildings requires special consideration, as stray currents from the circuit may occasionally reach the steel, resulting in the production of a voluminous type of rust which, in turn, may cause the concrete to burst, thus exposing the steel. Generally, suitable electrical precautions should prevent this.

6 Initiation of attack

A fundamental concept of the electrochemical theory is that, in addition to the presence of an electrolyte, there must be a *potential difference* which, in practice, may be either between two dissimilar metals or between areas of a single metal (that is, between anodes and cathodes). In the case of the single metal there are a number of factors which may determine the areas which becomes anodes and cathodes. Emphasis is placed here on a single metal as far as initiation of attack is concerned.

As already noted, attack is initiated, and subsequently proceeds, at the anodic areas. Although it is convenient to outline separately the main factors involved, it is important to note that the cause of corrosion, in any given circumstance, may be due to one or more of these factors. On the other hand, as corrosion proceeds, the factors responsible for attack may change. The fact that rust, for example, spreads from the original localized area of attack is due to the formation of concentrated cells beneath the rust.

(a) Non-uniformity of the metal

The whole of a metal, particularly its surface, will rarely follow the ideal metal lattice, while there may also be differences in the atoms. Boundary conditions between various grains of the metal may exist, giving rise to micro-cells, in which the boundary is usually the anode.

Intercrystalline corrosion may occur in a number of alloys (stainless steels and certain aluminium alloys, for example) as a result of precipitation in the grain boundaries. There may be differences in composition from place to place in an alloy, in which case the parts of the alloy which contain a greater concentration of the more noble phase (the component with the more positive potential) will be cathodic to the rest of the surface. The corrosion of aluminium–zinc alloys can be increased under such conditions. In general, departures from the ideal structure of a metal, and hence non-uniformity, are increased by impurities. Thus pure metals are likely to resist corrosion better than metals containing impurities. For example, pure aluminium is more resistant to attack than an aluminium alloy.

Electrically conducting materials (metallic or non-metallic) which may be included in a metal or in contact with the surface of a metal do often act as cathodes. The contact of dissimilar metals is an obvious example, but the corrosion of copper tubing as the result of carbon films left during manufacture is another. As regards the latter, it may be noted that BS 2871 Part 1 1971 requires that the internal surfaces of copper tubing shall be free of all deleterious films.

Areas of *unequal stress* or *deformation* give rise to different potentials, and are important in the production of galvanic cells with a single metal. In general, the more stressed parts are anodic, and corrode more readily. Variations in stress can be caused by many factors, such as strains or external stresses. Some examples of corrosion resulting from unequal stress or deformation include boilers, at bends in iron or steel, heads of rivets and cracking of brass.

An important point for the initiation of attack occurs where there is a breakdown of the protective oxide film on the metal. The breaks in the film cause the underlying metal to be exposed. It is the exposed area which becomes anodic, and hence attacked. Breakdown of the oxide film on aluminium often occurs. In the case of iron or steel, discontinuities of the mill scale are often responsible for intense localized corrosion of the underlying metal.

The phenomenon of rust creeping underneath damaged paint or other similar protective films present an anomaly because the corrosion is initiated by a breakdown in the film, yet the exposed metal is cathodic, rather then anodic. This can be explained by *differential aeration* corrosion (see (b) below). The exposed part of the metal has free access to oxygen, whereas that covered by the paint film has not. The latter therefore becomes anodic and corrodes (see Diagram D3.7/8).

(b) Non-uniformity of the liquid

Non-uniformity in the composition of the liquid in contact with the metal, or close enough to have an influence on the corrosion process, may lead to the setting up of currents. The requisite non-uniformity may arise due to concentrations of metal ion, salt, hydrogen ion (pH value significant), oxygen or oxidants, as shown in Diagram D3.7/9. It will be seen in the diagram that, with the exception of the neutral salt, areas of high concentration result in cathodes, and those of low concentration in anodes.

Some of the conditions under which *variations* in the liquid are likely to occur are worth noting. With the metal ion, concentration can occur in flowing waters, particularly if there are variations in the rate of flow. In the case of copper, for example, attack takes place at the area from which the copper ions are removed most readily, that is, where the flow is fastest. With salts, differences of chloride concentration are likely to occur when fresh and sea water meet (flowing of rivers into the sea).

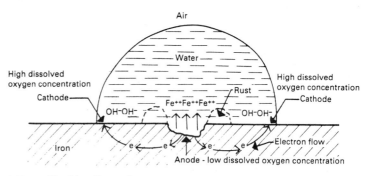

1 Drop of liquid on the surface

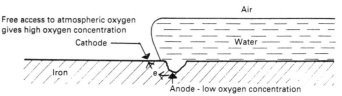

2 Partial coverage of surface by liquid

Diagram D3.7/8 *Three basic examples of corrosion cells formed due to differential aeration. (See also Diagram D3.7/12). Diagram illustrates the way in which the corrosion product may influence the distribution of attack – corrosion is initiated at the original anode but may spread, the attack taking place beneath the product where new anodes are formed, due to restricted oxygen access. In the case of other solids, e.g. dust, paint films, etc, the principle is similar except there is no original anode*

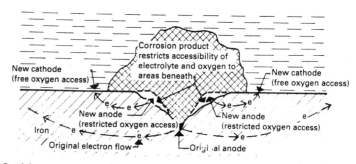

3 Partial coverage of surface solids

Variations in oxygen concentration require special consideration, because corrosion currents which arise as a result of the differences of oxygen distribution in the liquid account for a significant number of corrosion failures in practice (crevice corrosion is notable). Such corrosion is usually known as *differential aeration corrosion*. Cases in which there is likely to be a significant difference in oxygen concentration include partial immersion of a metal, in which only the exposed part has free access to oxygen. As already explained, that part devoid of oxygen becomes anodic, and thus attacked. Similar conditions occur when one part of the surface of a metal is covered with liquid and the other exposed (drops or small areas of liquid on the surface), or when one part is covered by solids (dust, corrosion

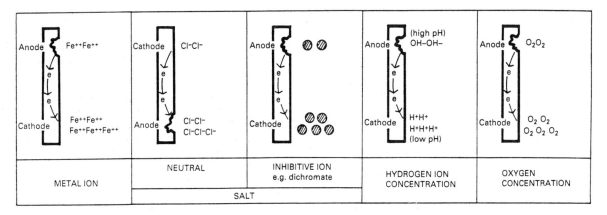

| METAL ION | NEUTRAL | INHIBITIVE ION e.g. dichromate | HYDROGEN ION CONCENTRATION | OXYGEN CONCENTRATION |

SALT

products, paint and similar protective films, etc.) and the other exposed to the atmosphere. (See also '8 Forms of corrosion', p. 476.) Some examples of differential aeration corrosion are included for convenience, in Diagrams D3.7/8 and D3.7/12.

Diagram D3.7/9 *Corrosion cells formed by non-uniformity in the composition of the liquid in contact with the metal. Note: Apart from the neutral salt, areas of high concentration result in the formation of the cathode; and those with low concentrations, the anode*

(c) Variations in physical conditions

Variations in physical conditions which may give rise to the setting up of currents include differences in temperature, stray currents and flow of water. The significance of stray currents is explained in 5 (c)(3), p. 470.

Differences in temperature usually result in the hotter part becoming anodic, although the nature of the liquid and metal involved is important. Thus, in sulphate, the hot part is the cathode on copper and lead. Temperature differences can lead to corrosion in boilers and refrigerating equipment.

The flow of water has two effects: first, stimulation of the cathodic process by inceasing the concentration of reactants at the metal surface and, second, stimulation of the anodic process by facilitating the removal of corrosion products.

7 Rate of attack

The fact that, in any given circumstance, the basic conditions necessary for corrosion to take place may exist gives no indication of the rate of attack which may occur. In practice, it is the rate of attack which is most important. It is perhaps axiomatic that this rate will be dependent on the strength of the electric current in any given galvanic cell. Although it may be possible to predict the strength of the current on the basis of potential differences between the anode and cathode on the open circuit, as given in Chart C3.7/1, there are three other factors, namely, polarization, the conductivity of the electrolyte and the relative areas of the anode and cathode, which must be taken into account. For convenience, the effects of pH value (hydrogen ion concentration) and dissolved matter (collectively referred to as 'composition') are included with conductivity. (Composition of water is also included in 3.2 *Exposure*, under 'Water supply', Table T3.2/3 p. 205.

(a) Polarization

Whatever the potential difference between a given anode and cathode may be on the open circuit, this is modified in an actual galvanic cell by displacement of the potential of the cathode towards the anode and vice versa. The shifts of potential are known as *polarization*. Thus the operative potential may be near that of the open circuit anode or cathode potential, or intermediate between the two. The importance of polarization lies in the fact that it may, in certain cases, result in extremely small potential differences, even though these differences on the open circuit are large. The precise value of the operative potential will, in turn, depend on whether the corrosion process is controlled by the anode, cathode or a combination of both.

In general, the dissolution of metal ions at the anode is a very fast reaction. The anodic reaction only controls the rate of corrosion with 'near' noble metals, such as copper, or with metals in the *passive state* (stainless steel notably) as the diffusion of ions through the oxide layer is slow.

Cathodic control is association with the control of the oxygen reduction reaction. This control, in which oxygen reduction or the effects of polarization are important, may be illustrated by two examples. On the basis of an open circuit, the potential differences between copper–aluminium and aluminium–stainless steel are the same. In galvanic cells, the copper is an efficient cathode, with the result that oxygen is readily reduced so that the aluminium is severely attacked. However, because of the passive film on the stainless steel, oxygen is not readily reduced, with the result that the stainless steel will be readily polarized to the aluminium potential so that the galvanic effect will be small or negligible.

(b) Conductivity and composition

The conductivity of electrolytes varies widely. However, those which are good conductors not only increase the rate of corrosion in general but also enable cathodes and anodes which are comparatively far apart from one another to take part in the corrosion process. With those electrolytes of poor conductivity the flow of current will be limited to the immediate areas of contact between the two areas.

The cathodic protection provided by zinc on galvanized steel when the steel is exposed through breaks in the zinc coating is dependent not only on the area of steel exposed but also on the conductivity of the electrolyte. When a small area of steel is exposed, the potential of the exposed steel is polarized to a more negative potential at which the ferrous ions can no longer leave the metal. Thus, the steel is cathodically protected, with oxygen reduction taking place on the iron cathode and an increase in corrosion of the zinc anode. When a large area of steel is exposed, protection is maintained only if the electrolyte has a high conductivity (brackish waters and sea water are notable). If the electrolyte offers a high resistance, then only the steel adjacent to the zinc will be protected.

The *pH value* of solutions is discussed in *3.2 Exposure*, under 'Water supply', p. 209. It is a measure of the acidity or alkalinity of aqueous solutions, in which a scale, ranging from 0 to 14, is used to express the hydrogen ion concentration. On this scale *pure water* has a value of 7.0 (at 25°C), *acid* solutions a value of *less than* 7.0, and *alkaline* solutions a value of *more than* 7.0. The way in which the pH value, that is, the hydrogen ion concentration, will influence corrosion rate depends on whether:

1. The metal is noble (noble metals are stable in both acid and alkaline solutions); or
2. Its oxide is soluble in acid or both acid and alkali.

Metals such as aluminium, lead, tin and zinc are soluble in both acid and alkaline solutions, their rate of corrosion being parabolic. The minimum rate of corrosion occurs at a pH value of 6.5 for aluminium, 8.0 for lead, 8.5 for tin and 11.5 for zinc. Most metals have oxides which are soluble in acids but insoluble in alkalis. Generally, the lower the pH value, the higher the rate of corrosion. In the case of iron, for example, the critical pH value is 3.0. Below this value the rate of corrosion increases. It is no longer necessary for oxygen to be present, as it is possible for hydrogen to be liberated. Between pH 4.0 and 9.0 the rate of corrosion is low and constant, falling to a minimum when a value of pH 12.0 is reached. In alkalis, iron forms a protective film. Above a pH of 12 the rate of corrosion increases. Variation in corrosion rate with pH value is shown in Diagram D3.7/10.

Dissolved salts may influence corrosion according to their concentration and type. Ions such as chloride, Cl^- and sulphate, SO_4^{--}, are highly aggressive, capable of breaking down or preventing the formation of protective films, while those such as carbonate and bicarbonate, CO_3^{--} and HCO_3^-, and calcium, Ca^{++}, have inhibiting properties and therefore are capable of restraining corrosion. The latter may be exemplified by the formation of a film of calcium carbonate, as

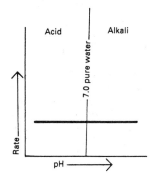

(1) Noble metals
(e.g. gold, platinum)

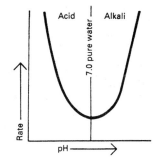

(2) Metals with amphoteric oxides
(e.g. aluminium, lead, tin. zinc)

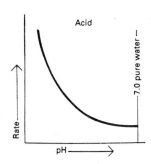

(3) Acid-soluable metals
(e.g. cadmium, copper, chromium, manganese, magnesium, nickel)

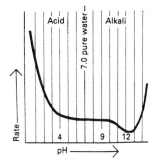

(4) Iron

Diagram D3.7/10 *The effects of the concentration in the pH value on the rate of corrosion of various groups of metals*

occurs in water supply systems carrying hard water. The aggressive or inhibiting properties are however, dependent on the concentration of the dissolved salts.

Dissolved gases, principally carbon dioxide and oxygen, may inflence corrosion. The effect of carbon dioxide in water is twofold. First, it lowers the pH value, in itself a stimulant to intensive attack. Second, free carbon dioxide makes water aggressive because it reverses the reaction which leads to the formation of carbonate scale, thereby dissolving scale that may be present. Oxygen, on the other hand, acts both as cathodic depolarizer and an anodic polarizer or passivator.

(c) Relative areas of cathode and anode

The importance of the relative areas of the anode and cathode, as far as rate of attack is concerned, depends on the amount of dissolved salts in water, that is, the conductivity of the electrolyte. The influence of the latter may be viewed in two different ways:

1. *Dissolved salts*: In waters with a relatively low amount of dissolved salts the area of the cathode will be unimportant, as the attack on the anode is restricted to the area immediately next to the junction of the anode and cathode (or, in the case of two metals, next to the bimetallic junction), as outlined in (b) above. However, in a highly conductive electrolyte such as salt water the attack will be controlled by the relative areas of the anode and cathode. Provided the cathodic reaction is unrestricted by polarization, the greater the area of the cathode relative to that of the anode, the greater the amount of attack. In fact, the amount of attack at the anode will be proportional to the area of the cathode. For example, an aluminium rivet in copper would suffer intense attack, whereas a copper rivet in aluminium, although undesirable, would suffer less.
2. *Strength of current*: The strength of the current flowing depends largely on the amount of oxygen reaching the cathode. Consequently, if the cathode is large, the current may be quite strong in relation to the size of the anodic area. In the case of iron or steel, breaks in the mill scale expose the bare metal which becomes the anode. Attack may therefore become very intense when the exposed portion of the iron is small. Intense attack is frequently associated with this combination of large cathode and small anode.

8 Forms of corrosion

Different types of corrosion yield different forms of corrosive attack. The more common, most of which are localized, are included below.

(a) Uniform attack

This, in a sense, is the ideal form of attack, as there is uniform thinning which can be allowed for in design. Such attack usually takes place in acid solutions, or in strongly alkaline solutions on some metals. It is also likely to occur in solutions with a high content of dissolved salts. Surfaces which have a uniform deposit are more likely to have a reasonably uniform type of attack, while deposits of calcium carbonate will reduce considerably the rate of attack. In practice, it is probably unwise, except in special circumstances, to rely on uniform corrosion taking place, as attack is more commonly localized.

(b) Grooving

Grooving is a form of localized attack in which thinning occurs where there has been particular concentration of an electrolyte. In condensers, for example, grooves are initiated where the steam first starts to condense, and consequently indicate the flow path. Such condensates are acid. Local concentrations may also occur from water run-off from roofs or merely from dripping. Grooving of *flashings* often takes place when the water is acidic.

(c) Pitting

Pitting is one of the most dangerous forms of localized attack. It is believed to be associated at its inception with a small anodic area and a large cathodic area, and may be due to variations in the metal, in the surface film (metallic or otherwise) or in the film–solution interface.

Breakdown of *mill scale* on iron and steel – a film laid down during manufacture – is a common cause of intense localized attack (see Diagram D3.7/11).

At one time severe pitting occurred with copper cold water pipes in hard waters, due, it is thought, to a film of carbon derived from the lubrication used in the drawing process. Such carbon films are now removed. In fact, BS 2871: Part 1: 1977, for copper tubing, requires that the inner surface shall be free from deleterious films. In rare cases pitting may be caused as a result of films by moorland waters.

Some metals, notably aluminium, may suffer from localized attack by copper-bearing waters. Such attack is induced if dissolved oxygen, calcium bicarbonate and chloride are present.

Pitting of aluminium, which may be due to breakdown in oxide film or dirt deposits, can usually be minimized by regular cleaning.

Sulphate-reducing bacteria also cause pitting, which may be hemi-spherical or elongated if individual pits run into one another. Stainless steels may be pitted (usually elongated pits) by salts (particularly chlorides) which break down the passivity of the surface.

(d) Waterline corrosion

Waterline corrosion is commonly associated with water tanks (Diagram D3.7/12), when localized corrosion takes place at the metal/water/air boundary. A similar type of attack can happen at the periphery of a droplet of water on a metal surface. Attack occurs along a line just beneath the level of the miniscus. This is an example of differential aeration corrosion, and as the area above the waterline is cathodic, it is completely unaffected by corrosion.

Susceptibility to waterline attack varies with the metal and the solution. Iron and steel are particularly vulnerable to most waters.

(e) Crevice corrosion

Crevice corrosion usually results from *bad design*, although it can also occur as the result of deposits of foreign matter on the surface. Invariably, the crevices formed are inaccessible, and so localized attack is often intense. Attack is more commonly due to differences in oxygen content (differential aeration), although the formation of a concentration cell may be the result of differences in salt or hydrogen ion.

One of the main reasons why atmospheric corrosion occurs rapidly in crevices is more likely due to the fact that moisture is retained in them long after moisture has dried up on the surface. Corrosion deposits in

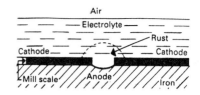

Diagram D3.7/11 *Breakdown of mill scale on iron or steel is a common cause of intense local corrosion*

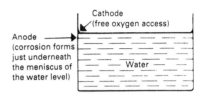

Diagram D3.7/12 *Waterline attack – an example of differential aeration corrosion. (Compare with Diagram D3.7/8)*

crevices can exert considerable forces, which may be sufficient to result in damage to surrounding material (see '4 Basic mechanisms', (b) 3, p. 464).

(f) Dezincification

Dezincification is the selective corrosion of *brass*, in which one of the constituents, usually zinc, is removed, leaving a weak porous residue. Such corrosion occurs more often in moving waters (differential aeration is significant) and so brass taps and valves are commonly affected. Soft waters and sea water are particularly aggressive.

Other *copper alloys* may be similarly affected. In the case of *aluminium bronzes* with 8% or more of aluminium, *de-aluminification* occurs.

(g) Corrosion in rapidly moving water

Corrosion of metals subjected to rapidly moving waters is usually the result of corrosion *and* mechanical abrasion (erosion). Attack may be caused by local breakdown of a protective layer by suspended particles (sand, for instance), or gas bubbles which impinge on the surface (usually continually in the same place), or by turbulence alone. The form of attack is usually localized, giving rise to pits. The conditions necessary for such attack arise at the entrance of pipes, at sharp bends, near deposits, and where the cross section of the flow stream changes abruptly. Metals most likely to suffer from corrosion/erosion attack are those which form poorly adhering corrosive products.

Changes in the rate of movement of water, resulting in intense local attack, are significant in the case of copper. For local attack to occur, water must pass rapidly over the surface at *one* point, while the water over the remainder of the surface is relatively stagnant. Under these conditions copper ions are removed from the point where rapid movement of water occurs, keeping the ionic concentration low, while the ionic concentration will be high elsewhere. The point of low ionic concentration will, therefore, be permanently anodic (see '6 Initiation of attack', (b), p. 471).

(h) Stress corrosion

Cracks which may develop in metals are often due to various forms of stress corrosion, in which only a small amount of metal is removed by corrosion. Basically, two types of stress corrosion may be identified:

- One results from the stresses due to applied loads which cause a fracture, with corrosion serving to break down obstructions which would otherwise retard the advance of the crack
- The other is essentially steady electrochemical destruction of metal (destruction of the grain boundaries is significant) with stresses due to applied loads serving to concentrate the corrosion on the tip of the crack. Stress corrosion of a kind may occur due to fatigue – this type of corrosion is more correctly known as fatigue corrosion.

The form of cracking or the agents responsible for cracking give rise to various descriptions. Some of these include:

1. *Season-cracking*, which occurs in brass, and so called because of the resemblance of the cracks to those in seasoned timber;
2. *Caustic cracking* (sometimes also *caustic embrittlement*), a type of intercrystalline attack initated by caustic alkali and found in boilers. The attack occurs in joints, seams or crevices in which water can leak and become sufficiently concentrated with caustic alkali;

3. *Nitrate cracking* is another example of intergranular attack which takes place on mild steel (not all types) at high temperatures in the presence of nitrates.

Effects of corrosion

In terms of building, the principal effects of the corrosion of a metal are as follows.

1 Structural soundness

Corrosion may affect the structural soundness of the component. It may generally be assumed that corrosion will impair the strength of the component. The extent to which a reduction in strength may be significant will depend on circumstances. In the case of uniform attack, the effects of corrosion on strength may be allowed for, as the corrosion will result in uniform reduction of thickness. The same is not necessarily true with the various forms of localized attack, which, on the whole, are probably more frequent. Other things being equal, localized reduction in strength can, of course, have more serious consequences than uniform reduction.

When considering the effects of corrosion on the structural soundness of a metal component it is always important to remember that all components are required to be strong enough to perform their primary functions, which in a great many cases may not be specifically related to the primary structural stability of a building. A simple example may clarify this important point.

The primary function of water pipes is to facilitate the flow of water. In order to do this, the pipes must be strong enough to resist the weight of water, including any pressures involved. A fracture or hole caused or initiated by corrosion would result in leakage of water. Other examples would include various fixing devices and related components used in pipework, engineering services, panelling or cladding and hinges, etc.

2 Distortion or cracking of other building materials

The products of corrosion are far more voluminous than the metal or alloy from which the products are formed (see 'General considerations 4 Basic mechanisms (b) (3)' p. 465). Apart from special cases, such as some forms of stress corrosion, corrosion is confined to the exposed surface of a metal. Consequently, the growth of corrosion products may cause distortion or cracking of other building materials in which the metal may either be embedded or with which it may be in contact. The failure of other building materials in this way may, in turn, lead to more rapid attack of the metal due to the freer access of water and/or oxygen to the metal and, consequently, 'renewed' damage on the associated building material. So the cycle may continue, making the deleterious effects of corrosion progressively worse. The fact that the rate of destruction is often increased is also extremely significant.

3 Entry of water into the building

The failure of the component may lead to entry of water into the building, as may occur with metal roof finishes, flashings, dpcs, gutters, pipes, engineering services, equipment, etc. In this it is important to note that the water which may gain entry is not restricted to rainwater, as outlined in *3.2 Exposure*, p. 212).

4 Changes in appearance

The changes in appearance associated with corrosion are generally unsightly. The surfaces affected may be either the metal or some other building material adjacent but underneath a corroding metal. Examples of the latter include the brown staining from iron and steel, or the green staining from copper and its alloys often seen on concrete, stonework and masonry. Other materials, including paint, may also be affected. Rainwater,or some other source of water, flowing over the surface of the corroded metal transfers some of the corrosion products onto the adjacent material. Indiscriminate storage of metals (steel reinforcement is often notable) on or near other building materials may cause staining of the latter. In almost all cases, particularly when porous materials are involved, the staining is extremely difficult to remove.

Exposure

1 Generally

The extent to which a metal or alloy may corrode is basically dependent on two interrelated factors:

- The properties of the metal or alloy, including the effects of deformation; and,
- The conditions to which the metal or alloy is exposed.

In buildings, metals may be exposed to a number of different conditions. In each case *the presence of water is fundamental,* and so all sources of water, including the extent to which the water may be polluted, as discussed in *3.2 Exposure,* must be considered. A point which needs to be emphasized is that the severity of exposure will be governed by the effects of the specific use of a metal. In this the actual form and exposure condition of a detail are extremely important. Thus, although exposure to the atmosphere, external or internal, does represent a basic condition of exposure, this type of exposure is seldom principally responsible for most cases of corrosion. Although it would be wrong to underestimate the significance of atmospheric exposure, it is important to note that failure nearly always occurs in some *detail* and not in general exposure to the atmosphere (even the external atmosphere). In view of the fact that metals are seldom, if ever, used in isolation, it is also necessary to include as a condition of exposure the effects of other building materials, including the juxtaposition of different metals or alloys.

In practice, it is important that consideration be given to all types of exposure, with due regard paid to *details* which may influence the severity of exposure in given circumstances. In general terms, it is possible to identify five types of exposure which may arise. Each of these is considered separately.

2 External atmospheres

The amount of corrosion of metals or alloys exposed to external atmospheres will be primarily dependent on the kind of metal or alloy and the effects of climatic factors. As regards the latter, it is not only the macro-climate which is important but, more precisely, the *micro-climate*, which is influenced to a large extent by the building in general and by details in particular. As previously emphasized, the length of *time* that moisture may be in contact with the surface of a metal or alloy is an important governing factor as far as the rate of corrosion is concerned. The aggressive nature of the moisture is also significant, for *both* rate *and* intensity of attack.

Climatic factors which influence conditions of exposure are:

- Humidity;
- Temperature;
- Rainfall;
- Wind; and
- Exposure to the prevailing wind and rain.

Although it is, in general, unlikely that corrosion will take place when the relative humidity is below 70% (the critical humidity – see 'General considerations 5(a) Corrosion classification' p. 467), cognisance must be taken of the fact that, under certain temperature/humidity conditions, *condensation* may take place. Temperature and wind influence significantly the rate at which water is evaporated from surfaces. The effects of seasonal variations are generally important (see *3.1 General considerations* '6 Prevailing wet and dry conditions', p. 140), but in the case of metals daily variations may be significant. Rainfall, particularly exposure to wind-driven rain, will influence which faces will be primarily subjected to water.

The aggressiveness of the water in contact with the surface of a metal, whether derived from atmospheric humidity (condensation notably) or from rain, will be dependent on atmospheric polution (see *3.2 Exposure*, 'Atmosphere', p. 168 ff). Further examples of the effect of pollution in the atmosphere on the rate of corrosion are given under 'Resistance to corrosion' later.

The extent to which water may be retained on the surface of a metal will be determined by the slope of the surface. In general, horizontal or near-horizontal surfaces will have the effect of increasing the severity of exposure, as removal of any retained water will be dependent on evaporation. Retention of water, and hence an increase in the condition of exposure, will occur if water is allowed to enter crevices (see also '5 Contact with conductive water' later).

3 Internal atmospheres

Nowadays the exposure of metals to internal atmospheres should be considered in two ways: first, the exposure of metals directly to the prevailing humidity conditions of the internal air and, second, the exposure of metals indirectly to the results of condensation on either exposed surfaces (i.e. from surface condensation) or concealed surfaces within the thickness of a construction (i.e. from interstitial condensation).

In both cases the prevailing conditions of humidity will be an important determinant. Such conditions are initially influenced by the external humidity conditions, including the degree of its pollution. However, and importantly, large amounts of moisture may be produced within buildings (for example, in kitchens and bathrooms in domestic buildings, in swimming pools, in commercial kitchens or in some factories). Unless removed, such moisture may add signifcantly to the internal humidity. The rate and/or amount of condensation will

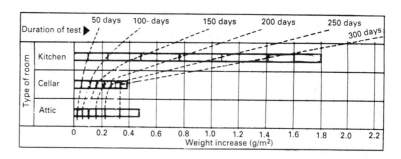

Chart C3.7/2 *Comparison of the effects of three different internal atmospheres on the corrosion of zinc. (Data from* Atmospheric corrosion resistance of zinc, *ZDA and AZ11, Fig. 9)*

be influenced either by the prevailing internal humidity conditions for surface condensation or by the external and internal humidity conditions in conjunction with temperature and the hygrothermal properties of the construction (see *Materials for building*, Vol. 4 'Condensation', p. 124).

With the use of more thermal insulation within constructions and higher levels of heating together with the increase in the amount of moisture produced within buildings it is now more important that the effects of all forms of atmospheric 'modifiers' are considered. In some cases air conditioning may reduce the severity of exposure; in others it may increase it.

Chart 3.7/2 illustrates, by way of a simple example, the effects of three different internal atmospheres on the corrosion of zinc. Not surprisingly, the greatest corrosion occurred in a kitchen. Bearing in mind that the data were collected well before the present levels of heating and humidity commonly found in buildings, the amount of corrosion in the attic is greater than in the cellar. Today the difference between the two would undoubtedly be greater.

Particularly corrosive atmospheres may be created in buildings by a number of different substances used every day (e.g. solvents, cleansing agents and aerosols). Gases in flues from the combustion of fuels may also be relevant. In all cases, condensation usually increases the aggressive nature of a corrosive atmosphere. (See also *3.6 Chemical attack*, 'Sulphate attack', '7 Domestic chimneys', p. 421.)

4 Other building materials

Metal components may be embedded in or in contact with a variety of building materials. The latter include mortars, plasters, concrete, floor compositions or wood.* The extent to which any of these materials may influence corrosion depends largely on how much water releases, or concentrates, corrosive agents from them. In this, water introduced during construction may be important. For convenience, the corrosive agents which may be derived from building materials and other sources, together with the possible effects of these agents, are included in Table T3.7/1.

Although not strictly a building material, the soil should be regarded basically as an aggressive material in which metals may be embedded or in contact. Conditions of exposure of pipes, and other metalwork buried under ground, is covered under 'General considerations', 5(c) p. 469.

*In the context of corrosion, wood includes both natural (i.e. untreated) and treated timber. Some species of timber have a marked acidity (e.g. the heartwoods of oak, sweet chestnut and Douglas fir). Whereas timber normally contains only a small amount of electrolyte, their conductivity can be increased greatly by the salts (derived from sea water), certain preservatives or flame retardants introduced into the timber. Bimetallic corrosion can occur with metals such as aluminium (for example, if the timber has been treated with a preservative containing copper).

5 Contact with conductive water

As mentioned previously in type 1 exposure, there is a greater risk of corrosion if water is retained in crevices between metal surfaces or between a metallic and some other material, than where a metal is simply exposed to the normal action of the weather. Another factor which may equally well increase the condition of exposure is the dripping of rainwater onto the surface of a metal. All these increases in the severity of exposure are, usually inevitably, brought about by details in building construction.

A further increase in the risk of corrosion arises when the water (rainwater, condensate or any other source of water) is made more conductive, that is, made a more efficient electrolyte, by dissolved acids, alkalis or salts. Such dissolved matter may, in general, be derived from the atmosphere or from material with which the metal comes into contact. There are also those particular cases, such as occur in chemical works or other industrial processes, where the water may be highly corrosive. In addition, account must sometimes be taken of the corrosive nature of some materials used during maintenance, including cleaning, or of leaking service pipes and equipment. Finally, supply water may also constitute an increased risk of corrosion.

6 Contact between dissimilar metals

(a) Difficulties of assessment

The contact between two unprotected dissimilar metals in the presence of moisture presents a complex condition of exposure. Assessment of the severity of exposure of any given bimetallic junction and, in particular, the risk of corrosion of the less noble metal under given conditions is made difficult because of the many variable factors involved. The potential difference between two dissimilar metals in contact can give some guidance as to the possible rate of corrosion. The galvanic series, described in 'General considerations', 4(c) p. 466, from which potential differences between any two metals may be estimated is based on an *open circuit*, whereas, in practice, potential differences result from a *closed circuit*, and these may be significantly different from those in the open circuit. Thus, the galvanic series does not serve as a reliable practical guide. Attempts have been made by a number of workers in the field to formulate, in a simplified way, tables showing the possibility of corrosion when two dissimilar metals are in contact, and based on practical experience. Such tables are usually the product of engineering experience, but they are, nevertheless, of value in building practice.

(b) Tabulated guidance

Compilers of tables giving the degree of corrosion at bimetallic junctions have, in order to simplify the tables, had to make generalizations. However, the tables are still of value, as they do at least give an indication of the basic condition of exposure, *provided* they are not regarded as infallible and that the variables that may alter the basic condition of exposure are taken into account. For convenience, Diagram D3.7/13 has been prepared from a table produced by the Admiralty, War Office and Ministry of Aviation Inter-service Metallurgical Research Council. The original table, based on pratical experience, is a useful summary of available information in a simplified form. The

Table T3.7/1   *Effects of corrosive agents on metals*

Corrosive agents			Effects[a]	
Source	Common occurrence	Type	Metal(s)	Description/Comments
Portland cement	All cement-based products, e.g. mortars, renders, plasters, screeds, concrete	Sodium and potassium hydroxide (alkaline – about pH 12.5)	Aluminium	Generally harmful but degree of attack partly dependent on type of alloy. Protection under damp conditions essential
				Anodizing coatings rapidly destroyed
				When embedded, unsightly salt efflorescence can occur above level of embedment. Cracking of the embedding medium may also occur
			Lead Zinc	Harmful. Protection under damp conditions essential
High-alumina cement	All cement-based products	Alkalis (alkalinity less than for Portland cement)	Aluminium Lead Zinc	Generally as for Portland cement, except effects are less marked as high-alumina cement is much less alkaline
High-calcium and magnesium limes	Mortars and plasters		Aluminium	Severely attacked. Protection under damp conditions essential
			Lead Zinc	Relatively slight effect
Salt accelerators	Gypsum plasters (Keene's, Parian anhydrous gypsum plasters)	Acid reaction	All unprotected metals	Some corrosion likely before plaster has dried out, i.e. while it is still damp. Normally, corrosion should not be troublesome once the plaster has dried out and provided there are subsequently no prolonged periods of dampness
	Reinforced concrete	Calcium chloride	Mild steel	Calcium chloride added must be evenly distributed (preferably added to mixing water) and must not exceed 2% calculated on the weight of cement
Smoke and flue gases	Flue terminals	Principally sulphur dioxide	Aluminium Copper Zinc (mainly in galvanized steel	Attack is rapid when metals are directly exposed to smoke and flue gases
Combustion of coal and other fuels, and from sea spray	Atmosphere Rainwater	Principally sulphur dioxide, carbon dioxide and/or corrosive salts, e.g. sulphates and chlorides	Aluminium	Generally, corrosion resistance is fairly high. Sometimes corrosion stifles itself and the corrosion rate falls to a low value, but with a few alloys corrosion may be continuing and severe
			Zinc	Protective film not sufficiently dense and adherent to prevent steady though slow attack
			Iron and steel	Corrosion greatly stimulated. Protection essential
Water supply	Pipes and related components	Principally chlorides and sulphates, but dissolved carbon dioxide also significant	Iron and steel (except stainless steels)	Dissolved carbon dioxide may influence rate of corrosion due to its acidic character, and, indirectly, prevent the formation of protective calcium carbonate
			Aluminium	Generally, not recommended for use with ordinary supply waters, but excellent for use in specially treated waters used industrially
			Copper	Corrosion appreciable in soft waters with appreciable dissolved carbon dioxide contents, causing green staining on plumbing fixtures
			Lead	Attack by soft waters sufficient to cause a physiologically dangerous lead concentration – 'lead poisoning'
Sea water	Immersed structures and pipe lines	Principally chlorides	Iron and steel	Rate of corrosion increased when compared with fresh water, but this rate is increased by movement of the sea water
			Stainless steels	Local attack and pitting usually marked in stagnant water

Table T3.7/1 *(Continued)*

	Corrosive agents		Effects[a]	
Source	*Common occurrence*	*Type*	*Metal(s)*	*Description/Comments*
Wood	*Softwoods*: Western red cedar and Douglas fir *Hardwoods*: Oak and sweet chestnut	Mainly organic acids, but also soluble salts	Aluminium Copper Lead Zinc	Corrosion unlikely if timber is well seasoned, maintained dry or the metal is isolated from the timber
				Copper has been partly affected by acidic rainwater run from cedar roofs
				Aluminium nails used for cedar shingles or sidings severely attacked if directly exposed to rainwater. No danger of attack if nails are protected by an overlapping shingle
				Zinc valley gutter in contact with hardboard underlay has been known to be attacked within two years
				Aluminium paint can be safely used on wood
Magnesium oxychloride cement	Flooring	Salt (hygroscopic)	Aluminium	Severely attacked whether in contact or embedded in flooring. Protection essential
Sulphates	Clay products, particularly brickwork	Calcium, magnesium and potassium (soluble salts)	Zinc	Zinc used as flashings, soakers, etc., partly attacked by soluble salts and partly by alkali from mortar
Algae, moss or lichen	Pitched roofs	Organic acids and carbon dioxide	Aluminium Copper Lead, zinc	Attack principally associated with gutters and flashings and where acidic rainwater drips on to the metal
Foaming agents (certain types only)	Foamed cement used for insulation of pipes	Small amounts of ammonia	Copper	Copper containing small amounts of phosphorus affected. In the absence of avoiding the use of ammonia-producing foaming agents, suitably annealed or phosphorus-free copper should be used
Acetic acid fumes	Industrial processes, e.g. breweries, pickle factories and sawmills	Acid	Lead	Use of lead for glazing bars, flashings, etc., best avoided as complete protection is rarely possible under the particular circumstances
Ashes or clinker	Made-up ground	Water-soluble matter, and unburnt carbonaceous matter	Iron and steel	Serious corrosion likely

[a] The effects included in this table are intended to cover those commonly experienced and taking into account common usage of the various metals in buildings.

value of the table (and Diagram D3.7/13) lies in the fact that it does show, at a glance, whether there is any danger of the corrosion of one metal being increased by contact with a second metal.* However, when use is made of the chart to assess severity of exposure, account must be taken of the following factors.†

(1) The electrolyte and its effects

Corrosion at bimetallic junctions only occurs while a continuous film or body of water joins the different metals so that a small electric current, an essential part of the corrosion process, can pass through this water. The circuit is completed by the metallic contact. The conductivity of the water, itself dependent on the quantity and nature of electrolyte dissolved in it, will influence the severity of corrosion in a given time.

*Evans Ulick R. and Rance, Vera E., *Corrosion and its Prevention at Bimetallic Contacts*, 3rd edn, HMSO London, 1963. The main part of this publication is the table setting out the 'Degree of Corrosion at Bimetallic Contacts'. Diagram D3.7/13 is an abridged version of the table in that those metals not commonly used in building practice have been omitted while a different form of presentation has been adopted.

†These factors have been included under 'General considerations' earlier, but are repeated here so that they may be read directly with Diagram D3.7/13.

486

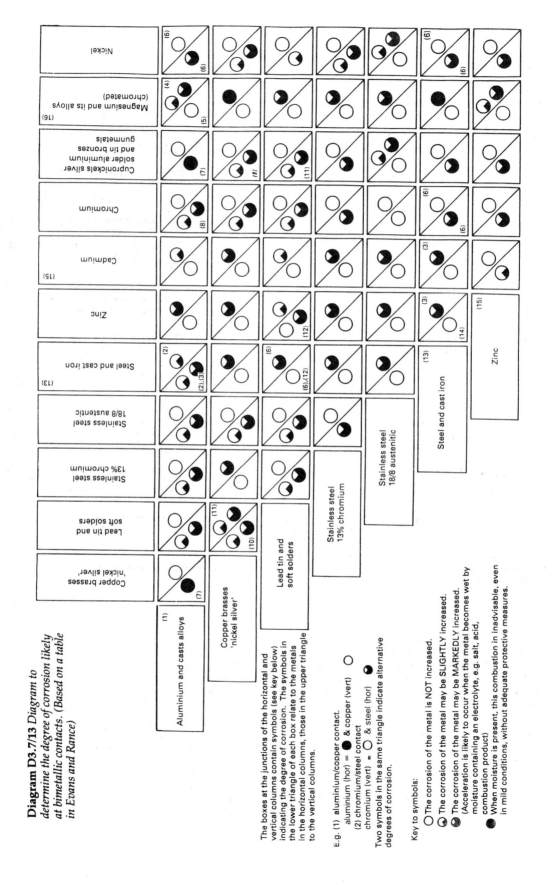

Diagram D3.7/13 Diagram to determine the degree of corrosion likely at bimetallic contacts. (Based on a table in Evans and Rance)

The boxes at the junctions of the horizontal and vertical columns contain symbols (see key below) indicating the degree of corrosion. The symbols in the lower triangle of each box relate to the metals in the horizontal columns, those in the upper triangle to the vertical columns.

E.g. (1) aluminium/copper contact.
aluminium (hor) = ● & copper (vert) ○
(2) chromium/steel contact
chromium (vert) = ○ & steel (hor) ◑
Two symbols in the same triangle indicate alternative degrees of corrosion.

Key to symbols:

○ The corrosion of the metal is NOT increased.

◔ The corrosion of the metal may be SLIGHTLY increased.

◕ The corrosion of the metal may be MARKEDLY increased. (Acceleration is likely to occur when the metal becomes wet by moisture containing an electrolyte, e.g. salt, acid, combustion product)

● When moisture is present, this combustion in inadvisable, even in mild conditions, without adequate protective measures.

Notes:

(1) Bimetallic contact of different alloys of aluminium can occur, because aluminium becomes more noble when alloyed with **appreciable amounts of copper and less noble with appreciable amounts of zinc.**

(2) Serious local corrosion through oxygen-screening or in other ways (even when the total destruction of metal is finished) may sometimes be caused when corrosion products from iron or steel reach aluminium, or vice versa.

(3) The corroded metal may provide an excellent protective coating for the steel or cast iron, the latter usually being electrochemically protected at gaps in the coating.

(4) Adverse galvanic affects will be minimized by using aluminium alloys containing little or no copper (0.1% max.).

(5) Aluminium may be attacked by alkali formed at the aluminium when magnesium corrodes in sea water or certain other electrolytes.

(6) The 'non-corroding metal' may be used as a protective coating to the 'corroded metal', provided the coating is continuous - good coatings may become discontinuous under abrasive conditions.

(7) For maximum protection of the aluminium when contact with copper or copper-rich materials is unavoidable, ensure that the copper-rich metal is first placed with tin or nickel and then with cadmium.

(8) ◐◕ For contact with thick chromium plate.
For contact with thin chromium plate.

(9) Corrosion of copper and brasses may sometimes be accelerated by contact with bronzes and gunmetals.

(10) In some immersed conditions the corrosion of copper or brass may be seriously accelerated at pores or defects in tin coatings

(11) In some immersed conditions there may be serious acceleration of soldered seams in copper and its alloys.

(12) When exposed to the atmosphere, lead in contact with steel or galvanized steel can be rapidly corroded where the access of air is restricted (narrow crevices notable).

(13) The corrosion of mild steel may sometimes be increased by coupling with cast iron especially when the exposed area of mild steel is small compared with the cast iron.

(14) In most supply waters at temperatures above 140°F (60°C) zinc may accelerate the corrosion of steel.

(15) The choice between cadmium and zinc may be influenced by the fact that the corrosive product on zinc is, in certain circumstances, more voluminous and less adherent than on cadmium.

(16) The behaviour of magnesium alloys in bimetallic contacts is particularly influenced by the environment, depending especially on whether an electrolyte can collect and remain as a bridge across the contact. The behaviour indicated in the chart refers to fairly severe conditions.

In practice, sea water obviously represents a strong electrolyte, but dissolved fuel combustion products in rainwater also give a water a relatively high conductivity. Intermittent contact through water results in the bimetallic effect being dependent on the total time of contact.

Conditions of exposure are most severe for metals immersed in water. On exposure to atmospheric conditions, corrosion is usually localized in the vicinity of the line of contact. Condensed moisture containing dissolved electrolyte from polluted atmospheres (industrial and marine) can cause an appreciable acceleration of corrosion.

Water containing complex ions may alter the data given in the diagram. For example, ammonia has no action on iron, but attacks copper (normally a more noble metal) in the presence of air.

At soldered joints certain fluxes, which leave hygroscopic residues and produce moist conditions at the joint, are a source of trouble at bimetallic contacts.

Finally, it should be noted that serious acceleration of corrosion can occur when the two metals concerned are not directly joined together. For example, dissimilar metals, exposed to the sea in the wooden hull of a ship, may be dangerous even though they are connected only remotely by a metallic path through the ship.

(2) Condition of the metal

The diagram disregards the fact that different alloys within one group may differ significantly in composition and may, therefore, give rise to bimetallic effects when in contact with each other. In the case of aluminium alloys, for example, bimetallic contact is also dependent on heat-treatment conditions.

In the case of stainless steel it may be noted that its more noble character is dependent upon the protective film on the metal and this film is, in turn, maintained only if oxygen is continuously present.

(3) Metallic coatings

Metals such as chromium, nickel and zinc are most commonly used as protective metal coatings on another metal. The electrochemical behaviour of a metallic coating is generally similar to that of the same metal in the passive state, *provided* that it is (1) resonably thick and (2) free from pores. Where the coating is thin and contains many pores the elctrochemical behaviour of the coated metal may approximate more to that of the basis metal, or to that of a relatively thick non-porous underlay. For example, the electrochemical behaviour of a coated metal with a thin decorative chromium on a good undercoat of nickel on steel or brass may approximate to that of the nickel.

Discontinuities of the coating may set up corrosion cells, with two possible effects. If the coating is more noble, the basis metal may be attacked; if the coating is less noble, the coating may be attacked.

Certain trimetallic contacts may need attention. Consider, for example, galvanized steel (zinc on steel) in contact with a more noble metal aluminium. Contact of the zinc with the aluminium may accelerate the corrosion of the zinc, leaving the steel (the base metal) exposed. In the absence of the zinc, contact between steel and aluminium (now less noble) may accelerate corrosion of the latter.

(4) Relative areas

In general, the danger of corrosion at a bimetallic junction is greatest if the area of the cathode (the nobler metal) is large compared to that of the anode (the less noble metal). Immersed conditions represent more severe exposure than do atmospheric conditions.

(5) Microgalvanic effects

In some cases microscopic bimetallic couples can be formed, resulting in accelerated corrosion (usually pitting) of the baser metal, when small amounts of the more noble metal are dissolved in water. Thus the two dissimilar metals concerned do not have to be in metallic contact, as normally implied. In copper–water systems, for example, small amounts of copper dissolved by the water may be deposited on less noble metals such as zinc, aluminium or ferrous alloys in the same system.

(6) Effects of graphite, coke or a carbonaceous film

These may, if left on a metal after fabrication, act as a noble electrode in a couple and accelerate corrosion.

(7) Other effects

When using Diagram D3.7/13 account should also be taken of the fact that corrosion may be accelerated due to other types of exposure previously outlined. Thus, for example, corrosion of an identical metal may be accelerated if there are crevices; deposits on the metal surface can cause differential aeration corrosion; certain materials may give off corrosive agents.

(c) Typical examples

Experience has shown that, in building practice, there is serious risk of attack at bimetallic junctions. Some typical examples of corrosion due to the contact of dissimilar metals which may occur in buildings are noted below. In the examples given, it has been assumed that *no precautions* have been taken to isolate the one metal from the other, while no attempt has been made to classify the degree of corrosion. For guidance here reference should be made to Diagram D3.7/13.

1. Copper flashings in contact with galvanized steel – the galvanized steel corrodes.
2. Copper parapet capping held in position by steel or galvanized steel nails – the nails corrode.
3. Aluminium window or door frames with copper or brass hinges and other exposed brass fittings – the aluminium corrodes.
4. Copper flashings, where the 'run-off' is on an aluminium roof – copper will be deposited on the surface of the aluminium, resulting in serious pitting of the aluminium.
5. Copper pipes in contact with galvanized steel water tanks – corrosion occurs at the point of contact and often at the water line of the tank as well as on the floor of the tank.
6. Brass valves in galvanized steel water pipes – corrosion products from the steel blocks the pipes.
7. Aluminium rainwater goods (particularly pipes) carrying rainwater from copper roofs – copper will be deposited on the surface of the aluminium, resulting in serious localized attack of the aluminium (perforation is usually rapid).

Resistance to corrosion

1 Scope

As the resistance to corrosion which a metal or alloy will provide is primarily dependent on its properties and the conditions to which it is exposed, it is more realistic to consider not only the basic composition of the various metals and alloys used in building practice but also the effects of the different basic conditions of exposure outlined previously under 'Exposure' p. 480. The resistances which metals will provide are summarized in Table T3.7/2.

There are two basic groups of metals, namely, the *ferrous metals* and the *non-ferrous metals*. The metals and alloys which fall under either of these groups are discussed separately.

2 Ferrous metals

(a) Generally

The term 'ferrous metals' is used to include those consisting largely of iron. In engineering and building pig iron from the blastfurnace is not used. Instead, the pig iron is subjected to refining treatment, including alloying, in which carbon is important, although other elements are also used. Consequently, it is the iron alloys which must be considered, and these are basically cast iron, wrought iron, ordinary steels* and stainless steels.

Apart from the stainless steels, ferrous metals are not chosen because of their resistance to corrosion but mainly due to their excellent mechanical properties, ease of fabrication, relative cheapness, wide availability of suitable ores and ease of extraction. In most environments, *iron* has a low resistance to corrosion when compared with other metals. This can be accounted for by the following factors:

● The ease with which cathodic reactions can proceed on its surface.
● The readiness with which concentration cells are formed; and,
● The poor protection afforded by corrosion products.

The basically poor resistance of iron can be increased by alloying, and in this the high resistance to corrosion generally offered by the stainless steels, that is, alloys of iron containing not less than 12% chromium, may be noted. In general, ferrous metals, apart from stainless steel, need some form of applied protection – the degree depending on the severity of exposure.

(b) Cast iron

Cast irons are ferrous alloys containing over 1.5% of carbon with significant amounts of phosphorus, silicon and possibly, deliberate additions of alloying elements. Cast irons include grey iron castings, iron castings in which the graphite is in a spheroidal or nodular form, and malleable iron castings. The familiar grey fracture associated with cast iron is due to the presence of free graphite, which, in the form of flakes, accounts for the brittleness of cast iron, while it is the important distinguishing characteristic between cast iron and mild steel.

Small percentages of alloying additions have little effect on the corrosion resistance of cast iron; larger percentages, on the other hand, have been made to develop cast irons of better mechanical and corrosion-resistant properties. Special alloys of this kind include the high-silicon irons (up to 18% silicon), the high-chromium irons (25–30% chromium) and austenitic irons (not less than 15% nickel and 7% copper). In general, these special alloys are not commonly used in buildings.

The presence of a *casting skin* which may be either white iron, if there has been chilling, or high in silica if moulded in sand, increases the initial corroding characteristics of cast iron. This skin, it may be noted,

*As explained later, the term 'ordinary' is used for convenience in order to distinguish mild and low-alloy steels from stainless steels.

adheres more firmly than the mill scale found on mild steel, and thus offers greater protection.

All cast irons have *good resistance to atmospheric corrosion*. Resistance is also good in distilled or fresh waters, but the rate of attack increases with salt concentration, aeration and temperature. In soft acidic waters and sea water *graphitization* of the cast iron occurs and, although the appearance and shape of the original iron is largely retained, it is much weaker mechanically (when used underground the corroded component is easily fractured by earth movement). Retention of appearance and shape is due to the fact that graphite is virtually unaffected by most corrosion processes and is often found *in situ* after the iron has been removed by corrosion. Graphitization results in a porous graphite residue ('graphite corrosion residue') impregnated with insoluble corrosion products (iron phosphide, the phosphorus in cast iron, which is also virtually unaffected by most corrosion processes and silcon, which is usually oxidized to silica or silicates, the latter helping to bind the other constituents). It may be noted here that, although the corrosion rates of cast iron and steel are much the same in soft acidic waters and sea water, cast iron appears to behave better due to its normally much thicker section and to the graphitization which retains the original form and thus enables a pipe, for example, to retain water.

Carbon dioxide dissolved in water influences corrosion for two reasons. First, it can increase the hydrogen evolution type of attack due to its acidic character (carbon dioxide decreases the pH of the solution). Second, and indirectly, concentration of protective calcium carbonate is not formed with increases in carbon dioxide content of water; instead, calcium carbonate remains in solution in the form of calcium bicarbonate.

The corrosion resistance of cast iron can be increased if *calcareous deposits* (hard waters notably) with a favourable type of crystalline structure can develop. Calcareous deposits of some protective value can also develop when cast iron (or steel) is immersed in sea water, particularly when cathodic protection is applied. The protective value of calcareous deposits can be minimized by the presence in sufficient quantities of chloride, which interfere with the deposition of the protective layers. The result is the formation of rust nodules, beneath which intensive attack and pitting can take place. The presence of sulphate ions may also interfere with the formation of protective deposits. It may be noted that only a small amount of saline matter (relatively low concentration) is sufficient to enable corrosion cells to function at or near their maximum efficiency. In *sodium chloride* solutions, for example, the corrosion rate for iron increases relatively little after the concentration has reached 1000 ppm sodium chloride.

The decomposition and leaching of dead vegetation yields organic acids which, in turn, produce *waters with a lower pH value* and hence increase the danger of corrosion. The presence of *sulphate-reducing bacteria* may also increase the risk of corrosion.

(c) Wrought iron

At present little, if any, wrought iron is made. Interest in its behaviour is therefore largely historic, although wrought iron components (pipes, chains, gates, ballustrades and railings, for example) are still is use. When required, wrought iron is often replaced by mild steel and, as service life often favours wrought iron, it is useful to appreciate the reasons for the difference in behaviour.

Wrought iron is an *iron alloy* of *very low carbon content* (between 0.02% and 0.03%) and in the case of British wrought irons, with a high slag

Table T3.7/2 *Resistance of metals to corrosion*

Metal	Resistance to			
	Atmosphere	Other building materials	Conductive water	Other metals (see also Chart C.3.7/3)
Aluminium and its alloys	1. Generally fairly high and dependent on type of alloy and degree of atmospheric pollution 2. Surface corrosion and pitting may take place when atmospheric pollution is high 3. Resistance against direct exposure to flue gases and smoke only fair. In flue terminals the metal cannot be expected to last very long	1. Resistance generally low when in contact with wet cement-based products with either Portland or high-alumina cement. Attack is severe when in contact with wet high-calcium and magnesium limes 2. Resistance low to magnesium oxychloride cement whether in contact or embedded 3. Attack likely by some timbers, e.g. western red cedar, Douglas fir, oak and sweet chestnut 4. Some attack likely before gypsum plasters (Keene's, Parian, anhydrous) have dried out. Subsequent attack unlikely unless the plaster becomes damp for prolonged periods	1. Resistance generally good against rainwater – water run-off from copper to be avoided 2. Resistance generally poor against waste water, particularly when water supply is carried in copper or when washing compounds are present in the waste water 3. Excellent resistance when used to carry specially treated waters – not generally recommended for ordinary water supplies 4. Resistance to sea water may be good with some alloys	1. Resistance generally fairly low when in contact with most common metals, including dissimilar alloys in contact 2. Resistance particularly low when in contact with copper and copper alloys – attack of aluminium is rapid 3. Good resistance when in contact with zinc or cadmium 4. Resistance to iron varies according to type of alloy and nature of electrolyte. Attack generally negligible even when in contact with stainless steels
Cast iron	1. Good resistance which is further enhanced if the casting skin on the iron is still intact 2. Corrosion resistance of nodular graphite cast iron is at least as good, if not better, as that of flake graphite cast iron	1. Good resistance relative to those materials with which cast iron normally comes into contact 2. Resistance to soils mainly governed by oxygen content. Resistance generally poor in anaerobic soils, e.g. clays (presence of sulphate-reducing bacteria important). Acid soils usually aggressive 3. Poor resistance in made-up ground containing ashes or clinker (presence of water-soluble and unburnt carbonaceous matter significant)	1. Good resistance in distilled or fresh waters. Rate of attack increased by salt concentration, aeration and temperature 2. Rate of corrosion generally increases in sea water, particularly with increases in movement of water 3. Graphitization occurs in soft acid waters and sea water 4. Calcarous deposits may help to stifle attack – may be significant when cathodic protection is used	1. Resistance considerably reduced when in contact with cupro nickels, aluminium-bronzes, gun-metals, copper, brasses, lead[a] and soft solders[a], stainless steels and chromium. Accelerated corrosion is likely to occur when highly conductive electrolytes (salt, acid, combustion products) are present 2. Resistance to corrosion may be moderately reduced when in contact with aluminium and its alloys
Copper and its alloys	1. Excellent resistance to most conditions of exposure 2. Flue gases and smoke will attack under severe conditions of exposure	1. Excellent resistance against most building materials except when ammonia is present as in certain types of foamed cement or when organic acids from certain timbers are present 2. Poor resistance when copper tubes are covered with a thin carbonaceous film	1. Generally excellent resistance against most waters. In roofs attack may be caused when water is made acidic by organic matter (algae, lichen or moss) and where this water drips onto the metal 2. Resistance low against soft waters with appreciable amounts of dissolved carbon dioxide	1. Resistance generally excellent when in contact with most common metals, including dissimilar alloys in contact 2. Resistance may be low in some immersed conditions when there are pores or defects in tin coatings to copper or brass
Lead	1. Very high resistance due to formation of basic lead carbonate or sulphate but certain conditions of environment or detail may reduce resistance considerably[b] 2. Resistance poor against acetic acid fumes	1. Resistance fairly low against alkali attack from wet cement-based products 2. Resistance low against contact with certain types of wood 3. Resistance in soils is generally high except sometimes in made-up ground containing ashes and in clay soils containing gypsum and perhaps chloride	1. Resistance against most waters is generally high 2. Resistance against soft waters is sufficiently low to enable a lead concentration to build up that is physiologically dangerous 3. Resistance is considerably lowered when rainwater run-off from roofs is made acidic by organic matter (algae, lichens, moss)	1. In some immersed conditions there may be serious acceleration of soldered seams in copper and its alloys 2. Rapid corrosion where the access of air is restricted (narrow crevices notable) when in contact with steel or galvanized steel and exposed to the atmosphere 3. Varied reduction in resistance when in contact with copper and its alloys, chromium, stainless steel and nickel
Stainless steel (includes a wide range of alloys – as many as 60)	1. Generally excellent, but dependent on composition. (Careful selection for given environment important) 2. Resistance may be lowered considerably, with resultant superficial localized attack, where debris occurs, in crevices or by marine spray	1. Generally excellent, including exposure to most soil conditions	1. Generally excellent to supply lake or river waters, even when these are relatively highly polluted, irrespective of temperature 2. In sea waters and saline solutions resistance is considerably reduced, depending on the ability of the chloride of breaking down the passive film. Stagnant waters usually result in marked localized attack and pitting	1. Generally unaffected by contact with other metals, except in sea water when contact with aluminium alloys, ordinary steel, zinc and sometimes copper may lead to attack, but relative areas of metals involved and movement of water important

Table T3.7/2   *(Continued)*

Metal	Resistance to			
	Atmosphere	Other building materials	Conductive water	Other metals (see also Chart C.3.7/3)
			3. Susceptibility to corrosion by strongly reducing mineral acids, but otherwise unaffected by acids and alkalis	
Steel (excluding stainless steels)	1. Generally poor but resistance is improved by small additions of copper (e.g. copper-bearing steels) or other alloying elements such as chromium and with relatively high phosphorus and silicon contents (e.g. low-alloy steels)	1. Unaffected by cement- or lime-based products, but resistance seriously reduced if embedded in concrete containing excessive amounts of calcium chloride or if the calcium chloride is not evenly distributed 2. Attacked by gypsum plasters, magnesium oxychloride cements and some organic materials (e.g. certain timbers and plastics that exude acids or other corrosive compounds) 3. Resistance to soils mainly governed by oxygen content. Resistance generally poor in anaerobic soils, e.g. clays (presence of sulphate-reducing bacteria important). Acid soils usually aggressive 4. Poor resistance in made-up ground of ashes or clinker (presence of water soluble and unburnt carbonaceous matter significant)	1. Resistance to fresh waters generally low but rate of attack increased by salt concentration, aeration and temperature 2. Rate of corrosion generally increases in sea water particularly with increases in movement of water 3. Rate of corrosion generally increased by soft acidic waters 4. Calcareous deposits may help to stifle attack – may be significant when cathodic protection is used	1. Resistance considerably reduced when in contact with cupronickels, aluminium-bronzes, gunmetals, copper, brasses, lead[a] and soft solders[a], stainless steels and chromium. Accelerated corrosion is likely to occur when high conductive electrolytes (salt, acid, combustion products) are present 2. In certain soft waters, change of potential occurs at tempeeratures above 60°C and steel is rapidly corroded by zinc
Zinc	1. Fair resistance as protective film is not dense nor firmly adherent. Resistance considerably reduced by presence of sulphur	1. Resistance fairly low against alkali attack from wet cement-based products, acid attack from acid plasters such as Keene's and salts from brickwork 2. Resistance low against contact with certain types of wood, including hardboard	1. Resistance varies considerably and is largely dependent on the type of dissolved salts and gases. Attack is generally also influenced by temperature – marked increase in attack at high temperatures	1. Resistance generally low to most metals except magnesium. Useful as cathodic protector for iron and steel 2. In certain soft waters, change of potential occurs at temperatures above 60°C and zinc will rapidly corrode iron or steel

[a] Lead or soft solders may be used as coatings without risk of corrosion, *provided* that continuity of the coating is good.
[b] Covered moist conditions with access to carbon dioxide as in cases of interstitial condensation are particularly notable.

content (varying between 1.0% and 4.0% but usually about 2.0%) and a characteristic laminated structure. The fundamental differences between wrought iron and steel are due to both processing and composition, with the former influencing the latter. In the case of processing, steel is cast into ingots, whereas wrought iron is removed from the furnace in a semi-molten plastic condition and is formed into bars or billets under a steam hammer. As far as composition is concerned, the presence of slag in wrought iron distinguishes it from steel. In addition, wrought iron has less carbon and manganese and usually more phosphorus. Although the silicon content of the two materials is roughly the same (between 0.10% and 0.20%) the state of the silicon is different.

The slag content plays an important, though by no means the only, part in the corrosion resistance of wrought iron. The slag performs a barrier-like action which impedes the penetration by corrosion. The effects of the slag content on the rate of corrosion may be illustrated by comparing the performance of Swedish wrought irons, which are almost slagless, with that of British wrought irons, which contain much more slag. The Swedish wrought iron rusts outdoors 25% *more* quickly than mild steel, while the British wrought irons rust about 25% *less* rapidly than mild steel.

In addition to slag content, the existence, in wrought irons, of two types of resistant zones and differences in the nature of the scale and its adhesion to the metal surface (the adhesion is good) account for the reported longer life of wrought iron tubes compared to ones of mild steel.

(d) Ordinary steel

(1) Terminology and groups

The term 'ordinary' is used here for convenience, in order to differentiate between those steels having a comparatively low percentage of alloying elements, other than carbon, and those, such as the stainless steels, having a high percentage. Two main groups of steel, commonly used, may be identified, namely, *mild steel* and *low-alloy steel*. Both are classified as low-carbon steels, that is, with a carbon content up to 0.25%. The classification of iron–carbon alloys, based on the percentage of carbon present, is usually as follows:

Low-carbon steel
  Low-alloy steel up to 0.20
  Mild steel up to 0.25
Medium-carbon steel 0.25–0.45
High-carbon steel 0.45–1.50
Cast iron 2.50–4.50

(2) Role of carbon

The *role of carbon* in steel is important as far as the hardness and strength of the material is concerned. Neither carbon not the small quantities of other elements, such as sulphur, phosphorus, silicon and manganese, normally present, have much effect on the general corrosion resistance of steel, particularly in neutral media, as the corrosion rate is controlled primarily by the transport of oxygen. However, the inclusion of small amounts of certain alloying elements such as copper, chromium and nickel, does increase the resistance of iron and steel outdoors.

(3) Differences in composition

The primary differences in composition between mild steel and low-alloy steels may be summarized as follows.

Mild steel has no deliberate alloy addition, and a carbon content not exceeding 0.25%. Although alloying elements are not deliberately included, mild steel usually contains small amount of sulphur, phosphorus, silicon, manganese and copper. Small percentages of these elements are inevitably introduced into steel through the scrap used in its manufacture. Copper contents, for example, of well over 0.03% are common in British steels.

The corrosion resistance of mild steel can be increased by the addition of copper. Such steels are generally known as *copper-bearing steels*. BS 4360: 1986, *Mild steel for general constructional purposes*, specifies two grades of copper-bearing steel which contain either 0.20–0.35% or 0.035–0.50% copper. The effect of the additional copper is marked, but there is a limit, which is dependent on the sulphur content. However, there is a progressive decrease in corrosion as the copper content is increased until the limit (about 0.1%) is reached. Thereafter, further additions of copper cause little additional improvement. (See Chart C3.7/3.)

Low-alloy steel generally has a carbon content not exceeding 0.20%, to which small percentages of alloying elements up to, say, 3.0% in all

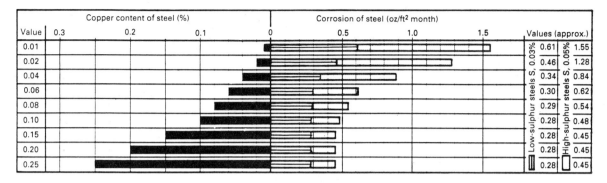

Value	Copper content of steel (%)				Corrosion of steel (oz/ft² month)				Values (approx.)			
	0.3	0.2	0.1	0	0.5	1.0	1.5		Low-sulphur steels S, 0.03%		High-sulphur steels S, 0.05%	
0.01										0.61		1.55
0.02										0.46		1.28
0.04										0.34		0.84
0.06										0.30		0.62
0.08										0.29		0.54
0.10										0.28		0.48
0.15										0.28		0.45
0.20										0.28		0.45
0.25										0.28		0.45

Chart C3.7/3 *Comparison of the effect of copper content on the corrosion of mild steel outdoors. Note: Included for guidance in connection with the text only, as the CP on which the values are based has now been replaced – see below. (From CP 2008, Protection of iron and steel structures: 1966, Fig. 4, p. 162. Now replaced by BS 5493: 1977)*

have been deliberately added. The effect of these additions may be to increase the mechanical properties or corrosion resistance of the steel, or both. Although the low-alloy steels and the copper-bearing steels are more resistant to corrosion than ordinary mild steel, they are not necessarily immune to corrosion and are, therefore, better considered as 'slow-rusting'. For example, steel with 1.0% chromium, 0.6% copper and relatively high phosphorus and silicon contents corrodes in the open atmosphere at one-third, or less, of the rate for ordinary mild steel. The relative performance of some mild, copper-bearing and low-alloy steels in an industrial atmosphere and in pure mountain air are shown in Diagram D3.7/14. Corrosion is most rapid during the first year or so, and the beneficial effects of low-alloy additions are of greater practical value in the more corrosive atmosphere.

The improved resistance of both copper and low-alloy steels outdoors has been attributed to the influence of alloying elements on the nature of corrosion product. The rust formed on these steels is more compact and less permeable and thus more protective than on ordinary mild steel. However, it may be noted that natural weathering appears to favour this because these alloyed steels show little or no advantage when exposed indoors or when sheltered from the rain.

Another contributing factor to the increased resistance provided by the alloy steels is that their *mill scale* is more firmly held than on mild steel. If, however, the scale is not removed the danger of localized attack is correspondingly greater than with mild steel. It may be noted that, in general, the presence of mill scale on the metal surface does not affect the overall corrosion but does increase significantly the localization of attack, due to breaks in the mill scale which frequently occur in practice. It has been shown that in sea water the penetration of pitting after one year was four times greater with steel with mill scale than with steel without mill scale. Over a period of time this difference decreases, although the effect was still evident after 15 years. (See Chart C3.7/4.)

Unlike cast iron, steel generally has poor resistance to both fresh water and sea water, although both corrode more rapidly in the latter than in the former. The effects of dissolved carbon dioxide, chlorides, sulphate ions, organic acids and calcareous deposits are similar to those described for cast iron. The effects of sulphate-reducing bacteria on steel are severe pitting.

Although *the effect of temperature* on the corrosiveness of a water varies with operating conditions, corrosion reactions do, in general, proceed more rapidly as the temperature rises. In moving waters there is a twofold increase initially for every rise of 10°C above atmospheric

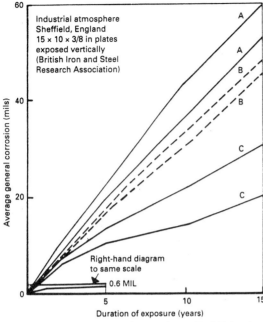

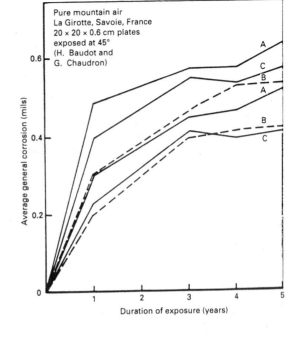

The curves show the limit of results observed by British and French investigators for groups of A mild steels: B copper-bearing steels: C low-alloy steels containing up to 1.0% chromium and 0.65% copper.

Diagram D3.7/14 *Corrosion/time curves for mild steels, copper-bearing steels and low-alloy steels exposed outdoors in an industrial environment (left-hand diagram) and pure mountain air (right-hand diagram).* Note: *The diagram is included for comparative purposes. CP 2008 on which it is based has now been replaced by BS 5493: 1977*

temperature. At higher temperatures this tendency is reversed, as oxygen becomes less soluble. Thus, in open systems from which liberated gases may escape, there is a maximum corrosion rate at 70–80°C followed by a decrease up to the boiling point. On the other hand, in a closed system there is no such maximum rate, as the oxygen is retained within them. However, in such cases corrosion generally ceases when the oxygen supply has become exhausted through the corrosion reaction.

(4) Effects of water movement
As with most other metals, *water movement* often increases the rate of corrosion by promoting the supply of oxygen to the metal surface, while high speeds may also prevent the adhesion of protective corrosion products or calcareous deposits. There may be a considerable increase in exposure when dissimilar steels are in contact. The scouring effect of sand and detritus may seriously aggravate corrosion in harbour and similar installations.

(5) Effect of cement
In general, *the effect of cement mortar and concrete* on iron and steel embedded in or in contact with them is dependent, in addition to the presence of moisture, on a number of factors, chief of which are:

- The highly *alkaline nature* of most cement mortars and concretes (Portland cement mixes have an alkalinity of pH 12.5) is normally sufficient to inhibit rusting. Such inhibition may be reduced if the leachate from the concrete contains appreciable amounts of calcium sulphate (gypsum) in addition to the calcium hydroxide.

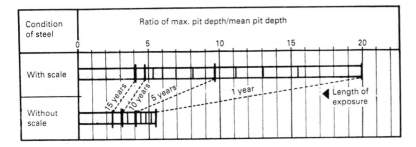

Chart C3.7/4 *Comparison of the effect of mill scale on pitting of steel in sea water. (Data from Butler and Ison, Fig. 19, p. 66)*

- *Carbonation* of the surface layers (the action of carbon dioxide makes these layers less alkaline), which results in an increase in volume of one-sixth, may effectively seal the concrete if it is of good quality; in poor-quality concrete such sealing does not occur. Thus with dense concrete and with proper depth of cover to the steel, sufficient uncarbonated material will remain uncarbonated to protect the steel for at least 50 years.
- The quality of concrete in which steel in embedded and the depth of cover are important if cracks, and thus the entry of water, is to be prevented. Entry of water will promote corrosion. In certain cases it may be necessary to prevent the occurrence of large shrinkage cracks (small hair cracks are normally insignificant).
- The presence of large amounts of *chlorides*, either through the use of salt-containing aggregates or as additives to accelerate hardening, will promote corrosion. In general, the amount of chloride used in reinforced concrete should not exceed 2.0%, calculated on the weight of cement.

(6) Effects of other factors
All types of low-alloy and straight carbon steels can be used together without danger of increased attack. Such steels are, however, anodic to the stainless steels containing large amounts of chromium and nickel. In sea water there is a risk of corrosion when mild steel is in contact with copper or stainless steel. Rate of movement of this type of water will influence the rate of attack – generally, the faster the flow, the faster the rate of corrosion. When considering protective metal coatings the relative position in the galvanic series with regard to iron is extremely important. *Anodic* coatings, such as zinc, provide good protection provided the area of exposed iron is not extensive (see 'General considerations', 4(b) p. 462). When cathodic coatings, nickel or chromium for example, are used, pitting may result at pores, cracks or areas of damage. As rivets are the sites of poor paint adhesion, they are frequently heavily corroded. Increased resistance may be provided if the rivets are alloyed with small additions of nickel and copper. Similarly, a weld of 2.0% nickel-steel suffers less corrosion than a carbon-steel weld, while corrosion or pitting of the mild steel plate is not accelerated.

(e) Stainless steel

(1) Definition and variable resistance
Stainless steel may be defined as a steel containing sufficient chromium, or chromium and nickel, to render it highly resistant to corrosion. The amount of chromium basically controls corrosion resistance, with a minimum of about 12% required; nickel is added to

improve ductility. Increased resistance to corrosion is provided if molybdenum is included.

Although stainless steels have a high resistance to corrosion, they are by no means completely immune from attack, and under certain conditions their resistance is far from good. In this, it is to be remembered that stainless steel does not imply a single alloy but a group of alloys consisting of something like *60 varieties*. Consequently, the term 'stainless' should not be taken too literally, while care in selection of the correct type is important.

Stainless steels commonly used in building applications include those containing 17% chromium (known as 17% stainless steel); 18% chromium and 8% nickel (known as 18/8 stainless steel); and 18% chromium, 10% nickel and 3% molybdenum (known as 18/10/3 stainless steel).

(2) Protective coating

The high resistance to corrosion provided by stainless steel is due to the presence of a thin protective oxide film, which reforms and heals itself spontaneously if the surface is damaged. With the protective oxide film the stainless steel is in the passive state, but passivation (for which 11.5% chromium content is required) occurs readily only in oxygenated environments. Consequently, stainless steel will be in active state, and thus lose its corrosion resistance, in non-oxidizing environments, which cannot repair the oxide film. When repair of the film is prevented by lack of oxygen access, severe corrosion can take place at a rate comparable with that of mild steel. Such conditions are likely to occur beneath debris, including sticky particles of sulphur or iron-bearing compounds and suspended salt particles, or in crevices. Adequate cleaning of the surface will minimize the deleterious effects of the debris (superficial local breakdown resulting in disfigurement by rust-staining), and careful design those effects associated with crevices. In this, it is important to note that the particular quality of stainless steel is important. In general, the more corrosion-resistant molybdenum-bearing quality provides good resistance to most industrial corrosive atmospheres.

(3) Corrosion resistance

When corrosion of stainless steel does occur it is usually localized. The presence of chlorides increases the susceptibility to attack. The corrosion resistance of stainless steel in sea water or when exposed to marine atmospheres is generally not high. Under these conditions the behaviour of the stainless steel is dependent on the ability of the chloride to break down the passive film. In sea water stainless steel may be attacked when in contact with aluminium alloys, ordinary steel and zinc; sometimes attack may also take place when it is in contact with copper. In this, the relative areas of the metals involved and the movement of water are important.

Stainless steel normally has a high resistance to corrosion when exposed to a wide range of supply, lake and river waters, even when pollution is relatively high, and irrespective of whether the water is hot or cold. Resistance to building materials is also high, although corrosion can take place when stainless steel is exposed to strongly reducing mineral acids.

3 Non-ferrous metals

(a) Generally

As a class, the non-ferrous metals (only aluminium, copper, lead and zinc are considered in detail here) offer greater resistance to corrosion than do the ferrous metals. This is due either to their self-passivation, as in aluminium, or to the formation of restraining films, as in zinc, or to their inherent nobility, as in copper. However, non-ferrous metals may be corroded, sometimes extensively or rapidly, when exposed to certain acids and alkalis, many of which may be derived from other building materials in contact with the metal or at bimetallic junctions.

Although non-ferrous metals may be preferred for a number of conditions of exposure because of their high resistance to corrosion, it is not always possible to use them due either to their inadequate mechanical properties or to their high cost. In some instances, as in galvanized steel or chromium plating, non-ferrous metals may be used as protective coatings to the less resistant ferrous metals. In this, it should be noted that the efficacy of the non-ferrous metal coating is, to a large extent, dependent on the thickness of the coating, while breaks in the coating may lead to bimetallic corrosion. As regards the latter, and depending on the metals involved, either the base metal or the 'coating' metal may suffer attack.

(b) Aluminium and its alloys

(1) Effects of alloying and treaments
A wide variety of aluminium alloys are available but, apart from some special alloys, are generally known simply as aluminium. The selection of a given aluminium alloy for a given condition of exposure is extremely important (see Photograph P3.7/1) because pure aluminium offers the greatest resistance to corrosion. The presence of impurities, alloying elements, the nature of the surface and heat treatments tend to lower, sometimes significantly, resistance to corrosion.

The high resistance to attack offered by aluminium and its alloys depends on the formation of a tightly adherent and protective surface film. *Passivation* of aluminium, that is, the formation of the protective surface silm, is produced both by the reaction of the metal with oxygen in the air or dissolved in water and by direct reaction with the water itself. If removed, the film immediately reforms.* It may be noted that in the absence of its protective surface film, aluminium is a very reactive chemical element. The film formed on aluminium is generally stable in the pH range 4.5–8.5. However, outside this range the aluminium

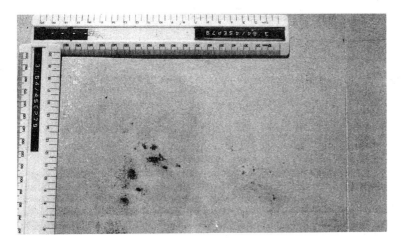

Photograph P3.7/1 *Pitting corrosion of aluminium long strip sheet in a fully supported roof due to the combined effects of iron deposited from scaffold poles and the corrosive nature of the rainwater – the exposed surface illustrated*

*It is perhaps axiomatic, but worth noting, that a continuing cycle of removal and reforming of the protective film leads to thinnning of the aluminium component.

becomes active. Aluminium dissolves particularly easily in alkalis but corrosion resistance above pH 8.5 is dependent on the source of the alkali, so that aluminium is stable in many alkaline waters. Thicker oxide films with greater resistance to corrosion and abrasion may be formed by anodic treatment – anodizing.

(2) Influencing factors

The resistance of aluminium and certain of its alloys is high to normal atmospheric exposure. In clean, dry air the appearance of the aluminium will remain unaltered for a long time. However, when exposed to normal outdoor conditions in the UK, surface corrosion does occur to a varying extent and should be taken into account. Surface corrosion, together with the deposition of dirt, tends to make the aluminium, unless specially finished, become duller and darker, and eventually roughened. Washing, either artifically or by rain, is beneficial as it removes dirt and other debris that accelerate corrosion. Although dulling of aluminium does indicate that corrosion has taken place, it does not imply that the metal is becoming progressively thinner, although some pitting of the surface may occur, particularly in sheltered areas. Corrosion is generally greatest with those aluminium alloys containing copper, and unless these are coated with pure aluminium they will also suffer severe loss of strength. Susceptibility to corrosion can also be reduced by extended heat treatment. From experience, both in the UK and in America, it would seem that most of the aluminium alloys likely to be used as building elements are, in general, capable of withstanding normal atmospheric conditions without serious deterioration. It may, however, be noted that although the structurally important alloys are slightly more susceptible to corrosion than the purer aluminium used in sheet form, the greater thickness of the former compensates for this. However, in heavily polluted industrial and marine atmospheres corrosion of the structural alloys is accelerated and protection of the metal – usually by painting – is required.

(3) Corrosive conditions

Highly corrosive conditions of exposure are likely to occur in poorly ventilated industrial buildings, particularly those with uninsulated single-skin roofs.* Acid vapours condensing on the underside of the aluminium roof sheets will attack the protective film and thus hasten corrosion of the surface. The use of open fires for space heating without adequate ventilation has, in certain cases, led to severe corrosion of the undersurface of aluminium roofing.

Waters contaminated with substances from neighbouring materials, or those present in industrial and marine atmospheres, may become highly corrosive. *Pools of water* remaining static for long periods on the surface of aluminium can lead to corrosion by limiting the oxygen supply. Small amounts of metallic salts present in domestic water supplies are liable to 'plate out' on the surface of aluminium supply lines and lead to corrosion. However, industrially, in specially treated waters, aluminium tubing, etc. is highly resistant to corrosion.

In *rainwater installations* alloys of the same type should be used throughout (except for brackets, etc.) as cast and wrought alloys have different surface appearance and corrosion characteristics.

Kitchen and bathroom wastes are liable to harm aluminium. Contact between aluminium and copper leads to corrosion of the aluminium. Thus, in plumbing, aluminium should not be used in any system containing copper, or copper-based materials, or in any situation where water passes over a copper surface *before* reaching the aluminium.†

*In light of current standards of insulation, this observation is included as an historic note. However, the likelihood of interstitial condensation becoming corrosive should be addressed.

†Two examples reported by the BRS (in 1958) in which rapid and severe corrosion of aluminium associated with copper in plumbing are worth noting. (1) An aluminium waste pipe (of H10 alloy) used in a cloakroom corroded extensively in three years. The supply water was carried to the basins in copper pipes. Traces of dissolved copper appear to have accentuated corrosion caused by the soapy wastes. (2) Severe corrosion of aluminium sinks (alloy NS3) after three and a half years occurred in three places: (a) the aluminium in contact with the plated brass fitting; (b) under the drip of a tap; (c) below the bowl standing in the sink.

Under damp conditions, *cement-based products* may give rise to corrosion, the degree of attack depending on the composition of the alloy (anodizing coatings are rapidly destroyed) and the richness of the cement (the richer the cement, the more intense the attack). Corrosion of embedded metal can lead to cracking of the embedding medium and unsightly salt efflorescence above the level of embedment. *Chlorides,* used as workability aids, contained in the mix may accelerate corrosion. *Lime mortars* affect aluminium in the same way as cement, but less severely. All types of gypsum plaster are liable to cause some corrosion either while they are still damp in the early stages or if prolonged dampness occurs after the gypsum has dried out. *Magnesium chloride* in magnesium oxychloride flooring is corrosive and, in addition, also absorbs water, thus intensifying attack.

Although generally resistant to organic acids found in most timbers, aluminium may be corroded, under wet conditions, by *acids from western red cedar, Douglas fir, oak and sweet chestnut.* However, aluminium nails may be used with cedar shingles *provided* the heads of the nails are covered by the overlapping shingle. It is important to note that aluminium may be corroded when in contact with or adjacent to wood treated with a copper preservative.

Aluminium is likely to be corroded by direct contact with *soil.* Aluminium *flue terminals* are often rapidly corroded.

Corrosion of aluminium occurs when in *contact with copper,* brass or bronze, iron or mild steel. In heavily polluted industrial or marine atmospheres contact with lead or stainless steel also leads to corrosion of the aluminium.

(b) Copper and its alloys

(1) Formation of protective films
Pure copper has a high chemical stability in many corrosive media. Resistance to corrosion is due to the fact that copper is a relatively noble metal. On exposure it forms relatively thin adherent protective films of corrosion products. Because it has only a weak tendency to passivation, the effect of unequal aeration is very slight. In moving water differential ion concentration may result in corrosion as discussed under 'General considerations', 8(g), p. 478.

(2) Effects of alloying
Several grades of copper are available. The purest, with 99.9% copper, has high conductivity and is important in electrical work (wires, switches, busbars, generators, etc.). Copper commonly used for building purposes may contain small percentages of other elements

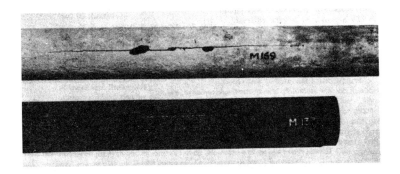

Photograph P3.7/2 *Stress-corrosion cracking of copper water pipes in foamed concrete. (From BRS Digest (First series) 110,* Corrosion of non-ferrous metals: I, *May 1958, Fig. 4, Building Research Establishment, Crown Copyright)*

such as phosphorus. These, although not as resistant to corrosion as pure copper, have, nevertheless, excellent resistance in most environments. The exceptions are important.

(3) Corrosive conditions

Copper may be severely attacked by *acids* particularly if these drip persistently onto the copper. In tiled roofing covered with algae, for example, rainwater run-off is made acidic, and where this drips onto copper, between the joints in the tiles, perforation of the copper has taken place (Photograph P3.7/3). Acid leached from cedarwood shingles may also lead to perforation.

The usually high corrosion resistance of copper tubing to water may be substantially reduced, with resultant pitting or stress corrosion cracking, if thin *carbonaceous films* (derived during manufacture but now generally removed) are present. In some rare cases, pitting has been caused by the formation of films by moorland waters.

In *soft waters* particularly those containing appreciable amounts of free *carbon dioxide,* and in carbonated waters in general, corrosion of copper does occur, but this is insignificant so far as durability and health risk is concerned. The formation of *green stains* on plumbing fixtures often implies that such waters may be corrosive enough to pick up sufficient copper to cause the stain. Hard waters are seldom corrosive to copper because of the rapid formation of a protective film of calcium compounds. Softening of waters with an appreciable temporary hardness can lead to corrosion, particularly if the water is heated above 140°C, and the calcium compounds necessary to form the protective film are absent, while sodium bicarbonate breaks down with the release of carbon dioxide.

Copper is *highly resistant to sea water,* although resistance may be significantly reduced in rapidly flowing sea water.

Copper, particularly if it contains phosphorus, is liable to corrode if *ammonia* is present. Small amounts of ammonia are released when a foaming agent is mixed with Portland cement to make foamed concrete.

(4) Brasses and bronzes

These are among the most important alloys of copper used in building. There are also copper–nickel alloys. *Brasses* are essentially alloys of copper with zinc (between 10% and 15% zinc), but often include other

Photograph P3.7/3 *Perforation of copper sheet by acid rainwater dripping from the roof. (From BRS Digest (First series) 110, Fig. 3, Building Research Establishment, Crown Copyright)*

components such as tin, iron, manganese, aluminium and lead. *Bronzes*, on the other hand, are generally the copper–tin series of alloys, although other components may be included. Corrosion behaviour of the alloys, although in many ways similar to copper, is largely dependent on their composition. Brass may suffer from a special form of corrosion, dezincification. (See 'General considerations' 8(f), p. 478).

(d) Lead

(1) Poor resistance of soluble films

The solubility and the physical properties of the lead compounds formed during exposure largely determine the corrosion resistance of lead. If the products are soluble, corrosion can proceed unhindered, but if the products are insoluble the metal surface has highly protective films. Exposure to slightly acidic conditions in the presence of carbon dioxide and organic acids, such as acetic acid, results in the formation of a whitish basic lead carbonate. Exposure to alkaline conditions (lime and cement mortars are notable) leads to the formation of a red-coloured lead oxide. Both these corrosion products allow corrosion to proceed unhindered.

(2) Importance of carbon dioxide

Lead is highly resistant to corrosion when exposed to the atmosphere and rainwater. This is because of the formation of a protective film of lead carbonate or sulphate. Both these films (sometimes also referred to as a 'patina') are stable (i.e. insoluble). There are, however, certain conditions when lead will corrode rapidly (sometimes in months rather than years) because it is starved of carbon dioxide. Condensation (better, interstitial condensation) is the most common cause of the rapid corrosion of lead, in insulated roofs in particular. In such roofs corrosion occurs when there is interstitial condensation on the hidden face of the lead (the internal face fronting onto its substrate). There are two reasons for this. First, the condensed water is distilled and lead will dissolve slowly in distilled water. Second, as there is usually insufficient free air movement under the lead (by virtue of the normal method of supporting the lead – see *3.4 Exclusion*, 'Fully supported roof membranes', p. 302) to provide the lead with the carbon dioxide it needs to form a stable protective film. The problem may be resolved, in principle, by constructing the element (roof or wall) so that there is a ventilation 'layer' between the underside of the substrate for the lead and the top of the layer of insulation. In other words, the principles of the cold deck roof design are adopted, where the risk of codensation is reduced by ventilation within the construction – see *3.4 Exclusion*, 'Methods of exclusion', '2 Roofs', p. 298.

(3) Corrosive conditions

Occasionally corrosion of lead gutters and weatherings occurs when these are associated with roofs covered with algae, moss or lichen. These produce organic acids and carbon dioxide and thus increase the acidity of rain water. Attack is generally slow, producing thinning, grooving and finally perforation, and so it can eventually be severe. Retention of moisture at the junction of lead and roof slates can produce corrosion at the edge, probably a form of crevice corrosion (Photograph P3.7/4).

Lead is relatively easily corroded by acetic acid fumes (premises such as breweries, pickling factories and saw mills are significant), even when the fumes are greatly diluted. Acids from timbers such as oak and deal are corrosive (Photograph P3.7/5). The resistance to corrosion of lead to alkalis is relatively poor, particularly under damp conditions.

Photograph P3.7/4 *Channelling and perforation of a lead valley gutter by acid rainwater from the roof: upper portion has been covered by slates. (From BRS Digest (First series) 110, Fig. 1, Building Research Establishment, Crown Copyright)*

Photograph P3.7/5 *Corrosion of lead from contact with wood. (From BRS Digest (First series) 110, Fig. 2, Building Research Establishment, Crown Copyright)*

Although generally resistant to most soil conditions, corrosion may occur in made-up ground containing ashes, and in clay containing gypsum and perhaps chloride.

The resistance of lead to corrosion in sea water is generally high. Soft waters generally attack lead significantly and in some cases sufficient quantities of a lead concentration (in excess of $0.1\,\mu g/cm^3$) to be physiologically harmful ('lead poisoning') can build up. However, lead is stable in soft waters containing calcium sulphate, calcium carbonate or silicic acid. Provided the water is not too soft, both carbonate and sulphate share in the formation of a protective layer.

(4) Alloys

Among important lead alloys are *soft solders*, lead–tin alloys of varying composition. Many of the failures which result from soldered joints (usually the metal being soldered) can be attributed to the action of the flux residues. Thus the choice of flux is important.

(e) Zinc

(1) Importance as coatings

Solid zinc is used to some extent in building constructions. However, its greater and most important use is in the corrosion control of steel components. The reasons for this are explained later under 'Protection, 3 Protective coatings', p. 510. Suffice it to say here that the corrosion resistance of the coatings, properly applied, is similar to that of solid zinc.

(2) Protective films

The corrosion resistance of zinc is dependent on the formation of protective films. These are generally not sufficiently dense and adherent to prevent steady though slow attack on the metal, particularly when it is exposed to the atmosphere. The variations in the corrosion of zinc in different environments at different seasons of the year are illustrated in Chart C3.7/5 and a comparison of the corrosion of zinc and iron in various environments is given in Chart C3.7/6.

Chart C3.7/5 *Comparison of the corrosion of zinc during different seasons at sites in the UK. (Data from* Atmospheric corrosion resistance of zinc, *Fig. 12)*

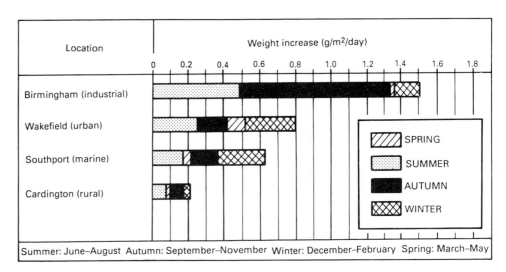

Place, Country and type of atmosphere		Value	Corrosion in one year expressed in micrometres (um) — INGOT IRON / ZINC	Value
Dry inland	BASRA, Iraq	10.2		0.7
	KHARTOUM, Sudan	0.7		0.6
Rural	ARO, Nigeria	8.6		1.4
	ABISKO, Sweden	4.0		0.9
	LLANWRTYO WELLS	48.0		3.0
Marine	APAPA, Nigeria	16.6		1.0
	CALSHOT, UK	53.4		3.1
	SINGAPORE, Malaya	13.1		1.2
Industrial	MOTHERWELL, UK	79.7		4.6
	SHEFFIELD, UK	98.6		14.6
	WOOLWICH, UK	88.2		3.7

Chart C3.7/6 *Comparison of corrosion rates of iron and zinc in different environments. (Data from Evans, U. R., An Introduction to Metallic Corrosion, 3rd edn, Table 4.3, p. 105, Edward Arnold, London, 1981)*

(3) Life expectancy

When exposed to normal atmospheric conditions the life of zinc is much less than that of other non-ferrous metals (40 years is the best that can be expected from the thicknesses of zinc commonly used). Thickness apart, the life is dependent on the degree of pollution of its environment. The greater the pollution, the shorter the life for any given thickness. It will be seen from Diagram D3.7/15 that a coating thickness of up to and including 200 µm is expected to last for 40 years in dry and damp interiors and non-polluted inland and coastal areas. If the latter are polluted, the expectancy drops to 30 and 20 years, even with an increased thickness (from 200 µm to just over 250 µm). Immersion in sea water gives an even lower life. Any of these life expectancies may, as in the case of other non-ferrous metals, be reduced by special conditions of severe exposure or by building details.

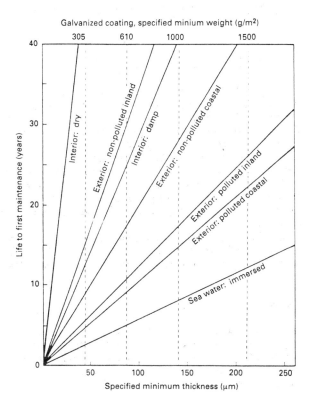

Diagram D3.7/15 *Typical lives of different thicknesses of zinc coating in different environments. (From BRE Digest 305, Zinc-coated steel, Fig. 1, p. 2, January 1986, Building Research Establishment, Crown Copyright)*

(4) Effect of acids and alkalis

Laboratory results indicate that zinc forms a stable protective film and therefore very low corrosion rates in the pH range from about 6 to 12.5. However, the rate of corrosion will be reduced markedly by strong acids and alkalis. Changes in temperature or aeration will also produce unstable films. Normal impurities in zinc include lead, cadmium and iron. A small iron content increases the resistance to attack of the zinc.

(5) Typical examples

- *Acid water from roofs* has a similar effect on zinc as on copper and lead.
- Zinc, like lead, is also *attacked by alkalis* released from lime or cement. Under damp conditions zinc is likely to corrode when in contact with all cement-based products, more so in those containing chlorides.
- Zinc is corroded significantly by *gypsum* plaster in moist conditions. The corrosion ceases once the plaster has dried out and remains dry. Corrosion of the zinc wil recommence if the plaster is subsequently dampened (by condensation, for example).
- Zinc *embedded in masonary* (e.g. wall ties) may be corroded because the alkaline protection provided by the lime or cement in the mortar is lost due to the action of atmospheric acid gases. The latter gain entry to the zinc because mortars tend to be porous and permeable to acid gases in the atmosphere. Concretes are less porous and more impermeable (see 'Protection, 6 Concrete', later, p. 524).
- *Provided* timber can be maintained continuously below a moisture content of about 20%, there is little, if any, risk of corrosion of zinc embedded in it (fixings, for example) in any combination of wood or preservative treatment.
- Whereas zinc is normally anodic to iron, a *reversal of the potential* of the zinc, making it cathodic to the iron, can occur in certain waters, particularly soft waters with an appreciable bicarbonate content, when the temperature of the water is above 60°C.

Protection

Although the corrosion resistance of a metal may be increased by modifications in its composition, this is, in a great many cases in practice, only a partial solution to the problem of reducing corrosion. Consequently, some form of protection is often required, particularly with the ferrous metals generally (stainless steels are possible exceptions). It should be emphasized, however, that, although methods of protection are more commonly associated with the ferrous metals generally, occasions do arise in the use of non-ferrous metals (bimetallic contacts or highly corrosive conditions of exposure caused by acids, alkalis and salts, for example) when some form of protection is advisable.

Methods of protection may be divided into *four* groups, namely: (1) treatment of the environment; (2) protective coatings; (3) concrete; and (4) cathodic protection. Each method is discussed below.

1 Treatment of the environment

As discussed earlier under 'General considerations, 5 Corrosion classification', three classes of environment, namely, atmospheric, immersed and underground, may influence the corrosion of a metal. In general, treatment of the environment is, when compared to other forms of protection, relatively restrictive.

(a) Atmosphere

Treatment of the atmosphere is necessarily limited to enclosed spaces, such spaces nowadays including those within a building and within a construction. In these, corrosion is often, but not exclusively, due to surface or interstitial condensation forming on the surfaces of metals – see explanation under 'Exposure, 3 Internal atmospheres', p. 481. Treatment of the internal atmosphere, essentially by reducing its moisture content through ventilation or air conditioning, has become more important than it was in the past. It may also involve the removal from inside a building of corrosive products emitted by certain manufacturing processes or by certain water treatments. (e.g. in swimming-pool halls).

(b) Immersed

The extent to which the immersed environment may be treated depends on the quanitiy of water involved, as it is the water which has to be treated, and when there are large quantities treatment is usually impracticable on economic grounds. If follows, therefore, that much can be done to reduce corrosion by water treatment in *enclosed* water systems.

Water treatment generally consists of the deliberate addition of chemicals to change or adjust its composition. In this, a distinction should be drawn between treatments which are specifically aimed to reduce or prevent corrosion and those such as filtration, chlorination, softening or distillation, which may either increase or decrease the corrosiveness of water. The important point with these forms of treatment is that they are not necessarily used in connection with specific corrosion problems but rather for health, convenience or specific industrial use.

Additives used in connection with corrosion may either *restrain* corrosion, that is, some corrosion may occur before protection is obtained, or *inhibit* it, that is, reduce or prevent attack at the outset. Most of the work on inhibitors has, it may be noted, been carried out with ferrous metals. The results of this work are also applicable to other common structural metals.

An important restrainer is a calcareous film deposited on the surface of a metal. The addition of lime to water enables such a film to be deposited. Although a high oxygen content and a certain bicarbonate hardness facilitates the formation of the calcareous film, it is generally desirable to remove dissolved gases from the water. When large quantities of water are involved it is often possible to carry out partial de-aeration for water conveyed in pipelines. De-aeration is effected by physical and thermal processes, which are supplemented by chemical treatments (lime to remove carbon dioxide and sodium sulphite or hydrazine to remove residual oxygen).

Chemicals added to water to inhibit corrosion require special care in selection and use. An inhibitor slows down the anodic and cathodic reaction or, in many cases, both. Most inhibitors affect both anodic and cathodic processes, but for convenience they are classified as anodic, cathodic or mixed, according to the processes which they primarily affect. Inhibitors useful for the protection of iron and steel include alkali chromates, nitrites, phosphates, benzoates, carbonates, borates, silicates and hydroxides.

The need for expert advice in the choice and concentration of a suitable inhibitor for a given circumstance is important for a number of different reasons. An inhibitor may prevent corrosion in one environment and increase it in another. The efficiency of inhibitors is affected by the composition of the water (pH value notably); temperature; the rate of flow; the presence of internal or applied stresses; the composition of the metal; and the presence of dissimilar metals. In addition, account must also be taken of loose scale, debris, crevices, etc. Finally, some inhibitors are toxic and precautions should be taken in their use.

(c) Underground

In the case of steel structures buried in highly corrosive soils, attempts have been made to reduce the attack on the steel by using special backfills. Pipelines, for example, have been surrounded by 150 mm or more of alkaline material such as chalk, limestone or calcareous sand. This procedure, by providing good drainage, temporarily converts anaerobic into aerobic soil conditions, but it is uncertain and has been known to increase corrosion.

2 Protective coatings – 1: Generally

(a) Categories

The covering used to protect a metal may consist of another metal or, as is probably more common, be formed of compounds, organic or inorganic, or aggregates of such compounds. There is, in fact, a bewildering variety of coatings, while each type of coating may be produced in more than one way. There are five broad categories of coatings; (1) metal coatings; (2) anodizing aluminium; (3) paint coatings; (4) vitreous enamelling; and (5) plastic coatings. Before considering each of these separately it is important to note those aspects common to all.

(b) Need for good adhesion

Although the inherent resistance to corrosion provided by coatings is liable to wide variation, the efficacy of any coating is primarily dependent on the fact that it *must adhere well* to the basis metal, while it should be free of pores or cracks and have acceptable mechanical properties.

  The requirements of good adhesion cannot be overemphasized if a protected coating is to give its maximum performance, particularly as many failures of protective coatings can be attributed to inadequate adhesion. The application of the most resistant coating to a poorly prepared surface can only ensure mediocre performance of the coating, and any departure from the best practice can only result in a loss of both time and money (and this can be annoying). *Cleanliness* is one most important requirement, if the maintenance of good adhesion is to be assured. In the case of metals to be protected, this means the proper removal of grease, dirt and oxide layers, including manufacturing scales. Pretreatment of metal surfaces in order to remove any undesirable matter that may affect adhesion may be undertaken by various methods, sometimes a combination of procedures being required.

(c) Pretreatment

Pretreatment methods may be classified as weathering, mechanical treatment, pickling and degreasing:

1. *Weathering* is the cheapest method of descaling steel, but normally requires a long period of exposure, varying with the weather conditions, the steel and the adhesion of the scale. After weathering, rust and any remaining mill scale can be removed fairly simply by wire brushing.
2. *Mechanical treatments* include sand and grit-blasting, flame cleaning, wire brushing and grinding (and polishing). Grinding and polishing are usually necessary before plating, the remainder being the main mechanical methods used before the application of paint or sprayed metal.
3. *Pickling* may be undertaken by chemical or electrolytic means. In the main, acids are used in chemical pickling, although molten salts may also be used prior to subsequent acid pickling. In electrolytic pickling, which is far quicker than chemical pickling, the metal is made cathodic and anodic in a suitable bath.
4. *Degreasing* is a most important part of the surface treatment. It may be carried out by physical or chemical methods. In painting work degreasing by solvents is adequate, but for work in aqueous solution (electroplating, for example) chemical or electrolytic methods are necessary, as the metal parts must be sufficiently clean to be wetted by water.

3 Protective coatings – 2: Metals

A number of different metals may be used as protective coatings. They may be applied in a number of different ways. It is more convenient to consider these two aspects separately.

(a) Types

It is important to draw a distinction between those metals used as coatings which may be cathodic and those which may be anodic to the basis metal. A cathodic coating, such as copper or nickel on steel, will not prevent corrosion of the basis metal if there are pin holes or cracks in the coating. On the other hand, an anodic coating, such as zinc or aluminium on steel, can protect the basis metal where it is exposed, provided the current density on the basis metal is sufficient. Thus in conferring protection the coating is destroyed. There are limits to which the anodic coating will provide such protection. Generally, protection ceases when the exposed area of the basis metal is large, although the conductivity of the electrolyte present is also important. (See also 'General considerations', 7 (b), p. 474.)

(1) Noble metals
Apart from their cost, noble metals, despite their high chemical resistance, are not generally suitable as protective coatings as they are strongly cathodic to metals such as iron, with the result that intensified attack may be expected at any discontinuity. At the same time, many (gold is notable) are soft and the thin economical coatings are easily damaged. Where cost permits, thick coats free from discontinuity can be obtained mechanically. Copper-clad steel plates are obtained in this way. Composite wires having a steel core sheathed in thick copper are also manufactured, the ductility of the copper allowing bending without exposing the steel. The use of brass plating on steel is of particular value in providing a bond for rubber, itself a valuable protective coating to steel – rubber fails to adhere to many metals, brass being the exception.

(2) Nickel and chromium

These two metals are usually employed as pairs in what is commonly known as 'chromium plating', and used mostly for the protection of steel or brass. The nickel deposits on the basis metal are covered with a thinner chromium coat. The nickel protects the basis metal against corrosion, while the chromium protects the nickel from fogging (dulling of the nickel on exposure to moist air containing sulphur dioxide); the hardness of chromium is also a further asset.

(3) Tin

Although the main use of tin is in the canning industry (tinplate–dead-mild steel carrying a tin layer) it is used to some extent for copper of lead water pipes and brass condenser tubes. A recent development is the application of a thin electrodeposited tin coating (0.8 µm thick or less) on steel before painting, to prevent the underrusting which would cause disfigurement or even subsequent peeling of the paint coat. The thin tin coatings are believed to provide superior protection to that given by phosphate films used for the purpose of increasing the life of paint coats.

(4) Tin–nickel alloy

This is notable, as it is a hard, bright deposit, resistant to atmospheres polluted with sulphur compounds, and withstands salt, mustard and lemon juice. It is, however, cathodic to steel and therefore fails to protect at gaps.

(5) Lead

The resistance of lead to sulphuric acid makes it valuable as a coating to steel sections suitable for the construction of pickling tanks. Lead has been used as a coating for metal window frames. It is mildly cathodic to steel under most conditions and a lead coat will not prevent rusting at pin-holes. However, it has been stated that the corrosion at pin-holes stifles itself (the inhibitive properties of lead compounds may be relevant), but this does not necessarily prevent disfigurement by rust spots – an outer coat of suitable paint is required to prevent the trouble.

(6) Zinc

A long-established and common method of protecting steel is by the application of a zinc coating. In building practice such protection is applied to a wide range of steel components including sections, sheets, wires, pipes, tanks, cisterns and fixing devices. Being anodic to steel, zinc will generally provide protection at a gap (an important exception occurs in hot water at temperatures, above 60°C when the potential difference between the zinc and steel is reversed, thus accelerating the corrosion of the steel – this reversal is notable with soft waters). In providing 'sacrificial' protection, the zinc is used up, so that if protection is to continue, thick coating are required. However, thick coatings are apt to be brittle while thin ones have a short life. Coatings of uneven thickness (thick in some parts and thin in others) combine the disadvantages of both, so that the aim should be to ensure uniformity of thickness of the coating, if the best advantage of the applied zinc is to be obtained. (The thickness and the uniformity of the coating – the latter in particular – are more important, as far as the life of the coating is concerned, than the process used to apply the zinc. Each coating method influences the use for specific purposes).

The thickness of the coating should be related to the corrosiveness of the environment, but where the coating cannot be increased in thickness, a paint or other coating is necessary. (Special care is required in the pretreatment of zinc to receive painting, as zinc reacts unfavourably with the ingredients of most paints, resulting in the disruption of the paint film.)

Although the life of a coat is, according to tests carried out on specially prepared samples with uniform coatings, proportional to the thickness (see Diagram D3.7/14 earlier and Chart C3.7/7 discussed below), this is not necessarily true under conditions of service. This is due mainly to the fact that, in practice, the coating is seldom completely uniform (the life would then be proportional to the average thickness). At the same time, account must be taken of the fact that accelerated corrosion will occur in crevices or where water drips onto the zinc.

For guidance, BS 5493: 1977. *Code of practice for protective coating of iron and steel structures against corrosion* (formerly CP 2008) gives typical life-to-first-maintenance of zinc coatings of various thicknesses and method of application in different types of environment. These are compared in Chart C3.7/7. It will be seen that life expectancy is directly proportional to the thickness of the coating for a given environment. Furthermore, the pollution of the environment reduces the life expectancy for a given thickness, in some cases considerably.

Chart C3.7/7 *Comparison of the life expectancy of zinc coatings of different thicknesses applied by different processes and exposed to different environments. (Data from BS 5493: 1977)*

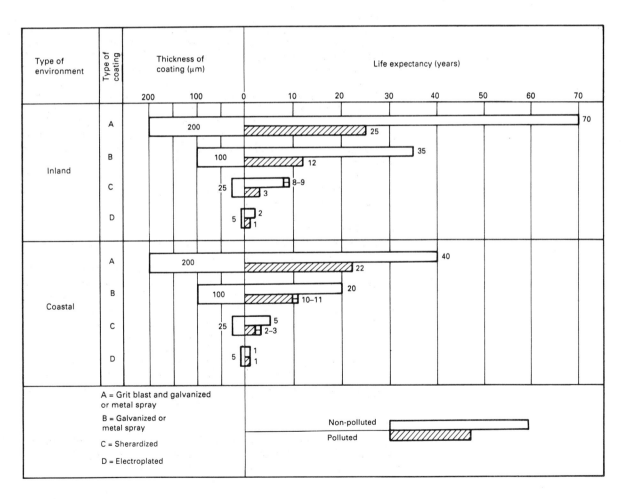

(7) Aluminium

The protection of steel with aluminium coatings is a relatively new technique. The performance of aluminium coatings has been good and protection of steel structures with aluminium coatings is used extensively. Like zinc, aluminium is anodic to steel, while the requirements of the coating in terms of thickness and uniformity are similar to those for zinc. When the thickness of the aluminium cannot be increased sufficiently to resist a given condition of exposure, a paint or other coating is necessary.

(8) Cadmium

The use of cadmium has not replaced zinc, as tests originally carried out which suggested superiority over zinc coatings did not represent service conditions. However, under storage conditions of high humidity and temperature, cadmium is superior to zinc, and so cadmium coatings are preferable for articles intended for the tropics.

(b) Methods

The main methods which may be used to apply metal coatings include dipping, electroplating, spraying, diffusion and cementation, vaporization and mechanical cladding. In all cases proper pretreatment to ensure good adhesion is imperative, as outlined earlier.

(1) Dipping

This is the oldest method of coating iron and steel. The basis metal is coated with the second ('coating') metal by dipping into a solution (usually the molten metal) of the latter. A pure solution of the 'coating' metal generally gives a thin coating, affording little protection; solutions containing deliberate additions of other elements give thicker, and hence more resistant, deposits in what are known as 'current-less' methods. With these methods an alloy layer forms between the basis and the deposited metal; it is often necessary for the formation of this alloy layer to be restricted. In the case of iron or steel, 'coating' metals are restricted to those with low melting points, such as lead, tin, zinc and aluminium. *Galvanized steel* is, strictly speaking, steel which has been protected by dipping into zinc. The term is, however, often loosely used to refer to all steel protected by a zinc coating whatever process may have been used.

(2) Electroplating

In this process the basis metal is coated by cathodic treatment in a solution containing a compound of the metal to be deposited. Electroplating has a number of advantages over dip coating. These include the range of metals that can be deposited (30 could be used, but about 15 are of technical interest), the saving in metal, the ability to produce thinner coats and the absence of any brittle intermediate layer. Many of the metals used are intended either as decoration or as protection against atmospheric corrosion. Plated metals are not generally used for immersed conditions (all types of water supply, for example). An important recent innovation is *brush plating*. This enables large areas of metal to be plated *in situ*. An anode, which is a tampon or brush, carries the electrolyte and is moved over the metal to be plated. In this process an extremely high current density is used.

(3) Spraying

The 'coating' metal is applied as a shower of tiny globules (often molten) which on striking the surface flatten to give a scaly porous coating. (Porosity is no great disadvantage if the coating is anodic to the basis metal.) The coating consists, therefore, of more or less separate particles strongly bonded together. Spraying of the globules is effected by means of a 'pistol', the metal fed into the pistol as wire or powder (liquid metal is seldom used now). Protection of steel by sprayed coatings of zinc or aluminium is common. Metals sprayed with aluminium may be referred to as 'metallized'.

(4) Diffusion and cementation
Cementation consists of heating the basis metal (usually mild steel) at high temperature in contact with a powder of the 'coating' metal (usually zinc and aluminium). Alloying of the two metals results in good adhesion, but the degree of alloying has to be limited if the coating is not to be hard and brittle. Two of the best known processes are *sherardizing*, diffusion coating with zinc, and *calorizing*, diffusion coating with aluminium. Sherardizing is generally confined to small articles only, as the coating tends to be fissured. However, coating with aluminium has wider application. Coating metals such as chromium, copper, tungsten, beryllium, tantalum, silicon, manganese, vanadium and titanium are more commonly applied by vaporization methods rather than by diffusion.

(5) Vaporization
Metal compounds are thermally decomposed on the metal surface. An important coating prepared in this way is that by chromium diffusion, *chromizing*.

(6) Mechanical cladding
The advantages of this method include a coating completely free of porosity and one which may be of any desired thickness. Various processes are used, but that most used is *roll-bonding*, in which the coating metal at the desired thickness is rolled onto the basis metal at the welding temperature. In another method, the coating metal is clad under heat and pressure with the aid of an intermediate layer of bonding material. Coating metals are mainly copper, nickel and aluminium, although highly resistant alloys such as stainless steel have also been used. An important example of cladding is that of an aluminium alloy with good mechanical properties clad with a thin sheet of pure aluminium which has superior resistance to corrosion.

(c) Anodizing aluminium

The process of anodizing a metal (also known as 'anodic oxidation') produces a thick oxide film on the surface; in effect the surface is 'pre-corroded'. The much more resistant film of oxide, which is very much thicker than that formed naturally on exposure to air, is formed by electrolytic anodic oxidation. A number of metals may be anodized, but the most common in building is aluminium.

Three conditions are necessary to produce thick oxide layers, namely: (1) the sollution must be able to attack the metal anode to some extent; (2) formation of capillaries in the oxide must take place; and (3) the solubility of the layer formed must be low in the electrolyte. The layer forms from the metal outwards and pores are necessary for the electrolyte to reach the metal so that the anodic process can continue. As the result of various reactions during the process the layer is not completely homogeneous, being hard and dense next to the metal and soft and less dense on the outside. Nevertheless, the film is chemically stable and relatively hard. The porosity of the film enables it to form an excellent base for paint or dyestuffs.

On the pure metal the film as formed is transparent but it is often coloured for decorative effect. After formation the film is sealed. The combined anodizing and sealing not only enhances the appearance but also increases resistance to corrosion and abrasion.

Anodized and sealed surfaces, although having a high resistance to atmospheric attack, can, like the naturally protected surfaces, be destroyed locally by corrosion resulting from the deposition of dust, dirt and soot. The film then becomes perforated and the aluminium pitted. As with ordinary aluminium, washing by rain water, or artificially, minimizes these effects.

The thickness of the film and the degree of sealing after anodizing or dyeing are the most important factors determining the durability of the finish. Sealing reduces the porosity of the oxide film and this may be carried out physically, by means of lacquers, oils or waxes, or chemically, by the action of pure water, steam or inhibitor solutions. The results of field tests have shown that, under identical conditions, the relationship between film thickness and corrosion resistance is not linear; doubling the thickness of the anodic film more than doubles the durability of the coating. However, it should be noted that because of the variable factors which influence the rate of corrosion, the relationship between the film thickness and durability has not been definitely established. For external use on buildings a film thickness of 25 µm is recommended, with thicker films in localities where air pollution is high. Thinner films can, however, be used on work that is regularly cleaned. The application of lacquers and waxes to the surface also help to reduce the frequency of maintenance.

4 Protective coatings – 3: Organic

(a) Types and uses generally

Organic coatings used to control corrosion include:

- *Paints and plastic coatings*, used mainly for protection against atmospheric corrosion. They provide the finish, so their appearance is another requirement to be considered.
- *Pitch, tar and bitumen* (sometimes loosely referred to as 'bituminous' products), as paints or tapes, are used either to separate the metal from other metals/building materials or for protection in underground or immersed conditions. Synthetic tapes may be used for the same purpose. As all these coatings will be hidden, appearance is of no importance.

The successful performance of all the coatings is dependent on the correct and thorough preparation of the metal before the coating is applied. As many are applied on-site, **creative pessimism** would caution that special care is taken in their specification and supervision.

(b) Paints – 1: Generally

(1) Durability and choice
A typical paint coating is between 25 and 100 µm thick. Despite their thinness, the protection they give is remarkable. Nevertheless, any coating has a relatively short life, even when properly applied. Durability and maintenance need to be considered whatever the coating used. Provision in design for subsequent recoating is important.

There is a wide variety of paint coatings from which to select. The technology of paint and its application is changing fast. Because of concern about the pollution of the environment and health hazards generally, injurious volatile solvents are being replaced by types that are less harmful. Dangerous lead additives have mostly been replaced by such materials as titanium dioxide.

Whatever new formulations and their method of application might be, the principles that should be adopted for successful performance of any paint coating will still be applicable. For this reason, details of paint coatings currently available have been excluded.*

(2) The paint film

A number of different factors will influence the choice of paint coatings suitable for given conditions. In this it is important to note that it is normal for a number of coats of paint to be used (multi-coat work), while what is normally referred to as a *paint film* is, in fact, the dried state not of one coat but rather of the various coats of paint used, each of which has a particular function to perform. The combination of the requisite coats for any given conditions is better referred to as a *paint system*.

(3) Influencing factors

The factors which will influence the choice of paint coatings, (the paint system) for given conditions may be viewed from different angles.

First, there are the *basic qualities* which the paint coating may possess, and these include:

- Spreading power;
- Elasticity;
- Resistance to the environment; and
- Compatibility with the metal being protected.

Second, there is the *method* (or methods) by which the basis metal is to be pretreated to receive the first coat of paint (usually the primer). In this a differentiation has to be drawn between these methods suitable for (1) factory† or (2) site application.

Third, *Methods of fabrication, storage, transport and erection.*

Fourth, *Maintenance schedule* and the ease with which maintenance can be carried out.

Fifth, *Compatibility* of the various coats of paint. (An intermediate coat may often be used to 'link' two otherwise incompatible coats.)

(4) Importance of pretreatment

The success of paint coatings is largely dependent on the thoroughness of pretreatment in order to ensure good adhesion and application and drying under favourable conditions.‡ As regards the latter, painting should not be carried out under damp conditions (or when the surface to be painted is damp); under conditions which are likely to retard drying out (air temperature should preferably be at least 5°C and relative humidity not more than 80%), or when significant amounts of dust, dirt and other debris are likely to be deposited on the surface, particularly prior to painting.

The basic characteristics of paint coatings, mechanisms of protection action and types of paint coatings, are more conveniently discussed under separate headings.

(5) The paint system

Normally a paint system consists of a *primer*, whose main functions are adhesion to and protection of the base; *undercoats*, which have a high

*Reference should be made in the first instance to the relevant British Standards or publications such as *Specification* (published annually by the Architectural Press) which contain authoritative summaries and references to British Standards.

†For new work, paticularly iron and steel, the adoption of factory painting to replace site painting is probably the best guarantee of durability that there is at present – factory painting is more efficient and avoids delays from bad weather and other difficulties on-site. But it is important to note that proper pretreatment is still essential if rapid paint failure and corrosion are to be prevented.

‡Outdoors, winter painting is very difficult; in the UK it has been shown, according to the BRE, that there is a greater risk of corrosion of iron and steel following November site painting than at any other time.

pigment content and a low gloss to cover and protect the primer; and *finishing coats* which provide further protection and finish. (In normal decorative work there is a clear distinction between undercoats and finishing coats, but in protective schemes this is not usually the case, and the undercoat is often a version of the finishing coat, sometimes differently tinted to ensure that the subsequent coat covers it fully.) In order to ensure that each of the coats performs its special function properly, it is essential that they should be mutually compatible. There are, it should be noted, certain systems as, for example, coal-tar/epoxy paints and some zinc-rich paints, in which primers are not used.*

In all multi-coat work it is essential that the paint system is considered as a whole.

(6) Paint composition

A paint consists of basically two components:

- *A vehicle*, a liquid constituent which gives the paint its fluidity (for spreading out thinly) and dries or evaporates (in time) to form a solid, adherent, dry film; and
- *A pigment*, a finely divided solid constituent, dispersed or suspended in the vehicle that controls the corrosion reaction or rate of diffusion of the reactants through the dry film.

A third component consists of :

- *Additives*, which accelerate the drying process or provide the dry film with improved resistance to the service environment.

The type of constituents used influence its:

- Spreading power
- Elasticity.
- Resistance to the environment, and
- Compatibility with the metal protected.

The vehicle may dry out in one of three ways:

- *By evaporation* of the solvent in the vehicle.
- *By chemical change*. This change is mainly oxidation of the liquid constituent of the vehicle (e.g. linseed oil). As the paint dries from the surface inwards, the layers of the coating are built up in a number of successive thin coats.
- *By polymerization*, which is a chemical reaction between the vehicle and a curing agent mixed with the paint just before it is applied to the metal. Paints drying in this way usually have the curing agent kept separate until required and are known as twin-pack systems. The curing process starts immediately the agent is added. Importantly the paint is suitable for application for a limited period after mixing. This is called the 'pot life'. As the paint dries throughout the film, it can be applied in thick layers.

The remaining solid portion of the vehicle in the dried film forms the *binder*. This holds the pigment in place, keys the film to the surface and provides resistance to the passage of water, oxygen and aggressive ions to the metal surface. Paints may be classified by the nature of the binder.

(7) Mechanisms of protective action

The basic mechanisms by which a paint film provides protection deserve special consideration. A paint film may effect the electrochemical corrosion reaction by inhibiting the cathodic and anodic reaction, or by providing a high resistive path between the anode and cathode

*Modern developments include systems that may not require primers and/or undercoats. The long term performance of them is not known, so they are excluded here.

(i.e. by insulation). Such paint systems are formulated to provide protection specifically by inhibiting the reactions (those intended for the protection of ferrous metals are particularly notable), while others have no inhibiting power and therefore protect entirely by providing a high resistive path.

In general, paint films are not completely impervious to water and oxygen. The extent to which a paint system will provide a resistive path is largely dependent on the *thickness* of the paint film, although other factors, such as the type of paint system, roughness of the metal and conditions of exposure, are equally important. However, the protective power of a paint film does increase with its thickness. Consequently, for those paint systems which provide protection entirely by insulating rather than inhibiting, the thickness of the paint film is extremely important. In this it is as well to note that marked differences in thickness (up to 100%) do occur from point to point, even when the paint is applied by a skilled painter, while equally great differences occur as between one skilled painter and another. Thus, discretion is needed in defining film thickness solely in terms of average value.

Inhibiting properties are achieved by the incorporation of inhibiting pigments,* and those commonly used include red lead,† calcium plumbate, zinc chrome, metallic lead and zinc dust. These are usually incorporated into priming paints, which, as already noted, must have good adhesive qualities while also providing a firm foundation for later coats. The efficiency of these pigments depends partly on the medium into which they are incorporated. Some pigments, it is important to note, although having an inhibiting effect on some metals may stimulate corrosion on others. A red lead primer, for example, should not be used on aluminium, as it stimulates corrosion. *Thickness* is also important when inhibiting primers are used, if their full potential is to be achieved. With one coat of primer it is usually impossible to achieve a continuous film of even thickness and, as important, free from pin-holes – the points at which corrosion begins. Apart from mild conditions of exposure, it is always advisable to use two coats of inhibiting primer, particularly in the protection of ferrous metals.

(8) Resistance to the environment

Apart from the mechanisms involved (see (7) above), the extent to which a paint film will provide protection depends largely on the durability of the film itself to any given conditions of exposure.‡ In this paint systems vary markedly in their resistance. In general, it is the *vehicle* which determines the durability of a paint film to given conditions of exposure, although the pigment may also be involved.

The characteristics and resistance of paint films grouped according to the chemical nature of the binder are summarized in Table T3.7/3. Brief notes on each group follow:

- *Oil-based paints*: These take a long time to dry; they need up to 48 hours between coats and up to 7 days before applying top coats to primers. The drying times can be shortened if zinc phosphate is added.
- *Oleo-resinous paints* (varnishes): These have shorter drying times than simple oil-based types. For structures immersed in water, tung oil with 100% phenolic resin can be used. The phenolic resin types do not tolerate damp surfaces during application; coal tar resin types do withstand some surface dampness.
- *Alkyds*: There are countless types of this widely used coating. The length of oil refers to the oil content, so the long oil-length types have a high oil content (usually about 65%) and short oil-length types have an oil content of less than 50%.

*The majority of pigments used in paints do not have any special inhibiting effects; they merely help to exclude damp and aggressive agents.

†The pigments red oxide and red lead are sometimes confused. Red oxide is an oxide of iron and has no specific corrosion-inhibiting properties. Certain types are used in the moderately effective oil primers and also in many quick-drying primers which are not always satisfactory. Red lead is an oxide of lead and has valuable corrosion-inhibiting properties. It is sometimes used with red oxide or graphite. It is not advisable to use it in paints other than primers for iron and steel.

‡Alkali resistance of paints is covered in *3.6 Chemical attack*, under 'Alkali action' (p. 393).

- *Epoxide resins*: There are two classes, air drying and stoving, each using a different type of curing agent. The air-drying varieties are in a twin-pack system; the stoving classes have the resin mixed in a single container. A dry film of over 250 μm thick can be formed in one application. The properties of expoxide vary over a wide range. Great care is needed when selecting a type to meet specific service conditions.
- *Coal tar epoxides*. The paint film is highly impermeable to water and most suited to immersed or very damp conditions.

Table T3.7/3   *Characteristics of paint systems based on type of binder*

Type of binder	Mode of drying	Acid resistance	Alkali resistance	Water resistance	Solvent resistance	Exterior weathering resistance
Raw linseed oil Boiled linseed oil	Air oxidative polymerization	Fair	Bad	Fair	Poor	Poor/fair
Oleoresinous varnishes	Air/stoving condensation/ oxidative polymerization	Fair	Bad	Fair/ good	Poor	Fair/ good
Long oil-length alkyd	Air oxidative polymerization	Fair	Bad	Fair	Poor	Very good
Medium oil-length alkyd	Air/stoving oxidative/ condensation polymerization	Fair	Poor	Fairly good	Fair	Very good
Short oil-length alkyd	Stoving condensation polymerization	Fair	Fair	Good	Fairly good	Very good
Urea formaldehyde alkyd blend	Stoving condensation polymerization	Fairly good	Fairly good	Very good	Good	Fair
Melamine formaldehyde alkyd blend	Stoving condensation polymerization	Fairly good	Fairly good	Very good	Good	Very good
Epoxide amino or phenolic resin blend	Stoving addition/ condensation polymerization	Good	Good	Very good	Very good	Good
Polyester/ polyisocyanate blend	Air/stoving addition polymerization	Fairly good	Good	Fairly good	Very good	Very good
Vinyl resin	Air solvent evaporation	Very good	Very good	Very good	Poor	Good
Chlorinated rubber	Air solvent evaporation	Good	Good	Very good	Poor	Good

Based on Trethewey, K. R. and Chamberlain, J., *Corrosion for Students of Science and Engineering*, Table 14.1, p. 247, Longman, Harlow, 1988.

- *Polyurethanes*: Air-drying systems come in two-packs, stoving varieties in one-pack. The properties of the paint depend on the ratio of alkyd to polyester in the resin. As a group, they do not tolerate high humidities or dampness during application or curing. The dry film has a high resistance to water and retains a high gloss for long periods.
- *Vinyls*: There are many different types of vinyls and drying time is short (2–5 minutes). The paints are difficult to apply by brushing so they are normally sprayed. Adhesion is impaired by surface moisture during application. Importantly, the solvents used are inflammable and toxic.
- *Chlorinated rubber*: Adhesion to the substrate and between coats is very good and damaged areas can be repaired easily. The paints are softened by many oils and greases.

(9) Priming paints

Given proper surface preparation, the priming coat (i.e. the coat applied to the prepared surface of the metal) is the most important determinant of the successful performance of the paint film. Its primary function is to provide a good base for subsequent coats of paint. It is the latter that provide most, if not all, of the corrosion control mainly by different degrees of impermeability. They cannot perform that function properly if they do not adhere adequately to the priming coat, with which they must, of course, be compatible. The priming coat must bond securely to the base metal and provide a surface to which the *first* overlying coat can bond securely. Needless to say, different metals need different primers. The surface preparation of the metal surface must therefore be appropriate for the primer (see '(c) Pretreatment' earlier, p. 508, for the surface treatments available).

To achieve good adhesion with the prepared surface of a metal the primer should 'wet' the surface readily. Since paint films are not completely impermeable to corrosive agents, priming paints should contain an inhibitive pigment. This is particularly important in the protection of iron and steel, which generally corrode comparatively easily.

Inhibiting pigments may be used in different vehicles. For example, red lead and calcium plumbate primers are normally made with oil-based vehicles which 'wet' the steel well; metallic lead and zinc chrome primers are often produced with more complex oleo-resinous vehicles; and zinc-rich primers are generally bound with chlorinated rubber, isomerized rubber, polystyrene or epoxide resin vehicles, in one- or two-pack systems. There are, in addition, what are known as pretreatment washes or etching primers. They are essentially vinyl resin based with additives. They come in one- or two-pack systems, their main function being to improve the adhesion of the paint to non-ferrous metals, although they can also provide temporary protection to blast-cleaned steel and sprayed metal coatings.

Summarized, the remaining anti-corrosion primers and their general characteristics are:

- *Red lead primer* to BS 2523 – three types. The binder is linseed drying oil. The types are the traditional primers for ferrous metals and are tolerant of indifferent surface preparation. As mentioned earlier (see notes on characteristics of binder accompanying Table T3.7/3), this type of binder hardens very slowly.
- *Red lead/red oxide primer*. The binder is a drying oil/resin type. It dries more quickly than a red lead primer (see above).
- *Zinc phosphate primer*. Drying oil/resin type binder with zinc phosphate as the inhibitor. This primer is non-toxic and is an

Photograph P3.7/6 *Paint film at the bottom of a painted mild steel baluster attacked by alkali in the concrete in which it is embedded. The result is localized rusting of the steel. (Forty-nine similar balusters were in various stages of rusting)*

alternative for those situations where a paint containing more than 5% of lead should not be used (e.g. on surfaces accessible to children). It dries fairly quickly and can provide protection without top coats for long periods, and is used mainly for ferrous metals.

- *Chromate primer*. Drying oil/resin binder with zinc chromate as the inhibitor, but red oxide may be incorporated. Also quick drying, but is used for both ferrous and non-ferrous metals.
- *Calcium plumbate primer* to BS 3698 in two types. Has a linseed oil binder but some types may have a drying oil binder. Used mainly on galvanized steel but not all paints adhere to the primers. Compatible undercoats/finishes are essential.
- *Metallic lead primer*. Has a drying oil/resin binder and at least 25% metallic lead pigment. It tolerates indifferent surface preparation to some extent. It is used for ferrous metals and should *be avoided* for use on aluminium alloys.
- *Zinc rich primer* to BS 4652 in three types, one in a one-pack and two in two-pack systems. Two are based on chlorinated rubber, the third on epoxy resin. Quick-drying primers for ferrous metals whose *surfaces must be well prepared*.

(10) Undercoats and finishes

The main functions of undercoats and finishes are to provide additional film thickness, water resistance and possibly decorative appearance (Photograph P3.7/6). Looked at in another way, undercoats and finishing coats may be said to serve to protect the primer and, when an inhibiting pigment is used, to enable the primer to retain its inhibiting effect. A highly decorative finish is rarely associated with maximum protection, although suitable colour coats for application over protective paints such as micaceous iron oxide paint are available.

Inhibitive pigments are unnecessary in undercoats or finishing coats. Inert pigments are generally more serviceable. Among the inert pigments are lamellar pigments, such as leafing aluminium, micaceous iron oxide and silica graphite, which orientate themselves in such a way as to reduce the permeability of the film and to retard its chemical degradation by sunlight. Other commonly used pigments are red iron oxide, white lead and rutile titanium oxide.

It may be noted that some undercoats act as links between incompatible priming and finishing coats.

(c) Paints – 2: Ferrous metals

Protection of ferrous metals is usually confined to iron and ordinary steel – stainless steel may need 'localized' protection at bimetallic junctions. Special care is needed in the selection of protective schemes for iron and steel due to their inherent corrosion-resistant weakness. In the UK they generally need some form of protection, whatever the exposure. Once a particular scheme has been selected, thorough pretreatment (particularly the importance of removing mill scale and rust) and application (a thick even film is significant) are essential. Primers with inhibiting pigments should always be used for ferrous metals, and, apart from mild conditions of exposure, two coats of primer are essential for full protection.

The choice of undercoats and finishing coats will be largely determined by conditions of exposure and, in some cases, quick drying as well. In some instances where conditions of exposure are severe, paint coatings may be inadequate, even if these are applied over non-ferrous metal coatings. Consequently, alternatives, either another type of coating or another material (metal or otherwise), may be essential.

Photograph P3.7/7 *Corrosion staining due to the lack of a bituminous paint coating where the lead is in contact with mortar*

Although choice of protective system will be determined by conditions of exposure, account should also be taken of the differences between site- and factory-application. The latter should provide greater durability, other things being equal.* It should also be noted that for good protection a total film thickness of at least 125 µm and often up to 250 µm is necessary. With air-drying paints this normally means at least four coats, including two priming coats; thixotropic and chemically cured paints may require fewer coats.

(d) Paints – 3: Non-ferrous metals

(1) Generally

The need for protective schemes for the non-ferrous metals is generally less important, particularly for ordinary atmospheric exposure, than in the case of the ferrous metals. However, it is often advisable to paint the surfaces of non-ferrous metals such as aluminium and zinc, particularly when these are used as coatings (see '3 (a) Metal coatings'). Because the non-ferrous metals are more corrosion-resistant than the ferrous ones the choice of a paint system is less critical except in highly corrosive atmospheres or in the special cases outlined later. Simpler paint systems, such as ordinary alkyd undercoats and gloss finishes, are usually adequate, but in chemically contaminated environments chlorinated rubber or epoxy paints may be needed. Although adequate thickness is important, this need not be as great as for ferrous metals.

Serious consideration of protective schemes for non-ferrous metals needs to be given either when the contact of dissimilar metals cannot be avoided or when non-ferrous metals are in contact with or embedded in other building materials (Photograph P3.7/7 and P3.7/8). As regards the latter, contact with materials likely to produce acidic or alkaline conditions (see 'Resistance to corrosion', p. 489) should be avoided.

Where the contact of two dissimilar metals cannot be prevented the aim should be to ensure that there is no electrically conducting path between them (**separate lives**). This can be achieved by *thick* coats (adequate thickness is important) of bituminous paint or hot applied bitumen, inhibitive primers or pastes. The use of bituminous felts and some plastics sheets, although not paint systems, are also acceptable treatments.

Resistance to alkaline solutions (concrete, cement or lime mortar) or to acidic solutions (gypsum plaster – e.g. Keene's) may be given by *thick* coats of bitumen. For alkaline surfaces, including contact with plaster, alkali-resisting primers may be used. Contact with those timbers which may affect non-ferrous metals (especially timbers such as western red

Photograph P3.7/8 *Corrosion of a zinc soaker in contact with brickwork. (From BRS Digest (First series) 111, Corrosion of non-ferrous metals: II, June 1958, Fig. 1, Building Research Establishment, Crown Copyright)*

*An important aspect of factory-applied paint coatings is that they are virtually impossible to make good on-site if they are damaged. This is particulary true of stove-enamelled finishes.

cedar, Douglas fir, oak and sweet chestnut) is better avoided by the use of bitumen or zinc chromate primers – wood primers may not be suitable.

The basic requirements for painting of the individual non-ferrous metals are summaraized below.

(2) Aluminium and its alloys

Paint treatments are the same for aluminium and all its alloys. Normally, the priming paint should contain corrosion-inhibiting pigments such as zinc chromate and other chromates constituting about 20% of the dry weight of the film – red lead must be avoided. Red oxide primers with about 5% chromate (these primers are cheaper) may be used for dipping or spraying where the alloy is resistant to corrosion and the exposure is not severe. Etch primers are particularly suitable for aluminium and its alloys – the two-pack being superior to the single pack. Sprayed aluminium coatings may be painted as for zinc (see later), or simpler paint systems, usually with an etch primer, can be used. Hot-dipped ('aluminized') steel should be painted as for aluminium sheet.

(3) Zinc-sheet, hot-dipped or electrolytic coatings

Corrosion-inhibiting primers are not necessary for zinc, but the primer must have good adhesion – zinc rapidly causes most oil-based paints to lose their adhesion. This can be obtained with either etch primer or calcium plumbate primer, and both require thorough degreasing – weathering or mordant washes are unnecessary. Calcium plumbate primers applied to a zinc coating which has weathered so as to expose the steel and rusting will help to prevent further corrosion. On weathered zinc, zinc dust/zinc oxide paints have proved successful. Red lead primer should not be used.

(4) Sprayed zinc coatings

Subject to the investigation of suitable paint systems, the best appears to be a 'modified' (water-resistant) etch primer plus a coat of zinc chromate primer. In mild conditions these may be followed by suitable decorative or protective top coats, but in the most severe conditions micaceous iron oxide paint is probably the best.

(5) Copper and its alloys

The chief difficulty in conventional painting is due to chemical reaction between the metal and the drying oil paints. Green stains or poor adhesion sometimes result. Roughening of the surface before painting can be achieved with fine abrasive paper, preferably used wet or with white spirit. Both aluminium pigmented primers and etch primers are satisfactory. Indoors it is possible to paint directly with one or two coats of alkyd gloss paint. Preservation of the original appearance and prevention of discoloration or a patina is intended with a recently produced special clear lacquer ('Incralac').

(6) Lead

New lead surfaces should be abraided or treated with phosphating solutions or phosphoric acid. Etch primers and many ordinary metal primers, provided they do not contain graphite, are satisfactory.

(7) Chromium, nickel and tin

These metals are usually only painted when corroded. Corrosion products should be removed with fine abrasive paper followed by an etch primer and any decorative paint system.

(8) Cadmium
Weathering to produce a paintable surface should not be used; instead, phosphating treatments are suitable, while an etch primer will provide a key for decorative paints.

(e) Plastics coatings

Plastics of various kinds are now used for coating metals, particularly steel, by brushing, dipping, spraying or by cladding with the aid of adhesives. Provided the coatings are adherent and non-porous, the resistance is that of the plastics itself. Most of the plastics used are reasonably resistant to a number of environments, although they may only withstand moderate temperature and are often soft enough to be easily damaged.

PVC was among the first of the plastics to be used for coating steel, mainly under factory conditions. The coated products were for external use. More recent developments have seen the emergence of a factory process known as *coil coating* for profiled metal cladding sheets of *galvanized* steel or aluminium. The finishing coats, which are based on a number of different organic 'vehicles', are applied under controlled conditions as either liquid or film coatings to a strip of the metal. The coated strip can then be profiled by roll forming or press forming. Like paint, the surfaces of the coatings may suffer in appearance from dirt deposition or colour change. Experience so far indicates that some degradation may be apparent, at worst, after about 10 years and, at best, after about 20. New developments are constantly taking place to improve the life expectancy of the coatings, of which there are many types. Work is also proceeding on finding acceptable ways of recoating the finish satisfactorily on-site.

On-site nylon may be applied by flame spraying. Some steel ironmongery and railings are given a dipped or powder-process plastics coating; in the latter form especially, epoxy resins give very good protection.

Window frames may be coated with extruded PVC rather than painted, to give the best protection in severely corrosive environments. Steel pipes for water and gas services may be clad with extruded PVC sheathing which, among other things, has excellent electrical resistance. Where high resistance to corrosion is required from without and within, PVC coatings can be applied to spirally welded steel piping internally or externally or both. Other types of plastics coatings include Neoprene and Hypalon.

(f) Special paints and tapes for immersed conditions

Where metals have to be immersed in water, permanently wet or very damp conditions there is a range of special paints and tapes available to provide the required corrosion control. Not surprisingly, none have the appearance of the finished paint or tape as a criterion. Corrosion-resisting characteristics are paramount. Thick coatings are normally essential. Products available include the following:

(1) Chemical-resistant paints
These include epoxy resins, polyurethane and polyester resin and are supplied as two-pack systems. As the coating cures by polymerization, the surfaces to which it is applied must be cleaned thoroughly.

(2) Hot-applied bituminous coatings
These are used for protecting pipelines. They are plasticized coal tar enamels or petroleum asphalt. Glass-fibre wraps may be used to reinforce the coating.

(3) Wrapping tapes
Three types are available:

- Petroleum-impregnated natural and synthetic fabric;
- Synthetic resin or plastics tapes (e.g. PVC or polythene); and
- Coal tar or bitumen tapes.

5 Vitreous enamelling

Vitreous enamel coats the surface of the metal with a continuous inorganic glaze which is highly resistant to corrosion. Corrosion resistance, particularly against acids, can be increased by additional amounts of silica and the introduction of titania. Accidental damage often causes chipping which renders the coating useless. This is particularly true of vitreous enamel on steel. However, vitreous enamel on aluminium – a comparative innovation in the UK, although used extensively in the United States – has distinct advantages in that it may, if successfully prepared, be sawn, drilled cut and punched without chipping. Nevertheless, care must be taken to seal cut edges with a corrosion-inhibitive paste to prevent unsightly staining by corrosion products.

The process of vitreous enamelling consists briefly of applying, to a suitably prepared surface, powdered glass of suitable composition, often suspended in a liquid, usually water. The particles, after being dried in warm air, are strongly heated to melt the glass, which under suitable conditions flows over the whole surface, producing a coherent coat. Generally, several coats are applied, the first chosen for adhesion and subsequent coats for resistance to corrosion or abrasion. Temperatures of between 650° and 850°C are generally required to fuse the silicate or borosilicate glass coatings. Because of the low melting point of aluminium, special frits, usually with lead-free formulations which can be bonded to the surface at around 560°C, are used.

The heating necessary to fuse the glass causes a reduction in strength of the metal. Consequently, vitreous enamelling should be confined to non-load-bearing components.

6 Concrete

(a) Generally

The use of concrete as a means of protecting steel (sometimes iron as well) is a secondary requirement in that the concrete is required primarily for other reasons: for structural purposes in the case of reinforced concrete or for fire resistance in encased steelwork. Nevertheless, due consideration must be given to the requirements for proper corrosion control of the steel by the concrete. If the steel corrodes sufficiently, the concrete will crack, thereby encouraging more corrosion and cracking (Photograph P3.7/9). In extreme cases the cracking may be sufficient to put at risk, if not reduce significantly, the structural and/or fire-resistance performance of the structure.

(b) The corrosion process

Hydraulic cements in concrete offer some corrosion protection to the steel. When hydrated, the cement component in concrete produces *alkaline* compounds. These raise the pH of the matrix to between 12.6 and 13.5, a range in which steel remains passive. The steel is given

Photograph P3.7/9 *Severe cracking of a reinforced concrete column due to the corrosion of the steel*

further chemical and physical protection by the cement matrix that is a barrier to the ingress of moisture and oxygen. This protection only remains effective while the high pH and the physical integrity of the cover is retained. Other factors are also involved, but these are less important in the present context.

The protective alkalinity starts to be lost at the exposed surface of the concrete. This surface is exposed to (i.e. it is in contact with) atmospheric carbon dioxide and sulphur that form highly acidic solutions in moist conditions. The alkaline materials are usually unstable when they react with the acid solutions. The reactions which result in the reduction of alkalinity near the surface of the concrete is known as *carbonation*. Among the several factors that govern the extent to which the protective alkalinity will be lost, the most important are:

- The permeability of the concrete;
- The extent and depth of cracking; and
- The humidity and the concentration of acid gases in the atmosphere.

The depth of carbonation increases with time and with rises in the permeability of the concrete. Its rate is greatest where the humidity is in the range 50–75% rh.

Diagram D3.7/16 compares (a) the fully protected steel, (b) corrosion of the steel in carbonated concrete and (c) corrosion of steel in cracked concrete. The significant features of each are:

1. A dense concrete results in a relatively thin carbonation zone at the exposed surface of the concrete. The density of the concrete is

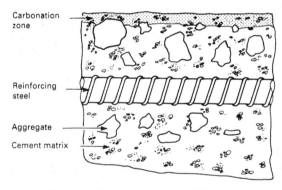

(a) **Steel protected from corrosion in partially carbonated concrete**

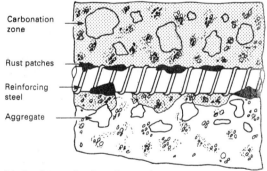

(b) **Steel corroding in carbonated concrete**

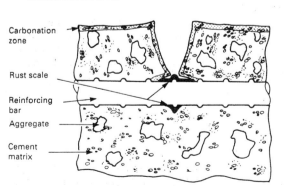

(c) **Steel corroding in cracked concrete**

Diagram D3.7/16 *Comparison of the effects of the corrosion of steel reinforcement of partial and full carbonation and cracks in its concrete cover. (From BRE Digest 263, Figs 4–6, p. 5, Building Research Establishment, Crown Copyright)*

Photograph P3.7/10 *An example of the cracking and spalling of concrete due to inadequate cover of the reinforcing steel in it – part of reinforced concrete columns (after 4 years)*

responsible for a low rate of carbonation. The steel remains protected from corrosion throughout the life of the structure.

2. The increased depth of carbonation results if the concrete is permeable. The atmospheric gases can therefore gain entry into the concrete, thereby causing a greater depth of concrete to carbonate. Once the carbonation zone has reached the steel, it begins to corrode, usually in localized places. The corrosion spreads in time.

3. Cracks in the concrete provides direct access of the atmospheric gases to the steel which then begins to corrode. It will be seen that the thin carbonation zone extends down the sides of the cracks. In view of the direct access of the atmospheric gases to the steel, such protection as the carbonation zone may provide elsewhere is of little significance.

The greater volume occupied by the corrosion products imparts internal tensile forces to the conrete cover causing cracking and spalling (Photograph P3.7/10). The atmospheric gases can then gain access to other parts of the steel, so the corrosion spreads. At the same time, the cross-sectional area of the steel will be reduced by the corrosion. Normally, structural weakening of steel reinforcement is unlikely to occur until some time after the cracking has begun.

(c) Applying the cover

The concrete cover may be applied to steel in one of two ways, namely, (1) by casting the concrete around the steel either on site or in a factory or (2) by spraying the concrete around the steel. The latter method is more commonly applied to structural steel or for lining steel bunkers and chimney stacks. Casting may be applied either in reinforced concrete work or in structural steelwork. There are some detailed differences in requirements between these two methods, which are more conveniently discussed separately. However, it should be noted that in either case all loose rust and all loosely adherent mill scale should be removed from the steel – other surface preparation is generally unnecessary.

(d) Using additional coatings

Inevitably, circumstances may be such that reliance cannot be placed on the steel being adequately protected by the concrete or there is a need to arrest further corrosion.* There are two ways of approaching the problem: coating the reinforcement with zinc and/or coating the surface of the concrete.

(1) Zinc-coated reinforcement
According to BRE, zinc-coated reinforcement has been used successfully to reduce the risk of corrosion. Importantly, long-term tests have shown that the initial attack on zinc by the alkalis released during hydration of the cement is not progressive. The coating can therefore by expected to have good durability. The zinc may be coated by:

- Electrochemical deposition;
- Sherardizing;
- Hot-dip galvanizing;
- Metal spraying; and
- Applying a zinc-rich paint.

The first four are explained in 3(b), p. 512; the last is included in 4(b), p. 514. On balance, hot-dip galvanizing is the most economical way of providing a satisfactory zinc-coated surface.

*In some cases it may, of course, be necessary to adopt remedial measures. A number of different techniques are available. These are outside the scope of this book, but see BRE Digests 263, 264 and 265, *The durability of steel in concrete*, the third dealing with the repair of reinforced concrete.

The *thickness* of the coating required in any given circumstance will be influenced by factors explained earlier, such as exposure, the resistance of the zinc to attack and the quality of the concrete. As the corrosion products of zinc are soluble, *low conditions of humidity* are needed for the zinc to have a long life. The alkalinity of the concrete causes a layer of zinc hydroxide to form initially and one of a complex *calcium zincate* compound subsequently. The latter is insoluble in the highly alkaline pore liquid in the concrete. Importantly, the chemical reaction produces a tight bond between the concrete and the zinc-coated steel, and this provides a barrier against further attack of the underlying zinc.

The zinc coating must be sufficiently thick to allow for:

- The formation of the protective zincate layer; and
- A margin for corrosive attack when the concrete environment around the steel has lost its alkalinity through carbonation.

A coating weight of 200–700 µm is advised by BRE – a 'rule' that may change in time. This range also gives adequate protection during short-term storage on-site.

It should be noted that much thicker zinc coatings would be required at any air/concrete/steel interface, but such interfaces are better avoided. If not, there is no alternative for the zinc coating to be protected with alkali-resistant paint.

(2) The concrete surface

The function of a surface coating on the exposed surface of the concrete is to control, if not eliminate, carbonation of the concrete. BRE has developed an accelerated test to determine the relative resistance of such coatings to carbon dioxide. From the results using this test a wide variety of coating types, properly applied, can provide the necessary resistance to carbon dioxide.

To fulfil its primary function, the coating needs to be relatively impermeable. The permeability of coatings generally increases with thickness. BRE advise that a coating should have a total mean thickness of at least 200 µm over the whole surface. To aid achievement of the latter, the concrete should be level and uniform. Special cementitious slurries may have to applied to make rough surfaces acceptable. Account must also be taken of the needs of good painting practice.

Pigments in the coating increase its resistance. Paint systems consisting solely of clear coating materials have been shown to have little resistance to carbon dioxide penetration because they cannot be applied thickly enough. Experience has also shown that clear coatings are not as durable as pigmented ones. Coating systems should therefore include a pigmented top coat.

(e) Cast concrete

The main requirements for concrete cast around steel in order to prevent corrosion of the embedded steel may be summarized as follows.

(1) Quality

In order to confer maximum protection, good-quality concrete is essential. The materials used should be carefully selected and due consideration given to correct water/cement ratios; correct cement/ aggregate ratios; correctly graded and good quality aggregates; correct mixing; adequate formwork; correct compaction; and correct curing. The main aim should be to ensure a concrete which is compact, that is, relatively impermeable, and free from cracks, other than fine haircracks. Construction joints designed to ensure the exclusion of water are also important.

(2) Cover

The cover required will depend on the grade of concrete used and the conditions to which it is exposed (Table T3.7/4). It will be seen from the table that small depths of cover are related to the stronger grades of concrete in locations completely protected from the weather. The lower grades of concrete are not suitable for the more exposed locations.

Note that the provision of adequate cover should not be confined to the main reinforcing bars but should also include stirrups, hooping and binding wire (see Diagram D3.7/21, p. 538). Care is required in placing and fixing of reinforcement, including all the 'accessories' to ensure that they do not 'migrate' towards the exposed surface, thereby reducing the designed cover during the placing or compaction of the concrete. Close supervision of the work in progress is essential.

(3) Additives

The use of additives such as calcium chloride to accelerate the hardening of concrete should be strictly limited in accordance with current codes of practice. When calcium chloride within the permitted limit is used, concrete of low permeability is essential, while it is equally important to ensure that the chloride is thoroughly mixed to avoid any uneven concentrations, as these can set up corrosion currents. To achieve an even distribution of the chloride it should be dissolved in the mixing water and not added as a dry powder.

Although cases of failure are rare when hydrochloride acid is used to wash precast concrete units, this practice should be closely controlled, as variations in absorption of chloride leads to increased risk of corrosion.

Table T3.7/4   *Nominal concrete cover for reinforcing steel for different conditions of exposure*

Condition of exposure	Nominal cover (mm)				
	Concrete grade				
	20	25	30	40	50 and over
Mild: e.g. completely protected against weather or aggressive conditions, except for brief period of exposure to normal weather conditions during construction	25	20	15	15	15
Moderate: e.g. sheltered from severe rain and against freezing while saturated with water. Buried concrete and concrete continuously under water	–	40	30	25	20
Severe: e.g. exposed to driving rain, alternate wetting and drying and to freezing while wet. Subject to heavy condensation or corrosive fumes	–	50	40	30	25
Very severe: e.g. exposed to sea water or moorland water and with abrasion	–	–	–	60	50
Subject to salt used for de-icing	–	–	50[a]	40[a]	25

[a] Applicable only if the concrete has entrained air.

From BRE Digest 263, *The durability of steel in concrete: Part 1 Mechanism of protection and corrosion*, p.4, July 1982, Building Research Establishment, Crown Copyright.

(f) Sprayed concrete

Concrete may be sprayed by means of a high-pressure cement gun. The normal mixture for such a gun working at 0.21–0.28 N/mm^2 is cement and sharp sand mixed dry. A cement/sand ratio of 1 : 3½ is satisfactory for most atmospheric conditions, but where water tightness is desired this should be reduced to 1 : 3. For resistance to heat, special semi-refactory materials are used. No foaming agent or other admixture is added and mixing with water takes place at the nozzle. In order to avoid the risk of shrinkage cracking the water/cement ratio is closely controlled.

For interior work the coating should generally be at least 40 mm but preferably 50 mm, while for exterior work exposed to corrosive conditions the coating should be at least 50 mm. Such coatings should be reinforced with welded wire mesh 4.0 or 3.50 mm thick, firmly anchored to the steel surface. These thick coatings are applied in several layers, and the cement/aggregate ratio should be the same throughout, while each layer should be allowed to harden for 3 hours before being overcoated. Proper curing of the completed coating is essential.

7 Cathodic protection

(a) Mechanism

During the corrosion process there is a flow of current from an anodic area into the electrolyte. Cathodic protection is the application of a stronger current in the opposite direction, flowing from a specially provided anode through the electrolyte to the metal which neutralizes the currents responsible for corrosion and renders the metal cathodic over its entire area. Under these conditions, the anodic process occurs at the specially provided anode.

(b) Sources of current

The countercurrent may be supplied by baser metals, that is, more electronegative ones (usually magnesium, aluminium and zinc) which undergo 'sacrificial attack' or, alternatively, by an impressed current drawn from a power supply using a transformer and rectifier to produce low-voltage d.c. (any other d.c. source may also be used). The principle upon which cathodic protection is based and the alternative methods available are illustrated in Diagram D3.7/17.

(c) Application

Cathodic protection is commonly applied to steel structures or containers (tanks, for example) that are immersed in water or buried in a damp environment. Continuous contact with a mass of electrolyte is generally required.

Although the principle of cathodic protection is fairly simple and straightforward, expert knowledge and experience is required in its application. The precise details of a system for a given circumstance (i.e. the current needed) cannot be accurately predicted without test, although rough estimates can be based on previous experience. The conductivity of the electrolyte is important – more current is needed with electrolytes of low conductivity (fresh water) than with electrolytes of high conductivity (sea water).

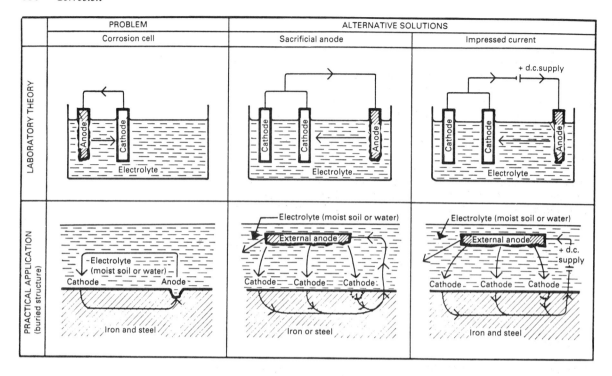

PROBLEM	ALTERNATIVE SOLUTIONS	
Corrosion cell	Sacrificial anode	Impressed current

Diagram D3.7/17 *The principle of cathodic protection*

As a general rule the *choice of system* will depend on the size of the structure, the conductivity of the electrolyte and the availability of power. Thus, sacrificial anodes may be preferred when the conductivity of the electrolyte is sufficiently high (notably, fresh-water tanks) for small temporary or isolated systems; the anodes may also be preferred where the passage of a large current through the soil may be objectionable. An impressed current system, on the other hand, is likely to be preferred for large structures where a large total current is necessary and where power is readily available.

The *choice of material* as the anode for impressed current and sacrificial anode systems will be based on different criteria. The basic requirements of each system are better considered separately.

(1) Impressed current systems
The requirements which the anodes should meet are:

● Reasonable working life;
● Ability to discharge the current to be used without unduly rapid disintegration of the anode connections; and
● Sufficiently low resistance to earth.

These requirements determine not only the choice of material but also the form of anode and its surface area. Where a large area is required to provide high conductivity scrap metal may be used, but often materials such as graphite (if impregnated to reduce porosity), high silicon-iron, lead alloys and platinized titanium may be specifically chosen to resist deterioration because their choice will be more economical. The geometric arrangement of anodes is also important. For the protection of buried structures it is common to use a number of buried anodes connected together to form an extensive groundbed.

(2) Sacrificial anode systems

The electrolytes in which the anodes are to operate govern the choice of material. In low-conductivity electrolytes the emf is the primary consideration; in electrolytes with high conductivity the anode should have a reasonable working life. An anode of *magnesium alloy* has the highest driving potential and is the most universally effective in protecting buried structures. A magnesium alloy is often used for the protection of fresh-water tanks. Its useful life, particularly in sea water, is reduced because if suffers natural corrosion in addition to the loss corresponding to the protection current. *Zinc and aluminium* have lower driving potentials. However, their use may be advantageous in high-conductivity environments in which magnesium anodes would give more current than necessary. In certain cases it is necessary to surround buried anodes with a backfill to reduce corrosion, minimize polarization and provide better contact with the soil.

(d) Side effects

The use of cathodic protection may, under certain conditions, produce undesirable side effects which could mitigate against the degree of protection expected or induce corrosion elsewhere. The main side effects may be summarized as follows:

(1) Stray currents

As a cathodic protection installation for buried structures causes a direct current to flow in the soil, adequate precautions are required in order to prevent this current from flowing along neighbouring structures, thus initiating their corrosion (see 'General considerations', 5(c)(3) p. 470). Susceptibility to corrosion by stray currents and the magnitude of their effect will depend on the total protection current; the quality of any coatings on the structures, the conductivity of the soil and the relative positions of the structure and the groundbed (the anodes). However, the effect can be minimized by care at the design stage, but at the same time it is advisable to notify other organizations likely to have pipes or cables buried or in contact with the soil nearby of the intention to install cathodic protection.

(2) Hydrogen and alkalinity

In some cases it may be necessary, in order to achieve protection at positions where the protective effect is least, to employ potentials more negative than the protection potential over the greater part of the surface. This can result in the evolution of hydrogen gas, which can disrupt coatings, or the production of a high degree of alkalinity, which can destroy paint by saponification (see *3.06 Chemical attack*; p. 392).

(3) Calcareous deposits

In sea water and in some soils the alkalinity causes the precipitation of an insoluble calcareous deposit. Although this may be beneficial in that it reduces the current needed for cathodic protection, it may be undesirable in some water installations where an excessive deposit may lead to blockage of flow channels or reduction of heat transfer.

Precautions

1 Generally

The problems associated with the corrosion of metals are varied and complex. The complexity of these problems and their possible solutions, as outlined in the preceding sections of this chapter, are mainly due to the many variables and interactive factors involved. In practice, the difficulties that are often encountered in establishing, for any given circumstance, the precautions which should reasonably be taken to minimize the risk and/or effects of corrosion, may be due to two closely interrelated factors:

• It is generally difficult to predict, with complete accuracy, the degree of corrosion that may occur in any given circumstance. Consequently, such precautions as are taken should, for safety, assume the very worst case **(creative pessimism)**.
• Cost is invariably important, but here, too, a course must be steered between the requirements of durability, on the one hand, and maintenance, on the other, in addition to other requirements such as strength.

Despite all these difficulties there has been sufficient experience in the use of metals in buildings which has led to the establishment of some definite and clear recommendations particularly in the use of iron or steel. Many of the precautions that should be taken to minimize the risk or the effects of corrosion have been described in various ways in the preceding sections of this chapter. However, these have been included to form part of the discussion aimed at explaining not only the causes and effects of corrosion but also possible remedies. Consequently for convenience this section brings together the more important precautions without the related reasons. These have already been covered earlier.

2 Selection

Selection relates to the approach to the problem and the form of protection. Depending on circumstances, the problems of corrosion prevention may be considered basically from one of two points of view: first, the use of a metal which is highly resistant to corrosion, or second, the use of some form of protection to a metal which is not resistant to corrosion. The latter course implies maintenance.

Cost will always influence which of these two approaches is adopted, but in a great many cases *strength* will be significant. However, eight other factors will influence selection. These summarized are:

• *Conditions of exposure*, that is, basic exposure plus the effects of design. In a great many cases exposure will be one of the most important factors influencing selection. Careful analysis of exposure for any given condition is essential.
• *Proposed life of structure or component.* In this a distinction must be drawn between short- and long-life buildings and/or life-to-first-maintenance.
• *Periods required between maintenance.*
• *Importance of structure or component.*
• *Difficulty of access for maintenance.* This applies to all metal work, including those components designed to be replaced (i.e. designed as short-life components).
• *Shape and size of structure or component.*
• *Protection during storage and transport.*
• *Fabrication methods.*

3 Design

Many details of design may seriously increase the risk of corrosion generally. Attention to such details is therefore important, particularly when metals of low corrosion resistance, such as ordinary steel, are being used. In the absence of taking the necessary steps to eliminate the use of 'corrosion-increasing' details it is necessary to ensure that some means of protection, or additional protection, depending on circumstances, is provided.

Design should include consideration of such factors as geometry, working conditions, contact between dissimilar metals or between metals and other building materials, and accessibility for maintenance. For convenience, only geometric factors and working conditions are discussed here; contact and accessibility being considered under '4 Contact' and '6 Maintenance', respectively.

(a) Geometric factors

Geometric factors may cause the initiation of corrosion, particularly local corrosion, for a number of different reasons in both atmospheric (external and internal) and immersed environments.

(1) Atmospheric environments
One of the most important precautions which should be taken with metals exposed to atmospheric environments is the avoidance of details which encourage the entrapment of moisture and dirt (crevices are notable). Typical examples of these and other details which should be avoided are illustrated in Diagram D3.7/18. The precautions may be summarized as follows:

- *Arrange* constructional and other members so that trapping of moisture and dirt is discouraged (Diagram D3.7/18(a)). Alternatively, adequate drainage holes (sufficient diameter important) may be provided, if rearrangement of members is impracticable.
- *Arrange* joints and fastenings to give free and uninterrupted lines (Diagram D3.7/18(b)). Welds are generally preferable to bolted joints, with butt welds better than lap welds. If the latter have to be used, then appropriate welding or filling with mastic may be necessary to avoid the entrapment of moisture and dirt.
- *Avoid* crevices, (Diagram D3.7/18(c)), as they also allow the retention of moisture and dirt and thus increase corrosion. Where it is not possible to do so, they should be filled by welding, using a filler or mastic.
- *Avoid* structural arrangements that prevent the free circulation of air through and arround them (Diagram D.3.7/18(d)) so as to reduce condensation risk, particularly where thermal insulation is impracticable. *Hollow structures* such as boxed and tubular sections should preferably be hermetically sealed.
- *Avoid* sharp corners and edges or rough surfaces to metals which are to receive a protective coating, particularly if this is paint (Diagram D3.7/18(e)). All *contours* should be as *rounded* as possible in order to ensure an even coating thickness, and also to avoid the possibility of mechanical damage. Paint coatings may also break down if there is a rapid change in contour (the heads of rivets and screws and crevices) in which pickling or pretreatment solutions can become trapped.
- *Avoid* features that allow water and condensation to drip or be blown back to other parts of a structure. In this the use of drips is important, but in some cases suitable drainage with downpipes may be essential.
- *Ensure*, where it is difficult to prevent sheltered steel surfaces, such as those under eaves, where evaporation of water is retarded, that the sheltered areas are given additional protection. Other metals in similar positions (aluminium and stainless steel notably) should be washed regularly.

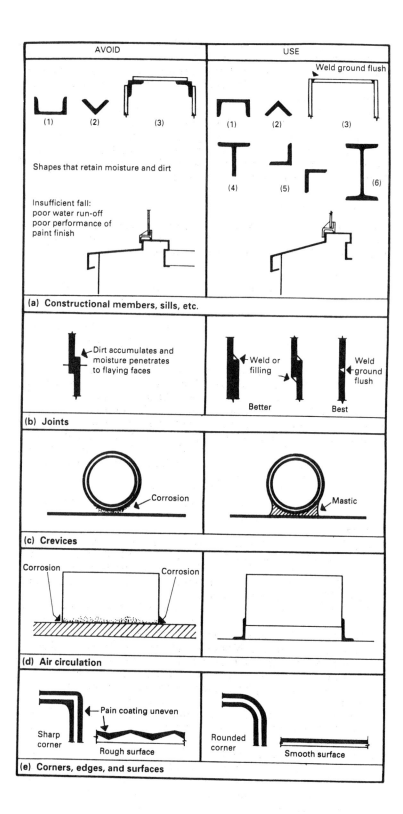

Diagram D3.7/18 *Some of the geometric factors that should be taken into account in design for exposure of metals to atmospheric environments*

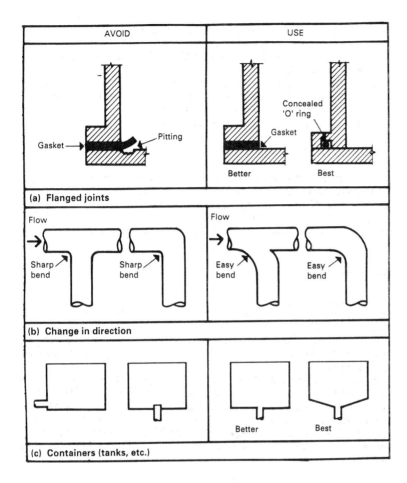

Diagram D3.7/19 *Some of the geometric factors that should be taken into account in design for exposure of metals to immersed environments*

(2) Immersed environments

Some of the points made under atmospheric environments above are also relevant to metals in contact with liquids. Those features which are specially applicable to liquids are set out below; some of these are illustrated in Diagram D3.7/19:

- *Consider* velocity of the liquid.
- *Avoid* crevices. These may be particularly pronounced at screwed joints. In flanged joints (Diagram D3.7/19(a)) the use of special gaskets is advised).
- *Avoid* sharp changes in direction (Diagram D3.7/19(b)), especially when high-velocity liquids are being transported.
- *Ensure* that water flows from anodic metals in the system to cathodic metals.
- *Avoid* the use of storage tanks and cylinders which are difficult to drain (Diagram D3.7/19(c)).

(b) Working conditions

It is important to ensure that constructional procedures do not cause conditions which are likely to lead to *stress corrosion*. The highest stresses are likely to be set up by welding, so adequate stress-relieving treatments are necessary. 'Force-fitting' of parts, for example, when bolt holes do not coincide, which results in the concentration of stresses should be avoided.

Despite the necessity for stress-relieving treatment when welded joints are used (such treatment is also necessary with other forms of jointing), these are preferable, as riveted joints and expansion fits, for example, will tend to loosen on heat treatment.

(c) Specification

It is essential that design intentions are made absolutely clear to those who have to undertake and supervise the installation of the relevant component. **Creative pessimism** would caution that a specification might also describe the working and environmental conditions/exposure and explain the need for the precautions described.

4 Contact

In general, it is safest to avoid contact between metals and absorbent building materials or between dissimilar metals. Where this is impracticable, then it is necessary to provide some form of isolation **(separate lives)**. This must, in most conditions, be electrically resistant and non-absorbent, consisting of suitable paints, tapes, linings, gaskets or washers.

When considering contact, it is always important to remember that this might also be caused when water flows from one surface onto another. Corrosion of aluminium gutters and pipes, for example, is often caused by copper-bearing water from roofs or pipes in a water system. The same also applies to water which first passes over building materials containing corrosive agents and then over a metal. Dripping water having corrosive agents can often be particularly damaging.

(a) Contact with other building materials

Conditions which are likely to lead to corrosion of metals when in contact with other absorbent building materials are included in 'Exposure 4. Other building materials' (p. 482).

Most non-ferrous metals (copper is a notable exception) are susceptible to attack by alkalis and acids. When these are in contact with alkaline concrete, cement or lime mortar, or with acidic gypsum plasters (e.g. Keene's) the metals should be protected by *thick* coats of bitumen. Alkali-resisting primers may be used on alkaline surfaces and should also suffice for plaster contacts.

Wood, especially western red cedar, Douglas fir, oak and sweet chestnut, may affect metals and contact may be prevented by the use of bitumen or zinc chromate primers. Wood primers may not be suitable.

(b) Contact between dissimilar metals

This is a particularly important cause of corrosion. The metals which are most vulnerable when in contact and typical examples of contact in building construction are included under 'Exposure, 6. Contact between dissimilar metals' (p. 483). Contacts that should always be avoided are those between copper, nickel, and their alloys (for example, brass and bronzes) with aluminium or zinc, and of aluminium or zinc components with steel.

In order to avoid contact it is important that the isolating medium is electrically resistant. Suitable treatments include *thick* coats of bituminous paint or hot-applied bitumen, inhibitive primers paints, or tapes, while bituminous felts and some sheet plastics are acceptable alternatives. When fixing devices are used, insulating gaskets, washers, etc. (see Diagram D3.7/20) may be more convenient – the use of coatings may be undesirable due to the ease with which they may be damaged.

In water systems it is essential to ensure that *all debris*, including steel drillings, etc., is removed from tanks, particularly galvanized steel tanks, before they are filled with water.

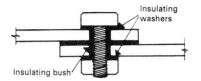

Diagram D3.7/20 *A method of electrically insulating dissimilar metals. (Based on Butler and Ison, Fig. 42, p. 235)*

(c) Soldering, brazing and welding

It is important to ensure that fluxes used with soldering or brazing material are thoroughly cleaned off. Weld metals which differ substantially from the base metal should be avoided, while the preparation, welding techniques and finishing should be carefully carried out so that not only is the final weld relatively smooth and well shaped but it also excludes the presence of porosity, voids, crevices, scale inclusions or changes in structure.

(d) Agressive water

In all types of water systems check the compatibility of the water with any metal it is proposed to use. Aggressive waters include those from water supplies which may be soft or those containing free carbon dioxide, while acids, alkalis and salts may be derived from a number of sources (see also Table T3.7/1, p. 484).

Galvanized steel should not be used for hot water cylinders, nor should the internal surfaces of steel pipes be galvanized if the temperature of the water is likely to exceed 60°C. Above this temperature the zinc coating may no longer protect the steel pipe.

(e) Flue gases

The use of boilers for heating and hot water makes it important to ensure that metals (especially flue linings) are selected for their resistance to the gases to be encountered in any particular situation. In some cases metal components may be completely inadequate.

5 Protection

The four main methods by which protection may be achieved, namely, (1) treatment of the environment, (2) protective coatings, (3) concrete, and (4) cathodic protection are covered in detail under 'Protection', p. 506–31. Treatment of the environment is excluded here.

(a) Coatings

A decision to use a coating to increase corrosion resistance means that the coating must be considered during design (see '1 Selection' earlier). The factors which need to be taken into account include the success of the application of a coating to a particular metal (some metals are restricted in the coatings they can receive successfully); possible damage to the coating during storage, transport and assembly; the probable life of the coating; and whether the coating can be renewed *in situ*. At the same time, consideration must also be given to the resistance of the coating to its 'working condition', which may include the need for resistance to abrasion, impact, chemicals, sunlight, etc., or thermal stability and effects on heat transfer (the latter would be important in water systems). Alkali softening of paint applied to steelwork, for example, is a common cause of corrosion.

In all cases, it is imperative that the metal to be coated is thoroughly clean, free from grease, dirt, mill scale, rust, etc. This requires good pretreatment. Correct application of the coating is also essential, while surfaces which are likely to lead to differing thicknesses of the coating (see Diagram D3.7/18(e)) should be avoided.

Ordinary steel, it is important to note, can seldom be used without some form of protection. Selection of the coating is important, but proper pretreatment and subsequent application should not be neglected (see 'Protection' 2(c), p. 508).

(b) Concrete

(1) General

Whenever steel is embedded in concrete (reinforced concrete work, encasing of structural steel members or pipes, for example), the following precautions are essential:

- There must be adequate depth of cover. In reinforced concrete work this means cover to all steel, including stirrups, hooping and binding wire, and not merely that to the main bars (Diagram D3.7/21). In those cases where it may be difficult to comply with the amount of cover required particularly in lightweight concrete work, consideration should be given to the use of zinc-coated reinforcement. Pipes encased in concrete below ground must not rest on the soil but on a concrete bed (Diagram D3.7/22).
- The concrete must be of good quality. This means attention must be given to the quality of the aggregates (shrinkable aggregates should be avoided); correct water/cement ratio; and correct compaction.
- Avoid the excessive use of chlorides (calcium chloride should not be added in excess of 2% by weight of cement). When chlorides are used, ensure that these are thoroughly mixed and that a concrete of low permeability is achieved by specifying well-graded aggregates, low water/cement ratios and thorough compaction.
- Additional precautions will be necessary if corrosive substances are present in ground waters and in marine and industrial environments. (Sulphate attack may be significant – see *3.6 Chemical attack*, p. 410).

Cover must include binding wire, etc.

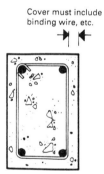

Diagram D3.7/21 *Adequate cover of the steel in reinforced concrete should include stirrups, binding wires, etc.*

Diagram D3.7/22 *Illustration of the need for complete encasement of steel pipes buried below ground*

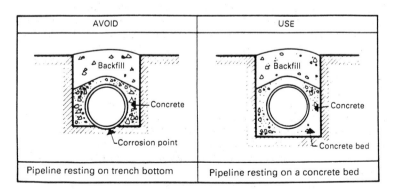

AVOID	USE
Backfill — Concrete — Corrosion point	Backfill — Concrete — Concrete bed
Pipeline resting on trench bottom	Pipeline resting on a concrete bed

- Ensure that cracks in the cover to the steel are not deep enough to allow moisture, etc., to penetrate to the steel. In this, quality control is vital, but also formwork should not be undersized or badly made, while it is important to check that a good bond is achieved at all contruction joints – badly made honeycombed joints are likely to lead to trouble.

(2) Reinforced concrete floors

Special consideration is required with reinforced concrete floors, particularly in the following circumstances:

- Where the finish itself is unaffected by corrosive liquors but does not provide sufficient protection of the underlying structure. Examples are: (1) concrete surfaces permeable to saline wash waters which do not affect the concrete itself; (2) epoxy and polyester floorings sold for use in corrosive situations and laid up to 6 mm thick. Many liquids that do not affect these floorings may leak through them, particularly when some wear has occurred. The risk may not be recognized when they are laid by firms with little chemical engineering experience.
- Where the finish, e.g., granolithic concrete, may distintegrate. Here the user might carry out repairs to the concrete yet not recognize that the steel below is corroding.
- Where the floor finish can disintegrate to produce a corrosive chemical. In this category there is only magnesium oxychloride. This flooring has been responsible for at least one collapse of a reinforced concrete floor caused by the action of magnesium chloride on the reinforcement. The code of practice for *in-situ* flooring, CP 204, points to the need for protecting the metalwork of gas, water and electrical services by 25 mm of dense concrete when this flooring is used *even in dry conditions*. In wet conditions the flooring should never be used because the reinforcement cannot be protected by bitumen or galvanizing.
- In corrosion-resistant flooring where detailing at channels, gulleys, upstands, etc., is defective.
- Where there is a change of use in a factory and it passes unnoticed that a corrosion-resistant floor *designed to resist other chemicals* may not be suited to the new use.
- Where, before laying epoxy resin flooring, washing with hydrochloric acid to remove laitance is recommended. (Hydrochloric acid may also be used to make concrete floors less slippery.)

(3) Exposure

The vulnerability of copings, sills and other projections and of large plane-surfaced walls of reinforced concrete must always be taken into account.

(c) Cathodic protection

Cathodic protection can be applied to any parts of a structure that are immersed in water or buried in a damp environment. The choice of system and details of design require expert knowledge and experience; it is therefore important that such advice is always sought.

Factors which need to be taken into account include the following:

- Continuous contact with a mass of electrolyte;
- Resistivity of the electrolyte to determine probable corrosion risk to structure;
- Field tests to determine groundbed locations, and the requirements of the groundbed and current supply;

- Provision of permanent and accessible insulated test leads at representative points to enable electrical tests to be made of the structure;
- Provision of insulation from any neighbouring structures. Other organizations likely to have pipes or cables buried or in contact with the soil near a cathodically protected structure should be notified;
- Provision should be made for subsequent inspection and maintenance.

6 Maintenance

Maintenance may, depending on circumstances, include regular washing and replacement of a protective system or of a metal or other component (notably sealants).

(a) Inspection

Maintenance is required whenever systems of protection against corrosion with short lives are used. However, it is always wise to ensure that, subsequent to the completion of a building, all metalwork, including the structures in which metal may be embedded or concealed, is regularly inspected.

(b) Accessibility

It is important at the design stages to ensure adequate accessibility. This would include physical access to the metalwork or structure as such, access by means of removable panels, etc. to conceal metalwork and, as important, access to enable coatings such as paint to be applied. As regards the last, details of design may seriously hinder this; some typical examples are given in Diagram D3.7/23. If it is impracticable to provide access, additional protection is important.

(c) Frequency

This will be largely governed by the durability of the metal used (if unprotected, that is) or the durability of the protective system. In this details of design as outlined earlier under '1 Design' can increase the intervals between maintenance.

(d) Washing

Regular washing of unprotected and sheltered metals, such as aluminium and stainless steel, can greatly increase their durability.

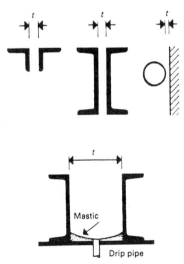

Detail shown above is difficult to maintain; detail shown below is preferable

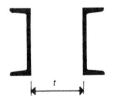

Note: In all cases *t* should provide adequate access for painting

Diagram D3.7/23 *Typical situations in which adequate provision for access is important for all painted metalwork*

Sources

Butler, G. and Ison, H. C. K., *Corrosion and its prevention in waters*, Leonard Hill, Glasgow, 1966

Evans and Ulick, *An introduction to metallic corrosion*, 3rd edn, Edward Arnold, London, 1981

Hendrick, T. W., *Corrosion and its Prevention*, a manual published by the Organization for Economic Cooperation and Development, April 1964

Alexander, William and Street, Arthur, *Metals in the Service of Man*, 8th edn, Penguin, Harmondsworth, 1982, reprinted with revisions 1985. This is an excellent book, written for the layman, on metals in general. It is included for its coverage on metals rather than corrosion

BRE
Note
The latest publications from BRE are included below. BRE Digests of the first and second series now superseded were included in *Materials for Building*, Vol. 3. These are now useful as background only and have been used here accordingly.
BRS Digest (first series) 110 and 111, *Corrosion of non-ferrous metals I and III*, May and June 1958
BRS Digest (second series) 59, *The protection against corrosion of reinforced steel*, June 1965
BRS Digest (second series) 29 and 30, *Aluminium in building – 1: Properties and uses* and *2: Finishes*, December 1962 and January 1983
BRS Digest (second series) 70 and 71, *Painting metals in buildings – 1: Iron and steel* and *2: Non-ferrous metals and coatings*, May and June 1966
Principles of Modern building, Vol. 1, 3rd edn, HMSO, London, 1959

Digests
  98, *Durability of metals in natural waters*, minor revisions 1977
109, *Zinc-coated reinforcement for concrete*, September 1969
121, *Stainless steel as a building material*, September 1970
263, *The durability of steel in concrete*: Part 1 *Mechanism of protection and corrosion*, July 1982
264, *Durability of steel in concrete*: Part 2 *Diagnosis and assessment of corrosion-cracked concrete*, August 1982
265, *The durability of steel in concrete*: Part 3 *The repair of reinforced concrete*, September 1982

Information paper
7/89, Davies H. and Rothwell, G.W., *The effectiveness of surface coatings in reducing carbonation in reinforced concrete*, May 1989

BSI
BS 5393: 1977, *Code of practice for protective coating of iron and steel structures*
BS 8110: Part 1: 1985, *The structural use of concrete*

Trade associations
Information and technical data sheets from the Aluminium, Copper, Lead and Zinc Development Associations

Frost action

Introduction

1 Definition and influencing factors

Frost action is used here to describe the cause and effects of water that freezes in porous materials. Such action cannot take place in non-porous ones. Frost action is associated with changes in temperature and the moisture content of the relevant materials. Two quite different problems are linked with it: first, the effects of low temperatures on materials and processes during construction* and, second, the durability of materials during the lifetime of the building.

Although frost action only occurs at temperatures below freezing, precautions should be taken during construction when the air and/or material temperature is falling or is expected to fall below 3°C.† Consequently, it is necessary to consider the factors that influence the way the temperature of the air around materials and of the materials themselves changes. The influencing factors, such as the heat transfer processes and mechanisms of heat flow, are explained in detail in *Materials for Building*, Vol. 4, under 'General considerations'. The factors that influence moisture content are covered in this book in 3.3 *Moisture content*, to which reference should be made as necessary.

2 Frost resistance

In practice, the frost resistance of materials is required. This property is dependent on the interrelationship of three factors:

- Properties of a material such as the strength, absorption, and saturation coefficient;
- The position in a construction in which the material is used; and
- Climatic conditions.

3 Deleterious effects

The deleterious effects of frost action are mainly:

- *During construction*, the breakdown of newly laid cement-based products such as mortars, renders and concretes being more usual although breakdown of units of materials can also occur; and
- *During the life of a building*, the splitting of the surface of a material (hence the term 'frost splitting' sometimes used to describe this effect) or heave (as might occur in ground-floor slabs).

In general, the plain wall surfaces of buildings are not subject to the deleterious effects of frost action; those parts heavily exposed to water (likely to have high moisture content for long periods of time) usually

*Also included is what is sometimes referred to as 'winter building', a description that is used later in this chapter.

†Hitherto, air temperature has been considered to be the criterion to use. However, in some cases the temperature of an element of construction can be just as, if not more, important.

are. Walls apart, roofs may be similarly exposed. It is therefore these parts to which special consideration should be paid.

4 Guidance and aids available

It would obviously be helpful if the frost resistance of materials were known, more so when existing or new materials are to be used in new ways or locations. In the past research workers have made attempts to devise laboratory tests for the determination of frost resistance of materials, of brick and stone in particular. Unfortunately, no reliable test has yet been devised.* Consequently, reliance has still to be placed on experience of the behaviour of materials in constructions over a period of time, usually at least 10 years (i.e. an empirical approach). Despite the unreliability of such laboratory tests as have been carried out for direct practical application, their results do offer some guidance or lessons to be learned on the mechanisms involved and the influencing factors in particular. It is for those reasons that the test results and their relevance in gaining an understanding of the problems associated with durability are included here.

The problems that were formerly encountered with winter building on-site have, to a large extent, been eased in recent years. This has been due partly to the availability of relatively inexpensive insulating materials in sheet (such as bubble polythene) and board or quilt forms to protect vulnerable parts of a construction. Another factor has been the availability of easy-to-use digital thermometers to measure both air and surface temperatures. Weather forecasting for the construction industry from the Meterological Office is probably another influencing factor.†

General considerations

1 Basic mechanisms

Unlike most other liquids, water ceases to contract just above freezing point (4°C) and begins to expand again. On conversion to ice at freezing point there is a further sudden expansion of 9% by volume (see 3.1 *General considerations*, 'Basic properties of water', p. 131 ff). Thus, weight for weight, ice is bulkier than water. However, the effects of pressure are equally significant. Ice will melt with the application of pressure, the degree of pressure needed being dependent on the temperature of the ice below freezing point – the lower the temperature, the higher the pressure required. Looked at in another way, the application of pressure reduces the freezing point of water.

If, before freezing takes place, water only partially fills a closed container, then it follows that there is room in which the expansion that occurs on freezing can be accommodated, as shown in Diagram D3.8/1(a). (In this it is assumed that the air can escape.) If, on the other hand, a closed container is already completely full of water before freezing takes place, then the subsequent expansion will result in pressure on the walls of the container, with one of three possible effects, depending on the strength properties of the walls, as shown in Diagram D3.8/1(b). Subject to the temperature involved, the walls of the container may be strong enough to resist the pressure. Alternatively, the walls may undergo deformation and, provided this is within the elastic limit, the shape of the container will be restored when the ice subsequently melts. Finally, the walls of the container may be broken.

*The unreliability of tests is no longer necessarily generally acceptable. For example, the 1990 revision of BS 402, *Specification for clay roof tiles and fittings*, now includes the BCRL's panel freezing test – see p. 552 – for the determination of the frost resistance of clay roof tiles. It seems likely that BSI will extend this to other ceramic products such as bricks. Readers should note changes in the relevant BSs. In the meantime, the notes on tests included later should still be of assistance in the understanding of the difficulties involved in testing.

†The Meterological Office offers, for a fee, a special quick-reply service, CLIMEST. This gives estimates of the monthly and yearly averages of frost and low temperatures (among other climatological factors) in the locality of the building (useful in the planning of contract periods). For work in progress pre-arranged *short-term forecasts* or *a warning service* can be provided for builders. Details are available from the Director General, Meteorological Office, London Road, Bracknell, Berks RG12 2SZ (telephone: 0344 20242, fax: 0344 422907).

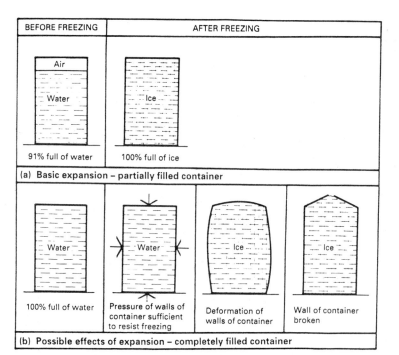

Diagram D3.8/1 *Illustration of the basic mechanisms and possible effects of water freezing in a container. In (a) it has been assumed that air can escape*

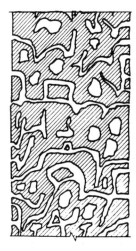

Diagram D3.8/2 *The complex pattern of containers that may occur in porous materials such as bricks*

2 Freezing in materials

(a) Scope

The pores in porous building materials are the containers which may be filled, in varying degrees, with water. However, the arrangement of the pores presents an extemely complex pattern of containers, as shown in Diagram D3.8/2. Whether or not expansion of the water within pores during freezing is likely to cause damage to the walls of the pores depends on a number of different factors, the most important of which include moisture content, moisture gradient, rate of freezing, temperature gradient, freezing/thawing cycles and pore structure. The relationship between these factors is extremely complicated. However, it is convenient to discuss the significance of each separately, indicating, as far as possible, the relationship between them.

(b) Moisture content

By definition, moisture content (see *3.3 Moisture content*, p. 224) is the moisture which a material contains at a given time and expressed as a percentage of its dry weight. Values of moisture content only indicate the amount of water that is present in a material and not how the moisture is distributed, nor the extent to which the pores are filled. However, in general terms, materials with high moisture contents (generally saturated) subjected to freezing are more likely to suffer from frost action than those under similar conditions with low moisture contents.

(c) Moisture gradient

Moisture gradient (*3.3 Moisture content*, p. 229) is the variation of moisture content between the outer and inner part of a piece of material (or construction). In winter in external wall elements there is generally a drop in moisture content across the thickness of the element, with high moisture contents near the exposed faces (**high > low**). Warmth from within the building may help to maintain low moisture contents at or near the internal face but the significance may be changed by thermal insulation (see (d) below). In general, but subject to pore structure, a moisture gradient is advantageous, in that there is room in the interior into which water being squeezed out by the expansion caused by freezing at or near the exposed external face may escape (Diagram D3.8/3).

In the more heavily exposed parts of a wall, such as parapets, there may not be a moisture gradient *across* the construction, although there may be a moisture gradient *down* the construction. In the absence of a dpc under a coping stone, saturation of a considerable part of the parapet may occur. Under such conditions the initial freezing of water in the coping may not result in any deleterious effect on the material of which the coping is made. However, it is possible for water to migrate from unfrozen to frozen parts, as is known to happen in the formation of ice lenses and the consequent heaving of soils, for example, under freezing conditions. Water drawn from the underlying masonry may then contribute to subsequent splitting or breakdown of the coping when this additional water also freezes (Diagram D3.8/4). In this the ease with which water will migrate from the underlying masonry will depend more on the bed joint than on the elusive differences in character between one unit and another. Nevertheless, it is important that dpcs are inserted under copings as *one* way of increasing the durability of the coping stone.

(d) Rate of freezing

Rate of freezing and the tensile strength of the walls of the pores in a material are closely interrelated. In this it is important to note that the tensile strengths of the materials normally subjected to frost action, such as bricks, tiles and stones, are, in general, low.

If the rate of freezing is *slow*, the pressure exerted is gradual and, as important, may be well within the tensile strength of the material. If this is so, then it is possible for the ice which may be formed to be *liquefied*. The surplus water may thus be extruded, either to the outside, that is, above the surface of the material, or to empty pores within the material. This liquefaction enables the material to relieve stresses.

If, on the other hand, the rate of freezing is *rapid*, and particularly when freezing occurs at low temperatures, the pressures exerted, in addition to being rapid, are often of a magnitude far in excess of the tensile strength of the material. Consequently, liquefaction and thus relief of stress is not possible, and this, in turn, leads to damage of the material.

(e) Temperature gradient

In winter there is a drop in the temperature gradient across a construction from inside to outside. The amount of the drop (i.e. angle of the slope of the gradient) will be governed by the relative internal/external air temperature, the distribution of the temperature being dependent on the thermal properties of the materials making up the construction (see *Materials for Building*, Vol. 4, *4.04 Thermal*

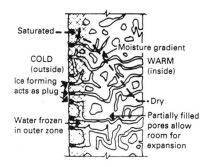

Diagram D3.8/3 *Illustration of moisture gradient and the advantage this may have in leaving room into which water squeezed out by expansion caused by freezing at or near the exposed face may escape*

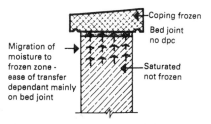

Diagram D3.8/4 *Illustration of the migration of water from unfrozen to frozen parts as may occur in a parapet. The durability of copings is usually increased by inserting a dpc under the coping. A dpc should be considered essential. It also helps to reduce the likelihood of damage to the parapet itself*

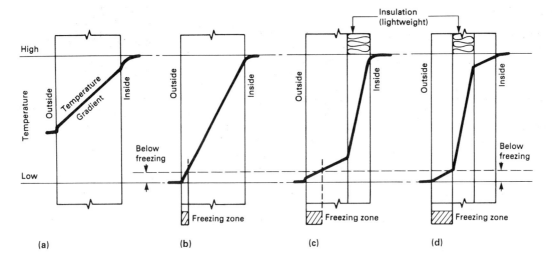

Diagram D3.8/5 *Illustration of temperature gradient and the effects of changes in temperature and the addition of insulation on the slope of the gradient and the size of the freezing zone. Steady-state conditions are illustrated – see text for effects of thermal capacity*

insulation, pp. 80–87). An illustration of the influencing factors is shown in Diagram D3.8/5. The temperature gradients shown in the diagram are what are known as *steady-state* gradients. This state rarely occurs in a construction in practice, but it is the state commonly used as a basis for analysis. The rate with which the different layers respond to changes in temperature and the way these store and then release heat are important when considering frost action. Generally, dense materials respond relatively slowly and store much heat, which they release relatively slowly; the opposite is true of lightweight materials (*Materials for Building*, Vol. 4, 4.05 Thermal capacity, pp.97ff).

Taking all these factors into account, the chief hygrothermal characteristics related to frost action in Diagram D3.8/5 are:

- (a): This illustrates the essential features of a temperature gradient in a homogeneous construction. The external temperature is assumed, for comparative purposes, to be above freezing. The temperature gradient slopes gradually from high to low.
- (b): This differs from (a) in that the external air temperature is assumed to be below freezing.

1. *Zone of freezing*: There is therefore a zone in which the construction itself is below freezing.
2. *The effects of thermal capacity*: Because of the relationship of the slope and the thickness of the construction, the zone is relatively narrow. If the air temperature drops further, the temperature gradient will not change immediately, and therefore nor will the zone of freezing within the construction because of the effects of the thermal capacity of the construction. If the latter is of a dense material, the change will be delayed for some time. The duration of the delay can be important if the external temperature is initially above freezing and then suddenly drops below it. The thermal capacity of a dense construction would tend to delay its zone of freezing for a relatively long time, and generally this would be an advantage. The same would not be true if the construction was lightweight. On the other hand, a dense construction would warm up slowly when the temperature rises, thereby maintaining any zone of freezing for a relatively long time.

3. *Moisture and interstitial condensation*: There will be some migration of moisture from the inside to the outside of a building but interstitial condensation is unlikely if the outer parts of the construction are not saturated – a saturated 'layer' within a construction acts as a vapour control layer. Such condensation as might occur will then be added to the moisture content of the outer 'layer'. If, however, the outer 'layer' of the construction is very damp or saturated there will be a tendency for the moisture to migrate towards the inner, drier, layers on the warmer side of the construction from which some of the moisture might evaporate (**high > low**). If this happens, and there is freezing within the outer exposed 'layer', the freezing water may be extruded into the drier parts, thereby reducing damage. This, among other factors, is said to account for the less frequent damage in wall constructions than might otherwise have been expected. As explained in (c) and (d) below, this 'safety valve' might not exist in well-insulated constructions (**creative pessimism**).

- (c) and (d): These have the same temperature drop as in (b). They differ in that they include a layer of insulation – on the inside in (c) and sandwiched in (d). For the present purposes the basic construction is assumed to be dense as the insulation is, in any case, lightweight.

1. *Effects of the insulation on freezing*: Because of the thermal properties of the insulation, the temperature gradient across it is very steep, leaving the remainder to drop across the remaining dense and less well-insulated layers. The effect of the insulation, wherever it is placed on the warmer side, is to render the outer, denser, layer much colder overall than is the case in (b). As will be seen, the effects of the precise location of the insulation does influence the width of the zone of freezing in the outer layer. As a rule, the closer the insulation is to the outer face, the wider the zone of freezing in the outer layer and the less the contribution of that layer to moderating changes in the external temperature.

2. *Effects of saturation*: The width of the zone of freezing may, of course, become significant and the risk of frost action increased if, in addition, the freezing zone is very damp or saturated. Such conditions may occur, depending on the amount of moisture entering the construction from inside and the amount of this moisture that might condense interstitially. Whether or not this is likely to happen depends, in the first instance, on the effectiveness or otherwise of any vapour control layer incorporated on the warm side. As explained in (b), above, a very damp or saturated outer layer of the construction may also act as a vapour control layer.

3. *Limited drying out*: Before the use of insulation in the amounts now common in wall constructions (as in other elements, of course) it was thought that warmth from the inside of a building would dry out the faces nearer that warmth, thereby providing a dry layer into which freezing water could escape. By its very nature, the insulation reduces such heating/drying considerably. The possibility of water freezing at or near the outer face being squeezed into areas of the construction with lower moisture contents near the warmer face as described in (b) (3) above is therefore also reduced considerably.*

(f) Freezing/thawing cycles

Severe damage may occur if water is trapped between two layers of ice. This kind of condition can be found in heavily saturated walls subjected to severe frosts between which there are partial thaws.

*It is important to note that the increased risk to be expected in modern well-insulated constructions is based on a theoretical consideration of influencing factors. At present there is no evidence to show that increases in the insulation of cavity walls, for example, have caused an increase in the amount of frost shattering of the masonry units or that there is a greater risk of such damage occurring. Pending hard evidence, **creative pessimism** would caution that extra care should be taken when selecting materials for wall constructions, the location of the insulation and the need or otherwise of a vapour control layer.

Brickwork, for example, may, under conditions of severe frost, freeze to a depth of 75–100 mm during one night. During the day following the frost, the sun may thaw the surface and moisture from melted snow trickle over it. There may be a severe frost during the night following the partial thaw. Thus water is trapped between two layers of ice at different depths.

(g) Pore structure

Pore structure plays an extemely important part in determining the resistance to frost action which a material may provide. The arrangement and size of pores influence two of the factors already outlined, namely, moisture content and moisture gradient. These, in turn, are associated with absorption and the saturation coefficient of a material, both of which have been used in the past as possible guides to the frost resistance of materials. As a result of work carried out in both the USA and the UK, a relationship between these has been shown to exist, but because of the wide variation in pore structure between materials of the same kind, this relationship cannot always be

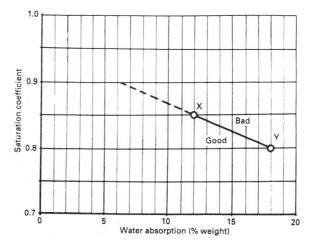

Diagram D3.8/6 *The relationship between frost resistance of bricks and their water absorption and saturation coefficients (see Diagram D3.8/7 for relationship of bricks and certain limestones). (From Butterworth, B. Bricks and Modern Research, Fig. 23, p. 95, Crosby Lockwood, London, 1948)*

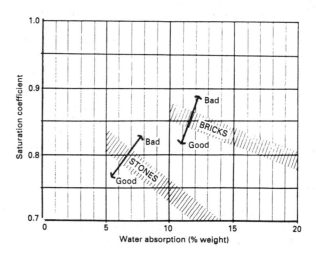

Diagram D3.8/7 *The different relationships between frost resistance and water absorption and saturation coefficients of bricks and certain limestones. (From Butterworth, Fig. 24, p. 96)*

universally applied in practice. However, it is of interest to outline this relationship. (Absorption and saturation coefficient are both covered in *3.3 Moisture content*, p. 231 ff.)

The amount of water absorbed by a material is usually expressed as a percentage of the weight of the material, while the saturation coefficient is the ratio of the volume of water absorbed and the total volume of voids in a material. Values of either do not, however, indicate the zones of a material which actually contain water, nor whether pores are completely or partially filled nor whether there is any room for expansion. However, it has been shown, in laboratory tests, that materials with high saturation coefficients were of poor durability, while those with low coefficients were of high durability. This, in fact, proved to be a generalization, as it was not possible to draw a line which clearly demarcated the good from the bad. Tests carried out on the basis of absorption and crushing strength tended to yield the same result. However, if all three tests were taken together, it was possible to separate the good from the bad. In the absence of reliable artificial frost tests, this method has provided a useful empirical generalization.

Simple charts based on the relationship between absorption and saturation coefficient alone (strength has been found to add little to the accuracy with which durability can be predicted) have been suggested, as shown in Diagrams D3.8/6 and D3.8/7. Diagram D3.8/6 illustrates the 'give and take' that was found possible in assessing the durability of a brick. For example, a brick with an absorption of 12% and a saturation coefficient of 0.85 is likely to be as durable as one with an absorption of 18% and a saturation coefficient of 0.80. Diagram D3.8/7 illustrates the different relations between frost resistance and water absorption and saturation coefficient for bricks and certain limestones. In this the pore structure between the materials is significant, although differences in strength should also be included (p. 556).

Photograph P3.8/1 *Random frost splitting of bricks in a relatively new cavity wall due to the combined effects of inadequate damp-proofing (at sill level) and blockages (mainly mortar droppings) in the cavity. Exposed site in a colder part of the UK – see also Photograph P3.8/2 for a detailed view of a shattered surface*

Practical considerations

Emphasis is given here to the problems related to the durability of materials, as the difficulties of winter building are dealt with separately later (p. 556).

1 Effects on materials

Damage due to frost action is generally confined to the shattering of the surface of materials (Photograph P3.8/1).* Under extremely severe conditions of freezing, as often occur in laboratory freezing tests, the whole or a substantial part of a material may be shattered (Photograph P3.8/2). The degree of shattering and the manner in which it takes place depend largely on the pore structure and strength of the material and the conditions during freezing, as outlined under 'General considerations' earlier. Materials such as stone with a laminated structure tend to fail at the boundaries of the various layers of which the material is composed. Effects such as these can be seen on slate roofs which have been subjected to frost damage. In stone walling, delamination can occur if the beds are laid parallel to the external face.

A rather different effect may result with sandy or chalky soils under concrete ground floors, particularly when the soil has been used as a

Photograph P3.8/2 *Detailed view of a sill from another part of the building in Photograph P3.8/1 showing the nature of splitting in a brick immediately under the (flush) sill without proper damp-proofing*

*The considerable expansion which accompanies freezing may result in the overall expansion of an element of construction. This is generally not seen to be a major cause of cracking but may, in some cases, be an important contributory cause. For this reason, shattering of materials is given as the chief deleterious effect of frost action on materials.

Photograph P3.8/3 *The result of frost heave of chalk used as hardcore under oversite concrete in an unoccupied and unheated house. Freezing of the chalk while waterlogged lifted the oversite concrete and brickwork above. (Building Research Establishment, Crown Copyright)*

fill. Heaving of the soil resulting from expansion may lift a building or part of a building differentially out of the ground, giving rise to cracking of the structure (Photograph P3.8/3). On thawing, the building may settle to its original position in the ground. However, this settlement does not repair the cracks formed during freezing, which means that the structure remains permanently weakened.

2 Likely positions

In general, damage as a result of frost action is mostly confined to those parts of a building which may be saturated with water when exposed to frost (Photograph P3.8/4). Saturation with water immediately before freezing is, of course, an extremely significant requirement if damage is to occur. Thus conditions from frost action in external walls are usually only found when there is exposure to damp conditions with little or no protection. Such conditions are likely to arise in parapets, parts of walls below dpc level, retaining and free-standing walls and especially horizontal surfaces such as copings, cornices, string courses and sills, as shown in Diagram D3.8/8.

Roofs finished in tiles or slates may suffer damage from frost action. Attack is generally more severe in low-pitched tiled roofs because the tiles remain wet for longer periods than when steeper pitches are used. In general, tiles or slates at or near the eaves are more susceptible to damage than those in the rest of the roof slope.

Although the incidence of frosts is an important climatic factor to be taken into account, the likelihood of frosts following immediately on a rainy spell also needs to be considered. In the UK it is unusual for sudden changes from heavy driving rain to severe frosts to occur. This, plus the fact that in relatively poorly insulated constructions warmth from within the building tended to encourage sufficient drying out (see (b)(3) above), explains, in part at any rate, why external walls between dpc level and eaves have remained unaffected generally by frost action. As explained in (c)(3) above, this might not continue to be the case in well-insulated wall constructions. The effects of added insulation apart, the important exceptions have been, and will continue to be, those parts of a wall which may be saturated by rainwater prior to freezing. It is with these parts that, because of the inherent vulnerability, care in the selection and use of materials is important, if not essential.

Photograph P3.8/4 *Frost shattering is more likely to occur in locations of walls that are saturated for relatively long periods of time such as parapets (a) or free-standing/retaining walls (b)*

(a)

(b)

3 Assessment of resistance

(a) Difficulties of measurement

As yet there is no numerical measure of frost resistance as there is for thermal or electrical resistance. Difficulties have been experienced by a number of workers in establishing a correlation between laboratory test results and field experience. To be meaningful, a measure of frost resistance needs to take into account the particular

- Form of construction;
- Location; and
- Climate of the location.

A broad distinction can be made between 'normal' and 'severe' exposure, the latter implying saturated conditions prior to freezing.* Despite these difficulties, an outline of the present position generally and that relative to bricks in particular should be useful. It is appropriate to consider the resistance of mortar in brickwork during its laying in cold weather in particular.

(b) Generally

(1) Laboratory tests

In the absence of reliable results from artificial tests, simple charts based on the relationship between absorption and saturation outlined earlier ('General considerations', 2(b), p. 544) provide some guidance on durability. For many years, attempts have been made to subject samples materials to freezing conditions in a laboratory, but experience showed that the tests did not simulate natural conditions with sufficient accuracy. One reason was the fact that samples of materials were used rather than samples of the constructional context in which they normally occur (i.e. brickwork in the case of bricks). The limitations of samples has been largely overcome by the BCRL panel freezing test – see (3) later. For completeness, a description of BRS Tray test is included in (2) below).

(2) BRS Tray test

In an attempt to overcome some of the limitations of the artificial tests, the BRS developed what has become known as the 'tray test'. In this, samples of materials, some laid in mortar, are immersed in water in shallow trays and exposed to natural climatic conditions. Observations extending over many years have demonstrated that it is possible to differentiate between materials that normally show good frost resistance in copings, sills and other exposed features from those that are susceptible to frost damage in similar conditions. The 'tray test' method also gives information on the broad relationship that exists between frost resistance and the more easily measured properties of water absorption, so that, when necessary, tentative deductions can be made on that basis. Obviously in making such deductions the possible effects of the inclusion of thermal insulation in the construction should also be taken into account.

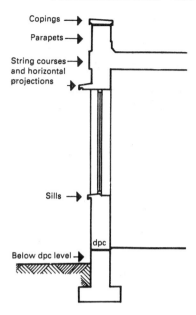

Parts of external walls

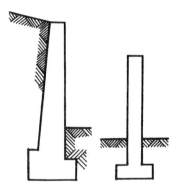

Retaining walls   Free standing walls

Diagram D3.8/8 *Illustration of the positions in which frost action is likely to occur. All these positions may be saturated with water when exposed to frost and are therefore better considered as severely exposed, irrespective of the climatic exposure of the site*

*BCRL has produced a *Freezing index*. This relates rainfall and air frosts during winters in the UK since 1959. This has highlighted those years when exposed brickwork has been exposed to an increased risk of damage. See Beardmore, C. and Ford R. W., *Winter weather records relating to potential frost failure of brickwork*, BRCL, Technical Note 372, 1988.

Table T3.8/1   *Susceptibility of materials to deterioration by frost action*

Class of material	How affected
Natural stone	Variable. Best stones unaffected. Some stones with pronounced cleavage along the bedding planes are unsuitable for copings or cornices
Clay products	Variable. Best bricks and tiles unaffected. Some products, insufficiently fired, or with flaws of structure originating in the machine, may deteriorate, especially bricks in copings, and tiles on flat-pitched roofs
Cast stone, concrete, asbestos cement	Material of good quality is rarely affected

Note: A laminar structure usually makes a material more liable to deterioration. From *Principles of Modern Building*, Vol. 1, 3rd edn, HMSO, London, 1959, Table 9.1, Crown Copyright.

(3) BCRL panel freezing test*
More recently, the results of tests on panels of bricks (or roofing tiles) subjected to a hundred freeze–thaw cycles are encouraging as a rapid test to establish the resistance of bricks (or roofing tiles) to frost damage. Briefly, the test consists of placing a panel of brickwork in an insulated jacket, saturating the panel and then subjecting it to freezing and thawing one hundred times. Clay bricks or roofing tiles in a panel surviving this test are considered to be frost resistant.

(4) Tabulated guidance
As a general guide, Table T3.8/1 summarizes the susceptibility of materials to deterioration by frost action.

(c) Bricks

Bricks varieties run to hundreds. They may be used in many locations with different types of mortar. Laboratory tests have been carried out on a large number of samples and the performance of bricks and brickwork has been monitored by research workers and brickmakers over many years. Yet successive editions of BS 3921† have been less specific as to the means by which frost-resistant bricks should be determined. The classification for frost resistance in the current code as to how the classes should be determined is as follows:

(F)   *Frost resistant.* Bricks durable in all building situations, including those where they are in a saturated condition and subjected to repeated freezing and thawing.
(M)   *Moderately frost resistant.* Bricks durable except when in a saturated condition and subject to repeated freezing and thawing.

*Peake, F. and Ford, R.W., *Brick freeze damage: site and lab results compared*, Building Technical File, 23, October 1988, pp. 21–24.

†The latest being BS 3921: 1985, *British Standard Specification for clay bricks*

(O) *Not frost resistant*. Bricks liable to be damaged by freezing and thawing if not protected as recommended in BS 5628: Part 3 during construction and afterwards (e.g. by an impermeable cladding). Such units may be suitable for internal use.

In the absence of any stated criteria upon which a frost-resistant brick (F) may be determined, as was the case previously*, it is now, by implication, left to individual manufacturers to classify their bricks on the basis of resistance in use over a number of years (usually at least three but preferably ten) and satisfy prospective specifiers accordingly. A rapid assessment of new products rapidly is only precluded until such times as a reliable laboratory method has been developed and accepted by the industry. The British Ceramic Research Laboratory's panel freeze test (see (b)(3) earlier) appears to be an answer to this. So far, the results of the test correlate well with experience in the field over 10 years, the period that BSI recommends as the minimum for 'real life' tests. It is of course noteworthy that BS 402: 1990 now includes the BCRL's test – see footnote on p. 543.

(d) Mortar

An important requirement of a mortar is that it should develop a good bond between it and the bricks (or other masonry units) to reduce the risk of rain penetration through the finished wall. Another is that the mortar must be able to resist frost and should develop its durability fairly quickly, particularly in winter. As the durability of a mortar is dependent on factors other than strength, early strength development is not essential. In some cases it might not be desirable (for example, in reducing cracking – see *2.3 Cracking*). A mortar should be no stronger than is necessary in particular circumstances. The assumption often made that 'the higher the strength, the better the mortar' is not justifiable.

In general, mortars should match the frost resistance of the masonry units. BS 5628: Part 3: 1985 includes five mortar groups (see Table T3.8/2). As will be seen, the higher the group number, the greater the strength of the related mortars but the less able are those mortars in accommodating movements. Within one group all the mixes have the same strength with the exception of group i, in which there is only one mix. For the present purposes it will be seen that frost resistance increases with mixes reading from left to right (for improvement in bond and rain resistance the reverse). When using the table the guiding principle is that the mortar should contain no more cement than is necessary to give adequate strength in the masonry unless there is a good reason for choosing a richer mix. One good reason would be working in winter, mainly because richer mixes develop their strength more quickly, usually sufficiently quickly to resist frost action. (Another reason would be where bricks have appreciable quantities of soluble sulphates and are used in exposed situations (see *3.6 Chemical attack*, 'Sulphate attack', p. 417.)

Table T3.8/2 gives a summary of the suitability of mortars of different groups according to the type of construction and masonry units used.

4 Precautions

Although there are relatively few situations in which there is a risk of frost damage occurring, it is nevertheless important that these are

*The criteria were given in the 1965 and 1974 editions. These are described in *Materials for Building*, Vol. 3, p. 93.

Table T3.8/2 *Durability of masonry in finished construction and the appropriate mortar designations required*

Constructional element	Quality of masonry units and appropriate mortar designation			
	Fired-clay units	Calcium silicate units	Concrete bricks	Concrete blocks
Internal walls and inner leaf of cavity walls	(F), (M) or (O) in (i), (ii), (iii) or (iv)[a]	Classes 2 to 7 in (iii) or (iv)[a]	$\geq 7\,\text{N/mm}^2$ in (iv)[a]	Any in (iii) or (iv)[a]
External walls above dpc – LRS	(F) or (M) in (i), (ii) or (iii)	Classes 2 to 7 in (iii) or (iv)	$\geq 7\,\text{N/mm}^2$ in (iii)	Any in (iii) or (iv)[a]
– HRS	(F) in (i) or (ii)	Classes 2 to 7 in (iii)	$\geq 15\,\text{N/mm}^2$ in (iii)	Any in (iii)
External walls below or near external ground – LRS (with/without freezing)	(F) or (M) in (i), (ii) or (iii)	Classes 3 to 7 in (iii) or (iv)	$\geq 15\,\text{N/mm}^2$ in (iii)	(a) of block density >1500 kg/m; or (b) Made with dense aggregate complying with BS 882 or BS 1047; or (c) Having a compressive strength >7 N/mm²; or (d) Most types of autoclaved block in (iii)
– HRS (*without* freezing)	(F) or (M) in (i) or (ii)	Classes 3 to 7 in (ii) or (iii)	$\geq 15\,\text{N/mm}^2$ in (ii) or (iii)	As in (a) to (d) above *but* in (ii) or (iii)
– HRS (*with* freezing)	(F) in (i) or (ii)	Classes 3 to 7 in (ii)	$\geq 20\,\text{N/mm}^2$ in (ii) or (iii)	As in (a) to (d) above *but* in (i)
Parapet walls – unrendered – LRS[b]	(F) or (M) in (i), (ii) or (iii)	Classes 3 to 7 in (iii)	$\geq 20\,\text{N/mm}^2$ in (iii)	As in (a) to (d) above *but* in (iii)
– HRS[c]	(F) in (i) or (ii)	Classes 3 to 7 in (iii)	$\geq 20\,\text{N/mm}^2$ in (iii)	As in (a) to (d) above *but* in (ii)
– rendered[d]	(F) or (M)[e] in (i) or (ii)	Classes 3 to 7 in (iii)	$\geq 7\,\text{N/mm}^2$ in (iii)	Any in (iii)
Cappings, copings and sills	(F) in (i)	Classes 4 to 7 in (ii)	$\geq 30\,\text{N/mm}^2$ in (ii)	As in (a) to (d) above *but* in (ii)
Chimneys (domestic) – unrendered – LRS	(F) or (M) in (i), (ii) or (iii)	Classes 3 to 7 in (iii)	$\geq 10\,\text{N/mm}^2$ in (iii)	Any in (iii)
– HRS	(F) in (i) or (ii)	Classes 3 to 7 in (iii)	$\geq 15\,\text{N/mm}^2$ in (iii)	As in (a) to (d) above *but* in (ii)
– Rendered	(F) or (M) in (i), (ii) or (iii)	Classes 3 to 7 in (iii)	$\geq 7\,\text{N/mm}^2$ in (iii)	Any in (iii)

Table T3.8/2   *(Continued)*

Constructional element	Quality of masonry units and appropriate mortar designation			
	Fired-clay units	Calcium silicate units	Concrete bricks	Concrete blocks
Free-standing walls				
– with coping	(F) or (M) in (i) or (ii)	Classes 3 to 7 in (iii)	$\geqslant$15 N/mm^2 in (iii)	Any in (iii)
– with capping	(F) (i) or (ii)	Classes 3 to 7 in (iii)	$\geqslant$20 N/mm^2 in (iii)	As for (a) to (d) above *but* in (ii)
Earth-retaining walls				
– with waterproofed retaining face and coping	(F) or (M) in (i) or (ii)	Classes 3 to 7 in (ii) or (iii)	$\geqslant$15 N/mm^2 in (ii)	As for (a) to (d) above *but* in (i) or (ii)
– with coping or capping but no waterproofing on retaining face	(F) in (i)	Classes 4 to 7 in (ii)	$\geqslant$30 N/mm^2 in (i) or (ii)	As for (a) to (d) above *but* in (ii)

Notes:
1. This table relates to the durability of masonry relative to frost action only. Changes in mortar designations and/or the use of sulphate-resisting cement may be advisable in certain situations (e.g. retaining walls, rendered parapets and chimneys – see *3.6 Chemical attack*, 'Sulphate attack'.
2. During construction, a frost hazard exists before mortar has hardened (usually 7 days after laying) or before the wall is protected against the entry of rain at the top.
3. Severe (i.e. saturated prior to freezing) conditions may exist irrespective of the climatic exposure of the building as a whole in locations such as parapets, sills and retaining walls – see text.

LRS = Low risk of saturation.
HRS = High risk of saturation.

[a] Where designation (iv) mortar is used it is *essential* to ensure that all masonry units, mortar and masonry under construction *are protected* from saturation and freezing.
[b] Some low parapets on some single-storey buildings.
[c] Where a capping only is provided for the masonry.
[d] Single-leaf walls should be rendered only on one face. All parapets should be provided with a coping with a dpc under it.
[e] Parapet walls of clay brick should not be rendered on both sides; if this is unavoidable, select mortar as though *not* rendered.

From BS 5628: Part 3: 1985, Table 13 and BRE Digest 160, p. 2.

carefully considered. The precautions that should be taken include the following.

(a) Selection of materials

(1) Exposure
Careful selection of materials is essential in all vulnerable positions, that is, those likely to be subjected to severe exposure by virtue of their location (e.g. in parapets, parts of walls below dpc level, retaining and free-standing walls) or plane of their 'horizontal' surface (e.g. copings, cornices, string cornices, sills and roofing units). **Creative pessimism** would caution that any part of a construction that is likely to become saturated should be considered, for the purposes of frost action, as being severely exposed.

(2) Experience as the guide

In general, experience of the performance of particular materials used over a number of years (at least ten) in particular positions and under specific climatic conditions is likely to prove the best guide. Special care is needed when using this guidance. Among other things, small changes in the nature of the construction (e.g. additional insulation and protective features) may be significant, in the long term in particular. Differences in climate over the country may also be important. A material that has performed satisfactorily in one part may not do so in another that is more heavily exposed to driving rain or is significantly colder in the winter.

(b) Choice of joints in masonry

Flush or bucket handle joints for masonry help to keep the mortar jointing relatively dry and thereby less likely to be damaged by frost action. Recessed joints, on the other hand, tend to encourage the entrapment of water and are therefore more prone to frost attack.

(c) Protection of tops of parapets

The tops of parapets should be protected with both copings and damp-proofing under them. The copings should preferably project beyond, and be so designed as to throw water clear of, the face(s) of the wall.

(d) Direction of bedding planes in stone

The bedding planes of stone should be laid at right angles to the external face of a wall.

(e) Roof pitch for slates and tiles

As a rule, the flatter the pitch of a roof, the better the quality of the slates or tiles should be.

(f) Fill of sandy soils and chalk

The use of sandy soils and chalk as a fill under concrete ground floor slabs is better avoided, particularly in poorly heated buildings, as these soils may be subjected to frost heave.

Winter building precautions

1 Background

(a) Effects of the weather

In the UK, bad weather conditions are not unusual at any time of the year. Generally, they may delay or otherwise affect the work on building sites. Low temperatures during the winter may, at best, delay the hardening of all *in-situ* cement-based products such as mortars,

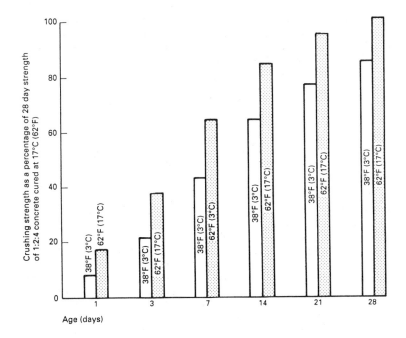

Chart C3.8/1 *Comparison of the effect of temperature upon the rate of hardening of a typical concrete made with ordinary Portland cement*

renders and concrete (see Chart C3.8/1) or, at worst, actually cause disintegration of those materials and of masonry units whether used externally or internally. The considerable amounts of water used in all the wet-trade processes (see *3.2 Exposure*, 'During construction', p. 200) make the relevant materials or forms of constructions particularly vulnerable to damage by frost action before they have developed sufficient strength. In addition, masonry and other units that do not require to be frost resistant in the finished building will be exposed to conditions that could cause them to be damaged should freezing occur.

(b) Basic requirements and need for preplanning

Put simply, the precautions that should be taken consist of providing various forms of protection, notably those that prevent the saturation of materials and those that keep them warm. An understanding of the vagaries of the weather in the UK is also important. Cold weather can and often does occur suddenly, sometimes after a period of disarming warm weather. Frosts at night are also not uncommon, even when winter has yet to begin or appears to have ended. These and other factors point to the need for a certain amount of preparation or preplanning. The requisite facilities for protection should be available on-site whenever low temperatures (i.e. from 3°C and below) can be expected.

(c) The risk of continuing to work

As a rule, brick- or blocklaying, rendering, screeding, concreting, floor laying, roof laying and even painting should be discontinued when the temperature of the work falls below 3°C or frost is imminent *unless* the operation is carried out in a permanently heated enclosure. Important-ly, the materials for the work should be stored and the work done within that enclosure. In general, the theoretical advantage of heating the materials (including water) for short-term working under cold conditions is seldom realized. Reliance on heating of materials is better avoided, even for short-term working.

2 *In-situ* cement-based products

(a) Generally

The precautions that should be taken with the various *in-situ* cement-based products such as mortars, plasters, renders, screeds and concretes differ in detail, but they do follow well-defined principles which may be summarized as follows:

- Avoid the use of accelerators, sometimes loosely referred to as 'anti-freeze' agents;
- Protect the finished work from rainwater with a waterproof material such as polythene (i.e. a 'raincoat' that does not absorb water) and in exposed conditions with side screens in addition;
- Insulate the work, the form of insulation depending on the conditions but, in any event, also waterproof for protection from rainwater. *Note*: Protection and insulation should be firmly secured against dislodgement by the wind.
- Encourage drying 'naturally' when conditions permit (e.g. by supporting covers clear of the work).

(b) Mortar

Mortars should only be as rich in cement as they need to be, depending on circumstances (see Table T3.8/2 and the related text). Resist the use of admixtures intended to reduce the freezing point of water.*

If the masonry units are unavoidably wet when laid, a plasticized cement:sand mortar is preferred. Alternatively, a masonry cement:sand mortar may be used, or the resisance to damage by freezing of cement:lime:sand mortars may be improved by adding a plasticizer and ensuring that the air content of the mortar is between 8% and 12% when used.†

(c) Materials generally

The guiding principle is that all materials stored on-site, whether aggregates or units, should not be allowed to become saturated by rainwater or from the ground and should not be permitted to become frozen. Saturation can be prevented by covering with a waterproof material (for rainwater) or stored on a raised platform (for groundwater), freezing by suitable insulation. Ideally, materials should be stored in an enclosure having a raised floor with facilities for moderate heating.

(d) In-situ concrete

In addition to the precautions generally described above, cements that set more rapidly (e.g. rapid-hardening) should be used in preference to calcium chloride. If the latter has to be added its amount must be carefully controlled. In any event, calcium chloride must not be added to extra-rapid-hardening Portland cement nor to high-alumina cement.

The acceleration of the setting and hardening of Portland cement is accompanied by an increase in the heat of hydration. Thus the concrete

*Calcium chloride in particular should be avoided. It and additives of this salt do not prevent frost damage to mortars; their use introduces the risk of corrosion of embedded metals (e.g. wall ties) and of permanent damage in the finished work.

†An air-entrained mortar that has set contains air spaces in it into which ice can expand when freezing occurs. However, such mortars will have reduced strengths and bond, both being influenced by the amount of air actually entrained.

attains its frost-resistant strength sooner and is warmer, particularly in its early stages.

3 Finishes

For convenience, the precautions that should be taken with the finishes are grouped together here. Heating and ventilation, a precaution that should be taken with each, is discussed first. This is followed by notes on the individual finishes.

(a) Heating and ventilation

Most (if not all) the finishing trades such as plastering, tiling (wall and floor), floorings and painting are executed when a building has been enclosed and/or fully protected from the weather. The heating may also be operating. It is always an advantage to provide heating and ventilation, both being important, to allow the finish to be applied safely and satisfactorily and to increase the rate of drying out. As to the latter it is essential that the level of heating and/or ventilation is sufficient to avoid localized drying out. This is particularly important with plastering.

As the humidity is likely to be considerable due to basically high levels in the winter that are supplemented by moisture during drying out, ventilation is important to reduce the risk of condensation occurring. Gas fuels for heating are better avoided, as these add considerable amounts of water, so increasing the risk of condensation.

(b) Plastering

(1) The problems
Plastering at low temperatures presents two problems. First, the rate of drying out is reduced considerably (due mainly to the inherently high humidities in winter that are increased by drying out) thereby delaying the following processes; and, second, there is the risk of the plaster freezing, in which the coldness of the background surfaces is significant.

(2) Choice of material
The plaster should be suitable for low-temperature working and special plasters are available that obviate the need to wet high-suction backgrounds. Gypsum-based plasters have rapid hardening properties and are therefore more suitable for low-temperature working than those containing cement and lime.

When cement and lime plasters must be used, ready-mixed mortars should be considered. These are made up under conditions free from frost. An aerated cement/sand undercoat will help to isolate the finishing plaster coat from any moisture in the wall. In some cases there may be no alternative but to use a dry lining technique.

(3) Enclosing the building
The part of the building to be plastered should be completely enclosed. If the glazing has not been completed, then some form of translucent sheet should be used over the window openings, ensuring that the sheet is properly sealed around the perimeter of the opening. All external door openings should be suitably covered. It helps if the number of entrances into the building are restricted. Once these precautions have been taken, consideration should be given to heating and ventilation as described in (a) above.

(c) Floor finishes

The appropriate dryness of its base must be determined before any floor finish is laid (see *3.1 General considerations*, 'Dampness', p. 152, for the use of appropriate measuring devices). The precautions to be taken with screeds should follow those previously described for cement-based products (p. 558).

(d) Painting

During the winter, painting may be seriously restricted or disrupted due to the presence of moisture on or under surfaces to be painted and to the delayed drying out of the paint itself caused by low temperatures. Nowadays with only a limited range of porous paints, the answer to early decoration lies either in selecting quick-drying paints or in the provision of appropriate heating and ventilation, or perhaps both. Whichever method is used, it is, of course, essential that the background to which the paint is to be applied is sufficiently dry. The use of appropriate measuring devices is advised (see *3.1 General considerations*, 'Dampness', p. 152 ff).

4 Flat roof work

An elaborate form of protective covering is required if flat roof decks are not to become wet before the waterproofing membrane is laid. Excessive moisture in all types of roof decks should be avoided for reasons other than frost action. As to the latter, decks of concrete with cement-based screeds contain, even without rainfall, considerable quantities of water (see *3.2 Exposure*, 'Entrapped water', p. 214 ff). Consequently, unless suitably insulated they are especially vulnerable to frost action. In addition, they do take a long time to dry out, even if protected from rainfall.

Sources

BRE
Digests
160, *Mortars for bricklaying*, December 1973
164 and 165 *Clay brickwork: 1 and 2*, minor revisions 1980 and May 1974
269, *The selection of natural building stone*, January 1983

National Building Studies
Some common defects in brickwork, National Building Studies, Bulletin No. 9, HMSO, London, 1950

Reports
Schaffer, R., *The weathering of natural building stones*, Building Research Special Report No. 18, HMSO, London, 1932 (facsimile reprint 1972 with a new Appendix II on the cleaning of external building surfaces and colourless treatments for masonry)
Principles of modern building, Volume 1, 3rd edn, HMSO, London, 1959
Lacy, R.E., *Climate and building in Britain*, BRE Report, HMSO, London, 1977
Winter building, interim review by Committee on Winter building, HMSO, London, 1963

Brick Development Association
Harding, J.R. and Smith, R.A., *Brickwork durability*, BDA Design Note 7, August 1983

Harding, J.R. and Smith, R.A., *Brick laying in winter conditions*, BDA Building Note 3, BDA, January 1986
Concreting and bricklaying in cold weather, National Building Studies, Bulletin No. 3, HMSO, Cement and Concrete Association
Monks, William and Ward, *External rendering, Appearance matters 2*, Cement and Concrete Association, 1982

BSI
BS 5262: 1976, *Code of practice for external rendered finishes*
BS 5534: 1978 (1985) *Code of practice for slating and tiling, Part 1: Design*
BS 5628: 1985, *British Standard Code of practice for use of masonry, Part 3. Materials and components, design and workmanship*

British Ceramic Research Ltd
Peak, F. and Ford, R. W. *Brick freeze damage: site and lab results compared*, Building Technical File, 23, October 1988, pp 21–24

Flow and changes in appearance

Introduction

1 Definitions

The change in appearance, deterioration and decay of building materials due to the effects of the surrounding environment is generally described as *weathering*. Of the factors involved in the weathering of materials, the flow of water over facades is of first significance. Other factors such as frost action are covered elsewhere (see *3.8 Frost action*). This chapter is concerned specifically with the changes of appearance that occur as a result of the flow of water over different materials.

When the flow of water over materials in uncontrolled, staining often results. Disfigurement arises from irregular redistribution of dirt on the surface, erosion, uneven drying out and therefore the creation of conditions suitable for the growth of certain lichens and moulds. It is the uneven and uncontrolled nature of the distribution of the agents and the effects of change – water, dirt and biological growths – that makes the distinction between an attractive 'patina of age' and ugly disfigurement. Weathering in some form is unavoidable: to make it add to the quality of the building, designers must be able to control it.

Exactly how water travels over the facade of a building is highly complex and dependent on a large number of variables. Due to the lack of research on this subject, this chapter can only consider these variables in qualitative terms. It is hoped, nevertheless, that even a broad discussion of the relevant factors will be useful both to designers and to those trying to understand the causes of the changing appearance of materials in building.

2 Principles relevant

Since the external environment cannot be controlled, weathering in some form is a fact of life. The process of weathering involves several of the basic principles described at the beginning of this book. Flow of water, of course, occurs because of the effects of gravity, the most elementary of the **high>low** principles. In the effects on materials of the flow of water, *reversion* is predominant (which, it will be recalled, is the fact that materials always try to return to their natural, disordered state). In terms of the practical control of staining, the objectives of **continuity** and **balance** apply, since, as will be seen, it is discontinuity in materials and their geometry and imbalances in exposure to environmental conditions that cause disfiguring flow patterns.

In order to design so as to avoid staining it is necessary to understand the mechanisms at work.

Basic mechanisms of flow and staining

1 Flow of water on surfaces

The term 'flow' refers to a number of ways in which water travels on the surface of materials. The first is the downward flow of water under the effect of gravity: this may be thought of as primary flow. The second – secondary flow – is the splashing back of water against a surface, either from rainfall striking the ground or a horizontal surface next to it or where water in primary flow strikes a horizontal surface such as a string course. Two other ways in which water reaches external materials should be mentioned. Strictly, they are not flow in the dynamic sense but they can play a part in the staining. The first is the deposition of water on a surface by condensation: externally, this occurs on a daily basis as dew, to varying extents according to the season, orientation, exposure and external conditions. A less common kind of static deposition takes place where there are hygroscopic or deliquescent salts on the surface. Sodium chloride is one such deliquescent salt, so this effect occurs close to the sea, where the salt in the air is deposited on facades. This, in turn, creates damp conditions on surfaces which may, as a result, be more prone to biological growths.

The basic factors affecting the rate of flow of water over a surface (w_r) are the quantities of incident water (w_i), the height of the surface (h) and its absorptivity (Diagram D3.9/1). The rate at which rain hits the surface is governed by its exposure. The absorptivity of the material regulates the proportion of water absorbed by the material (w_a) and the proportion which flows over the surface (w_r). Both the absolute absorptivity and the rate of absorption are relevant, and have to be related to the duration of the incident rain.

Consider a wall made of an absorbent material – London stock brickwork, for example. In light rainfall one can say from observation that the rate of absorption will initially be the same as the rate of incidence, and the overall absorptivity is such that the water is completely absorbed for several hours. If, however, the light rain persists for a long period a point will be reached where the surface is saturated and the rate of absorption is less than the rate of incident rain. The excess water will then begin to flow down the facade. In heavy rain this point is reached sooner. This relationship is shown in Diagram D3.9/2.

In fact, the quantity and rate of incident rain is not constant on all surfaces of the building. It clearly varies with orientation, but also in a more complex way with wind. This is discussed at length in *3.2 Exposure*, p. 181. Variations in wind speed affect the *amount* of water thrown against a facade whereas the wind vortices and modelling of the facade influence the direction and rate of flow of water. As a general rule, the tops and corners of walls are subject to a greater incidence of water than other parts of the facade, since it is along these lines that air flows must concentrate as they pass over and around the building.

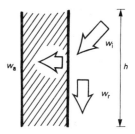

Diagram D3.9/1 *Factors affecting the rate of flow of water over a surface*
w_i: *water incident on the surface*
w_a: *water absorbed by the surface*
w_r: *water flowing over the surface*
$w_r = w_i - w_a$

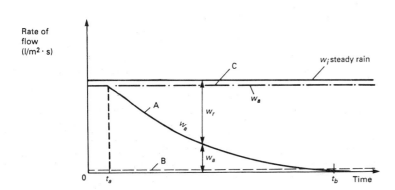

Rate of flow ($l/m^2 \cdot s$)

Diagram D3.9/2 *Simplified diagram showing the relationship between incident rain, run-off and absorption. Steady rainfall is assumed. Curve A (moderately absorbent material such as concrete). From t_o to t_a: $w_a = w_i$. All rain is absorbed and there is no run-off. From t_a to t_b: $w_a < w_i$. Some rain is absorbed but an increasing quantity runs off. From t_b onwards: $w_a = 0$. All rain runs off. Curve B (non-absorbent material such as glass): $w_a = 0$ so that from $t = 0$ all incident water runs off. Curve C (absorbent material such as porous brickwork). Here, during the measured time, $w_a = w_i$, so that there is no run-off*

Photograph P3.9/1 *A typical example of partial washing on a wall of moderate exposure. On the left-hand side the tide marks show that there is a greater flow of water from the sill-less windows than from the more absorbent brickwork above. On the other hand, the large window above the sign has a projecting sill which both shelters the wall below it (shown by the darker grey compared with the rest of the stonework) and concentrates the run-off at each end*

Photograph P3.9/2 *This shows the opposite effect to Photograph P3.9/1. Here the run-off from the brickwork over the stone string course is greater than from the windows, which in this case have projecting sills. This elevation faces north and is in a dirty environment. On the west-facing side it can be seen that the exposure to the prevailing weather has kept the stone completely free of streaks*

*In the excellent C&CA publication by Frank Hawes, *The weathering of concrete buildings* Wexham Springs, 1986.

It is because the different parts of the facade receive varying amounts of incident water that the phenomenon of partial flow occurs. This is very important to the problem of staining. What appears to happen is that the most exposed parts become saturated first, so that flow starts from them onto drier areas below. These absorb a proportion of the water until they too have a rate of incidence which exceeds the rate of absorption. In a given spell of rain there is a physical point on the surface where the run-off stops due to absorption. The position of this point depends on the factors described above. What is crucial from the point of view of staining is that it is near this point that any dirt picked up by the run-off water is redeposited. Whether this area of redeposition is indeed a hard line or a soft zone seems to depend on the absorption characteristics of the material (Photographs P3.9/1 and 2).

Rainfall, of course, occurs for variable periods, so that there must exist a critical relationship between the rate of absorption of a particular material, typical periods of incident rain, and its tendency to reach surface saturation and thus lead to run-off. *3.2 Exposure* p. 187, discusses the distinction between the annual and spell indices of driving rain. The important point to note here is that the annual index is the measure that affects the long-term changes in appearance, whereas the spell index indicates the susceptibility of a wall to saturation (and therefore, in the case of masonry, to rain penetration). This is because the annual index is an indication of the relative *washing* power of rain from various points of the compass.

Hawes* points out that, even on smooth materials, a complete film of streaming water on a surface requires a considerable quantity of incident rain. What usually happens is that the flow separates into separate streams, determined by fine variations in the absorption and geometry of the surface. In a similar way to streams and rivers, these patterns become self-reinforcing, particularly on soluble materials. The speed of flow – on porous materials, at least – is surprisingly slow: on concrete it rarely exceeds 0.5 m/min,

From the point of view of staining it is the random pattern of broken streams with a slow rate of flow that cause the worst disfigurement. A flowing continuous film will cause much more even washing.

2 Dirt and washing

Water flow alone might cause a change in appearance as a result of erosion (weathering in the geological sense), but in buildings whose life is always short compared with the geological time scale this factor plays only a small part. Materials which are soft enough to be weathered by running water would, in any case, have serious shortcomings in terms of basic durability.

For staining to occur there must also be a further medium on the surface . This is referred to here as *dirt*, a heading which includes fine atmospheric pollution – airborne dust, grit and soot (see *3.2 Exposure*, 'Pollution', p. 168) – loose material on the ground carried against the base of the building by splashback, or dirt of a biological source ranging from bird droppings to plant growths and their waste products.

Without the effects of rainfall it may be assumed (and this is an assumption based on the experience of what happens internally) that such dirt would be distributed fairly evenly over vertical surfaces of the facade, with higher concentrations on surfaces closer to horizontal such as window sills, copings and the tops of string courses and mouldings. Even though its quantity may be disfiguring (as is the case with buildings completely blackened by soot), it is unlikely that the

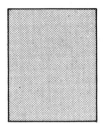

(a) **Even deposit of dirt**

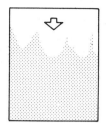

(b) **Water from above picks up some dirt...**

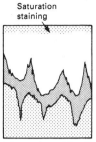

Saturation staining

(c) **...redepositing it at the limit of flow**

Diagram D3.9/3 *The basic mechanisms of staining*

distribution would ever naturally be so uneven as to constitute staining.

However, with the effects of water on the facade a different picture emerges. Dirt is picked up by water movement and redeposited elsewhere, so that its distribution becomes uneven. The original picture is upset: the areas which are most exposed to run-off and flow of water are washed cleaner than they would otherwise have been, whereas those where the dirt is redeposited have a greater accumulation. This basic mechanism is illustrated in Diagram D3.9/3 and Photograph P3.9/3.

An analogy was made with geological weathering earlier, and this can be used again. Factors which affect this washing and redeposition process are the rate of flow of water over the surface (faster flow being more effective at removing dirt of all sizes), the size of the particles of dirt, and their adhesion to the surface. These factors operate in a similar way to streams and rivers which carry or deposit solid particles of various sizes according to their rate of flow. From observation it does seem that there is a relationship between the size of dirt particles, their adhesion to the surface and the rate of flow needed to dislodge them. Smaller particles of dirt adhere more tenaciously than larger ones, and for larger particles to become lodged the surface needs to be rougher. It is evident from inspection of windows or metal claddings that large quantities of fine dirt will adhere to very smooth surfaces, and that therefore the idea of 'self-washing' is a myth. Once adhered, the small particles need a rapid rate of flow to dislodge them.

3 Drying out

As a general rule, the rate of absorption of the surface of a material may be assumed to be the inverse of its rate of drying out. In addition, parallel to the effect of orientation on exposure to driving rain, the orientation and colour of the surface affect the amount of solar radiation and therefore the speed of drying out. This is significant, because surfaces which remain damp for long periods are more likely to be colonized by lichens and other plant growths. Different species have various predilections for sunshine and therefore orientation – many prefer to be away from the sun. This, coupled with a slower rate of drying out, makes north-facing surfaces particularly prone to biological disfigurement. The fact that plants like constant dampness is vividly illustrated by the green patches on walls where leaking rainwater pipes have not been repaired.

Photograph P3.9/3 *The rough texture of these light grey aggregate-faced precast concrete cladding panels does little to disguise the effects of partial washing. The wall is washed more thoroughly at the top where it is more exposed and the dirt washed off is redeposited lower down. The effect of differences in exposure to rain is seen by the varying lengths of the tide marks below the projecting window sills*

Photograph P3.9/4 *This stooled Portland stone sill concentrates run-off at each end, showing that the general grey tone of the stone is due to an even deposit of dirt. Some of the dirt washed off is deposited to each side of the path of flow, but most is taken further down the wall. The light colour of the stone highlights the effect*

4 Differential flow

The flow of water on a facade also depends on the arrangement of materials with different absorption characteristics, together with its geometrical modelling. These geometrical factors are analysed at some length below. The key point to note is that while the variations in washing and deposition of dirt are the primary causes of staining, most of the former arise from differential flow (Photograph P3.9/4).

Factors affecting flow and staining

Having considered in outline the mechanisms at work, the individual factors affecting flow and staining are now discussed.

1 The surface

(a) Properties of the material

(1) Absorption and absorptivity

Absorbency is the ability of materials to take in liquids – in this case, water. Absorption is the quantity of liquid absorbed, measured as a percentage of the maximum possible, which constitutes saturation. Absorptivity, on the other hand, is the power to absorb, and therefore is a measure of the surface's rate of absorption. Absorption is the product of the absorptivity and the duration of exposure and is limited by saturation. These relationships are illustrated in Diagram D3.9/4.

Absorbency is the *ability* of a material to absorb liquid
Absorption is the percentage quantity of liquid
that could be absorbed

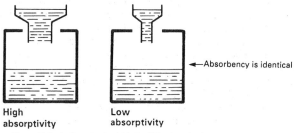

Absorpitivity is the *rate* at which liquid is absorbed

Diagram D3.9/4 *Graphical definition of absorbency and absorptivity*

Both absorption and absorptivity are governed by the porosity of the material, in terms of strict porosity (the percentage of volume taken up by pores) and the pore size and pore structure. These factors are of fundamental importance in determing the susceptibility of a material to surface flow and thus to staining, since the greater the absoptivity of the surface, the fewer are the conditions in which flow can occur.*

(2) Dirt retention
As noted above, the range of substances included under the general heading 'dirt' is fairly wide. The main properties of the dirt particles which affect their retention on a surface are their size and adhesion. Similarly, the relevant properties of the surface are therefore its texture, which should be considered at both the macroscopic scale (see below) and the microscopic scale – in other words, the pore size – and its specific adhesion to typical dirt particles. It is probable that porous materials with an open pore structure will retain more dirt than non-porous ones.

As was discussed in *2.2 Strength and the use of materials*, adhesion takes two forms, mechanical and specific. The dirt retention caused by the microscopic texture of the surface's porosity is, in effect, mechanical adhesion, an interlocking of the dirt particle with the 'key' of the surface. Specific adhesion will take place between two surfaces, one of which at least can 'wet' the other. 'Wettability' is governed by the free surface energy of the material, so that dirt and surfaces with high surface energies will tend to cling together more than those with low surface energies. Soot is actually sticky, and therefore adheres to most building materials. Although no longer the problem it was 30 years ago as a source of dirt, it is still being cleaned off facades by abrasion and high-pressure water techniques where the flow of rainwater alone has failed to shift it. The adhesion of dirt particles to surfaces increases with their physical wetness.

(3) Texture
From the point of view of staining, a strongly textured surface has both a benefit and a disadvantage. The benefit is that texture tends to break up and disperse the flow of water over the surface, thereby reducing the contrast between washed and unwashed areas. It also disguises staining, in as much as it does occur. The disadvantage is that it increases the dirt retention of the surface so that, overall, the surface is dirtier.

Some architects have used heavily textured exposed aggregate as a way of modelling precast concrete to control staining. In terms of staining, the results have often been successful, but at the cost of a very coarse finish, which may not always be considered desirable.

(4) Colour
Two aspects of colour are significant. First there is the constrast between the natural colour of the material and the areas of concentrated dirt. As a general rule, light-coloured materials show up staining more prominently than mid- or dark-coloured ones. The best example of this is the near-white Portland stone used in many of London's public buildings which, due to an unusually marked difference in dirt retention, have a strong contrast between washed and unwashed areas (Photograph 3.9/6). Where the unwashed areas coincide with parts in shadow, such as the underside of window sills, the result is often architecturally satisfactory. However, most materials do not show differential dirt concentrations to a pleasing effect, concrete being one of the most common examples.

Photograph P3.9/5 *These Portland stone columns illustrate the effect of differential washing caused by orientation*

Photograph P3.9/6 *The tops of the columns shown in Photograph P3.9/5 show the effects of concentrated differential flow by a marked contrast in colour between washed and unwashed areas. This is characteristic of Portland stone in urban areas. The lack of throatings on the underside of the abacus and architrave allows water to track back some way before dripping off*

*In theory, it should be possible to measure porosity and absorptivity of different materials and relate these to driving rain data to give a measure of the susceptibility of different surfaces to run-off. As far as can be ascertained, the research has never been carried out, so that the relationship between porosity, absorptivity, flow and staining remains empirical.

1. *Clay brickwork*

General	Has good weathering characteristics
Absorption	Relatively high: run-off only occurs under conditions of heavy rainfall
Dirt retention	Generally moderate to high, depending on type of brick. Engineering bricks much smoother than sand-faced bricks
Texture	Tends to be such as to resist concentrated dirt retention
Colour	Most brick colours disguise dirt retention. Mid-range colours in browns and red-purples disguise dirt most effectively
Saturation staining	Effects of differential weathering often stand out. Erosion of mortar on parpets and sills causes different tone. Absorbent bricks and mortar sustain biological growths

2. *Sandlime brickwork*

General	Has poor weathering characteristics. Tends to reveal strong contrast between washed and unwashed areas
Absorption	Is such that there are often conditions when run-off occurs but is absorbed at some point down the wall
Texture	Fine to moderate texture allows ready washing of large particles of dirt while enabling smaller particles to be held
Colour	Both pale and dark colours (for which these bricks have been favoured) tend to highlight effects of differential washing
Saturation staining	Prone to marked contrast between exposed and sheltered areas. Erosion of mortar joints as for clay brickwork. Sandlime bricks sustain biological growths

3. *Concrete*

General	Has very poor weathering characteristics. Usually shows a marked contrast between washed and unwashed areas
Absorption	Absorption is low so that in most conditions of rain there is run-off. In many conditions the run-off is insufficient to wash the whole of the wall surface
Texture	Varies with design: smooth textures 'from the forms' show contrast between washed and unwashed areas most markedly. Exposed agregate disguises dirt retention and breaks up flow
Colour	Light colour is normal, which highlights dirt retention
Saturation staining	Tops of parapets stain dark with retained dirt and biological growths

Chart C3.9/1 *The weathering characteristics of materials*

4. *Render*

General   Weathering varies from good to poor depending on finish and detail. Painted smooth render (or 'stucco') weathers well *as long as* it is frequently repainted

Texture   With unpainted render, the rougher the finish, the better the weathering

Colour   Light colours highlight differential washing to ill effect

5. *Stone*

General   Stone generally weathers well, not least because it is seen most on older buildings which with all their historic and cultural associations set the standard of what is acceptable. Further, it is used for buildings where the detail is more forgiving of differential washing than is the case in modern architecture

6. *Glass and vitreous enamel*

General   Weathering is good, both for clear glass and vitreous enamel. This is helped by regular window washing usually being part of routine maintenance

Absorption   None, so that there is a high level of run-off which removes the larger particles of dirt

Dirt retention   Very fine dust and sooty particles are retained and can cause staining if details create differential washing (Photograph P3.9/13)

Colour   Light colours used for vitreous enamel highlight differential dirt retention

Saturation staining   None

7. *Plastics*

General   Weathering generally poor because appearance only gets worse with age. The same notes as for glass apply except that solar radiation causes plastics to degrade, causing surface crazing which leads to increased dirt retention. Colour dyes begin to fade, causing dulling of the finish

8. *Metal sheets*

General   Profiled metals sheets and metal cladding panels tend to weather well, with low dirt retention and good cleaning by run-off. Deterioration of finish is an effect of solar radiation and corrosion rather than water run-off. Run-off from corroded areas can cause staining on areas below. Weathered mill-finish aluminium has moderate dirt retention due to increased roughness

Photograph P3.9/7 *This parapet to a wall of concrete blockwork illustrates saturation staining. The staining to the top course is emphasized both by the fact that there is no break in the walling material and that the exposed sloping face is visible from the ground. The darker colour is due to a mixture of dirt and biological growth*

Photograph P3.9/8 *An example of saturation staining on a brick parapet facing south-west. The colour change is more gradual than in the blockwork example in Photograph P3.9/7, and this change is most pronounced in the mortar joints. On a similar parapet facing north one would expect to see more biological growth*

The second aspect of colour is the effect of absorbed water on it. Invariably, surface saturation leads to a darkening of the material, even with those which are dark when dry. The reason for this derives from the way that light is reflected. When saturated, some materials absorb a greater proportion of the light that falls on them than when they are dry. Different materials have a different colour response to saturation. Materials which reach surface saturation quickly (such as concrete) show a rapid change of colour when wet, accentuated further if they start with a light colour. This explains the different rate of colour change between, say, limestone and concrete. It goes without saying that this is a property of porous materials only.

(5) Solubility

This is a factor which is limited to only a few materials, but in these cases the effect is often quite marked (see *3.2 Exposure*, p. 211 and *3.6 Chemical attack* p. 385). A problem occurs where water running over one material dissolves small quantities which are then deposited as stains on a material elsewhere. This can be seen where run-off from concrete leads to white streaking on brickwork below (the so-called 'spilt-milk', 'soot and whitewash' effect), or where the run-off from copper roofs or even brass nameplates has created patterns of green staining. Corrosion products of other metals, particularly iron, are considerably more unsightly.

Having set out the general parameters of materials which affect flow and staining, properties of particular materials are compared and contrasted in Chart C3.9/1. Note that most of this information is qualitative simply because the relevant quantitative data do not exist.

(b) Geometry of the surface

(1) Size and proportion

Observation of streaking patterns shows that it is difficult to generalize about the effects of the actual size of facades. Given two facades of identical porous materials and exposure conditions, one being taller than the other, one would expect the taller of the two to show more effects of partial washing, and indeed it is normal to see this pattern on tall buildings. Nevertheless, because of factors which vary in each case, streaking on narrow fascias is not uncommon.

(2) Orientation

Orientation of the surface affects the incidence of driving rain (see typical rain rose diagrams in Digram D3.2/18) and thus the frequency and rate of washing on the different faces of the building. This contrast between two adjoining faces is often very marked, particularly with materials such as concrete or Portland stone, which highlight different quantities of dirt retention (Photograph P3.9/5). Orientation, as already noted, also contributes to a surface's susceptibility to biological growths.

(3) Angle of inclination

The steeper the surface, the greater the rate of run-off, and the shallower the slope, the greater the exposure to rainfall of all inclinations and orientations. The tops of walls of absorbent materials such as masonry, concrete and brickwork differ from the general wall surface in two respects. On the one hand, the greater exposure makes them better washed, the dirt which is washed off being deposited further down the wall, and, on the other, they are particularly vulnerable to saturation and thus biological staining (Photographs P3.9/7 and 8).

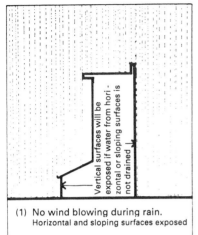
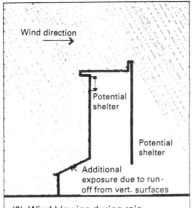

(1) No wind blowing during rain.
Horizontal and sloping surfaces exposed

(2) Wind blowing during rain.
Horizontal sloping and some vertical
(facing wind) surfaces exposed

Diagram D3.9/5 *Representation of the basic effect of wind during rain relative to the exposure of horizontal sloping and vertical surfaces using the section through a hypothetical building*

Photograph P3.9/9 *Run-off from the non-absorbent metal cladding is concentrated on the abutting dwarf wall*

(c) Architectural features and their effect on differential flow

(1) Protection

The amount of water striking the facade (and therefore the amount that can potentially lead to staining) is usually reduced if the top of the wall is protected by a large overhang (see Diagram D3.9/5). Note, however, that on large buildings where there are upwards air currents as the wind flows over and around it, overhangs can be a point of concentration for water.* Differential deposition of dirt and washing occurs wherever one part of the wall is protected more than another, either by overhangs or indeed by the shape of the plan form (Photographs P3.9/9, 10 and 11).

(2) Arrisses

It can be observed that arrisses generally receive the greatest concentrations of washing water and tend to be cleanest. The leading corners of a building are subjected to highest wind speeds and scouring effects of eddies and vortices as the wind is cleaved by the corner.

The surface tension of water also contributes to its behaviour on arrisses. On materials with an angle on contact less than 90°† water will spread as a film rather than form globules. In fact, water 'wets' most building materials, which is why throatings and drips are needed on soffits.

An overriding reason for water running down arrisses is that, in the absence of other forces, it will always take the shortest route to earth. When it is being driven by wind across a facade towards an external corner the arris represents a change of conditions, with a drop in the driving force on the leeward side. It is at that point, therefore, just beyond the effect of the wind, that the water flows vertically down the facade (see Diagram D3.9/6).

(3) Surface modelling

Modelling of the surface may concentrate or disperse the flow of water. Examples are given in Diagram D3.9/7.

Photograph P3.9/10 *Sandlime bricks always seem to highlight the effects of partial and differential washing. In this wall of black bricks the projecting window sills provide a marked rain shadow where dirt is concentrated*

*See BRE CP 81/74, *Observation of the behaviour of weather protective features on external walls*

†When a droplet of liquid rests on the surface of a solid the *angle of contact* is the angle measured between the droplet surface and the solid surface. This angle depends on the balance of the forces of attraction between the molecules within the liquid and between the molecules within the liquid and between the molecules of the liquid and the solid. The angle is constant for any two materials.

Diagram D3.9/6 *Washing of arrisses due to wind effects*

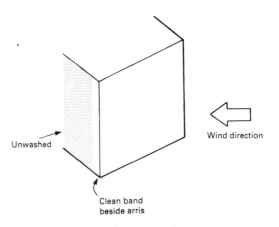

Unwashed

Wind direction

Clean band
beside arris

Photograph P3.9/11 *This example shows both a rain shadow below the continuous projecting sill and a short zone immediately below which appears to be more thoroughly washed, perhaps by water which drips off the sill at the throating being driven back against the wall. The panel material is a green slate*

Diagram D3.9/7 *Some examples of the effects of surface modelling*

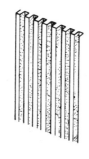

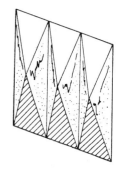

(a) Vertical ribbing on concrete can control flow. Attention must be paid to what happens at the top and bottom of the ribs

(b) Rough-textured surfaces disperse flow

(c) Variations in plane can lead to washed and unwashed areas accentuating light and shade. This is difficult to control

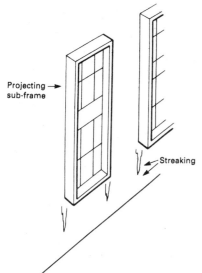

Projecting →
sub-frame

Streaking

Diagram D3.9/8 *Vertical features concentrate run-off*

(4) Vertical features

These tend to *concentrate* water in a flow. This is because where water is streaming on a surface it is often being blown across it until it reaches a channel, such as a vertical joint, or an obstacle, against which it runs down. Research has shown that the flow of water in vertical joints is much greater than the average flow of water over the wall (see *3.4 Exclusion*, p. 293). This is particularly marked on materials which are non-absorbent or have low absorptivity such as glass or metal claddings, where streaming conditions occur in normal conditions of rainfall (Photographs P3.9/12, and 13). Nevertheless, it can also be observed where precast concrete window sub-frames stand proud of brick walls – a common feature of 1950s architecture. These provide ideal vertical channels for run-off, causing streaking over the brickwork below (see Diagram D3.9/8).

(5) Horizontal features

Horizontal projections (such as string courses, window sills, copings) and recesses interrupt the flow of water down a facade and offer a means of controlling staining. Where horizontal projections do effectively throw water off a facade (such as at window sills) they have two effects. First, they create an unwashed zone immediately below the projection. This often coincides with the sun-cast shadow, but may still be seen as an unsightly disfigurement. Second, in orientations where the water that is thrown off is driven back against the facade lower down there will be a zone which is more fully washed by the concentration of run-off. There may then be a concentration of dirt at some further point below.

Unless horizontal features have a throating on their underside set just in from the bottom edge, water will track back along the soffit. With shallow projections such as sills this water may aggravate staining below the sill, or with large projections such as the underside of balconies ugly staining patterns may occur, clearly visible from the underside (Diagram D3.9/9).

The pattern of flow resulting from horizontal projections is greatly affected by the profile of the front edge. The extent to which water will be thrown off the face of a building is dependent on:

1. The rate of run-off over the projection (the faster the run-off, the more effectively will it be thrown over the edge);
2. The slope of the top surface. This affects the rate of run-off;
3. The size of the projection;
4. The profile of the leading edge. For the same rate of flow over the projection the sharper its edge, the more effectively it will throw water off (see Diagram D3.9/10).

What happens at the ends of horizontal projections of short length, such as sills, is also important, as shown in Diagram D3.9/11. Water streaming down the arrises of the reveal may hit the sill and stream over the edge.

(6) Joints

Joints in horizontal projections represent potential and – all too often – real discontinuities in the flow pattern. It is often surprising what an effect in terms of staining an open joint a mere 6 mm wide can have.

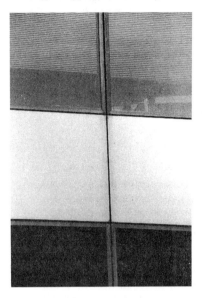

Photograph P3.9/12 *This open-jointed cladding system with an off-white panel shows a concentrated deposit of dirt at the edge of each panel. Run-off of dirty water from the glazing above is concentrated on the line of the mullions*

Diagram D3.9/9 *Illustration of uncontrolled horizontal flow of water along soffits (a). Two possible methods of control are shown in (b) and (c). Note: in the basic condition no account has been taken of the effects of the surface characteristics of the soffit: the worst condition is shown*

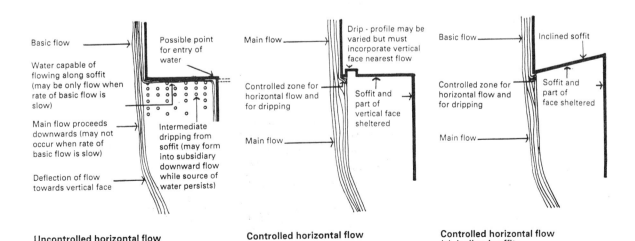

Basic flow

Water capable of flowing along soffit (may be only flow when rate of basic flow is slow)

Main flow proceeds downwards (may not occur when rate of basic flow is slow)

Deflection of flow towards vertical face

Possible point for entry of water

Intermediate dripping from soffit (may form into subsidiary downward flow while source of water persists)

Uncontrolled horizontal flow
(a) Basic condition

Main flow

Controlled zone for horizontal flow and for dripping

Main flow

Drip - profile may be varied but must incorporate vertical face nearest flow

Soffit and part of vertical face sheltered

Controlled horizontal flow
(b) Use of drip

Basic flow

Controlled zone for horizontal flow and for dripping

Main flow

Inclined soffit

Soffit and part of face sheltered

Controlled horizontal flow
(c) Inclined soffit

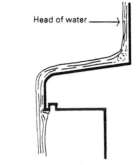

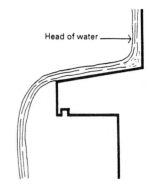

Diagram D3.9/10 *The effect of the profile of the top front edge of a projection on the flow of water*

(a) **Rounded edge**

(b) **Sharp edge**

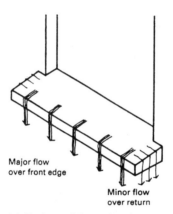

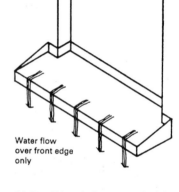

Diagram D3.9/11 *The ends of sills may allow water to run over the edge and stain the wall unless suitably detailed*

(a) **Basic condition and problem**

(b) **Possible solution-tray principle**

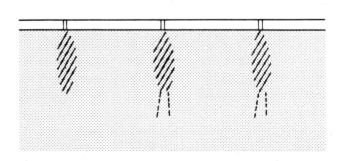

Diagram D3.9/12 *Characteristic staining from joints in the coping to a porous wall*

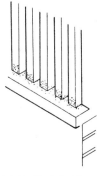

(a) **Profiled cladding
at a sill flashing**

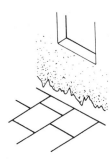

(b) **Ground level**
Note that a
net washing
effect may
be observed

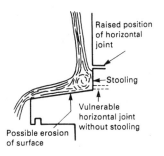

(c) **Masonry sill**

Diagram D3.9/13 *Some examples of
splashback*

Below such joints the staining may take the form of a linear streak or a
balloon type stain, depending on the absorptivity of the material (see
Diagram D3.9/12 and Photograph P3.9/14).

(7) Splashback
This occurs most frequently at the base of a building where it meets the
ground, but the same principle applies to water striking horizontal
ledges higher up the building. This effect is seen particularly on
projecting flashings at the base of metal claddings (being non-
absorbent, the run-off is high – see Diagram D3.9/13) and as
discoloration around the arrises of the panels in open-jointed
claddings (Diagram D3.9/14).

2 The environment

There are a number of different aspects to consider here – the exposure
of the building to rain, wind, sunshine, pollutants and sources of
vegetation. Climatic exposure is discussed in detail in *3.2 Exposure.*

(a) *Exposure to rain and wind*

The idea of wind-driven rain is very important because, as already
noted, the difference in the amount of washing between facades facing
the prevailing wind and those that are sheltered can be considerable.
The new methods for assessing the exposure to driving rain of
particular buildings can now take into account the local features which
generate a microclimate.

A further, finer, level of detail is the cryptoclimate of the building
which is the climatic sheath of up to 1 m thickness that surrounds the
building. This is the range of pressure variations and shading-
generated temperature variations that exist because of the specific
surface modelling. Although the level of detail is fine, the actual
difference in conditions can be large, with accompanying differences in
behaviour of water and dirt collection on identically detailed parts of
the same building. Being able to predict the cryptoclimate of a building
at the design stage is almost impossible: there are simply too many
unknowns.

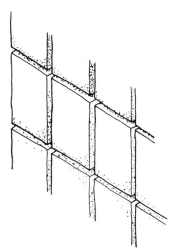

Diagram D3.9/14 *Staining at the
edges of joints in open-jointed claddings
caused by splashback*

Photograph P3.9/13 *This white vitreous enamel cladding faces the prevailing weather and is also heavily exposed to urban grime in its position beside a busy main road. The subtle staining patterns tell a story of the local effect of joints and fixings on the redistribution of fine particles of dirt by the incident rainfall*

Photograph P3.9/14 *Run-off causing staining on this sand-faced fletton brickwork is concentrated at minor joints in the metal verge trim. Even joints this small can be enough to cause concentrated flow of water*

The scouring effect of wind-driven rain is a major factor in removing dirt from surfaces. This is most marked at the extremities of facades and arrisses where wind speeds are greatest. Exposure to sunshine affects the rate of drying out and therefore susceptibility of porous materials to prolonged dampness and growth of plant matter.

(b) Exposure to dirt

Airborne pollutants are discussed in *3.2 Exposure*. The following will affect accumulation of dirt on facades:

- Proximity to industrial plant with smoke emissions;
- Proximity to main roads, since exhaust fumes and dust generated by traffic is a source of substantial quantities of dirt; and
- Proximity to trees which can supply vegetable matter and shading from the drying effect of the sun.

At ground level, splashback from rain hitting the ground adjacent to the building causes dirt from the ground to be picked up and deposited for a short distance up the wall. The rule of thumb for this height is 150 mm.

Practical precautions

1 Objectives

The discussion of practical precautions which follows is based on the observation that weathering of all materials will happen in some form, and that this weathering can add to or diminish the architectural quality of the building. It follows that an objective of the design should be to control weathering so that the inevitable ageing process looks intended and adds to the quality of the building, rather than something which leads only to degrees of shabbiness.

In discussing practical techniques for limiting staining and disfigurement it is useful to consider saturation staining separately from general staining, since the precautions to be adopted differ.

2 Saturation staining

Parts of the building which become saturated for long periods stain differentially to others. Saturation may arise from direct exposure to rainfall or concentrated run-off. The disfiguring effects of saturation are always worse on north-facing facades, where drying out is slower and conditions favour colonization by biological growths.

(a) Parapets

The tops of buildings, particularly parapet walls, are the most exposed and are therefore the most vulnerable to saturation. Where parapets are made of the same material as the lower part of the wall – as is usual – the difference in weathering is marked (Photographs P3.9/7 and 8). Where the wall is of an absorbent material, measures should be adopted which, first, keep the wall dry and, second, throw water away from the wall. A coping plays an important part in meeting these objectives, and, to be effective, it must be laid on a dpc. If the coping is of a different material to the main body of the wall then when it weathers differently this need not look unintentional.

In the discussion about the need for overhangs on copings it is often said that a drip is of no value at that point, since water dripping off the edge is likely to be thrown back against the wall lower down. While this is true, it does seem that a drip prevents a staining pattern of the jagged edge kind occurring below the top of the wall at the same time as causing a shadow area of unwashed wall below. The primary purpose of the coping is to keep the wall dry, and the need for an overhang should be assessed by considering exposure to prevailing winds, dirt retention and the susceptibility of the walling material to staining.

Photograph P3.9/15 *An example of splashback from the projecting flashing of a wall of profiled metal cladding*

(b) Sills

The problem with sills are similar to those of parapets, but here exposure to direct rainfall is supplemented by run-off from the window. Exposure varies considerably with orientation and the height of the window. Tall windows or vertical strips of window with non-absorbent spandrel panels increase the run-off over sills at the base.

In a similar way to copings, sills should be laid on dpcs to keep the wall below dry (unless they are continuous and impermeable, such as well-made precast concrete, slate or metal). Brick sills should be made of relatively non-absorbent bricks laid in a hard mortar.

The need for overhangs again depends on the amount of run-off and the exposure to dirt. With a large run-off it is wise to shed the water with a drip. In dirty conditions, however, drips lead to an unwashed line of dirt below the sill. Where the material of the wall below is vulnerable to partial washing for the conditions of exposure, the lack of a drip is likley to lead to streaking below the sill, which is better controlled by providing one.

Where sills are used they should follow the general rule and have an effective throating *on all three sides*, or else there is a danger of streaking run-off from the ends of the sill onto the adjacent wall. With both parapets and sills it is essential that the joints be properly filled so that staining does not occur at the position of the joints.

(c) Saturation near ground level from splashback and rising damp

Dpcs must be provided to prevent rising damp causing staining of walls near the ground. The use of a plinth of harder non-absorbent material disguises the effects of splashback. Cases can occur of splashback causing a zone of cleaner material near the ground.

3 General staining

Because staining and disfigurement is affected by so many variables it is difficult to set down hard-and-fast rules about specific materials and situations. What can be done is to put the factors in a broad order of importance and to set out the major relationships between them. As with so many aspects of building performance, this is an area of risk assessment in which trends rather than absolutes can be predicted. For each of the factors which have been discussed above a quantitative statement has been made.

(a) Exposure to dirt

The most important factor affecting disfigurement is exposure to dirt. Dirt on its own may disfigure surfaces even if there are no washing effects. By contrast, in perfectly clean conditions no staining could occur. However, such conditions never exist, and dirt in the sense used here is invariably present to some extent.

The following broad statements can be made:

- Grit-type dirt increases with polluted atmospheres (for example, proximity to major roads, industrial works, building sites, chemical works, etc.).
- Biological-type dirt increases:
  In rural environments where biological growth is not poisoned by pollution;
  On north-facing elevations where drying out is slower and the sunless environment provides suitable habitats for moulds and lichens;
- Dirt derived from the material itself increases:
  With soluble materials such as limestones;
  With run-off from corroding metals.

(b) Exposure to rain

The second most important factor is exposure to rain (and it is a close second). This factor is intimately connected with the properties of the exposed material. Since staining occurs because of partial washing, and the flow of water over materials depends on their absorptivity, the same exposure will cause different washing effects on different materials. Nevertheless, a number of broad statements can be made:

- The tendency for full washing to occur increases with exposure to driving rain–and the corollary:
- The tendency to staining increases as conditions allow partial washing.

In terms of materials:

- Smooth non-absorbent materials have good run-off in most conditions of exposure. However, the rate of run-off sufficient actually to *wash* off dirt increases with exposure to driving rain;
- Moderately absorbent materials such as sandstone and limestone are fully washed only when exposed to driving rain;
- Highly absorbent materials such as stock bricks and certain stones usually only receive sufficient water for partial washing to occur where there is heavy run-off from architectural features such as window sills.

The susceptibility of the surface to disfigurement is the other side of the equation. The various factors discussed in section 3 are assessed in turn.

(c) Properties of the surface material

(1) Absorption

For a given exposure susceptibility to partial washing is greatest where the absorptivity of the material allows partial run-off. Staining is least when the absorption is high enough to prevent run-off in all but extreme conditions, and is reduced where run-off is high enough to promote good regular washing under normal conditions. This factor is the least easy to control, and design decisions should be based on close observation of the behaviour of the material being considered under similar conditions.

(2) Dirt retention
Risk of staining increases with the susceptibility of the surface to retain dirt. Note, though, that this factor may be offset by the tendency of rough, dirt-retaining textures to disguise staining.

(3) Texture
Risk of disfiguring staining increases with the smoothness of materials which reveal the established paths of water run-off. Texture can be used both to disperse run-off and, by the effect of positive modelling of the surface, disguise differences in the concentration of dirt.

(4) Colour
Staining is disguised by darker colours and mottled or brindled tones. Very dark colours can show lighter staining, particularly the 'spilt-milk' effect of run-off from concrete and limestones. Black or dark grey sandlime bricks are particularly vulnerable.

(5) Solubility
The tendency to staining on materials further down the facade increases with solubility of the materials higher up: the effects of run-off should always be considered. On the other hand, soluble materials themselves tend to be more effectively washed since, above a critical level of run-off, the surface is removed with the dirt, leaving a fresh, clean surface.

(d) Geometry of the surface

(1) Orientation
This has a major and very marked effect on the exposure to driving rain and the prevailing wind. Observation of a wide range of building materials suggests that full washing is more likely with orientations facing the prevailing wind, and that the risk of staining increases on sheltered elevations. The incidence of biological growth, as already noted, increases on north-facing facades.

(2) Angle of inclination
The risk of staining due to partial washing increases with cut-back walls, parapets or curved and coved cornices. That due to dirt retention and saturation staining increases on the tops of inclined surfaces.

(3) Height
The risk of partial washing increases with height for a given exposure. However, exposure itself increases with height. High-rise buildings 20 floors high often show macro-staining effects with full washing for, say, half their height and then increasing dirtiness towards the ground. On the other hand, fascias or sills only inches high can show disfiguring stains if the rate of run-off is slow and there is a source of dirt to pollute the run-off nearby.

(e) Architectural features

(1) Generally
Strong modelling, particularly detailing from a clearly read and understood period such as the Gothic, seems of its own accord to disguise what would be disfiguring staining on a modern building. The reason for this is open to interpretation. Features and joints are all two-edged swords in the fight against staining: they can be used to control it or, if mishandled, they can aggravate the problem.

(2) Horizontal features
These will only reduce the risk of staining if they are effective in actually shedding water. If they are not (i.e. are shallow and do not have throatings) they can accentuate partial washing and actually cause staining.

(3) Vertical features
These are quite good at controlling flow and, thereby, staining. Careful thought must be given to what happens to the run-off at the base of the feature.

(4) Arrisses
These appear to be particularly vulnerable points for the concentration of run-off (as discussed above, this is due to the pressure gradients across the surface). This occurs on the scale of the entire building (the effect can be seen on multi-storey blocks of flats where there are flank walls of a single material) or on individual window openings or columns. For this to be avoided, careful design of vertical features which prevent the flow of water across the facade to the arris would be required.

4 Summary

To summarize, there are three basic steps which should be taken to control the disfigurement of buildings at the design stage:

1. Consider the likely pattern of the flow of water over the facade, paying particular attention to the possibility of differential flow arising from concentrations of water or partial washing. Exposure and the architectural features of the building will be the key aspects here.
2. Take steps to control the flow, primarily by geometrical means, although choice of materials is also important.
3. Control the *effects* of the flow, usually by choice of materials and their colour and texture.

Sources

Simpson, J.W. and Horrobin, P.J. (eds), *The Weathering and Performance of Building Materials*, MTP, Aylesbury, 1970
Hawes, F., *The weathering of concrete buildings*, C&CA, Wexham Spring, 1986
Marsh, P., *Concrete as a visual material*, C&CA, London, 1974
Morris, A.E.J., *Pre-cast Concrete in Architecture*, The Whitney Library of Design, 1978, pp. 286–309
White, R.B., *The Changing Appearance of Buildings*, HMSO, London, 1967
Hamilton, S.B., Bagenal, H. and White, R.B., *A qualitative survey of some buildings in the London area*, National Building Studies Special Report 33, HMSO, London, 1964
White, R.B., *Qualitative studies of buildings*, National Building Studies Special Report 39, HMSO, London, 1966
Principles of Modern Building, Vol 1, HMSO, London, 1959
Principles of Modern Building, Vol 2, HMSO, London, 1961
CP 01/74 *Observation on the behaviour of weather protective features on external walls*

Index

Abrasion, 120–22, 144
Absorbency, 566
Absorption, 151, 181, 231, 566–7
Absorptivity, 566–7
Acid deposition, 175
Acid gases, effect of:
  on brickwork, 389
  on cement-based products,
    390–91
  on stonework, 385–9
Acid rain, 169
Acidity:
  ground water, 199–200
  water supply, 209–11
Adhesion, 103, 111–12, 161
  effects of movement, 108
  loss of, 361
  precautions, 109–11
  principles, 108
  see also Bond formation
Adhesives:
  bonding techniques, 115
  classification, 112–13
  for plywood, 257
  materials and properties, 112
  selection of, 115
  special, 111–14
Aerobic soils, 469
Aggregates:
  alkali attack, 399–401
  impurities in, 383
  reactive, 396–401
  salt content, 359
  types of, 264–5
Air constituents, 170
Air pressure gradients, 23, 285–6
Alkali aggregate reaction, 396
Alkali carbonate reaction, 396
Alkali silica reaction, 396–401
Alkali etc. Works Regulation Act 1906,
  180
Alkalinity, water supply, 209–11
Alkalis, attack by, 391
  precautions, 401
Aluminium:
  coating, 512
  paint treatment for, 522
  use of, 499–501
Anaerobic soils, 469
Anchor fixings, 102, 104–6
Angular distortion, 71–2
Anodic coatings, see Anodizing
Anodizing, 497, 513–14
Arch principle, 22

Architectural features, effect on
    differential flow, 571–5, 579–80
Arrisses, 91, 571
Ashes and clinker (as aggregate), 360
Assembly, building, 25
Atmosphere, salts content, 365
Atmospheric exposure, 165, 480–82
  condensation, 167–8
  corrosion, 467
  humidity, 165
  pollution, 168–80
  variations in, 166
Attrition, 144

Backing materials, salts content, 363
Baffles, 117
Balance principle, 19, 28–30
Basements, water exclusion, 325–8
Bearing capacities, soil, 73
Below ground level structures, see
    Basements
Bending, 90–94
Binder, paint, 516
Biological attack, 381
  effects on weathering, 148
Bituminous coatings, 523
Blistering, 220
Block making, 95
Blow moulding, 95
Bond formation, adhesive, 116
  design of joints, 114
  techniques, 115–16
Bound moisture, 234–6
Bowing deformation, 48
Brasses, 502–3
BRCL panel freezing test, 552, 553
Bricks, clay:
  firing, 358
  frost resistance, 552–3
  making, 95, 384–5
Brickwork:
  acid gases on, 389
  calcium silicate, 272
    precautions, 272
    rules, 272
  clay, 271
    precautions, 271
    rules, 271
  efflorescence removal, 375–6
  moisture expansion, 271
  salts content, 363
  staining, 403–4
  sulphate attack, 417–19
  water exclusion, 315–16

British Board of Agrément (BBA), 6
Bronzes, 502–3
BRS Tray test, 551–2
Building principles, 15, 66
   defined, 17
   rules and precautions, 19
   see also under individual names
Bulb rivets, 102

Cadmium:
   coating, 512
   paint treatment, 523
Calcium carbonate, 361
Calcium silicate, 272
Calcium sulphate, 355, 386–8
Calorizing, 513
Capacity/capacitance meter, 153
Capillarity, 134, 197, 230–1
Carbide meter, 153
Carbon dioxide, 146, 361
Carbon monoxide, 175
Carbonation, 146, 392, 525
Case hardening, 229
Cast iron, 490–11
Cathodic coatings, 497
Cathodic control/protection, 474, 529–
   31, 539–40
Cathodic reactions, 463–4
Cavities, roof, 28
Cavity fixings, 107
Cavity walls, 8–9, 292, 317–21
   jamb compensation, 32–3
Cement, 263, 265
   ordinary Portland, 360–1, 398
      resistance of, 408–10
      and steel, 496–7
   sulphate resisting, 398, 409
Cement-based products, 262
   influencing factors, 263–9
   precautions, 270
Cementation, 513
Certification schemes, 7
Change, inherent nature of materials
   to, 9
   delaying, 13
Chemical attack, 378–81
   see also under specific chemicals
Chimneys, sulphate attack on, 421–2
Chipboard, 259–61
   precautions, 260–61
Chlorine compounds (from fuel
   burning), 175
Chromium:
   paint treatment, 522
   plating, 510
Chromizing, 513
Cladding, 314–15
   impact resistance, 125
   moisture exclusion, 322–5
Clay-nature, 75
Clay shrinkage potential, 76
Clay soils, 73, 74–6
   behaviour, 76–8
   precautions, 80
   UK categories, 79

Clean Air Act 1956, 168, 180
Cleaning *see under* Maintenance and
   cleaning
Cleavage loading, join, 114
Climatic data, 160
   factors, 481
   modification, 4
Cold bridges, 168
Cold deck design, 28–9, 300–1
Colliery shale, 87, 360
Colloidal constituents, 235
Compaction, floor screeds, 275
Compensation (of principles), 30–34,
   54
Complexity, growth, 5
Composite constructions, 28–30
Composite materials, 28
Compression (of fill), 83–4
Compression moulding, 94
Concrete, 264–8
   admixtures, 412
   autoclaved aerated, 265n
   cracking, 397
   curing, 267
   floor screeds, 273–7
   mix proportioning, 265–7
   pre-cast, 269
   as protective coating, 524–9, 538–9
   sprayed, 529
   and steel, 496–7
   sulphate attack, 410–16
   winter building precautions, 558
Concrete blocks/masonry, 272
Concrete floors, 415–16
Condensation:
   dampness, 151, 167–8
   interstitial, 218n, 547
   roof, 300
Conductivity meter, 153
Consolidation:
   of fill, 85
   of soil, 72–4
Construction water, 151, 200–202,
   369
Continuity principle, 19, 27–8, 279
Contraction movements/restraints,
   44–6
Conversion (of timber), 243–5
Copper and its alloys, 501–2
   paint treatment, 522
Corrosion, 144, 457–67
   classification, 467–70
   effects of, 479–80
   forms of, 476–9
   initiiation of, 470–3
   precautions against, 531–40
   by rain water, 170
   rate of attack, 473–6
   resistance to, 489–506
Corrosion cell, 462–5, 465–7
Corrosion products, 464–5
Corrosive agents, 145–6, 484–5
Cracking, 37–40
   classification, 44–50
   corrosion, 478–9

Cracking – *continued*
　initiation of, 46
　precautions against, 54, 65–9
　variations effects, 50–65
Creative pessimism principle, 19, 25–
　6, 414
Creep settlement, 83
Crevice corrosion, 477
Crocodile cracking, 65, 420
Curing, floor screeds, 275
Curtain walling, 314–15, 325
Cutting process, 89–90

Damage classification:
　floor slabs, 42
　from impact, 123–4
　walls, 41
Damp proof coursing (dpc), 317–20,
　330, 332–4
Damp proof membrane (dpm), 328–32
Dampness, 150–1
　and adhesion, 161
　causes, 151–2
　measurement of, 152–4
Degradation, radiation, 453–6
Degrees Clarke 207–8
Denudation, 144
Deviations, dimensional, 97
　induced, 97, 99, 315
　inherent, 97, 99, 315
　seating, 117
Dew, 168
Dezincification, 478
Differential aeration corrosion, 471,
　472
Differential drying shrinkage, 56–7
Differential flow, 566, 571–5
Differential movements, 24
Differential settlement, 85
Dimensional fluctuations, moisture
　content, 162
Dipping (coating), 512
Dirt staining, 564, 567, 576
Discontinuity principle, 27
Drips, use of, 156–7
Driving rain index, 160–61, 181, 187
Driving rain spell index, 161, 187
Dry-bulb temperature, 165
Dry rot, 431–2
Drying out, 200, 202, 230–1, 375, 565
　flat roofs, 220
　floor screeds, 275
　masonry mortars/renders, 271
　water movement, 236
Durability, 9–14
　differential, 25
　metals, 458

Eddies, 158
Efflorescence, 199, 348–54
　physical processes, 365–71
　precautions, 373–5
　removal of, 375–6
　sources of salts, 356–62, 362–4

Electrochemical theory, corrosion,
　460, 461–7
Electroplating, 512
Equilibrium moisture content, 225–6
　timber, 246–51
Erosion, 144
Evaporation, 138, 139, 141–3, 181
Exfoliation, 56, 148, 387
Expansion fixings, 102, 105–6
Expansion movements/restraints, 44–
　6
Exposure, 236
　changes in, 64
　effectof, 480–89
　liquid water effect, 164, 237
　staining effects, 575
　water vapour effect, 238–9
Extrusion forming, 94

Factors of safety, 26
　fixings, 103
Failures:
　lessons from, 15
　risk of, 6
Fasteners *see* Fixings, mechanical
Felspar, 146
Ferrous metals, 490–8
Ferrous sulphate, 356
Fibre building board, 257–9
　precautions, 259
Fill, 82–5
Finishing coat (paint), 516, 520
Finishing processes, 95
Fixings, mechanical:
　factors of safety, 103
　loading on, 103
　principles, 102–3
　strength of, 107
　theory, 103
　types of, 104–7, 305
Flat roofs:
　compensation, 33–4
　moisture entrapment, 216, 217–22
　winter building precautions, 560
Floor finishes, 442
Floor screeds, 273–7
Floors:
　abrasion on, 121
　concrete, 539
　ground moisture exclusion, 328–32
　impact damage to, 125
　precautions, 276–7
　rules, 276
Flow of water, 155–8, 562–6
Flue gases, 168
Flue linings, 422
Fluorine compounds (from fuel
　burning), 175
Former (support), 91
Foundation types, 74
Fractile point, 103
Framed and clad buildings, 7
Freezing, 135, 543–9
　and thawing cycles, 547
Frictional grip, 102

Frost action, 148
    damage, 549–51
    heave, 80
    precautions, 555–6
    resistance to, 551–4
    see also Winter building precautions
Fuel burning pollutants, 171–5
    dispersion of, 175–7
    distribution of, 178
Fumigation, 177
Fungi attack, 422–9
    precautions, 438–50

Gaskets, 117, 120
Gels, 235
Geological sculpturing, 144
Glass, alkali attack on, 404
Glass fibre-reinforced cement, 404
Glazing, impact resistance, 126
Gradients:
    stress-inducing, 54–7
    thermal, 20–21
Grinding process, 95
Grit and dust (from fuel burning), 174
Grooving (corrosion attack), 477
Ground/groundwater, exposure to,
    196–200

Hardboard, 258, 437
Hardcore, 85–7
Hardness, water, 207–9
Hardness differential, 89
High > low principle, 17, 20
High-strength friction-grip (HSFG)
    bolts, 107
High tech construction, 26n
Hogging deformation, 49
'Hot-house' effect, 135
Humic acid, 145
Humidity, 152, 165
Hydration, 146
'Hydrolic' cycle, 137
Hydrolysis, 146
Hygrometer, 152, 153
Hygroscopicity, 368
    control of, 251–4
    materials, 235n
Hygrothermal conditions, 267, 546

Immersed corrosion, 468
Impact, 122–6
    performance categories, 124
    precautions, 126
Impervious materials, 281, 299
Incompatibility of materials, 25
Industrial processes:
    salts content, 365
    use of water, 213
Inhibitors, corrosion, 507, 517
Injection moulding, 95
Innovation process, 5–7
Insect attack, 429–30
    precautions, 430
Insulating boards, 258
Intercrystalline corrosion, 471

Inversion (weather conditions), 177
Iron:
    as mineral, 146
    see also Cast iron, Steel, Wrought
        iron
Iron sulphide (pyrites), 357

Joinery, moisture content, 251
Jointing/joints, 279
    definitions, 96
    deviations allowance, 97
    function of, 97–8
    leakage through, 181
    need for, 97
    sealing of, 117–18
    theory of, 95–6
    types, 67–8, 99–101
    water exclusion, 292–7
    water ingress, 447
Junctions, moisture exclusion, 279,
    295–7, 334–45

Keying hold, 102
Knotting, 450

Lap joints, 115
Leaching effects, pollutants, 211
Lead:
    corrosion resistance, 503–4
    paint treatment of, 522
Lead coating, 510
Life cycle costs, 14
Lime:
    effect on paint films, 392
    preparation of, 382
    in rocks, 146
    salt content, 360, 363
Limestone, 146
    staining, 349, 371–2, 402–3
    precautions, 372
Liquid limit, 75–6
Local cycle (water), 139–40

Magnesium sulphate, 355, 384, 407–8
Maintenance and cleaning, 122
    dampness from, 152
    of protective systems, 540
    salts content, 365
    use of water, 212–13
Malleability, 91, 123
Map cracking, 65, 420
Masonry, 269–71
    concrete, 272
        precautions, 270, 273
    movement joints in, 68
    water exclusion, 315–16
Materials characteristics, 50–52, 373,
    568–9
    life, 13
Medium boards, 258
Metallic coatings, 488
Membranes, impermeable, *see*
    Tanking

Metals:
  bending, 93
  casting, 94
Minerals, decomposition of, 146
Mixing ratio, moisture, 165
Moisture, entrapped, 214
  flat roof procedure, 217–22
  precautions, 221–2
  reasons for, 215–17
  sources of, 214–15
Moisture content, 165, 223
  absorption, 231–3
  capillarity, 230–31
  in cement-based products, *see*
    Cement-based products
  definitions, 224–30
  effect of exposure, 236–9
  forms of, 234–6
  freezing, 544
  joinery, 250–51
  permeability, 233
  saturation coefficient, 233
  in timber, *see* Timber
  in timber-based products, *see*
    Timber-based products
Moisture gradient, 229
  freezing, 545
Moisture movement, 226–9, 237–43
  precautions, 253
Moisture transmission, 219, 230, 231,
  282
  see also Water exclusion
Mortars, 269, 374
  defects, 384
  frost resistance, 553
Mosaics, *see* Tiling
Mould growth, 428
Moulding, 94–5
Movement accommodation factor,
  118
Movement joints, 67–8
  in clay brickwork, 271
  in calcium silicate, 272
  in concrete masonry, 272
Multi-layer construction, 7–9, 30–31

Nails, 102, 104–5
Nickel, 510
Nitrogen oxides (from fuel burning),
  175
Non-ferrous metals, corrosion
  resistance, 498–506
  see also under individual names
Nut and bolt fixings, 106–7

'100 foot' rule (masonry), 68
Organic acids, 390
Oven drying, 152
'Overcoat' principle, 287
Oxidation:
  direct, 461
  photo-, 456
Oxygen (as weathering agent), 146

Paint coatings/systems, 252, 515–20
  behaviour of, 312–13
  chemical resistance, 523
  for ferrous metals, 520
  for immersed conditions, 523
  for joinery, 447
  for non-ferrous metals, 521–2
  for woodwork, 448–50
Paint composition, 516
Paint films, 391–2
  precautions, 393–6
  resistance of, 393–4
  wall moisture content, 395
Painting, winter building precautions,
  560
Panel walls, movement joints, 68
Particle board, *see* Chipboard
Peel loading, joint, 115
Permeability, 233, 281
pH value, 474–5
Photo-oxidation, 456
Pitched roof, moisture entrapment, 216
Pitting (corrosion attack), 477
Plaster:
  defects, 383
  efflorescence removal, 376
Plastering:
  winter building precautions, 559
Plastic coatings, 523
Plastic limit, 75–6
Plasticity, soil, 75
Plasticizers, 363
Plastics:
  bending, 94
  moulding, 94–5
Plywoods, 256–7
  durability, 437
  painting, 449
Polarization, 474
Polishing, 95
Pollutants/pollution, 168–9
  atmospheric, 162
  during construction, 202
  in ground water, 198–200
  industrial, 179–80
  leaching effect, 211
  smoke, 171, 173–4
  of water, 147
Polymers, use of, 451, 456
Pore solution, alkalinity of, 399
Pore structure:
  effects of, 366–7
  freezing, 548
Pot life (paint), 516
Potassium sulphate, 355
Precautions, 20
  against cracking, 66–9
  see also under individual materials/
    elements
Pressure effects, 4
Pressure release/unloading, 148
Primer (paint), 515, 519–20
Protective coatings, 252, 312–13, 508–
  29, 537–90
  see also under Paint coatings/systems

Pulverised fuel ash (pfa), 87, 398, 408, 409

Quality control, 6
Quicklime, 382–3

Rain/rainfall, 138, 141–3, 148, 152, 180
    efflorescence, 369
    penetration by, 160–1, 181, 282–3
    site exposure, 186–95
    wind significance, 181–6
'Rain rose' technique, BRE, 187–8
'Raincoat' principle, 287–9, 315
'Rainscreen' principle, 281, 287, 322–4
Reinforcement coating, 526–7
Rendering/renders, 273, 313
    efflorescence removal, 376
    sulphate attack, 420–21
    precautions, 273
    rules, 273
Resilience, 122
Resin anchors, 103
Restrainers (corrosion), 507
Restraint, degrees of, 44–6, 53
Reversion concept, 14, 23
Rising damp, 151
Risk of failure, 5
Rivets, 107
Rocks, decomposition of, 146
Roll-bonding, 513
Roof ventilators, 220, 221
Roofs:
    fixings, 305
    flat design, 300–304
    insulation, 305–307
    pitched, 216
    timber, 445
    water exclusion, 298–311
Rotating deformation, 50
Rules, 20
Rusting, 149

Safe working load, 103
Safety factors, *see* Factors of safety
Sagging deformation, 48
Sand, sea, 359
Sandstone, contour scaling, 370
Sandy ground, 73, 75
Saponification, 392
Sarking felt, 292, 310
Saturation coefficient, 233
Science:
    application of, 31
    interpretation of, 2
    Scientific approach, 6
Screws, 103, 105
Sculpturing, geological, 144
Sea water effects, 359, 408
Sealing/sealants, 117–18
    gaskets, 120
    precautions, 118
    types of, 118, 283–4
Self-jigging, 90
Separate lives principle, 18, 21, 24–5
Services, faulty, 212

Settlement, 46, 72–4, 85
Sewage, effects of, 390
Shaping process, 88–95
    by cutting, 89–90
    by bending, 91–4
    by moulding, 94–5
Shear loading joint, 114
Shearing deformation, 49
Sherardizing, 513
Shingles, *see* Tiles
Site exposure conditions, 186–95
Slags, 398
Slaking process, lime, 382–3
Slates, *see* Tiles
Smoke Control Order, 169
Smoke pollution, 171, 173–4
Sodium sulphate, 355
Softboard, 258
Soffits, flow along, 156–7
Soil:
    cohesive, *see* Clay soils
    granular, *see* Sandy ground
    index tests, 75
    salts content, 364
    types and variations, 71
Soil movements, 70–72
    clay, 74–80
    freezing, 80
    heave, 78
    lateral, 81
    loss of ground, 81
    mining subsidence, 80
    precautions, 80
    settlement, 72–4, 85
Soluble salts:
    nature of, 354–6
    sources, 356–65
Sooty deposits, 173–4
Spalling, 55
Spraying (coatings), 512
Staining effects:
    pollutants, 211
    precautions, 576–80
    water flow, 564, 567–76
Steel, 494–7
    low-alloy, 494–5
    mild, 494–5
    stainless, 497–8
Stone, natural, 359
Stonework:
    attacks on, 386–8
    efflorescence removal, 376
Stray current corrosion, 470
Stress concentrations, 57–9
    application of principles, 66
    precautions, 66–9
    typical locations, 59–65
Stress corrosion, 478, 535
Stress relaxation, 53–4
Stretching deformation, 49
Structural weakening, cracking, 40
Sulphate attack, 147, 162, 350, 406–10
    brickwork, 417–19
    concrete, 410–16
    domestic chimneys, 421–2

Sulphate attack – *continued*
  ground water, 198–9
  precautions, 412–15, 418–19, 420, 421
  rendering, 420–21
Sulphate-resisting bacteria, 464, 469–70
Sulphate skins, 386–8
Sulphates, 405–6
  naturally occurring, 410
  site classification, 414
Sulphur dioxide, 146n, 169, 171, 174–5, 361
Surface films, 461
  cooling, 55
  heating, 55
Surface tension, water, 134
Systems approach, 24

Tanking, 327
  precautions, 327
Temperature gradients, 23, 545
Tensile forces, 22
  joint loading, 114
Thatch, 311
Thermal effects, 43–4, 148, 455
Thermal insulation, 8
Thermal pumping, 301–2
Through-fixing, 102, 106–7
Tiles/tiling, 307–10
  clay, 370
  slates, 388–9
Timber:
  bending, 92–3
  coatings, 252
  conversion effects, 243–5
  drying, 246–8
  durability grading, 433
  fungal attack, 432–4
  hygroscopicity control, 251–4
  moisture content, 249–50
  moisture movements, 239–43
    precautions, 253–4
  preservative treatment, 434–7, 439, 449
  selection of, 439–40
  service specification, 249
  storage, 440–41
Timber-based products, 239, 254–5
  see also under specific names
Timber floors, fungal attack, 441–5
Timber framing, 445
Timber roof, 445
Tin coating, 510
Toggle fasteners, 102
Toughness, 123
Traditional construction, 26n
Trees, effect on clay soils, 76, 78
Twin-pack paint systems, 516

Ultra violet radiation, 381, 451–2
  action and effects, 453–6
  exposure to, 452–3
  reduction of effects, 456

Undercoat (paint), 515, 520
Underground corrosion, 469–70

Vacuum forming, 95
Vanadium salts, 356
Vaporization (coating), 513
Vapour control layers, 168
Vapour pressure, 23, 165, 285–6
Variations and their effects, 2, 50
Ventilation:
  roof, 220, 221
  underfloor, 443–4
Vibration loading, 43, 107
Vitreous enamelling, 524
Vortex formation, wind, 182, 184–5

Waferboard, 259
Wallboards, durability of, 437
Walls:
  moisture entrapment, 216–17
  moisture exclusion, 311–25, 333–4
Warm deck designs, 29–30, 300–301
Waste water, 213
Water, 129–30
  basic properties, 130–35
  capillarity, 134
  chemical combination, 234
  chemical formula, 131
  in construction, 151, 200–202, 369
  instability, 132–4
  flow:
    shedding, 22
    drainage, 23
  free, 234
    transport of, 134
  sorbed, 234–6
Water/cement ratio, 236n, 266, 275
Water cycle, 137–40
  significance of, 143–4
Water exclusion, 278–81
  joints and junctions, 292–7
  roofs, 298–311
  walls, 311–25
Water flow/movement, 22–3, 280–81
Water-repellants, 312–13
Water supply, 203
  contamination of, 204
  corrosivity, 210–11
  hardness, 207–9
  pH values, 209–10
  substances in, 136, 203–7
  treatment, 203–4
Water vapour, 197–8, 238–9
Waterfall effects, 147
Waterline corrosion, 477
Weather conditions, 140–43
  classification, 158–63
  forecasting, 543
Weathering, 144–5
  biological agencies, 148
  chemical, 145–7
  materials characteristics, 568–9
  physical, 148
  significance of, 148–9
  spheroidal, 146

Wet-bulb temperature, 165
Wet and dry conditions, 139, 140–43
Wet rot, 431
Wind effects, 148
    loading on fixings, 107
    with rain, 181–6
Wind pressure, 23
Winter building precautions,
    556–60

Wood-rots, 428–9, 431–2
Woodwool slabs, 438
Work hardening, 93
Wrapping tapes, 513, 524
Wrought iron, 491–3

Zinc:
    paint treatment, 522
    coating, 504, 510